Natural Products Isolation

METHODS IN BIOTECHNOLOGY™

John M. Walker, SERIES EDITOR

21. **Food-Borne Pathogens**, *Methods and Protocols,* edited by *Catherine Adley, 2006*
20. **Natural Products Isolation**, *Second Edition,* edited by *Satyajit D. Sarker, Zahid Latif, and Alexander I. Gray, 2005*
19. **Pesticide Protocols**, edited by *José L. Martínez Vidal and Antonia Garrido Frenich, 2005*
18. **Microbial Processes and Products**, edited by *Jose Luis Barredo, 2005*
17. **Microbial Enzymes and Biotransformations**, edited by *Jose Luis Barredo, 2005*
16. **Environmental Microbiology: *Methods and Protocols*,** edited by *John F. T. Spencer and Alicia L. Ragout de Spencer, 2004*
15. **Enzymes in Nonaqueous Solvents: *Methods and Protocols*,** edited by *Evgeny N. Vulfson, Peter J. Halling, and Herbert L. Holland, 2001*
14. **Food Microbiology Protocols**, edited by *J. F. T. Spencer and Alicia Leonor Ragout de Spencer, 2000*
13. **Supercritical Fluid Methods and Protocols**, edited by *John R. Williams and Anthony A. Clifford, 2000*
12. **Environmental Monitoring of Bacteria**, edited by *Clive Edwards, 1999*
11. **Aqueous Two-Phase Systems**, edited by *Rajni Hatti-Kaul, 2000*
10. **Carbohydrate Biotechnology Protocols**, edited by *Christopher Bucke, 1999*
9. **Downstream Processing Methods**, edited by *Mohamed A. Desai, 2000*
8. **Animal Cell Biotechnology**, edited by *Nigel Jenkins, 1999*
7. **Affinity Biosensors: *Techniques and Protocols*,** edited by *Kim R. Rogers and Ashok Mulchandani, 1998*
6. **Enzyme and Microbial Biosensors: *Techniques and Protocols*,** edited by *Ashok Mulchandani and Kim R. Rogers, 1998*
5. **Biopesticides: *Use and Delivery*,** edited by *Franklin R. Hall and Julius J. Menn, 1999*
4. **Natural Products Isolation**, edited by *Richard J. P. Cannell, 1998*
3. **Recombinant Proteins from Plants: *Production and Isolation of Clinically Useful Compounds*,** edited by *Charles Cunningham and Andrew J. R. Porter, 1998*
2. **Bioremediation Protocols**, edited by *David Sheehan, 1997*
1. **Immobilization of Enzymes and Cells**, edited by *Gordon F. Bickerstaff, 1997*

METHODS IN BIOTECHNOLOGY™

Natural Products Isolation

SECOND EDITION

Edited by

Satyajit D. Sarker

Pharmaceutical Biotechnology Research Group
School of Biomedical Sciences
University of Ulster at Coleraine
Coleraine, Northern Ireland
United Kingdom

Zahid Latif

Molecular Nature Limited
Plas Gogerddan, Aberystwyth
Wales, United Kingdom

Alexander I. Gray

Phytochemistry Research Lab
Department of Pharmaceutical Sciences
University of Strathclyde
Glasgow, Scotland, United Kingdom

HUMANA PRESS ✱ TOTOWA, NEW JERSEY

999 Riverview Drive, Suite 208
Totowa, New Jersey 07512

www.humanapress.com

This publication is printed on acid-free paper. ∞
ANSI Z39.48-1984 (American Standards Institute)

Permanence of Paper for Printed Library Materials.

Cover design by Patricia F. Cleary

For additional copies, pricing for bulk purchases, and/or information about other Humana titles, contact Humana at the above address or at any of the following numbers: Tel.: 973-256-1699; Fax: 973-256-8341; E-mail: orders@humanapr.com; or visit our Website: www.humanapress.com

Printed in the United States of America. 10 9 8 7 6 5 4 3 2 1

eISBN 1-59259-955-9

Library of Congress Cataloging-in-Publication Data

Natural products isolation. – 2nd ed. / edited by Satyajit D. Sarker, Zahid Latif, Alexander I. Gray.
p. cm. – (Methods in biotechnology; 20)
Includes bibliographical references and index.
ISBN 1-58829-447-1 (acid-free paper) – ISBN 1-59259-955-9 (eISBN)
1. Natural products. 2. Extraction (Chemistry) I. Sarker, Satyajit D. II. Latif, Zahid. III. Gray, Alexander I. IV. Series.

QD415.N355 2005
547′.7–dc22 2005017869

Preface

The term "natural products" spans an extremely large and diverse range of chemical compounds derived and isolated from biological sources. Our interest in natural products can be traced back thousands of years for their usefulness to humankind, and this continues to the present day. Compounds and extracts derived from the biosphere have found uses in medicine, agriculture, cosmetics, and food in ancient and modern societies around the world. Therefore, the ability to access natural products, understand their usefulness, and derive applications has been a major driving force in the field of natural product research.

The first edition of *Natural Products Isolation* provided readers for the first time with some practical guidance in the process of extraction and isolation of natural products and was the result of Richard Cannell's unique vision and tireless efforts. Unfortunately, Richard Cannell died in 1999 soon after completing the first edition. We are indebted to him and hope this new edition pays adequate tribute to his excellent work.

The first edition laid down the "ground rules" and established the techniques available at the time. Since its publication in 1998, there have been significant developments in some areas in natural product isolation. To capture these developments, publication of a second edition is long overdue, and we believe it brings the work up to date while still covering many basic techniques known to save time and effort, and capable of results equivalent to those from more recent and expensive techniques.

The purpose of compiling *Natural Products Isolation, 2nd Edition* is to give a practical overview of just how natural products can be extracted, prepared, and isolated from the source material. Methodology and know-how tend to be passed down through word of mouth and practical experience as much as through the scientific literature. The frustration involved in mastering techniques can dissuade even the most dogged of researchers from adopting a new method or persisting in an unfamiliar field of research.

Though we have tried to retain the main theme and philosophy of the first edition, we have also incorporated newer developments in this field of research. The second edition contains a total of 18 chapters, three of which are entirely new. Our intention is to provide substantial background information for aspiring natural product researchers as well as a useful reference guide to all of the available techniques for the more experienced among us.

Satyajit D. Sarker
Zahid Latif
Alexander I. Gray

Preface to First Edition

Biodiversity is a term commonly used to denote the variety of species and the multiplicity of forms of life. But this variety is deeper than is generally imagined. In addition to the processes of primary metabolism that involve essentially the same chemistry across great swathes of life, there are a myriad of secondary metabolites—natural products—usually confined to a particular group of organisms, or to a single species, or even to a single strain growing under certain conditions. In most cases we do not really know what biological role these compounds play, except that they represent a treasure trove of chemistry that can be of both interest and benefit to us. Tens of thousands of natural products have been described, but in a world where we are not even close to documenting all the extant species, there are almost certainly many more thousands of compounds waiting to be discovered.

The purpose of *Natural Products Isolation* is to give some practical guidance in the process of extraction and isolation of natural products. Literature reports tend to focus on natural products once they have been isolated—on their structural elucidation, or their biological or chemical properties. Extraction details are usually minimal and sometimes nonexistent, except for a mention of the general techniques used. Even when particular conditions of a separation are reported, they assume knowledge of the practical methodology required to carry out the experiment, and of the reasoning behind the conditions used. *Natural Products Isolation* aims to provide the foundation of this knowledge. Following an introduction to the isolation process, there are a series of chapters dealing with the major techniques used, followed by chapters on other aspects of isolation, such as those related to particular sample types, taking short cuts, or making the most of the isolation process. The emphasis is not so much on the isolation of a known natural product for which there may already be reported methods, but on the isolation of compounds of unknown identity.

Every natural product isolation is different and so the process is not really suited to a practical manual that gives detailed recipe-style methods. However, the aim has been to give as much practical direction and advice as possible, together with examples, so that the potential extractor can at least make a reasonable attempt at an isolation.

Natural Products Isolation is aimed mainly at scientists with little experience of natural products extraction, such as research students undertaking natural products-based research, or scientists from other disciplines who find they wish to isolate a small molecule from a biological mixture. However, there may also be something of interest for more experienced natural products scientists who wish to explore other methods of extraction, or use the book as a general reference. In particular, it is hoped that the book will be of value to scientists in less scientifically developed countries, where there is little experience of natural products work, but where there is great biodiversity and, hence, great potential for utilizing and sustaining that biodiversity through the discovery of novel, useful natural products.

Richard J. P. Cannell

In memory of Richard John Painter Cannell—b. 1960; d. 1999

Contents

Contributors

RUSSELL A. BARROW • *Microbial Natural Product Research Laboratory, Department of Chemistry, The Australian National University, Canberra, Australia*
RICHARD J. P. CANNELL • *Formerly, Glaxo Wellcome Research and Development, Stevenage, Herts, UK*
LAURENCE DINAN • *Inse Biochemistry Group, Hatherly Laboratories, University of Exeter, Exeter, Devan, UK*
DAVID G. DURHAM • *School of Pharmacy, The Robert Gordon University, Aberdeen, Scotland, UK*
ALASTAIR J. FLORENCE • *Department of Pharmaceutical Sciences, University of Strathclyde, Glasgow, Scotland, UK*
SIMON GIBBONS • *Centre for Pharmacognosy and Phytotherapy, The School of Pharmacy, University of London, London, UK*
ALEXANDER I. GRAY • *Phytochemistry Research Laboratories, Department of Pharmaceutical Sciences, University of Strathclyde, Glasgow, Scotland, UK*
WAEL E. HOUSSEN • *Marine Natural Products Laboratory, Chemistry Department, Aberdeen University, Aberdeen, Scotland, UK*
MARCEL JASPARS • *Marine Natural Products Laboratory, Chemistry Department, Aberdeen University, Aberdeen, Scotland, UK*
ANDREA JOHNSTON • *Department of Pharmaceutical Sciences, University of Strathclyde, Glasgow, Scotland, UK*
WILLIAM P. JONES • *College of Pharmacy, Medicinal Chemistry and Pharmacognosy, University of Illinois at Chicago, Chicago, IL*
DAVID A. KAU • *Cubist Pharmaceuticals (UK) Ltd, Berkshire, UK*
A. DOUGLAS KINGHORN • *College of Pharmacy, Medicinal Chemistry and Pharmacognosy, Ohio State University, Columbus, OH*
ZAHID LATIF • *Molecular Nature Limited, Plas Gogerddan, Aberystwyth, Wales, UK*
BO LI • *Kunming Institute of Botany, Chinese Academy of Science, Kunming, China*
STEVEN M. MARTIN • *Cubist Pharmaceuticals (UK) Ltd, Slough, Berkshire, UK*
JAMES B. MCALPINE • *Ecopia BioSciences Inc., Frederick Banting, Saint Laurent, Quebec, Canada*

PATRICK MORRIS • *Ecopia BioSciences Inc., Frederick Banting, Saint Laurent, Quebec, Canada*

LUTFUN NAHAR • *School of Life Sciences, The Robert Gordon University, Aberdeen, Scotland, UK*

HIDEAKI OTSUKA • *Department of Pharmacognosy, Graduate School of Biomedical Sciences, Hiroshima University, Minami-ku, Hiroshima, Japan*

RAYMOND G. REID • *Phytopharmaceutical Research Laboratory, School of Pharmacy, The Robert Gordon University, Aberdeen, Scotland, UK*

SATYAJIT D. SARKER • *Pharmaceutical Biotechnology Research Group, School of Biomedical Sciences, University of Ulster at Coleraine, Coleraine, Northern Ireland, UK*

VÉRONIQUE SEIDEL • *Phytochemistry Research Laboratories, Department of Pharmaceutical Sciences, University of Strathclyde, Glasgow, Scotland, UK*

NORMAN SHANKLAND • *Department of Pharmaceutical Sciences, University of Strathclyde, Glasgow, Scotland, UK*

YUZURU SHIMIZU • *Department of Biomedical and Pharmaceutical Sciences, University of Rhode Island, Kingston, RI*

STEPHEN K. WRIGLEY • *Cubist Pharmaceuticals (UK) Ltd, Slough, Berkshire, UK*

1

Natural Product Isolation

An Overview

Satyajit D. Sarker, Zahid Latif, and Alexander I. Gray

Summary

There has been a remarkable resurgence of interest in natural product research over the last decade or so. With the outstanding developments in the areas of separation science, spectroscopic techniques, and microplate-based ultrasensitive *in vitro* assays, natural product research is enjoying renewed attention for providing novel and interesting chemical scaffolds. The various available hyphenated techniques, e.g., GC-MS, LC-PDA, LC-MS, LC-FTIR, LC-NMR, LC-NMR-MS, CE-MS, have made possible the preisolation analyses of crude extracts or fractions from different natural sources, isolation and on-line detection of natural products, chemotaxonomic studies, chemical finger printing, quality control of herbal products, dereplication of natural products, and metabolomic studies. While different chapters in this book are devoted to a number of specific aspects of natural product isolation protocols, this chapter presents, with practical examples, a general overview of the processes involved in natural product research, starting from extraction to determination of the structures of purified products and their biological activity.

Key Words: Natural products; secondary metabolite; extraction; isolation; bioassay.

1. Introduction

Products of natural origins can be called "natural products." Natural products include: (1) an entire organism (e.g., a plant, an animal, or a

From: *Methods in Biotechnology, Vol. 20, Natural Products Isolation, 2nd ed.*
Edited by: S. D. Sarker, Z. Latif, and A. I. Gray © Humana Press Inc., Totowa, NJ

microorganism) that has not been subjected to any kind of processing or treatment other than a simple process of preservation (e.g., drying), (2) part of an organism (e.g., leaves or flowers of a plant, an isolated animal organ), (3) an extract of an organism or part of an organism, and exudates, and (4) pure compounds (e.g., alkaloids, coumarins, flavonoids, glycosides, lignans, steroids, sugars, terpenoids, etc.) isolated from plants, animals, or microorganisms *(1)*. However, in most cases the term natural products refers to secondary metabolites, small molecules (mol wt <2000 amu) produced by an organism that are not strictly necessary for the survival of the organism. Concepts of secondary metabolism include products of overflow metabolism as a result of nutrient limitation, shunt metabolism produced during idiophase, defense mechanism regulator molecules, etc. *(2)*. Natural products can be from any terrestrial or marine source: plants (e.g., paclitaxel [Taxol®] from *Taxus brevifolia*), animals (e.g., vitamins A and D from cod liver oil), or microorganisms (e.g., doxorubicin from *Streptomyces peucetius*).

Strategies for research in the area of natural products have evolved quite significantly over the last few decades. These can be broadly divided into two categories:

1. Older strategies:
   a. Focus on chemistry of compounds from natural sources, but not on activity.
   b. Straightforward isolation and identification of compounds from natural sources followed by biological activity testing (mainly *in vivo*).
   c. Chemotaxonomic investigation.
   d. Selection of organisms primarily based on ethnopharmacological information, folkloric reputations, or traditional uses.

2. Modern strategies:
   a. Bioassay-guided (mainly *in vitro*) isolation and identification of active "lead" compounds from natural sources.
   b. Production of natural products libraries.
   c. Production of active compounds in cell or tissue culture, genetic manipulation, natural combinatorial chemistry, and so on.
   d. More focused on bioactivity.
   e. Introduction of the concepts of dereplication, chemical fingerprinting, and metabolomics.
   f. Selection of organisms based on ethnopharmacological information, folkloric reputations, or traditional uses, and also those randomly selected.

A generic protocol for the drug discovery from natural products using a bioassay-guided approach is presented in **Fig. 1.**

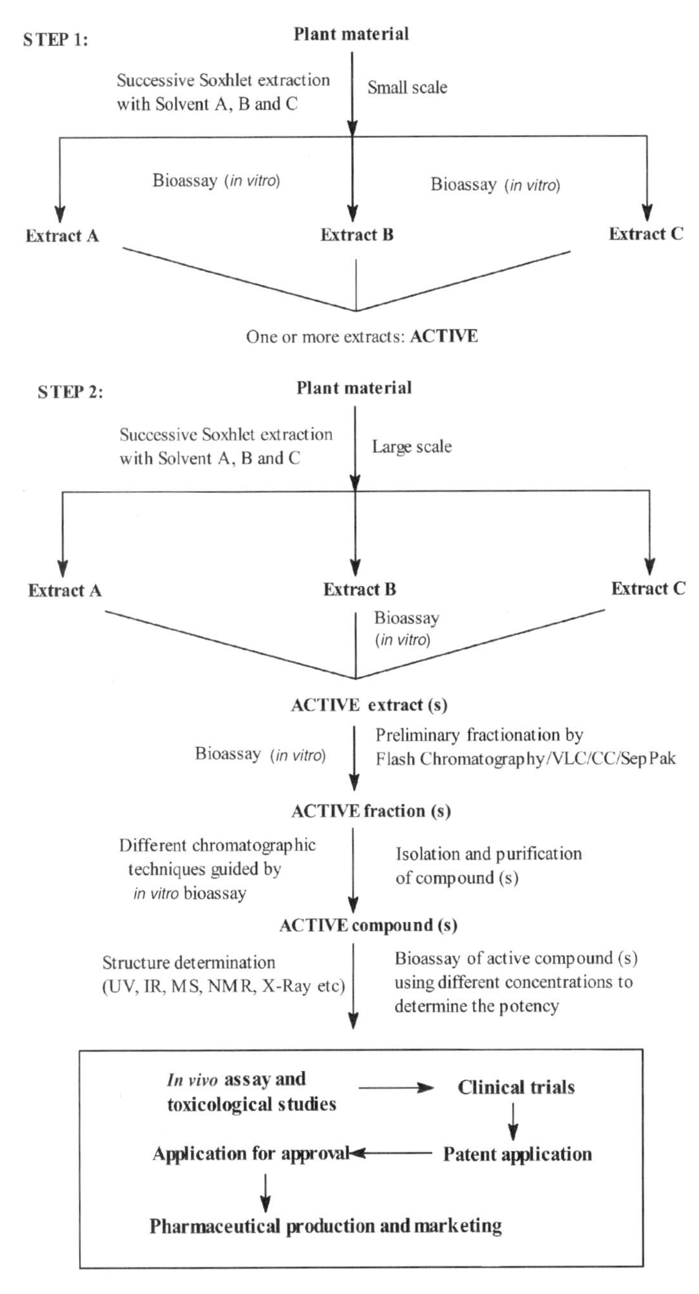

Fig. 1. An example of natural product drug discovery process (bioassay-guided approach).

2. Natural Products: Historical Perspective

The use of natural products, especially plants, for healing is as ancient and universal as medicine itself. The therapeutic use of plants certainly goes back to the Sumerian civilization, and 400 years before the Common Era, it has been recorded that Hippocrates used approximately 400 different plant species for medicinal purposes. Natural products played a prominent role in ancient traditional medicine systems, such as Chinese, Ayurveda, and Egyptian, which are still in common use today. According to the World Health Organization (WHO), 75% of people still rely on plant-based traditional medicines for primary health care globally. A brief summary of the history of natural product medicine is presented in **Table 1.**

3. Natural Products: Present and Future

Nature has been a source of therapeutic agents for thousands of years, and an impressive number of modern drugs have been derived from natural sources, many based on their use in traditional medicine. Over the last

Table 1
History of Natural Product Medicine

Period	Type	Description
Before 3000 BC	Ayurveda (knowledge of life) Chinese traditional medicine	Introduced medicinal properties of plants and other natural products
1550 BC	*Ebers Papyrus*	Presented a large number of crude drugs from natural sources (e.g., castor seeds and gum arabic)
460–377 BC	Hippocrates, "The Father of Medicine"	Described several plants and animals that could be sources of medicine
370–287 BC	Theophrastus	Described several plants and animals that could be sources of medicine
23–79 AD	Pliny the Elder	Described several plants and animals that could be sources of medicine
60–80 AD	Dioscorides	Wrote *De Materia Medica*, which described more than 600 medicinal plants
131–200 AD	Galen	Practiced botanical medicines (Galenicals) and made them popular in the West
15th century	Kräuterbuch (herbals)	Presented information and pictures of medicinal plants

century, a number of top selling drugs have been developed from natural products (vincristine from *Vinca rosea*, morphine from *Papaver somniferum*, Taxol® from *T. brevifolia*, etc.). In recent years, a significant revival of interest in natural products as a potential source for new medicines has been observed among academia as well as pharmaceutical companies. Several modern drugs (~40% of the modern drugs in use) have been developed from natural products. More precisely, according to Cragg et al. ***(3)***, 39% of the 520 new approved drugs between 1983 and 1994 were natural products or their derivatives, and 60–80% of antibacterial and anticancer drugs were from natural origins. In 2000, approximately 60% of all drugs in clinical trials for the multiplicity of cancers had natural origins. In 2001, eight (simvastatin, pravastatin, amoxycillin, clavulanic acid, azithromycin, ceftriaxone, cyclosporin, and paclitaxel) of the 30 top-selling medicines were natural products or their derivatives, and these eight drugs together totaled US $16 billion in sales.

Apart from natural product-derived modern medicine, natural products are also used directly in the "natural" pharmaceutical industry, which is growing rapidly in Europe and North America, as well as in traditional medicine programs being incorporated into the primary health care systems of Mexico, the People's Republic of China, Nigeria, and other developing countries. The use of herbal drugs is once again becoming more popular in the form of food supplements, nutraceuticals, and complementary and alternative medicine.

Natural products can contribute to the search for new drugs in three different ways:

1. by acting as new drugs that can be used in an unmodified state (e.g., vincristine from *Catharanthus roseus*).
2. by providing chemical "building blocks" used to synthesize more complex molecules (e.g., diosgenin from *Dioscorea floribunda* for the synthesis of oral contraceptives).
3. by indicating new modes of pharmacological action that allow complete synthesis of novel analogs (e.g., synthetic analogs of penicillin from *Penicillium notatum*).

Natural products will certainly continue to be considered as one of the major sources of new drugs in the years to come because

1. they offer incomparable structural diversity.
2. many of them are relatively small (<2000 Da).
3. they have "drug-like" properties (i.e., they can be absorbed and metabolized).

Only a small fraction of the world's biodiversity has been explored for bioactivity to date. For example, there are at least 250,000 species of higher plants that exist on this planet, but merely 5–10% of these have been investigated so far. In addition, reinvestigation of previously studied plants has continued to produce new bioactive compounds that have drug potential. Much less is known about marine organisms than other sources of natural products. However, research up to now has shown that they represent a valuable source for novel bioactive compounds. With the development of new molecular targets, there is an increasing demand for novel molecular diversity for screening. Natural products certainly play a crucial role in meeting this demand through the continued investigation of the world's biodiversity, much of which remains unexplored ***(4)***. With less than 1% of the microbial world currently known, advances in technologies for microbial cultivation and the extraction of nucleic acids from environmental samples from soil and marine habitats will offer access to an untapped reservoir of genetic and metabolic diversity ***(5)***. This is also true for nucleic acids isolated from symbiotic and endophytic microbes associated with terrestrial and marine macroorganisms.

Advent, introduction, and development of several new and highly specific *in vitro* bioassay techniques, chromatographic methods, and spectroscopic techniques, especially nuclear magnetic resonance (NMR), have made it much easier to screen, isolate, and identify potential drug lead compounds quickly and precisely. Automation of these methods now makes natural products viable for high-throughput screening (HTS).

4. Extraction

The choice of extraction procedure depends on the nature of the source material and the compounds to be isolated. Prior to choosing a method, it is necessary to establish the target of the extraction. There can be a number of targets; some of these are mentioned here.

1. An unknown bioactive compound.
2. A known compound present in an organism.
3. A group of compounds within an organism that are structurally related.
4. All secondary metabolites produced by one natural source that are not produced by a different "control" source, e.g., two species of the same genus or the same species grown under different conditions.
5. Identification of all secondary metabolites present in an organism for chemical fingerprinting or metabolomics study (*see* Chap. 9).

It is also necessary to seek answers to the questions related to the expected outcome of the extraction. These include:

1. Is this extraction for purifying a sufficient amount of a compound to characterize it partially or fully? What is the required level of purity (*see* **Note 1**)?
2. Is this to provide enough material for confirmation or denial of a proposed structure of a previously isolated compound (*see* **Note 2**)?
3. Is this to produce as much material as possible so that it can be used for further studies, e.g., clinical trial?

The typical extraction process, especially for plant materials (*see* Chap. 13), incorporates the following steps:

1. Drying and grinding of plant material or homogenizing fresh plant parts (leaves, flowers, etc.) or maceration of total plant parts with a solvent.
2. Choice of solvents
   a. Polar extraction: water, ethanol, methanol (MeOH), and so on.
   b. Medium polarity extraction: ethyl acetate (EtOAc), dichloromethane (DCM), and so on.
   c. Nonpolar: *n*-hexane, pet-ether, chloroform ($CHCl_3$), and so on.
3. Choice of extraction method
   a. Maceration.
   b. Boiling.
   c. Soxhlet.
   d. Supercritical fluid extraction.
   e. Sublimation.
   f. Steam distillation.

The fundamentals of various initial and bulk extraction techniques for natural products are detailed in Chapters 2 and 3.

5. Fractionation

A crude natural product extract is literally a cocktail of compounds. It is difficult to apply a single separation technique to isolate individual compounds from this crude mixture. Hence, the crude extract is initially separated into various discrete fractions containing compounds of similar polarities or molecular sizes. These fractions may be obvious, physically discrete divisions, such as the two phases of a liquid–liquid extraction (*see* Chap. 10) or they may be the contiguous eluate from a chromatography column, e.g., vacuum liquid chromatography (VLC), column chromatography (CC), size-exclusion chromatography (SEC), solid-phase extraction (SPE), etc. (*see* Chaps. 5,

13–15). For initial fractionation of any crude extract, it is advisable not to generate too many fractions, because it may spread the target compound over so many fractions that those containing this compound in low concentrations might evade detection. It is more sensible to collect only a few large, relatively crude ones and quickly home in on those containing the target compound. For finer fractionation, often guided by an on-line detection technique, e.g., ultraviolet (UV), modern preparative, or semipreparative high-performance liquid chromatography (HPLC) can be used.

6. Isolation

The most important factor that has to be considered before designing an isolation protocol is the nature of the target compound present in the crude extracts or fractions. The general features of the molecule that are helpful to ascertain the isolation process include solubility (hydrophobicity or hydrophilicity), acid–base properties, charge, stability, and molecular size. If isolating a known compound from the same or a new source, it is easy to obtain literature information on the chromatographic behavior of the target compound, and one can choose the most appropriate method for isolation without any major difficulty. However, it is more difficult to design an isolation protocol for a crude extract where the types of compounds present are totally unknown. In this situation, it is advisable to carry out qualitative tests for the presence of various types of compounds, e.g., phenolics, steroids, alkaloids, flavonoids, etc., as well as analytical thin-layer chromatography (TLC), (*see* Chap. 4) or HPLC profiling (*see* Chaps. 5, 8, and 9). The nature of the extract can also be helpful for choosing the right isolation protocol. For example, a MeOH extract or fractions from this extract containing polar compounds are better dealt with using reversed-phase HPLC (RP-HPLC). Various physical properties of the extracts can also be determined with a small portion of the crude extract in a series of small batch-wise experiments. Some of these experiments are summarized below.

1. *Hydrophobicity or hydrophilicity:* An indication of the polarity of the extract as well as the compounds present in the extract can be determined by drying an aliquot of the mixture and trying to redissolve it in various solvents covering the range of polarities, e.g., water, MeOH, acetonitrile (ACN), EtOAc, DCM, $CHCl_3$, petroleum ether, *n*-hexane, etc. The same information can be obtained by carrying out a range of solvent partitioning, usually between water

and EtOAc, $CHCl_3$, DCM, or *n*-hexane, followed by an assay to determine the distribution of compounds in solvent fractions.

2. *Acid–base properties:* Carrying out partitioning in aqueous solvents at a range of pH values, typically 3, 7, and 10, can help determine the acid–base property of the compounds in an extract. It is necessary to adjust the aqueous solution or suspension with a drop or two of mineral acid or alkali (a buffer can also be used), followed by the addition of organic solvent and solvent extraction. Organic and aqueous phases are assessed, preferably by TLC, for the presence of compounds. This experiment can also provide information on the stability of compounds at various pH values.
3. *Charge:* Information on the charge properties of the compound can be obtained by testing under batch conditions, the effect of adding various ion exchangers to the mixture. This information is particularly useful for designing any isolation protocol involving ion exchange chromatography (*see* Chap. 6).
4. *Heat stability:* A typical heat stability test involves incubation of the sample at ~90°C for 10 min in a water bath followed by an assay for unaffected compounds. It is particularly important for bioassay-guided isolation, where breakdown of active compounds often leads to the loss or reduction of biological activity. If the initial extraction of natural products is carried out at a high temperature, the test for heat stability becomes irrelevant.
5. *Size:* Dialysis tubing can be used to test whether there are any macromolecules, e.g., proteins, present in the extract. Macromolecules are retained within the tubing, allowing small (<2000 amu) secondary metabolites to pass through it. The necessity of the use of any SEC in the isolation protocol can be ascertained in this way.

The chromatographic techniques used in the isolation of various types of natural products can be broadly classified into two categories: classical or older, and modern.

Classical or older chromatographic techniques include:

1. Thin-layer chromatography (TLC).
2. Preparative thin-layer chromatography (PTLC).
3. Open-column chromatography (CC).
4. Flash chromatography (FC).

Modern chromatographic techniques are:

1. High-performance thin-layer chromatography (HPTLC).
2. Multiflash chromatography (e.g., Biotage®).
3. Vacuum liquid chromatography (VLC).
4. Chromatotron.
5. Solid-phase extraction (e.g., Sep-Pak®).

6. Droplet countercurrent chromatography (DCCC).
7. High-performance liquid chromatography (HPLC).
8. Hyphenated techniques (e.g., HPLC-PDA, LC-MS, LC-NMR, LC-MS-NMR).

Details about most of these techniques and their applications in the isolation of natural products can be found in Chapters 4–9 and 13–16. A number of isolation protocols are presented in **Figs. 2–6**.

6.1. Isolation of Spirocardins A and B From Nocardia sp

An outline of the general protocol described by Nakajima et al. *(6)* for the isolation of diterpene antibiotics, spirocardins A and B, from a fermentation broth of *Nocardia* sp., is presented in **Fig. 2**. The compounds were present in the broth filtrate, which was extracted twice with EtOAc (half-volume of supernatant). The pooled EtOAc fraction was concentrated by evaporation under vacuum, washed with an equal volume of water saturated with sodium chloride (NaCl), and further reduced to obtain an oil. This crude oil was redissolved in a minimal volume of EtOAc and subjected to silica gel CC eluting with *n*-hexane containing increasing amounts of acetone. It resulted in two fractions containing spirocardin A and spirocardin B, respectively, as the main components. Further purification was achieved by silica gel CC and RP-HPLC. For silica gel CC at this stage, an eluent of benzene–EtOAc mixture was used. Nowadays, benzene is no longer in use as a chromatographic solvent because of its carcinogenicity.

6.2. Isolation of Cispentacin From Bacillus cereus

Konishi et al. *(7)* presented an isolation protocol (**Fig. 3**) for an antifungal antibiotic, cispentacin, from a fermentation broth of *B. cereus*. This is an excellent example of the application of ion-exchange chromatography in natural product isolation. The broth supernatant was applied directly onto the ion-exchange column without any prior treatment. The final step of the isolation process employed CC on activated charcoal to yield cispentacin of 96% purity, which was further purified by recrystallization from acetone–ethanol–water.

6.3. Isolation of Phytoecdysteroids From Limnanthes douglasii

A convenient method (**Fig. 4**) for the isolation of two phytoecdysteroid glycosides, limnantheosides A and B, and two phytoecdysteroids,

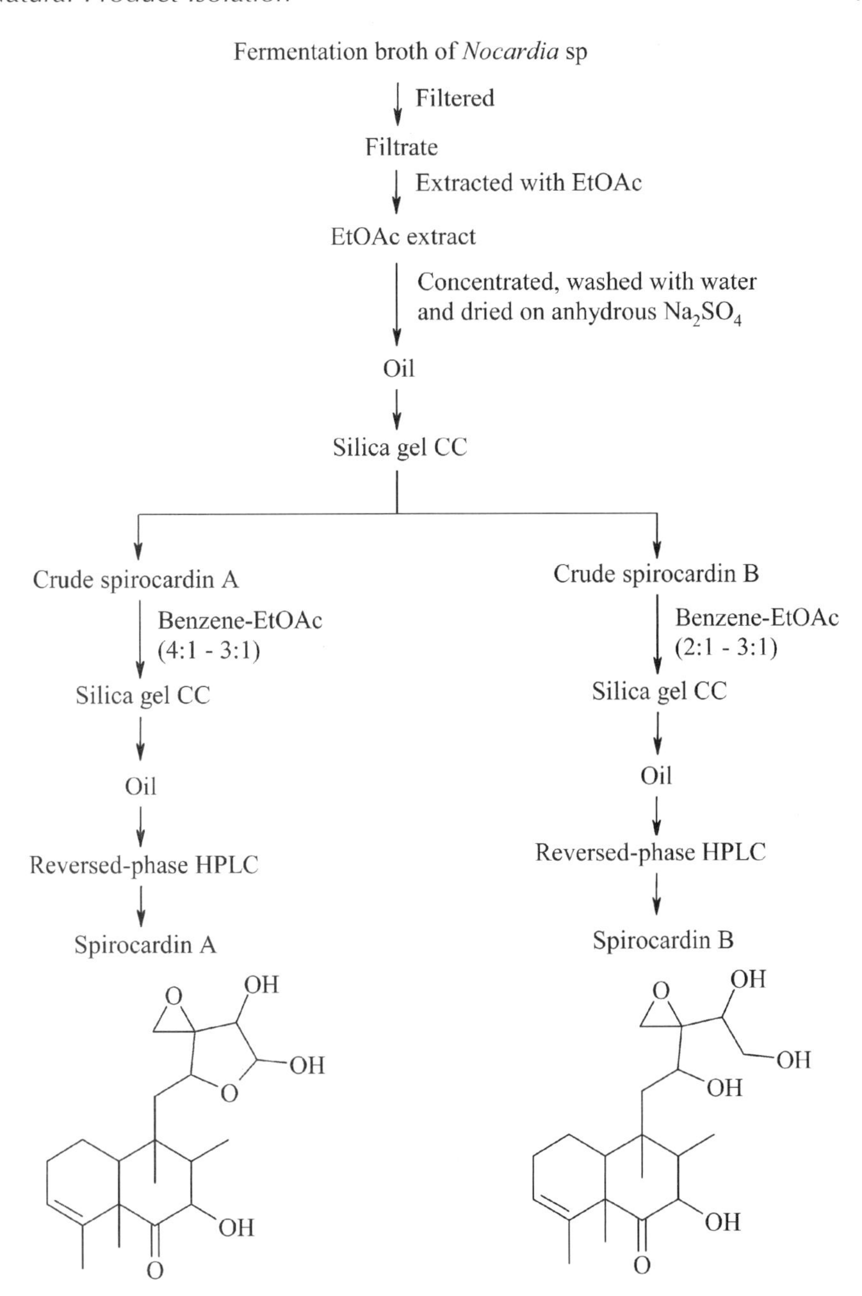

Fig. 2. Isolation of microbial natural products: spirocardins A and B from *Nocardia* sp.

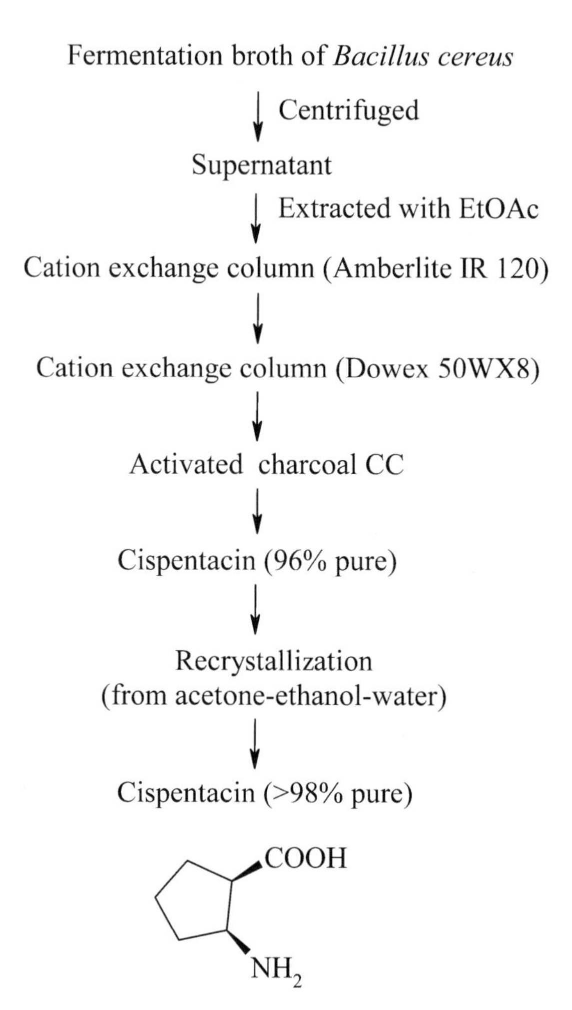

Fig. 3. Isolation of microbial natural products: cispentacin from *B. cereus*.

20-hydroxyecdysone and ponasterone A, using a combination of solvent extraction, SPE, and preparative RP-HPLC, was outlined by Sarker et al. ***(8)***. Ground seeds (50 g) were extracted (4×24 h) with 4×200 mL MeOH at 50°C with constant stirring using a magnetic stirrer. Extracts were pooled and H_2O added to give a 70% aqueous methanolic solution. After being defatted with *n*-hexane, the extract was concentrated using a rotary evaporator. SPE (Sep-Pak fractionation) of the concentrated extract (redissolved in 10% aq MeOH) using MeOH–H_2O step gradient, followed

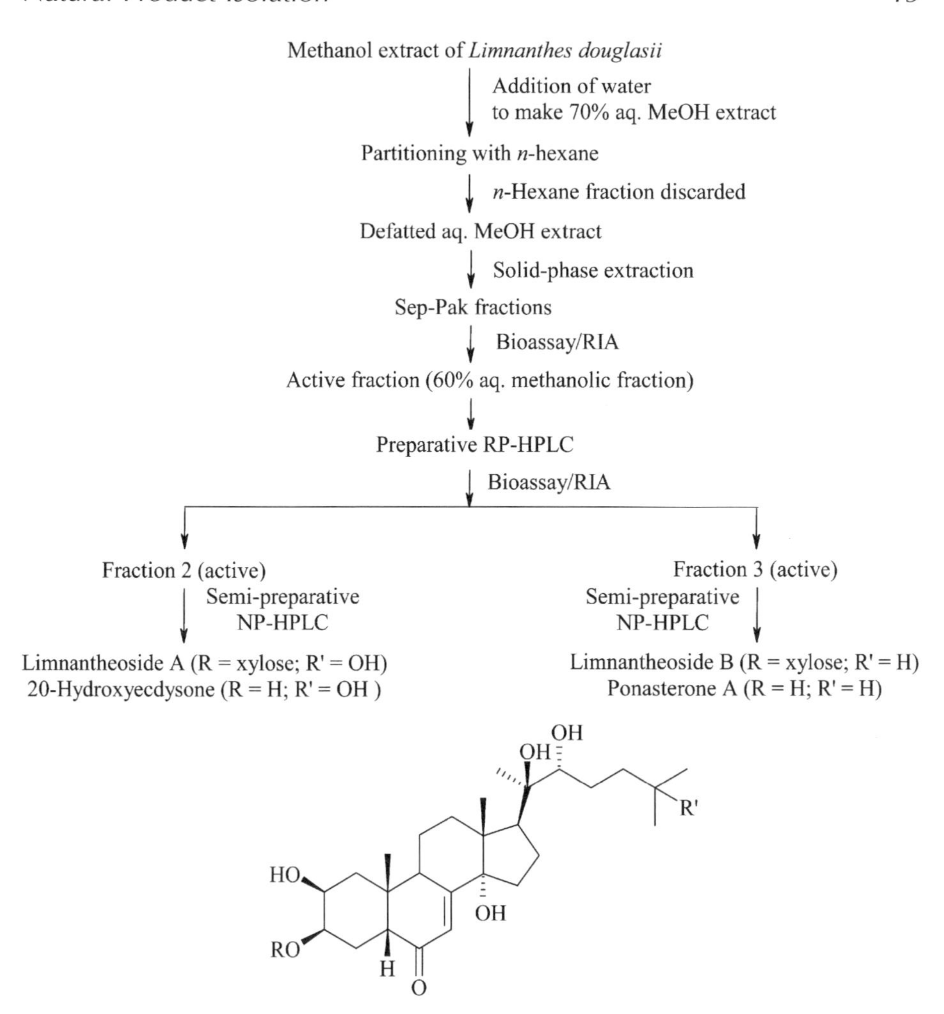

Fig. 4. Isolation of plant natural products: phytoecdysteroids from *L. douglasii.*

by ecdysteroid bioassay/RIA revealed the presence of ecdysteroids in the 60% MeOH–H_2O fraction, which was then subjected to HPLC using a preparative RP-column (isocratic elution with 55% MeOH–H_2O, 5 mL/min) to yield five fractions. Fractions 2 (R_t 18–20 min) and 3 (R_t 33–36 min) were found to be bioassay/RIA positive. Further NP-HPLC analyses of fraction 2 on NP-semiprep diol column (isocratic elution with 6% MeOH in DCM, 2 mL/min) produced 20-hydroxyecdysone (purity > 99%, R_t 13.1 min) and limnantheoside A (purity > 99%, R_t 19.2 min).

Similar purification of fraction 3 yielded ponasterone A (purity >99%, R_t 5.2 min) and limnantheoside B (purity >99%, R_t 10.8 min).

6.4. *Isolation of Moschatine, a Steroidal Glycoside, From* Centaurea moschata

Moschatine, a steroidal glycoside, was isolated from the seeds of *C. moschata* *(9)*. The isolation protocol (**Fig. 5**) involved successive Soxhlet extraction of the ground seeds with *n*-hexane, $CHCl_3$, and MeOH, followed by preparative RP-HPLC (C_{18} preparative column, isocratic elution with 55% MeOH in water, 5 mL/min). Final purification was carried out by RP-HPLC using a semipreparative C_6 column, eluted isocratically with 45% MeOH in water, 2 mL/min, to yield moschatine with a purity of >98%.

6.5. *Isolation of Saponins From* Serjania salzmanniana

The isolation of antifungal and molluscicidal saponins (**Fig. 6**) from *S. salzmanniana* involved the use of silica gel CC followed by countercurrent chromatography *(10)*. An unconventional feature of the final preparative TLC stage was the use of water as a nondestructive visualization "stain." The TLC plate turned dark (wet) when sprayed with water, except those regions represented by the sapogenins, which because of their hydrophobicity, remained white (dry).

7. Quantification

The yield of compounds at the end of the isolation and purification process is important in natural product research. An estimate of the recovery at the isolation stage can be obtained using various routine analytical techniques that may involve the use of a standard. In bioassay-guided isolation, the compound is monitored by bioassay at each stage, and a quantitative assessment of bioactivity of the compound is usually carried out by serial dilution method (*see* **Note 3**). Quantitative bioactivity assessment provides a clear idea about the recovery of the active compound(s) and also indicates whether the activity results from a single or multiple components. During the isolation process, if the activity is lost or reduced to a significant level, the possible reasons could be as follows:

1. The active compound has been retained in the column.
2. The active compound is unstable in the conditions used in the isolation process.

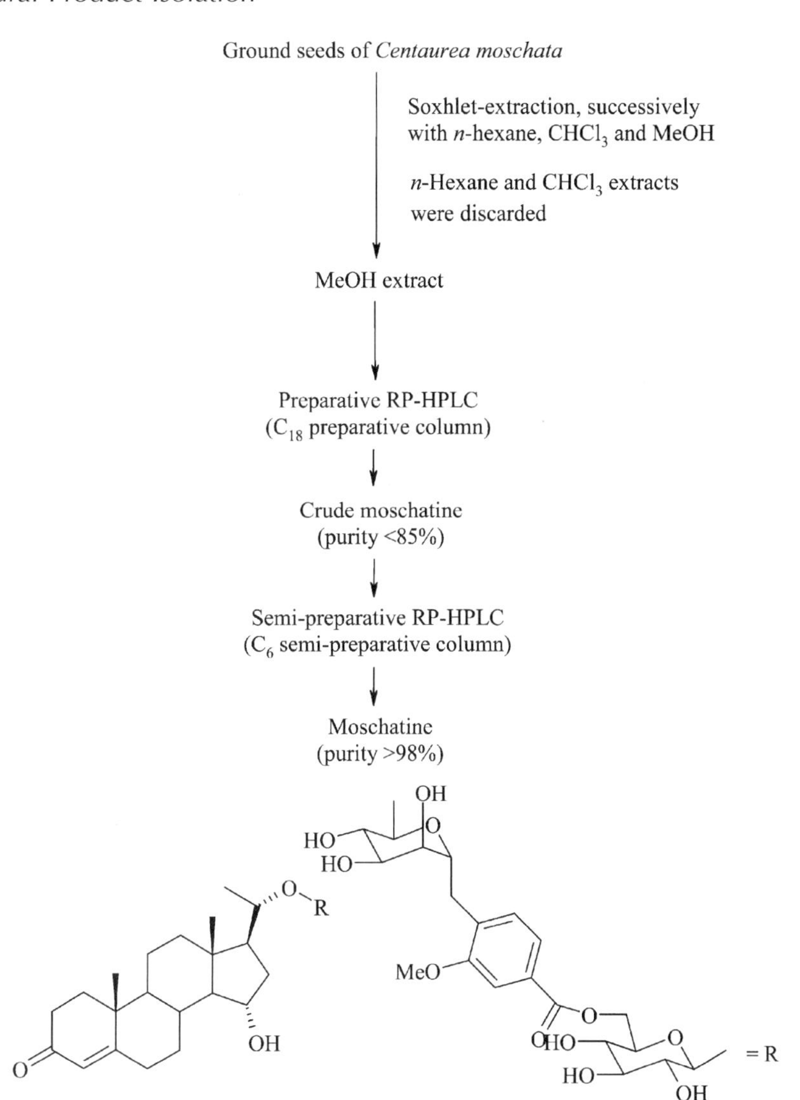

Fig. 5. Isolation of plant natural products: moschatine, a steroidal glycoside from *C. moschata*.

3. The extract solution may not have been prepared in a solvent that is compatible with the mobile phase, so that a large proportion of the active components precipitated out when loading on to the column.

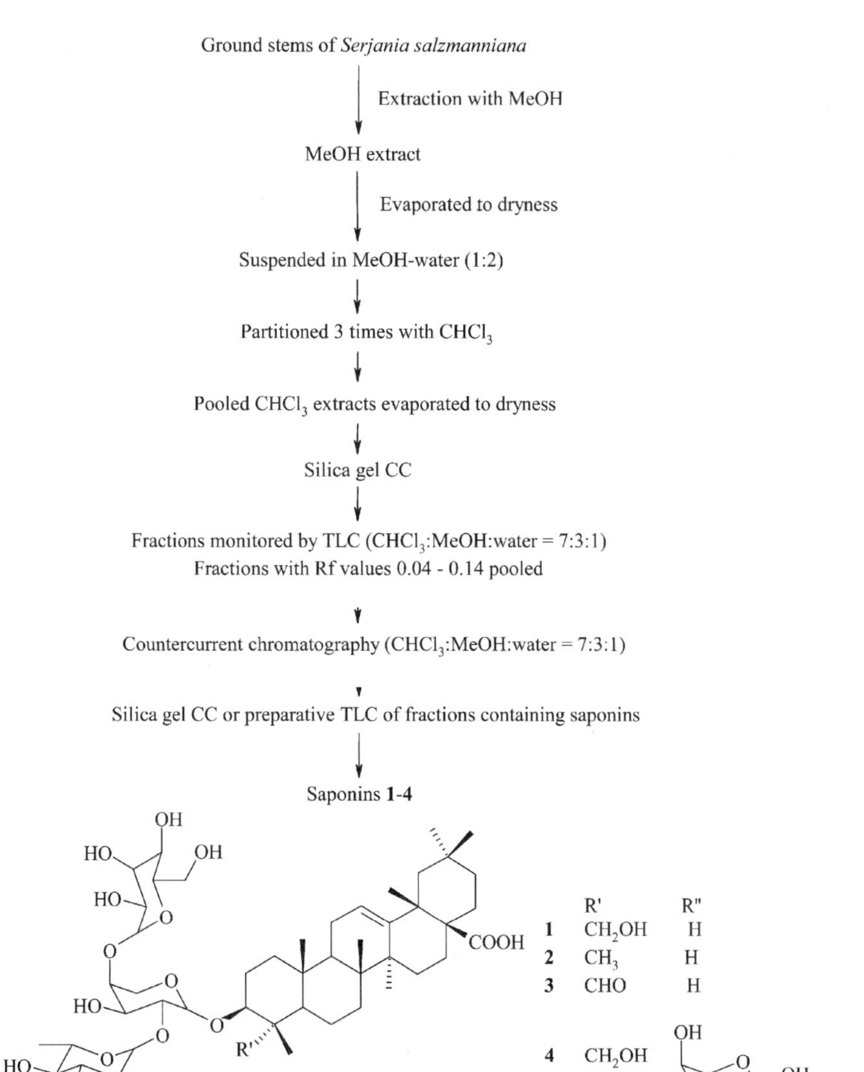

Fig. 6. Isolation of plant natural products: saponins from *S. salzmanniana*.

4. Most of the active component(s) spread across a wide range of fractions, causing undetectable amounts of component(s) present in the fractions.
5. The activity of the extract is probably because of the presence of synergy among a number of compounds, which, when separated, are not active individually.

8. "Poor-Yield" Problem

Poor yield or poor recovery is one of the major problems in natural product isolation. For example, only 30 g of vincristine was obtained from 15 t of dried leaves of *V. rosea* (or *C. roseus*) ***(11)***. Similarly, to obtain 1900 g of Taxol®, the felling of 6000 extremely slow-growing trees, *Taxus brevifolia*, was necessary to produce 27,300 kg of the bark. To tackle this poor-yield problem, especially in the case of Taxol®, a meeting was organized by the National Cancer Institute in Washington, D.C., in June 1990, where four suggestions were made:

1. Finding a better source for the supply of Taxol®, such as a different species or a cultivar of *Taxus*, or a different plant part or cultivation conditions.
2. Semisynthesis of Taxol® from a more abundant precursor.
3. Total synthesis of Taxol®.
4. Tissue culture production of Taxol® or a close relative.

Out of these four ways, the most successful one was semisynthesis. While three successful total syntheses of Taxol® have been achieved, they have not been proven to be economically better than the semisynthetic approach.

9. Structure Elucidation

In most cases of extraction and isolation of natural products, the end point is the identification of the compound or the conclusive structure elucidation of the isolated compound. However, structure elucidation of compounds isolated from plants, fungi, bacteria, or other organisms is generally time consuming, and sometimes can be the "bottleneck" in natural product research. There are many useful spectroscopic methods of getting information about chemical structures, but the interpretation of these spectra normally requires specialists with detailed spectroscopic knowledge and wide experience in natural product chemistry. With the remarkable advances made in the area of artificial intelligence and computing, there are a number of excellent automated structure elucidation programs available that could be extremely useful ***(12,13)***.

If the target compound is known, it is often easy to compare preliminary spectroscopic data with literature data or to make direct comparison with the standard sample. However, if the target compound is an unknown and complex natural product, a comprehensive and systematic approach involving a variety of physical, chemical, and spectroscopic techniques is required. Information on the chemistry of the genus or the family of plant or microbe under investigation could sometimes provide additional hints regarding the possible chemical class of the unknown compound. The following spectroscopic techniques are generally used for the structure determination of natural products:

1. *Ultraviolet-visible spectroscopy (UV-vis):* Provides information on chromophores present in the molecule. Some natural products, e.g., flavonoids, isoquinoline alkaloids, and coumarins, to name a few, can be primarily characterized (chemical class) from characteristic absorption peaks.
2. *Infrared spectroscopy (IR):* Determines different functional groups, e.g., —C=O, —OH, —NH_2, aromaticity, and so on, present in a molecule.
3. *Mass spectrometry (MS):* Gives information about the molecular mass, molecular formula, and fragmentation pattern. Most commonly used techniques are: electron impact mass spectrometry (EIMS), chemical ionization mass spectrometry (CIMS), electrospray ionization mass spectrometry (ESIMS), and fast atom bombardment mass spectrometry (FABMS).
4. *NMR:* Reveals information on the number and types of protons and carbons (and other elements like nitrogen, fluorine, etc.) present in the molecule, and the relationships among these atoms ***(14)***. The NMR experiments used today can be classified into two major categories:
   a. One-dimensional techniques: ^{1}HNMR, ^{13}CNMR, ^{13}CDEPT, ^{13}CPENDANT, ^{13}C *J* mod., nOe-diff., and so on.
   b. Two-dimensional techniques: ^{1}H-^{1}H COSY, ^{1}H-^{1}H DQF-COSY, ^{1}H-^{1}H COSY-lr, ^{1}H-^{1}H NOESY, ^{1}H-^{1}H ROESY, ^{1}H-^{1}H TOCSY (or HOHAHA), ^{1}H-^{13}C HMBC, ^{1}H-^{13}C HMQC, ^{1}H-^{13}C HSQC, HSQC-TOCSY, and the like.

In addition to the above-mentioned spectroscopic techniques, X-ray crystallographic techniques provide information on the crystal structure of the molecule, and polarimetry offers information on the optical activity of chiral compounds.

10. Assays

Chemical, biological, or physical assays are necessary to pinpoint the target compound(s) from a complex natural product extract. At present, natural product research is more focused on isolating target compounds (assay-guided isolation) rather than trying to isolate all compounds present in any extract. The target compounds may be of certain chemical classes, have certain physical properties, or possess certain biological activities. Therefore, appropriate assays should be incorporated in the extraction and isolation protocol.

The following basic points should be borne in mind when carrying out assays of natural products ***(2)***:

1. Samples dissolved or suspended in a solvent different from the original extraction solvent must be filtered or centrifuged to get rid of any insoluble matter.
2. Acidified or basified samples should be readjusted to their original pH to prevent them from interfering with the assay.
3. Positive and negative controls should be incorporated in any assay.
4. Ideally, the assay should be at least semiquantitative, and/or samples should be assayed in a series of dilutions to determine where the majority of the target compounds resides.
5. The assay must be sensitive enough to detect active components in low concentration.

Physical assays may involve the comparison of various chromatographic and spectroscopic behaviors, e.g., HPLC, TLC, LC-MS, CE-MS LC-NMR, and so on, of the target compound with a known standard. Chemical assays involve various chemical tests for identifying the chemical nature of the compounds, e.g., $FeCl_3$ can be used to detect phenolics, Dragendorff's reagent for alkaloids, 2,2-diphenyl-1-picrylhydrazyl (DPPH) for antioxidant compounds ***(15,16)***, and so on.

Bioassays can be defined as the use of a biological system to detect properties (e.g., antibacterial, antifungal, anticancer, anti-HIV, antidiabetic, etc.) of a crude extract, chromatographic fraction, mixture, or a pure compound. Bioassays could involve the use of *in vivo* systems (clinical trials, whole animal experiments), *ex vivo* systems (isolated tissues and organs), or *in vitro* systems (e.g., cultured cells). *In vivo* studies are more relevant to clinical conditions and can also provide toxicity data at the same time. Disadvantages of these studies are costs, need for large amount of test compounds/fractions, complex design, patient requirement, and difficulty in mode of action determination. *In vitro* bioassays are faster (ideal for HTS), and small amounts of test compounds are needed, but might not be relevant to clinical conditions. The trend has now moved from *in vivo* to *in vitro*. Bioassays available today are robust, specific, and more sensitive to even as low as picogram amounts of test compounds. Most of them can be carried out in full or semiautomation (e.g., using 96- or 384-well plates). There are a number of biological assays available to assess various activities, e.g., *Drosophila melanogaster* B_{II} cell line assay for the assessment of compounds with ecdysteroid (*see* **Note 4**) agonist or antagonist activity ***(17)***, antibacterial serial dilution assay using resazurin as indicator of cell growth *(18,19)*, etc. Most of the modern bioassays are microplate-based and require a small amount of extract, fraction, or compound for the assessment of activity. While it is not the intention of this chapter to discuss at great length various assays presently available, a summary of two typical assays used in natural product screening, the DPPH assay and antibacterial serial dilution assay using resazurin as indicator of cell growth, is presented here as an example. Details on various types of bioassays used in the screening of natural products are available in the literature *(20)*.

10.1. DPPH Assay for Antioxidant Activity

DPPH (molecular formula $C_{18}H_{12}N_5O_6$) is used in this assay to assess the free radical scavenging (antioxidant) property of natural products *(15,16)*. Quercetin, a well-known natural antioxidant, is generally used as a positive control. DPPH (4 mg) is dissolved in MeOH (50 mL) to obtain a concentration of 80 µg/mL. This assay can be carried out both qualitatively and quantitatively using UV-Vis spectrometer.

10.1.1. Qualitative Assay

Test extracts, fractions, or compounds are applied on a TLC plate and sprayed with DPPH solution using an atomizer. It is allowed to develop for 30 min. The white spots against a pink background indicate the antioxidant activity.

10.1.2. Quantitative Assay

For the quantitative assay, the stock solution of crude extracts or fractions is prepared using MeOH to achieve a concentration of 10 mg/mL, whereas that for the test compounds and positive standard is prepared at a concentration of 0.5 mg/mL. Dilutions are made to obtain concentrations of 5×10^{-2}, 5×10^{-3}, 5×10^{-4}, 5×10^{-5}, 5×10^{-6}, 5×10^{-7}, 5×10^{-8}, 5×10^{-9}, 5×10^{-10} mg/mL. Diluted solutions (1.00 mL each) are mixed with DPPH (1.00 mL) and allowed to stand for 30 min for any reaction to take place. The UV absorbance of these solutions is recorded at 517 nm. The experiment is usually performed in triplicate and the average absorption is noted for each concentration. The same procedure is followed for the standard (quercetin).

10.2. Antibacterial Serial Dilution Assay Using Resazurin as an Indicator of Cell Growth

Antibacterial activity of extracts, fractions, or purified compounds can be assessed and the minimal inhibitory concentration (MIC) value determined by this assay ***(18,19)***. Sufficient amounts of dried crude extracts are dissolved in dimethyl sulfoxide (DMSO) to obtain stock solutions of 5 mg/mL concentration. For purified compounds, the concentration is normally 1 mg/mL. Ciprofloxacin or any other broad-spectrum antibiotic could be used as a positive control. Normal saline, resazurin solution, and DMSO were used as negative controls. The antibacterial test is performed using the 96-well microplate-based broth dilution method, which utilized resazurin solution as an indicator of bacterial growth. All tests are generally performed in triplicate.

10.2.1. Preparation of Bacterial Species

The bacterial cultures are prepared by incubating a single colony overnight in nutrient agar at 37°C. For each of the bacterial species, 35 g of the bacterial culture is weighed into two plastic centrifuge tubes using aseptic techniques. The containers are covered with laboratory parafilm. The bacterial suspension is then spun down using a centrifuge at 4000 rpm for 10 min. The pellets are resuspended in normal saline (20 mL). The bacterial culture is then centrifuged again at 4000 rpm for another 5 min. This step is repeated twice to obtain a "clean" bacterial culture for the purpose of the bioassay. The supernatant is discarded and the pellets in each of the centrifuge tubes are resuspended in 5 mL of normal saline. The two bacterial suspensions of the same bacteria are added aseptically to a sterile universal bottle, thereby achieving a total volume of 10 mL. The optical density is measured at a wavelength of 500 nm using a CE 272 Linear Readout Ultraviolet Spectrophotometer, and serial dilutions are carried out to obtain an optical density in the range of 0.5–1.0. The actual values are noted and the cell-forming units are calculated using equations from previously provided viability graphs for the particular bacterial species ***(19)***. The bacterial solution is diluted accordingly to obtain a concentration of 5×10^5 CFU/mL.

10.2.2. Preparation of Resazurin Solution

One tablet of resazurin is dissolved in 40 mL sterile distilled water to obtain standard resazurin solution.

10.2.3. Preparation of 96-Well Plates and Assay

The top of the 96-well plates is labeled appropriately. For evaluating the activity of two different extracts, 100 μL of the extracts in DMSO, ciprofloxacin, normal saline, and resazurin solution is pipetted into the first row. The extract is added to two columns each, while the controls to one column each. Normal saline (50 μL) is added to rows 2–11. Using fresh sterile pipet tips, 50 μL of the contents of the first row is transferred to the second row. Serial dilutions are carried out until all the wells contain 50 μL of either extracts or controls in descending concentrations. Resazurin solution (10 μL) is added, which is followed by the addition of 30 μL of triple-strength broth (or triple-strength glucose in the case of *Enterococcus faecalis*) to each of the wells. Finally, 10 μL of bacterial solution of 5×10^5 CFU/mL concentration is added to all the wells starting with row 12. The plates are wrapped with clingfilm to prevent bacterial dehydration, and then incubated overnight for 18 h at 37°C. The presence of bacterial growth is indicated by color change from purple to pink.

11. Conclusion

Currently, there are a number of well-established methods available for extraction and isolation of natural products from various sources. An appropriate protocol for extraction and isolation can be designed only when the target compound(s) and the overall aim have been decided. It is also helpful to obtain as much information as possible on the chemical and physical nature of the compound(s) to be isolated. For unknown natural products, sometimes it may be necessary to try out pilot extraction and isolation methods to find out the best possible method. At the time of choosing a method, one should be open-minded enough to appreciate and weigh the advantages and disadvantages of all available methods, particularly focusing on their efficiency and, obviously, the total cost involved. Continuous progress in the area of separation technology has increased the variety and variability of the extraction and isolation methods that can be successfully utilized in the extraction and isolation of natural products. For any natural product researcher, it is therefore essential to become familiar with the newer approaches. In most cases, extraction and isolation of natural products are followed by structure determination or confirmation of the purified components. With the introduction of various hyphenated techniques (*see* Chap. 9), it is now possible to determine the structure of the compound as separation is carried out, without isolation and purification *(21)*. Because of the phenomenal progress made in the area of MS and NMR in the last few decades, it has now become possible to deduce the structure of a compound in microgram amounts *(22–24)*, thereby further blurring the boundaries between analytical and preparative methods.

12. Notes

1. The conclusive structure determination of an unknown complex natural product using high-field modern 1D and 2D NMR techniques requires the compound to be pure, >90%. The known structure of a compound can be deduced from a less pure one. In X-ray crystallographic studies, materials are required in an extremely pure state, >99.9% pure. For bioassays, it is also important to know the degree of purity of the test compound. The most reliable assay result can be obtained with a compound of ~100% purity, because it excludes any possibilities of having activities resulting from minor impurities.
2. If the extraction is designed just to provide enough material for confirmation or denial of a proposed structure of a previously isolated compound, it may require less material or even partially pure material, because in many cases this does not require mapping out a complete structure from scratch, but perhaps simply a comparison with a standard of known structure.
3. Approximate quantification can be performed by assaying a set of serial dilutions of every fraction at each stage of the separation process. To detect the peaks of activity, it is often necessary to assay the fractions at a range of dilutions, which approximately indicate the relative amounts of activity (proportional to the amount of compound present) in each fraction. Thus, the fraction(s) containing the bulk of the active compounds can be identified, and an approximate estimation of the total amount of activity recovered, relative to starting material, can be obtained.
4. Ecdysteroids, invertebrate steroidal compounds, are insect-molting hormones, and have also been found in various plant species.

References

1. Samuelsson, G. (1999) *Drugs of Natural Origin: A Textbook of Pharmacognosy*. 4th revised ed. Swedish Pharmaceutical Press, Stockholm, Sweden.
2. Cannell, R. J. P. (1998) How to approach the isolation of a natural product, in *Natural Products Isolation*. 1st ed. (Cannell, R. J. P., ed.), Humana Press, New Jersey, pp. 1–51.
3. Cragg, G. M., Newmann, D. J., and Snader, K. M. (1997) Natural products in drug discovery and development. *J. Nat. Prod.* **60,** 52–60.

4. Cragg, G. M. and Newman, D. J. (2001) Natural product drug discovery in the next millennium. *Pharm. Biol.* **39,** 8–17.
5. Cragg, G. M. and Newman, D. J. (2001) Medicinals for the millennia—the historical record. *Ann. N. Y. Acad. Sci.* **953,** 3–25.
6. Nakajima, M., Okazaki, T., Iwado, S., Kinoshita, T., and Haneishi, T. (1989) New diterpenoid antibiotics, spirocardins A and B. *J. Antibiot.* **42,** 1741–1748.
7. Konishi, M., Nishio, M., Saitoh, K., Miyaki, T., Oki, T., and Kawaguchi, H. (1989) Cispentacin, a new antifungal antibiotic I. Production, isolation, physicochemical properties and structure. *J. Antibiot.* **42,** 1749–1755.
8. Sarker, S. D., Girault, J. P., Lafont, R., and Dinan, L. (1997) Ecdysteroid xylosides from *Limnanthes douglasii. Phytochemistry* **44,** 513–521.
9. Sarker, S. D., Sik, V., Dinan, L., and Rees, H. H. (1998) Moschatine: an unusual steroidal glycoside from *Centaurea moschata. Phytochemistry* **48,** 1039–1043.
10. Ekabo, O. A., Farnsworth, N. R., Henderson, T. O., Mao, G., and Mukherjee, R. (1996) Antifungal and molluscicidal saponins from *Serjania salzmanniana. J. Nat. Prod.* **59,** 431–435.
11. Farnsworth, N. R. (1990) The role of ethnopharmacology in drug development, in *Bioactive Compounds from Plants* (Chadwick, D. J. and Marsh, J., eds.), John Wiley and Sons, New York, pp. 2–21.
12. Blinov K. A., Carlson D., Elyashberg M. E., et al. (2003) Computer assisted structure elucidation of natural products with limited 2D NMR data: application of the StrucEluc system. *Magn. Reson. Chem.* **41,** 359–372.
13. Steinbeck, C. (2004) Recent developments in automated structure elucidation of natural products. *Nat. Prod. Rep.* **21,** 512–518.
14. van de Ven, F. J. M. (1995). *Multidimensional NMR in Liquids: Basic Principles and Experimental Methods*, Wiley-VCH, New York, USA.
15. Takao, T., Watanabe, N., Yagi, I., and Sakata, K. (1994) A simple screening method for antioxidants and isolation of several antioxidants produced by marine bacteria from fish and shellfish. *Biosci. Biotechnol Biochem.* **58,** 1780–1783.
16. Kumarasamy, Y., Fergusson, M., Nahar, L., and Sarker, S. D. (2002) Biological activity of moschamindole from *Centaurea moschata. Pharm. Bio.* **40,** 307–310.
17. Dinan, L., Savchenko, T., Whiting, P., and Sarker, S. D. (1999) Plant natural products as insect steroid receptor agonists and antagonists. *Pestic. Sci.* **55,** 331–335.
18. Drummond, A. J. and Waigh, R. D. (2000) *Recent Research Developments in Phytochemistry.* vol. 4 (Pandalai, S. G., ed.) Research Signpost, India, pp. 143–152.

19. Sarker S. D., Eynon E., Fok K., et al. (2003) Screening the extracts of the seeds of *Achillea millefolium*, *Angelica sylvestris* and *Phleum pratense* for antibacterial, antioxidant activities and general toxicity. *Orient. Phar. Exp. Med.* **33,** 157–162.
20. Hostettmann, K. and Wolfender, J.-L. (2001) Application of liquid chromatography and liquid chromatography/NMR for the on-line identification of plant metabolites, in *Bioactive Compounds from Natural Sources* (Tringali, C., ed.), Taylor and Francis, New York, USA, pp. 31–68.
21. Viletinck, A. J. and Apers, S. (2001) Biological screening methods in the search for pharmacologically active natural products, in *Bioactive Compounds from Natural Sources* (Tringali, C., ed.), Taylor and Francis, New York, USA, pp. 1–30.
22. Neri, P. and Tringali, C. (2001) Applications of modern NMR techniques in the structure elucidation of bioactive natural products, in *Bioactive Compounds from Natural Sources* (Tringali, C, ed.), Taylor and Francis, New York, USA, pp. 69–128.
23. Peter-Katalinic, J. (2004) Potential of modern mass spectrometry in structure elucidation of natural products. International Conference on Natural Products and Physiologically Active Substances (ICNOAS-2004), Novosibirsk, Russia.
24. Peter-Katalinic, J. (1994) Analysis of glycoconjugates by fast-atom-bombardment mass-spectrometry and related ms techniques. *Mass Spectrom. Rev.* **13,** 77–98.

2

Initial and Bulk Extraction

Véronique Seidel

Summary

Currently, there is a growing interest in the study of natural products, especially as part of drug discovery programs. Secondary metabolites can be extracted from a variety of natural sources, including plants, microbes, marine animals, insects, and amphibia. This chapter focuses principally on laboratory-scale processes of initial and bulk extraction of natural products from plant and microbial sources. With regard to plant natural products, the steps required for the preparation of the material prior to extraction, including aspects concerning plant selection, collection, identification, drying, and grinding, are detailed. The various methods available for solvent extraction (maceration, percolation, Soxhlet extraction, pressurized solvent extraction, ultrasound-assisted solvent extraction, extraction under reflux, and steam distillation) are reviewed. Further focus is given on the factors that can influence the selection of a method and suitable solvent. Specific extraction protocols for certain classes of compounds are also discussed. Regarding microbial natural products, this chapter covers issues relating to the isolation of microorganisms and presents the extraction methods available for the recovery of metabolites from fermentation broths. Methods of minimizing compound degradation, artifact formation, extract contamination with external impurities, and enrichment of extracts with desired metabolites are also examined.

Key Words: Solid–liquid extraction; extraction methods; initial extraction; bulk extraction; maceration; percolation; Soxhlet extraction; ultrasonification; pressurized solvent extraction; extraction under reflux; steam distillation; infusion; decoction; broth fermentation.

From: *Methods in Biotechnology, Vol. 20, Natural Products Isolation, 2nd ed.*
Edited by: S. D. Sarker, Z. Latif, and A. I. Gray © Humana Press Inc., Totowa, NJ

1. Introduction

The natural products of interest here are small organic molecules (mol wt <2000 amu approx.), which are also frequently called secondary metabolites and are produced by various living organisms. The natural material (or biomass) originates from several sources including plants, microbes (e.g., fungi and filamentous bacteria), marine organisms (e.g., sponges, snails), insects, and amphibia. Unlike the ubiquitous macromolecules of primary metabolism (which are nutrients and factors fundamental for survival, (e.g., polysaccharides, proteins, nucleic acids, lipids), secondary metabolites comprise a range of chemically diverse compounds often specific to a particular species, which are not strictly essential for survival. Nevertheless, there is a growing interest in their study (particularly as part of drug discovery programs) as they represent a formidable reservoir of potentially useful leads for new medicines.

Prior to any isolation and purification work, natural products have to be extracted (or released) from the biomass. This could be with a view to isolate a known metabolite or to isolate and characterize as many compounds as possible (some of unknown structure) in the context of a systematic phytochemical investigation. An initial extraction is performed typically on a small amount of material to obtain a primary extract. This can be as part of a pharmacological study or to gain preliminary knowledge on the exact nature and amount of metabolites present in the material. Once specific metabolites have been identified in the initial extract, it may then become desirable to isolate them in larger quantities. This will involve either recollecting a larger amount of plant material or increasing the scale of the fermentation. In both cases, a bulk or large-scale extraction should follow.

Since natural products are so diverse and present distinct physicochemical properties (e.g., solubility), the question to address is how can these metabolites be extracted efficiently from the material under investigation. Solvent-extraction methods available for the initial and bulk laboratory-scale extraction of natural products from plant and microbial sources (solid–liquid extraction mainly) are presented in this chapter. Focus is also made on particular procedures, useful for removing unwanted interfering contaminants and enriching the extract with desired metabolites. Various other available natural product extraction methods are discussed in Chapters 3, 10, and 13–16.

2. Method

2.1. Extraction of Plant Natural Products

Plants are complex matrices, producing a range of secondary metabolites with different functional groups and polarities. Categories of natural products commonly encountered include waxes and fatty acids, polyacetylenes, terpenoids (e.g., monoterpenoids, iridoids, sesquiterpenoids, diterpenoids, triterpenoids), steroids, essential oils (lower terpenoids and phenylpropanoids), phenolics (simple phenolics, phenylpropanoids, flavonoids, tannins, anthocyanins, quinones, coumarins, lignans), alkaloids, and glycosidic derivatives (e.g., saponins, cardiac glycosides, flavonoid glycosides).

Several approaches can be employed to extract the plant material. Although water is used as an extractant in many traditional protocols, organic solvents of varying polarities are generally selected in modern methods of extraction to exploit the various solubilities of plant constituents. Solvent-extraction procedures applied to plant natural products include maceration, percolation, Soxhlet extraction, pressurized solvent extraction, ultrasound-assisted solvent extraction, extraction under reflux, and steam distillation.

2.1.1. Preparation of Plant Material

2.1.1.1. Selection

Any plant species and plant parts, collected randomly, can be investigated using available phytochemical methods. However, a more targeted approach is often preferred to a random selection. The plant material to be investigated can be selected on the basis of some specific traditional ethnomedical uses (*see* **Note 1**). Extracts prepared from plants and used as traditional remedies to treat certain diseases are more likely to contain biologically active components of medicinal interest. Alternatively, the plant can be selected based on chemotaxonomical data. This means that if species/genera related to the plant under investigation are known to contain specific compounds, then the plant itself can be expected to contain similar compounds. Another approach is to select the plant with a view to investigate a specific pharmacological activity. Additionally, work can be carried out on a particular group of natural products, a plant family, or on plants from a specific country or local area. Some plants can be selected following a combination of approaches. The use of literature databases (*see* Chap. 12) early in the selection process can provide some preliminary information on the type of natural products already isolated from the plant and the extraction methods employed to isolate them.

2.1.1.2. Collection and Identification

The whole plant or a particular plant part can be collected depending on where the metabolites of interest (if they are known) accumulate. Hence, aerial (e.g., leaves, stems, flowering tops, fruits, seeds, bark) and underground (e.g., bulbs, tubers, roots) parts can be collected separately. Only healthy specimens should be obtained, as signs of contamination (fungal, bacterial, or viral) may be linked to a change in the profile of metabolites present. Collection of plant material can also be influenced by other factors such as the age of the plant and environmental conditions (e.g., temperature, rainfall, amount of daylight, soil characteristics, and altitude). In some cases, it can be challenging, if not hazardous. This is particularly true if the targeted plant is a species of liana indigenous to the canopy (60 m above ground level!) of a remotely accessible area of the rain forests. It is important to take these issues into account for recollection purposes to ensure a reproducible profile (nature and amount) of metabolites.

It should be stressed that the plant must also be identified correctly. A specialized taxonomist should be involved in the detailed authentication of the plant (i.e., classification into its species, genus, family, order, and class). Any features relating to the collection, such as the name of the plant, the identity of the part(s) collected, the place and date of collection, should be recorded as part of a voucher (a dried specimen pressed between sheets of paper) deposited in a herbarium for future reference. More details on this particular aspect can be found in Chapter 13.

2.1.1.3. Drying and Grinding

If the plant is known to contain volatile or thermolabile compounds, it may be advisable to snap–freeze the material as soon as possible after collection. Once in the laboratory, the collected plants are washed or gently brushed to remove soil and other debris. Frozen samples can be stored in a freezer (at –20°C) or freeze-dried (lyophilized) (*see* **Note 2**). It is usual to grind them subsequently in a mortar with liquid nitrogen. Extracting the pulverized residue immediately or storing it in a freezer to prevent any changes in the profile of metabolites ***(1,2)*** is advisable.

It is, however, a more common practice to leave the sample to dry on trays at ambient temperature and in a room with adequate ventilation. Dry conditions are essential to prevent microbial fermentation and subsequent degradation of metabolites. Plant material should be sliced into small pieces and distributed evenly to facilitate homogenous drying. Protection from direct sunlight is advised to minimize chemical reactions (and the formation of artifacts) induced by ultraviolet rays. To accelerate the drying process (especially in countries with high relative humidity), the material can be dried in an oven (*see* **Note 3**). This can also minimize enzymatic reactions (e.g., hydrolysis of glycosides) that can occur as long as there is some residual moisture present in the plant material. The dried plant material should be stored in sealed containers in a dry and cool place. Storage for prolonged periods should be avoided, as some constituents may decompose.

The aim of grinding (i.e., fragmentation of the plant into smaller particles) is to improve the subsequent extraction by rendering the sample more homogenous, increasing the surface area, and facilitating the penetration of solvent into the cells. Mechanical grinders (e.g., hammer and cutting mills) are employed conveniently to shred the plant tissues to various particle sizes. Potential problems of grinding include the fact that some material (e.g., seeds and fruits rich in fats and volatile oils) may clog up the sieves and that the heat generated may degrade thermolabile metabolites.

2.1.2. Range of Extraction Methods

A number of methods using organic and/or aqueous solvents are employed in the extraction of natural products. Supercritical fluid extraction (which uses carbon dioxide in a supercritical state as the extractant), a solvent-free and environment-friendly method of extraction, is discussed in Chapter 3.

Solvent extraction relies on the principle of either "liquid–liquid" or "solid–liquid" extraction. Only the latter is described here, and theoretical and practical aspects related to liquid–liquid extraction are covered in Chapter 10. In solid–liquid extraction, the plant material is placed in contact with a solvent. While the whole process is dynamic, it can be simplified by dividing it into different steps. In the first instance, the solvent has to diffuse into cells, in the following step it has to solubilize the metabolites, and finally it has to diffuse out of the cells enriched in the extracted metabolites. In general, extractions can be facilitated by grinding (as the cells are largely destroyed, the extraction relies primarily on the solubilization of metabolites) and by increasing the temperature (to favor solubilization). Evaporation of the organic solvents or freeze-drying (of aqueous solutions) yields dried crude extracts (*see* **Note 4**).

2.1.2.1. Maceration

This simple, but still widely used, procedure involves leaving the pulverized plant to soak in a suitable solvent in a closed container at room temperature. The method is suitable for both initial and bulk extraction. Occasional or constant stirring of the preparation (using mechanical shakers or mixers to guarantee homogenous mixing) can increase the speed of the extraction. The extraction ultimately stops when an equilibrium is attained between the concentration of metabolites in the extract and that in the plant material. After extraction, the residual plant material (marc) has to be separated from the solvent. This involves a rough clarification by decanting, which is usually followed by a filtration step. Centrifugation may be necessary if the powder is too fine to be filtered. To ensure exhaustive extraction, it is common to carry out an initial maceration, followed by clarification, and an addition of fresh solvent to the marc. This can be performed periodically with all filtrates pooled together.

The main disadvantage of maceration is that the process can be quite time-consuming, taking from a few hours up to several weeks *(3)*. Exhaustive maceration can also consume large volumes of solvent and can lead to the potential loss of metabolites and/or plant material (*see* **Note 5**). Furthermore, some compounds may not be extracted efficiently if they are poorly soluble at room temperature. On the other hand, as the extraction is performed at room temperature, maceration is less likely to lead to the degradation of thermolabile metabolites.

2.1.2.2. Ultrasound-Assisted Solvent Extraction

This is a modified maceration method where the extraction is facilitated by the use of ultrasound (high-frequency pulses, 20 kHz). The plant powder is placed in a vial. The vial is placed in an ultrasonic bath, and ultrasound is used to induce a mechanical stress on the cells through the production of cavitations in the sample. The cellular breakdown increases the solubilization of metabolites in the solvent and improves extraction yields. The efficiency of the extraction depends on the instrument frequency, and length and temperature of sonication. Ultrasonification is rarely applied to large-scale extraction; it is mostly used for the initial extraction of a small amount of material. It is commonly applied to facilitate the extraction of intracellular metabolites from plant cell cultures *(4)*.

2.1.2.3. Percolation

In percolation, the powdered plant material is soaked initially in a solvent in a percolator (a cylindrical or conical container with a tap at the bottom) (*see* **Note 6**). Additional solvent is then poured on top of the plant material and allowed to percolate slowly (dropwise) out of the bottom of the percolator. Additional filtration of the extract is not required because there is a filter at the outlet of the percolator. Percolation is adequate for both initial and large-scale extraction. As for maceration, successive percolations can be performed to extract the plant material exhaustively by refilling the percolator with fresh solvent and pooling all extracts together. To ensure that percolation is complete, the percolate can be tested for the presence of metabolites with specific reagents (*see* Chap. 4).

There are several issues to consider when carrying out a percolation. The extent to which the material is ground can influence extracts' yields. Hence, fine powders and materials such as resins and plants that swell excessively (e.g., those containing mucilages) can clog the percolator. Furthermore, if the material is not distributed homogenously in the container (e.g., if it is packed too densely), the solvent may not reach all areas and the extraction will be incomplete. Both the contact time between the solvent and the plant (i.e., the percolation rate) and the temperature of the solvent can also influence extraction yields. A higher temperature will improve extraction but may lead to decomposition of labile metabolites. The other disadvantages of percolation are that large volumes of solvents are required and the process can be time-consuming.

2.1.2.4. Soxhlet Extraction

Soxhlet extraction is used widely in the extraction of plant metabolites because of its convenience. This method is adequate for both initial and bulk extraction (*see* **Note 7**). The plant powder is placed in a cellulose thimble in an extraction chamber, which is placed on top of a collecting flask beneath a reflux condenser. A suitable solvent is added to the flask, and the set up is heated under reflux. When a certain level of condensed solvent has accumulated in the thimble, it is siphoned into the flask beneath.

The main advantage of Soxhlet extraction is that it is a continuous process. As the solvent (saturated in solubilized metabolites) empties into the flask, fresh solvent is recondensed and extracts the material in the thimble continuously. This makes Soxhlet extraction less time- and solvent-consuming than maceration or percolation. However, the main disadvantage of Soxhlet extraction is that the extract is constantly heated at the boiling point of the solvent used, and this can damage thermolabile compounds and/or initiate the formation of artifacts.

2.1.2.5. Pressurized Solvent Extraction

Pressurized solvent extraction, also called "accelerated solvent extraction," employs temperatures that are higher than those used in other methods of extraction, and requires high pressures to maintain the solvent in a liquid state at high temperatures. It is best suited for the rapid and reproducible initial extraction of a number of samples (*see* **Note 8**). The powdered plant material is loaded into an extraction cell, which is placed in an oven. The solvent is then pumped from a reservoir to fill the cell, which is heated and pressurized at programmed levels for a set period of time. The cell is flushed with nitrogen gas, and the extract, which is automatically filtered, is collected in a flask. Fresh solvent is used to rinse the cell and to solubilize the remaining components. A final purge with nitrogen gas is performed to dry the material. High temperatures and pressures increase the penetration of solvent into the material and improve metabolite solubilization, enhancing extraction speed and yield ***(5)***. Moreover, with low solvent requirements, pressurized solvent extraction offers a more economical and environment-friendly alternative to conventional approaches ***(6)***. As the material is dried thoroughly after extraction, it is possible to perform repeated extractions with the same solvent or successive extractions with solvents of increasing polarity. An additional advantage is that the technique can be programmable, which will offer increased reproducibility. However, variable factors, e.g., the optimal extraction temperature, extraction time, and most suitable solvent, have to be determined for each sample.

2.1.2.6. Extraction Under Reflux and Steam Distillation

In extraction under reflux, plant material is immersed in a solvent in a round-bottomed flask, which is connected to a condenser. The solvent is heated until it reaches its boiling point. As the vapor is condensed, the solvent is recycled to the flask.

Steam distillation is a similar process and is commonly applied to the extraction of plant essential oils (a complex mixture of volatile constituents). The plant (dried or fresh) is covered with water in a flask connected to a condenser. Upon heating, the vapors (a mixture of essential oil and water) condense and the distillate (separated into two immiscible layers) is collected in a graduated tube connected to the condenser. The aqueous phase is recirculated into the flask, while the volatile oil is collected separately. Optimum extraction conditions (e.g., distillation rate) have to be determined depending on the nature of the material being extracted (*see* **Note 9**). The main disadvantage of extraction under reflux and steam distillation is that thermolabile components risk being degraded.

2.1.3. Selection of an Extraction Method and Solvent

The ideal extraction procedure should be exhaustive (i.e., extract as much of the desired metabolites or as many compounds as possible). It should be fast, simple, and reproducible if it is to be performed repeatedly. The selection of a suitable extraction method depends mainly on the work to be carried out, and whether or not the metabolites of interest are known.

If the plant material has been selected from an ethnobotanical point of view, it may be worthwhile reproducing the extraction methods employed traditionally (if they are reported) to enhance the chances of isolating potential bioactive metabolites. Traditional methods rely principally on the use of cold/hot water, alcoholic, and/or aqueous alcoholic mixtures to obtain preparations that are used externally or administered internally as teas (e.g., infusions, decoctions). Boiling solvent can be poured on the plant material (infusion) or the plant can be immersed in boiling solvent (decoction). If a plant has already been investigated chemically, a literature search can indicate the extraction methods employed previously. However, this does not exclude the possibility of choosing an alternative method that may yield different metabolites. If a plant is being investigated for the first time, the lack of information on suitable extraction methods leaves the choice to the investigator. The selection will be governed by the nature and amount of material to be extracted. If large amounts are to be extracted, the ease of transfer from initial to bulk scale must also be considered.

Extraction processes can employ water-miscible or water-immiscible solvents. The solvent selected should have a low potential for artifact formation, a low toxicity, a low flammability, and a low risk of explosion. Additionally, it should be economical and easily recycled by evaporation. These issues are particularly important in the case of bulk extraction where large volumes of solvents are employed. The main solvents used for extraction include aliphatic and chlorinated hydrocarbons, esters, and lower alcohols (**Table 1**) (*see* **Note 10**).

Table 1
Physicochemical Properties of Some Common Solvents Used in Natural Products Extraction

Solvent	Polarity index	Boiling point (°C)	Viscosity (cPoise)	Solubility in water (% w/w)
n-Hexane	0.0	69	0.33	0.001
Dichloromethane	3.1	41	0.44	1.6
n-Butanol	3.9	118	2.98	7.81
iso-propanol	3.9	82	2.30	100
n-Propanol	4.0	92	2.27	100
Chloroform	4.1	61	0.57	0.815
Ethyl acetate	4.4	77	0.45	8.7
Acetone	5.1	56	0.32	100
Methanol	5.1	65	0.60	100
Ethanol	5.2	78	1.20	100
Water	9.0	100	1.00	100

Extractions can be either "selective" or "total." The initial choice of the most appropriate solvent is based on its selectivity for the substances to be extracted. In a selective extraction, the plant material is extracted using a solvent of an appropriate polarity following the principle of "like dissolves like." Thus, nonpolar solvents are used to solubilize mostly lipophilic compounds (e.g., alkanes, fatty acids, pigments, waxes, sterols, some terpenoids, alkaloids, and coumarins). Medium-polarity solvents are used to extract compounds of intermediate polarity (e.g., some alkaloids, flavonoids), while more polar ones are used for more polar compounds (e.g., flavonoid glycosides, tannins, some alkaloids). Water is not used often as an initial extractant, even if the aim is to extract water-soluble plant constituents (e.g., glycosides, quaternary alkaloids, tannins) (*see* Chap. 16). A selective extraction can also be performed sequentially with solvents of increasing polarity. This has the advantage of allowing a preliminary separation of the metabolites present in the material within distinct extracts and simplifies further isolation *(7)*.

In an extraction referred to as "total," a polar organic solvent (e.g., ethanol, methanol, or an aqueous alcoholic mixture) is employed in an attempt to extract as many compounds as possible. This is based on the ability of alcoholic solvents to increase cell wall permeability, facilitating the efficient extraction of large amounts of polar and medium- to low-polarity constituents. The "total" extract is evaporated to dryness, redissolved in water, and the metabolites re-extracted based on their partition coefficient (i.e., relative affinity for either phase) by successive partitioning between water and immiscible organic solvents of varying polarity (*see* Chap. 10) ***(8,9)***.

Specific protocols during which the pH of the extracting aqueous phase is altered to solubilize selectively groups of metabolites (such as acids or bases) can also be used. For instance, these are applied to the extraction of alkaloids (which occur mostly as water-soluble salts in plants). On treating the plant material with an alkaline solution, the alkaloids are released as free bases that are recovered following partition into a water-immiscible organic solvent ***(10)***. Subsequent liquid–liquid extractions and pH modifications can be performed to separate the alkaloids from other nonalkaloidal metabolites (*see* Chap. 10). Alternatively, alkaloids can be extracted from the plant material in their salt form under acidic conditions ***(11)***. Acidic extraction is also applied to the extraction of anthocyanins ***(12)***. However, one drawback of the acid–base treatment is that it can produce some artifacts and/or lead to the degradation of compounds ***(13–15)***.

Finally, single solvents or solvent mixtures can be used in extraction protocols. When a solvent mixture is necessary, a binary mixture (two miscible solvents) is usually employed. In a Soxhlet extraction, it is preferable to use a single solvent simply because one of the solvents in the mixture may distill more rapidly than another. This may lead to a change in the solvent proportions in the extracting chamber.

2.2. Extraction of Microbial Natural Products

Microorganisms are also a valuable source of chemically diverse and potentially useful metabolites. To date, mostly filamentous bacterial species of the genus *Streptomyces* (Actinomycetes) and fungal species of the genera *Penicillium* and *Aspergillus* have been used for the extraction and isolation of their metabolites, which have important medical applications (e.g., antibiotics, immunosuppressants, hypocholesterolemic and anticancer agents). The search for novel microbial metabolites has been driven by the need for new antibiotics to combat the ever-increasing number of pathogenic microbes that are resistant to current antimicrobial agents. Aspects related to the selection, culture, and extraction of the microbial biomass are presented below. Further details on the purification and characterization of microbial metabolites are provided in Chapter 15.

2.2.1. Isolation and Fermentation

Because of the enormous diversity of the microbial world, it is not a simple task to select, identify, and culture pure strains that produce potentially bioactive metabolites. As many microorganisms are found in the soil, the investigation of microbial metabolites usually starts with the collection of soil samples. A wide variety of environments (e.g., soils of unusual composition or those from different climatic areas) can be explored to search for novel strains. The sample collected is typically prepared as a suspension in water, and appropriate dilutions of the supernatant are plated on a solid (agar) medium. *Streptomyces* species are widely found in the soil and will grow well on normal nutrient agar. The isolation of other species usually requires the use of selective media (e.g., MacConkey's medium for Gram-negative bacteria), the use of antibacterial/antifungal agents (e.g., nystatin to inhibit the growth of molds and fungi), and/or particular incubation conditions (e.g., thermophilic strains require incubation at 50°C) (*see* **Note 11**). Once individual colonies are obtained, they are subcultured several times on different media until they display purity (morphologically and microscopically). Pure strains are commonly stored in liquid nitrogen or freeze-dried in the presence of a cryoprotective agent (*see* **Note 12**). To enable cell growth and metabolite production, the isolated strains are transferred from stock to liquid broth (*see* **Note 13**).

The culture (or fermentation) is carried out initially in flasks containing a liquid medium before the strain is transferred to small fermenters (stainless steel closed vessels) and the whole process is scaled up. In flask fermentation, the culture is grown in a nutrient broth dispensed in flasks that are sealed and placed on a rotary shaker at a defined temperature (*see* **Note 14**). This allows a relatively good set-up to monitor both the growth rate of the biomass and the production of metabolites. It also provides a means of carrying out initial studies to optimize culture conditions and increase metabolite production (*see* **Note 15**). When performing studies in a small fermenter, the culture is grown under controlled conditions (*see* **Note 16**). The process can be scaled up once the effects of other important parameters for metabolite production have been optimized (e.g., aeration, stirring speed, temperature, pH, oxygen, and carbonic acid concentration), and the absence of external contamination has been ascertained. It is important that the growth of the producing strain be consistent to ensure reproducible productivity. This may not be true in cases where the morphology of the culture is different while growing in fermenters as opposed to flasks (e.g., actinomycetes and filamentous fungi can grow as two different morphologies, hyphae or pellets).

2.2.2. Selection of Extraction Methods

When selecting an extraction procedure for microbial metabolites, the following considerations should be borne in mind. Microbial metabolites are often produced in low yields, and one strain can yield a complex mixture of compounds. The metabolites may be completely or partially excreted by the cells into the (extracellular) medium or they may be present within the cells (intracellular). If metabolites of a certain type are expected, it is possible to refer to previously published protocols. The situation is more difficult when the strain is new, the metabolites are hitherto unknown, or the aim is to extract as many metabolites as possible. As for plant material, water-miscible and immiscible organic solvents, e.g., ethyl acetate (EtOAc), dichloromethane (DCM), *n*-butanol, methanol (MeOH), and so on, are used for the extraction of microbial metabolites (**Table 1**).

A variety of approaches can be employed in the recovery of microbial metabolites from fermentation broths. If the metabolites of interest are not only associated with the cells but are also present in the medium, a whole-broth solvent extraction is usually required to solubilize both intra- and extracellular compounds. In some cases, the fermentation broth may be freeze-dried prior to the extraction *(16,17)*. Alternatively, it can be clarified first by separating the microbial cells from the liquid medium prior to extraction. Clarification is achieved by filtration or centrifugation depending on the broth's physical properties (e.g., consistency) and the morphology and size of cells (*see* **Note 17**). Extraction is simpler if the metabolites are either entirely in the liquid medium, or adsorbed onto or located within the cells. For metabolites associated with the cells, it is advisable to perform the clarification (removal of media constituents and other contaminants) prior to the extraction *(18)*. The preliminary removal of physical "impurities" (e.g., cells, cell debris, insoluble medium components) is also advantageous if the metabolites of interest are principally extracellular. The extraction of compounds is then performed by partitioning the medium (aqueous phase) between a water-immiscible organic solvent *(19,20)*. Changing the pH of the aqueous phase and selecting a solvent into which the desired metabolites partition efficiently can extract metabolites selectively depending on their pK_a and partition coefficients. Adsorption procedures can also be employed in the extraction of metabolites from the medium. These exploit the fact that most secondary metabolites will be retained if the medium (aqueous solution) is passed through a column packed with a hydrophobic adsorbent. Following washes with water to elute inorganic salts and highly polar material (desalting step), elution with organic solvents (e.g., MeOH, acetone) or aqueous mixtures of organic solvents yields an extract enriched in the metabolites (*see* **Note 18**) ***(21,22)***. The adsorbent can sometimes be added directly to the fermentation broth to "trap" metabolites as they are produced ***(23)***.

In most cases, it will be necessary to obtain larger amounts of the microbial metabolites identified in the initial extract. This may be to carry out biological tests and/or to design structural analogs to investigate structure activity relationships. In such cases, not only is a larger volume of fermentation required but also good overall yield of metabolites is necessary for subsequent bulk extraction.

3. Conclusion

Plants and microorganisms produce complex mixtures of natural products, and the selection of the best protocol for an efficient extraction of these substances is not a simple task. "Classic" solvent-based procedures (e.g., maceration, percolation, Soxhlet extraction, extraction under reflux, steam distillation) are still applied widely in phytochemistry despite the fact that they lack reproducibility and are both time- and solvent-consuming. This is principally because they only require basic glassware and are convenient to use for both initial and bulk extraction. Accelerated solvent extraction is a newer instrumental technique. While it offers some advantages over conventional methods (mainly efficiency and reproducibility), it is best suited for initial rather than bulk extraction. It has found a wider application in industry (where large numbers of extracts have to be produced in an efficient and reproducible way) rather than in academia.

To date, mainly plant and microbial sources have been investigated for their metabolites. However, it is important to remember that researchers are only beginning to explore other biotopes (e.g., the marine environment, insects) and that many plants and microorganisms have not yet been characterized. Moreover, several species among the bacteria known are yet to be cultured under laboratory conditions ***(24)***. This leaves much scope for the potential discovery of novel and/or useful natural products in the future.

4. Notes

1. Several databases and journals (such as the *Journal of Ethnopharmacology*) containing some ethnobotanical, chemical, and biological information can be consulted on the Internet. These include the following links:

   http://ukcrop.net/perl/ace/search/PhytochemDB
   http://ukcrop.net/perl/ace/search/MPNADB
   http://ukcrop.net/perl/ace/search/EthnobotDB
   http://www.ars-grin.gov/duke/
   http://www.leffingwell.com/plants.htm

2. If fresh plant material is used, enzymes can be denaturated (or deactivated) by soaking the material in an organic solvent (e.g., methanol or ethanol).
3. Temperature below 30°C is recommended to avoid the loss/degradation of thermolabile compounds (e.g., the volatile constituents of essential oils).
4. Solvent removal should be done immediately after extraction to minimize the loss of compounds unstable in solution. Prolonged exposure to sunlight should also be avoided because of the potential for degradation. For organic solvents, the extract is concentrated by evaporation under reduced pressure (using a rotary evaporator) at a temperature below 40°C to minimize the degradation of thermolabile compounds. Precautions are required if extracts contain some metabolites that foam (e.g., saponins), as these may spill into the solvent collecting flask. Small volumes of solvents (<5 mL) can be evaporated under a gentle stream of nitrogen gas. If an organic/aqueous mixture was used as the extractant, the sample is evaporated under reduced pressure and then freeze-dried (lyophilized). Freeze-drying relies on the removal of water from a frozen sample by sublimation under vacuum. The freeze-dried extract is best stored in a sealed container in a freezer (–20°C) until required to minimize degradation at room temperature.
5. The plant can be left to macerate within a muslin bag to facilitate further filtration and addition of fresh solvent.
6. A metallic container is recommended if hot solvent is used. Details of the apparatus can be found in a variety of pharmacopeia monographs. The plant is left to soak initially for up to 24 h in the percolator. Only coarsely fragmented material that passes through a 3 mm sieve is adequate for percolation.
7. Initial and bulk Soxhlet extraction can be performed for up to 72 h. An initial extraction uses 200–500 mL on average, while a bulk extraction uses 2.5–5 L of solvent. For safety purposes, it is recommended the Soxhlet apparatus be in a walk-in type of fume cupboard.
8. Accelerated solvent extraction is a registered process (ASE®) developed by Dionex (Sunnyvale, CA, USA). Extractions are carried out under high

pressure (100–200 bar) within 12–20 min, using 15–45 mL of solvent, and at temperatures ranging from ambient to 200°C. Three systems are currently available. The ASE 100 has a single loading cell (10–100 mL) and is best suited for low-throughput initial extractions. Higher-throughput extractions can be achieved with the ASE 200, which has a rack of 24 cells (1–33 mL). Samples of up to 30 g can be extracted. The ASE 300 (12 cells, 34–100 mL) is ideal for sample sizes above 30 g.

9. In steam distillation, the plant material should not be powdered too finely. Coarsely fragmented or crushed material is preferable. Some glycerol can be added to the water to facilitate the extraction of tough material (e.g., barks, seeds, roots). Xylene may be added to the graduated receiver to trap the distilled volatile oil produced. Both the description of the distillation apparatus and the steam distillation procedure are found in a variety of pharmacopeia monographs.
10. Toxic solvents and those detrimental to the environment (e.g., benzene, toluene, and carbon tetrachloride) must not be used. Diethyl ether should be avoided as it is highly flammable and can lead to the formation of explosive peroxides. Aliphatic hydrocarbons (either pure alkanes or crude petroleum fractions) are employed to defat the plant material as they selectively solubilize waxes and fats. A disadvantage is that they are highly flammable. Dichloromethane (DCM) is preferred to chloroform ($CHCl_3$), the latter being more toxic. Both can however produce artifacts ***(25)***. Acetone is not commonly used (it may also give rise to artifacts under acidic conditions). In general, an aqueous-alcoholic mixture is the solvent of choice for an extraction. However, it has been shown that MeOH may produce artifacts ***(26)***. General-purpose grade solvents (available commercially in plastic containers or in plastic-stoppered Winchesters) frequently contain plasticizers (additives used in the manufacture of plastic). Minimizing contamination with plasticizers is especially important when bulk extraction is carried out and large volumes of solvent are used. Dioctylphthalate ester is the most frequently encountered contaminant of plant extracts. It can be detected by the presence of an intense purple-pink spot on a silica gel TLC plate (R_f value = 0.4 in petroleum ether–EtOAc, 95:5) following detection with anisaldehyde–sulfuric acid reagent and incubation for 5 min at 110°C. Spectroscopic data UV λ_{max} 275 nm (log ε 3.17), 282 nm (sh); ^{1}H NMR (δ, $CDCl_3$) 7.70 (2H, dd, H-3 and H-6 aromatic protons), 7.52 (2H, dd, H-4 and H-5 aromatic protons), 4.20 (4H, dd, H-1′ and H-1″ 2-ethylhexyl moiety), 1.2–1.8 (14H, m, CH and CH_2s of 2-ethylhexyl moiety), 0.90 (12H, CH_3 groups); EIMS (m/z) 279, 167, and 149 (100%). Distilling solvents prior to extraction and using glass containers for storage can eliminate contamination with plasticizers. Another possible contaminant is high-vacuum grease (a silicone-based lubricant) used as a seal and to prevent glassware

joints from seizing. Mass spectrometry can be used to detect contamination; silicone grease presents a typical mass fragmentation pattern (m/z 429, 355, 281, 207, and 133), which differs from aliphatic hydrocarbon greases (fragmentation every 14 mass units intervals) *(27)*. When leaves are extracted, the extract will contain high levels of chlorophyll pigments. These can be removed by gel filtration chromatography with Sephadex LH-20 material.

11. For the selective isolation of actinomycetes, soil samples are air-dried for a few days to reduce contamination with nonsporulating bacteria (especially Gram-negative organisms). An isolation medium without peptone limits the germination of *Bacillus* spores (which are resistant to the drying process). By placing a membrane filter on the agar plate and inoculating its surface with the soil suspension, the filamentous hyphae of actinomycetes will penetrate through the filter. Once the filter is removed and the plate incubated, actinomycetes will grow on the agar almost exclusively. A useful method to enrich the population of various actinomycetes (other than *Streptomyces*) is to air-dry and then heat the soil sample at 100–120°C for 1 h. For the selective isolation of Gram-positive *Bacillus* strains, soil samples are heated at 70°C or suspended for 1 h in 50% ethanol. A selective isolation medium with peptone and other amino acids can favor the germination of *Bacillus* spores (which are heat- and disinfectant-resistant). The optimum incubation temperature for the preferential isolation of fungi is 20–25°C.
12. A stock can also be prepared by storing aliquots (e.g., 1 mL) of the culture at −80°C. Glycerol (20%) can be used as a cryoprotective agent.
13. Some colonies are transferred from the stock onto a suitable solid medium before transferring to liquid medium. If the transfer to broth is not immediate, regular subculturing onto solid medium can be performed. Subcultures can be kept for up to several months at 4°C. Repeated subculturing is not recommended for bacteria (e.g., actinomycetes), whose phenotypes can be altered over several generations (genetic instability).
14. Typically 50–100 mL of liquid medium is dispensed into flasks (500 mL capacity). The small volume of medium relative to the size of the flasks is necessary to increase aeration. Sterile cotton plugs are used to avoid external contamination and allow gaseous exchange. Flasks (up to 50–100) are placed in a shaking unit in a temperature-controlled room. A temperature-controlled shaking incubator may be more convenient.
15. Liquid media are often very complex. They are likely to contain carbon sources (e.g., sugars, fatty acids, organic acids), nitrogen sources (e.g. yeast, peptones, meat-based extracts), and mineral elements (e.g., iron, cobalt, phosphorus, sulfur, potassium). The most suitable medium composition (variable amounts and proportions of different nutrients) has to be selected for an optimal metabolite production.

16. Small fermenters have a 5–10 L capacity. Mini-fermenters (up to 2 L) can provide an intermediate stage in the scale-up process.
17. Microbial cells include bacteria (1×2 μm), yeasts (7×10 μm), and fungal hyphae (1×10 μm). The latter can grow as a filamentous mycelium.
18. Solid phase extraction (SPE) is carried out using commercially available 'cartridges' (adsorbing material packed in a small syringe), which are also used for chromatographic purification (*see* Chap. 5). Adsorbents used include reverse phase (C_{18} silica gel), ion-exchange, size exclusion (Sephadex G25), and non-ionic styrene–divinylbenzene polymeric (e.g., Amberlite XAD-16, XAD-2, XAD-1180 [Rohm and Haas], and Diaion HP20 [Mitsubishi]) material. For stronger binding, a brominated polystyrene resin (Sepabeads® SP-107 [Mitsubishi]) can be employed.

References

1. Schliemann, W., Yizhong Cai, Y., Degenkolb, T., Schmidt, J., and Corke, H. (2001) Betalains of *Celosia argentea*. *Phytochemistry* **58,** 159–165.
2. Brown, G. D., Liang, G.-Y., and Sy, L.-K. (2003) Terpenoids from the seeds of *Artemisia annua*. *Phytochemistry* **64,** 303–323.
3. Takahashi, H., Hirata, S., Minami, H., and Fukuyama, Y. (2001) Triterpene and flavanone glycoside from *Rhododendron simsii*. *Phytochemistry* **56,** 875–879.
4. Mohagheghzadeh, A., Schmidt, T. J., and Alfermann, A. W. (2002) Arylnaphthalene lignans from in vitro cultures of *Linum austriacum*. *J. Nat. Prod.* **65,** 69–71.
5. Waksmundzka-Hajnos, M., Petruczynik, A., Dragan, A., Wianwska, D., Dawidowicz, A., and Sowa, I. (2004) Influence of the extraction mode on the yield of some furanocoumarins from *Pastinaca sativa* fruits. *J. Chromatogr. B* **800,** 181–187.
6. Benthin, B., Danz, H., and Hamburger, M. (1999) Pressurized liquid extraction of medicinal plants. *J. Chromatogr. A* **837,** 211–219.
7. Cottiglia, F., Dhanapal, B., Sticher, O., and Heilmann, J. (2004) New chromanone acids with antibacterial activity from *Calophyllum brasiliens*. *J. Nat. Prod.* **67,** 537–541.
8. Lin, L.-C., Yang, L.-L., and Chou, C.-J. (2003) Cytotoxic naphthoquinones and plumbagic acid glucosides from *Plumbago zeylanica*. *Phytochemistry* **62,** 619–622.
9. Akhtar, M. N., Atta-ur-Rahman, Choudhary, M. I., Sener, B., Erdogan, I., and Tsuda, Y. (2003) New class of steroidal alkaloids from *Fritillaria imperialis*. *Phytochemistry* **63,** 115–122.

10. Zanolari, B., Guilet, D., Marston, A., Queiroz, E. F., Paulo, M. Q., and Hostettmann, K. (2003) Tropane alkaloids from the bark of *Erythroxylum vacciniifolium*. *J. Nat. Prod.* **66,** 497–502.
11. Zhang, X., Yea, W., Zhaoa, S., and Che, C.-T. (2004) Isoquinoline and isoindole alkaloids from *Menispermum dauricum*. *Phytochemistry* **65,** 929–932.
12. Toki, K., Saito, N., Shigihara, A., and Honda, T. (2001) Anthocyanins from the scarlet flowers of *Anemone coronaria*. *Phytochemistry* **56,** 711–715.
13. Kiehlmann, E. and Li, E. P. M. (1995) Isomerisation of dihydroquercetin. *J. Nat. Prod.* **58,** 450–455.
14. Sulaiman, M., Martin, M. T., Pais, M., Hadi, H. A., and Awang, K. (1998) Desmosine, an artefact alkaloid from *Desmos dumosus*. *Phytochemistry* **49,** 2191–2192.
15. Salim, A. A., Garson, M. J., and Craik, D. J. (2004) New Alkaloids from *Pandanus amaryllifolius*. *J. Nat. Prod.* **67,** 54–57.
16. Funayama S., Ishibashi M., Ankaru Y., et al. (1989) Novel cytocidal antibiotics, glucopiericidinols A1 and A2. Taxonomy, fermentation, isolation, structure elucidation and biological characteristics. *J. Antibiot.* **42,**1734–1740.
17. Cao S., Lee A. S. Y., Huang Y., et al. (2002) Agonodepsides A and B: two new depsides from a filamentous fungus F7524. *J. Nat. Prod.* **65**, 1037–1038.
18. Chinworrungsee, M., Kittakoop, P., Isaka, M., Maithip, P., Supothina, S., and Thebtaranonth, Y. (2004) Isolation and structure elucidation of a novel antimalarial macrocyclic polylactone, menisporopsin A, from the fungus *Menisporopsis theobromae*. *J. Nat. Prod.* **67,** 689–692.
19. Machida K., Trifonov I. S., Ayer W. A., et al. (2001) 3(2H)–Benzofuranones and chromanes from liquid cultures of the mycoparasitic fungus *Coniothyrium minitans*. *Phytochemistry* **58**, 173–177.
20. Liu, Z., Jensen, P. R., and Fenical, W. (2003) A cyclic carbonate and related polyketides from a marine-derived fungus of the genus *Phoma*. *Phytochemistry* **64,** 571–574.
21. Shiono, Y., Matsuzaka, R., Wakamatsu, H., Muneta, K., Murayama, T., and Ikeda, M. (2004) Fascicularones A and B from a mycelial culture of *Naematoloma fasciculare*. *Phytochemistry* **65,** 491–496.
22. Yun, B.-S., Lee, I.-K., Cho, Y., Cho, S.-M., and Yoo, I.-D. (2002) New Tricyclic Sesquiterpenes from the fermentation Broth of *Stereum hirsutum*. *J. Nat. Prod.* **65,** 786–788.
23. Gerth, K., Bedorf, N., Irschik, H., Hofle, G., and Reichenbach, H. (1994) The soraphens: a family of novel antifungal compounds from *Sorangium cellulosum* (Myxobacteria) I. Soraphen A1α: fermentation, isolation, biological properties. *J. Antibiot.* **47,** 23–31.
24. Demain, A. L. (1999) Pharmaceutically active secondary metabolites of microorganisms. *Appl. Microbiol. Biotechnol.* **52,** 455–463.

25. Phillipson, J. D. and Bisset, N. G. (1972) Quaternisation and oxidation of strychnine and brucine during plant extraction. *Phytochemistry* **11,** 2547–2553.
26. Lavie, D., Bessalle, R., Pestchanker, M. J., Gottlieb, H. E., Frolow, F., and Giordano, O. S. (1987) Trechonolide A, a new withanolide type from *Trechonaetes laciniata. Phytochemistry* **26,** 1791–1795.
27. Banthorpe, D. V. (1991) Classification of terpenoids and general procedures for their characterisation, in *Methods in Plant Biochemistry,* vol. 7 (Dey, P. M. and Harborne, J. B., eds.) Academic, New York.

Suggested Reading

1. Bruneton J. (1995) *Pharmacognosy, Phytochemistry, Medicinal Plants.* Springer-Verlag, Berlin.
2. Evans W. C. (2002) *Trease and Evans' Pharmacognosy*, 15 ed, Bailliere Tindall: London.
3. Heinrich M., Barnes J., Gibbons S., and Williamson E. M. (2004) *Fundamentals of Pharmacognosy and Phytotherapy*. Churchill Livingstone, Edinburgh.
4. Kaufman, P. B., Cseke, L. J., Warber, S., Duje, J. A., and Brielman, H. L. (1999) *Natural Products From Plants*, CRC Press, Boca Raton
5. Lancini G. and Lorenzetti R. (1993) *Biotechnology of Antibiotics and Other Bioactive Microbial Metabolites*. Plenum, New York.
6. List P. H. and Schmit P. C. (1989) *Phytopharmaceutical Technology*. Heyden & Son Ltd, London.
7. Williamson E. M., Okpako D. T., and Evans F. J. (1996) Selection, preparation and pharmacological evaluation of plant material, in *Pharmacological Methods in Phytotherapy Research*, vol. 1. John Wiley and Sons, Chichester.
8. Zygmunt, B. and Namiesnik, J. (2003) Preparation of samples of plant material for chromatographic analysis. *J. Chromatogr. Sci.* **41,** 109–116.

3

Supercritical Fluid Extraction

Lutfun Nahar and Satyajit D. Sarker

Summary

Supercritical fluids (SCFs) are increasingly replacing organic solvents, e.g., *n*-hexane, dichloromethane, chloroform, and so on, that are conventionally used in industrial extraction, purification, and recrystallization operations because of regulatory and environmental pressures on hydrocarbon and ozone-depleting emissions. In natural product extraction and isolation, supercritical fluid extraction (SFE), especially that employing supercritical CO_2, has become the method of choice. Sophisticated modern technologies allow precise regulation of changes in temperature and pressure, and thus manipulation of solvating property of the SCF, which helps the extraction of natural products of a wide range of polarities. This chapter deals mainly with the application of the SFE technology in the natural product extraction and isolation, and discusses various methodologies with specific examples.

Key Words: Supercritical fluid extraction (SFE); supercritical fluid (SCF); supercritical carbon dioxide; natural products.

1. Introduction

The ever-increasing concern about the environmental pollution attributed to several chemical wastes has paved the way for the introduction of "green chemistry." Chemists are now becoming more and more careful about the use of chemicals and solvents, and are putting significant efforts in designing environment-friendly research protocols. Extraction and isolation of natural products from various sources conventionally generate

From: *Methods in Biotechnology, Vol. 20, Natural Products Isolation, 2nd ed.*
Edited by: S. D. Sarker, Z. Latif, and A. I. Gray © Humana Press Inc., Totowa, NJ

large amounts of waste organic solvents. An eco-friendly alternative to the use of organic solvents in natural product extraction is the application of supercritical fluid extraction (SFE) protocol. The emerging stricter environmental regulations concerning the use of common industrial solvents, most of which are hazardous to human health, have led to the increasing popularity and growth of the SFE technologies, especially those employing supercritical CO_2.

Cagniard de la Tour discovered critical point in 1822. The critical point of a pure substance is defined as the highest temperature and pressure at which the substance can exist in vapor–liquid equilibrium. At temperatures and pressures above this point, a single homogeneous fluid is formed, which is known as supercritical fluid (SCF). SCF is heavy like liquid but has the penetration power of gas (**Table 1**). These qualities make SCFs effective and selective solvents. SCFs are produced by heating a gas above its critical temperature or compressing a liquid above its critical pressure (**Fig. 1**). Under these conditions, the molar volume is the same, irrespective of whether the original form is a liquid or a gas. SFE can be used to extract active ingredients or analytes from various plants and microbial samples, and can be particularly useful in the extraction of unknown natural products or preparation of the whole-organism extracts for chemical and biological assays in any screening program. While the most commonly used SCF is supercritical CO_2, a number of other SCFs, e.g., ethane, butane, pentane, nitrous oxide, ammonia, trifluoromethane, and water, are also used. The main advantages of using SCFs for extractions are that they are inexpensive, contaminant-free, selectively controllable, and less costly to dispose safely than organic solvents. Oxidative and thermal degradation of active compounds is much less likely in SFE than in conventional solvent extraction and steam distillation methods. One of the best examples of the utilization of SFE in natural product extraction

Table 1
A Comparison of Properties of Gas, Liquid, and SCF

Property	Density (kg/m^3)	Viscosity (cP)	Diffusivity (mm^2/s)
Gas	1	0.01	1–10
SCF	100–800	0.05–0.1	0.01–0.1
Liquid	1000	0.5–1.0	0.001

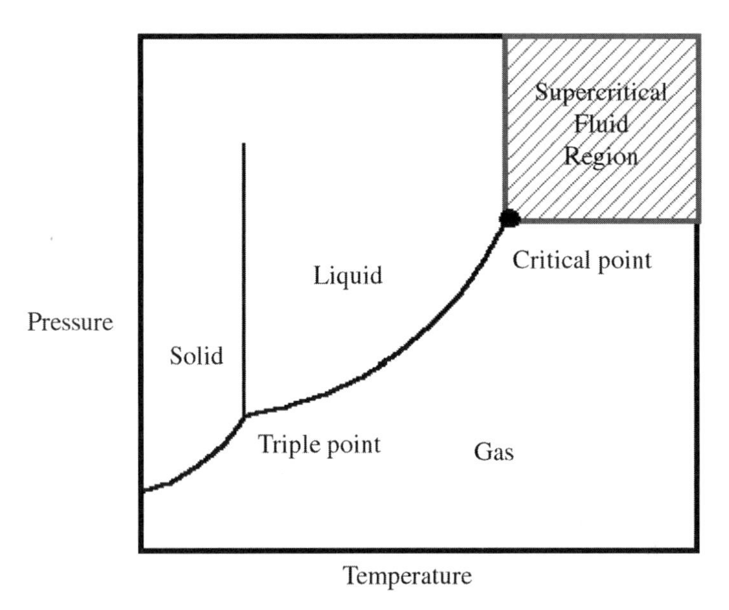

Fig. 1. Phase diagram.

is the use of supercritical CO_2 to extract caffeine from coffee. SCFs can have solvating powers similar to organic solvents, but with higher diffusivities, lower viscosity, and lower surface tension (**Table 1**). The solvating power can be adjusted by changing the pressure or temperature, or adding modifiers to the SCF. A common modifier is methanol (MeOH), typically 1–10%, which increases the polarity of supercritical CO_2.

While a number of books on SFE, which cover the history, principles, instrumentation, and methodologies, are now available ***(1–10)***, this chapter focuses on the application of the SFE technology in the natural product extraction and isolation, and discusses various methodologies with specific examples.

2. Principle of Solvent-Free Extraction Process: A Typical Supercritical CO_2 System

Most of the currently available SFE systems utilize CO_2, which is generally considered as safe for solvent-free extraction processes. The fundamental steps involved in SFE are as follows:

(1) Liquid CO_2 is forced into supercritical state by regulating its temperature and pressure.

(2) Supercritical CO_2 has solvent power and extracts predominantly lipophilic and volatile compounds.
(3) Gaseous CO_2 returns to CO_2 tank. After a full round, the new extraction starts with circulating CO_2.

The whole process is summarized in **Fig. 2**. The essential parts of a typical SFE system include a CO_2 source, a pump to pressurize the gas, an oven containing the extraction vessel, a restrictor to maintain high pressure in the extraction line, and a trapping vessel or analyte-collection device. Fully integrated and automated SFE systems of various sizes are nowadays available from a number of commercial suppliers (**Table 2**). Analytes are trapped by letting the solute-containing SCF decompress into an empty vial, through a solvent, or onto a solid sorbent material. Extractions can be carried out in three different modes: dynamic, static, or combination modes. In a dynamic extraction, the SCF continuously flows through the sample in the extraction vessel and out of the restrictor to the trapping vessel. In the static mode, the SCF circulates in a loop containing the extraction vessel for some time before being released through the restrictor to the trapping vessel. In the combination mode, a static extraction is performed for some period of time, followed by a dynamic extraction.

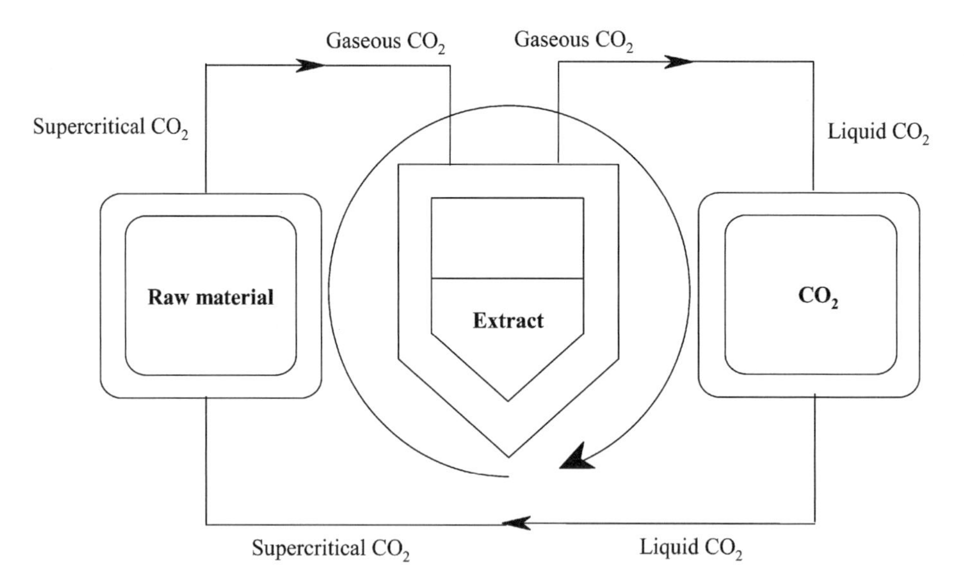

Fig. 2. SFE using CO_2 (continuous circulation of supercritical CO_2 in a closed hermetic system).

Table 2
Vendors for SFE Systems and Accessories

Vendor	Address	Web site
Applied Separations, Inc.	930 Hamilton Street, Allentown, PA 18101, USA	http://www.appliedseparations.com/about_asi.htm
Supercritical Fluid Technologies, Inc.	One Innovation Way, Suite 304, Newark, DE 19711, USA	http://www.supercriticalfluids.com/about.htm
Thar Technologies, Inc.	100 Beta Drive, Pittsburgh, PA 15238, USA	http://www.thartech.com/systems/extraction/
Teledyne Technologies, Inc.	12333 West Olympic Boulevard, Los Angeles, CA 90064, USA	http://www.isco.com/default.asp
ESEL TechTra Inc.	1382-4, Kuweol-3dong, Namdongku, Incheon, Korea	http://www.labkorea.com/index.html
JASCO	18 Oak Industrial Park, Chelmsford Road, Great Dunmow, Essex CM6 1XN, UK	http://www.jasco.co.uk/secure/sfc.asp
Agilent Technologies	395 Page Mill Road, Palo Alto, CA 94306, USA	http://www.agilent.com/about/index.html
Foss Tecator AB	730 Birchwood Boulevard, Birchwood, Warrington WA3 7PR, Cheshire, UK	http://www.foss.dk/c/p/default.asp?width=800
CDS Analytical	465 Limestone Road, PO Box 277, Oxford, PA 19363–0277, USA	http://www.cdsanalytical.com/
Supelco	The Old Brickyard, New Road, Gillingham, Dorset SP8 4XT, UK	http://www.sigmaaldrich.com/Brands/Supelco_Home.html
Perkin-Elmer	45 William Street, Wellesley, MA 02481–4078, USA	http://www.perkinelmer.com/

2.1. Solvent

Supercritical CO_2, because of its low critical parameters (31.1°C, 73.8 bar), low cost, and nontoxicity, is the most widely used supercritical solvent for SFE. Generally, cryogenic-grade CO_2 is not pure enough for most SFE operations. High-purity CO_2, which must be free from water, hydrocarbons, and halocarbons, is necessary for SFE. SFE-grade CO_2 is available commercially. However, the costs are significant. Also, contaminant build-up may not be sufficiently mitigated when long dynamic SFE analyses are necessary or when performing SFE coupled to gas chromatography (SFE-GC). Using an in-line catalyst-based purification system, minimum purity "bone dry" CO_2 can be purified to the level of the highest purity, the most expensive CO_2 available. Significant cost savings can be realized using the various commercial purification systems with inexpensive, low-purity CO_2 in place of much more expensive, high-purity CO_2.

Apart from CO_2, several other SCFs can also be used (**Table 3**). Organic solvents are usually explosive; hence, an SFE unit using organic solvents must be explosion-proof. Chloroflurocarbons (CFCs) are good solvents for SFE because of their high density. However, the industrial use of chlorofluorohydrocarbons is restricted because of their effect on the ozonosphere. More recently, the use of water in SFE has become quite popular. One of the unique properties of water is that above its critical point

Table 3
Various Supercritical Solvents and Their Critical Conditions

Fluid	Critical temperature (K)	Critical pressure (bar)
Carbon dioxide	304.1	73.8
Ethane	305.4	48.8
Ethylene	282.4	50.4
Propane	369.8	42.5
Propylene	364.9	46.0
Trifluoromethane	299.3	48.6
Chlorotrifluoromethane	302.0	38.7
Trichlorofluoromethane	471.2	44.1
Ammonia	405.5	113.5
Water	647.3	221.2
Cyclohexane	553.5	40.7
n-Pentane	469.7	33.7
Toluene	591.8	41.0

(374°C, 218 atm), it becomes an excellent solvent for organic compounds and a poor solvent for inorganic salts. This feature allows water to be used in SFE for extracting inorganic and organic components sequentially.

2.2. Pressure System

The pump in SFE, which is used to pressurize CO_2 for achieving supercritical state, must be capable of generating high pressure and delivering reproducible volumes at a constant flow rate. In SFE, the analyte is collected for a finite time before it is analyzed further. Therefore, the total volume of SCF passed through the extraction chamber is crucial. It may be necessary to modify the conventional pumps to make them suitable for handling SCFs, since the compressibility of these fluids is different from the liquids that are designed for the pumps. There must be a provision for pump head cooling to maintain CO_2 in its liquid state. Customized pumps are also available commercially nowadays. For example, the ISCO Series D pumps or JASCO SFC/SFE PU-1580-CO_2 pumps are well-suited for use as SCF pumps.

In SFE operations, two types of pumps, which provide the required flow and pressure, are generally used: reciprocating and syringe pumps. Reciprocating pumps have an 'infinite' reservoir and supply a continuous flow of SCF. Modifiers are normally added using doped cylinders or a second pump with proportioning valves. To pump liquid CO_2, the pump head must be cooled using low-cost cryogenic CO_2.

Syringe pumps provide pulseless flow and can easily be filled with liquid CO_2. These pumps do not need to be cooled because CO_2 is liquefied by pressure, not temperature. However, because of the limited volume, the syringe must be filled and repressurized when the pump cylinder is emptied. When changing modifiers, it is necessary to flush the pump head thoroughly to prevent carryover. Ideally, these pumps should be able to deliver fluid at 0.5–4 mL/min flow rates.

2.3. Sample Preparation

In an SFE system, an extraction vessel holds the matrix to be extracted. Natural product matrix could be of various origins, e.g., plants, microbes, and so on, and of several physical forms. For bulk matrix, it is essential to carry out some preliminary sample preparation including grinding, sieving, drying, mixing, pH adjustment, or wetting, depending on the physical form of the matrix. For nonporous or semiporous material, a smaller particle size allows more efficient and faster extraction. In case of semisolid, gel, or liquid matrix, it is necessary to immobilize the matrix on a solid support because SFE is generally not suitable for liquid samples (*see* **Note 1**). Such matrix is applied to a piece of filter paper, a solid support such as diatomaceous earth, or to a drying agent to carry out the extraction efficiently and to prevent the matrix being swept out of the extraction cell. Water removal is essential for wet matrices to enhance the recovery and reproducibility of the extraction process. Addition of sodium sulfate and diatomaceous earth to make a free-flowing powder often gives good results.

Most of the currently available extraction cells, which must be inert and able to withstand the pressure generated by the pump, are stainless steel (SS) tubes with compression end fittings. However, more sophisticated and expensive extraction cells are automatic sealing thimbles. Typically, most SFE samples have masses <10 g (*see* **Note 2**). The size of commercial analytical SFE extraction cells ranges from 100 to 50 mL. The amount of SCF required for a typical extraction depends on a number of parameters. As a general guideline, the amount of SCF is at least three extraction cell volumes. The extraction cell is usually in an oven to control the temperature because any fluctuation in the temperature results in a change in SCF density and solvating power. Higher temperature may also increase the solubility of any added modifier causing equilibrium complications.

2.4. Extraction

Static, dynamic, or recirculation modes can be used to perform SFE. In the static mode, the extraction cell is pressurized with SCF, and allowed to equilibrate prior to removal of the analyte for collection. In the dynamic mode, which appears to be the most popular among the three, SCF is passed through the extraction cell, and the analyte is collected continuously. In the recirculation mode, the same fluid is pumped through the sample for a while before it is repeatedly pumped to the collection device.

Addition of a modifier can bring about enhanced efficiency of the SFE process. Modifier addition can be accomplished in three ways:

(1) By adding it (usually 500 μL or less) directly to the sample in the extraction cell or by mixing it with the sample prior to loading in the extraction cell.
(2) By using CO_2 gas cylinders containing premixed amounts of added modifier.
(3) By adding it in a continuous fashion with an external modifier pump.

Among these options, the modifier pump approach appears to be the most desirable and convenient way.

It is necessary to maintain the extraction cell under pressure to reach the supercritical state. Once the supercritical CO_2 and analyte leave the extraction vessel, the supercritical CO_2 must be depressurized to change from an SCF state to a gas state. This can be achieved by introducing a fixed or variable restrictor, the former being used more frequently. Early instruments used fixed restrictors, essentially fused silica capillary tubes. The diameter and the length of this tube can provide the appropriate back pressure. Thus, the pressure and SCF density can be changed by varying the flow rate through the restrictor. The restrictor needs to be replaced to regulate the system pressure to achieve desired change in the density of SCF at constant flow rate. Also, during method development, fixed restrictors must be changed between extractions. Fixed restrictors tend to plug and have limited lifetimes. In addition, flow rate variations between restrictors can cause irreproducible results.

Variable restrictors, which do not need to change during method development, are designed to regulate pressure independent of flow rate by mechanically controlling the size of a small opening. Now all major vendors have automatic variable restrictors that can open and close to maintain flow rates at desired levels. The decompression control offered by automatic variable restrictors has been a major factor contributing to the reduction of clogging. This is particularly important in the analysis of high-fat foods.

2.5. Sample Collection

As the supercritical CO_2 passes through the restrictor, the change in pressure here causes the pressure of the SCF to decrease, and eventually gaseous CO_2 may form. This process is called depressurization. After passing through the restrictor, the supercritical CO_2 reverts to CO_2 gas, and the analyte is deposited in a collection vial, which may be filled with glass wool or an organic trapping solvent. Some vendors' instruments have the capability to cool the collection vial, which for some analytes has been shown to improve recoveries.

Some instruments allow for alternative analyte trapping techniques. For example, cryogenically controlled solid-phase extraction (SPE) traps are beneficial for some applications. When SPE traps are used, the analyte is deposited in the trap, and the trap is then rinsed with a small volume (usually 1–2 mL) of *n*-hexane, $CHCl_3$, iso-octane, ether, or some other organic solvent. Alternatively, the solid-phase trap can be filled with glass beads, alumina, or other types of solid supports, depending on the specific application. Once the analyte is recovered, its level in the natural product matrix can be determined by simple gravimetric analysis, or a portion of the extract can be removed with a syringe and analyzed by gas chromatography (GC) or high-performance liquid chromatography (HPLC). This type of analysis is referred to as off-line analysis or off-line collection. Off-line collection is simple, and the collected samples can be analyzed by several methods. Other advantages of this approach are accommodation of larger sample sizes, feasibility of multiple analyses from a single extraction, and accommodation of a wider range of analyte concentrations.

In on-line collection, the analytical instrument is usually coupled to the extraction device. The analyte is trapped at the head of a GC, SFC, or HPLC column, and then analyzed. On-line analysis is a more advanced technique, which involves coupling the GC or HPLC to the restrictor of the SFE, so that the analyte can be transferred directly into the analytical instrument. It is advantageous for trace analysis. Since the analyte is transferred to the analytical instrument, on-line SFE can be 1000 times more sensitive than off-line SFE. Off-line analysis, however, is far simpler to perform. On-line analysis will undoubtedly be the next major SFE area that will grow in importance as vendors create instrument accessories to simplify the cumbersome complexities of the technique. Furthermore, the development of more reliable, nonplugging SFE instruments increases the likelihood of coupling SFEs to other instruments. The disadvantage of on-line analysis is that the analyte is directed into the analytical instrument, and none can be analyzed by other techniques without carrying out another extraction. On-line extraction can also deposit many contaminant(s) onto the column head. There is also a risk of column overloading.

3. Properties of SCFs and Solubility of Solids in SCFs

A number of important properties of a pure liquid can be changed dramatically by increasing the temperature and pressure approaching the thermodynamic critical point. For example, under thermodynamic equilibrium conditions, the visual distinction between liquid and gas phases, as well as the difference between the liquid and gas densities, disappears at and above the critical point. Applications of SCF include recovery of organics from oil shale; separations of biological fluids; bioseparation including natural products; petroleum recovery; crude de-asphalting and dewaxing; coal processing; selective extraction of fragrances, oils, and impurities from agricultural and food products; pollution control; combustion; and many other applications. SFE is based on the fact that near the critical point of the solvent, its properties change rapidly with only slight variations of pressure. Above the critical temperature of a compound, the pure, gaseous component cannot be liquefied regardless of the pressure applied. The critical pressure is the vapor pressure of the gas at the critical temperature. In the supercritical environment, only one phase exists. The SCF, as it is termed, is neither a gas nor a liquid and is best described as intermediate to the two extremes. This phase retains solvent power approximating liquids as well as the transport properties common to gases. A comparison of typical values for density, viscosity, and diffusivity of gases, liquids, and SCFs is presented in **Table 1**.

4. Description of a Typical Supercritical Apparatus

The SCF apparatus, which was designed and constructed by Dr Arthur T. Andrews ***(11,12)***, is presented in **Fig. 3**. The extraction–circulation-sampling system is located in a GCA Precision Model 18EM forced convection oven ($46 \times 33 \times 48$ cm), modified by the installation of a temperature controller to ensure the isothermal operation (±0.2°C). The primary circulation loop consists of a modified Milton Roy extraction vessel, a magnetically coupled gear pump, and a sampling-monitoring system parallel to the main flow loop. All primary components are made of stainless steel (SS). Tubing (0.64 cm) in the main loop is rated for a pressure of 10,000 psi. To minimize volume, the sample loop is made of 0.16 and 0.32 cm steel tubing. Valves and fittings manufactured by Autoclave Engineers and Crawford are used. The overall limits on operating temperature and pressure are 25–100°C and 4000 psig, respectively.

An eductor-equipped cylinder contains liquid CO_2, which passes through a filter (7 μL) and is delivered to a Milton Roy Model 396–89 piston pump. For continuous flow desorption runs, to prevent cavitations, the pump head is cooled via an external cooling bath. An adjustable Haskel Model 53379–4 pressure control valve (PCV) and safety rupture disk (10,000 psig rated) prevent overpressurization. The basic unit, comprising the O-ring sealed cap and the lock ring, is from an original Milton Roy screening apparatus. A 0.16-cm SS sheath thermocouple, type K, is inserted into the vessel for measuring bed temperature. The primary O-ring is made of high-pressure CO_2-compatible Buna N synthetic rubber.

Supercritical CO_2 is circulated by a magnetically coupled gear pump (Micropump Model 183HP-346), modified for operation up to 5000 psi. A filter (40 μm, FI) is installed upstream of the pump, and a sintered metal filter (2 μm) is located downstream to prevent solids from entering the

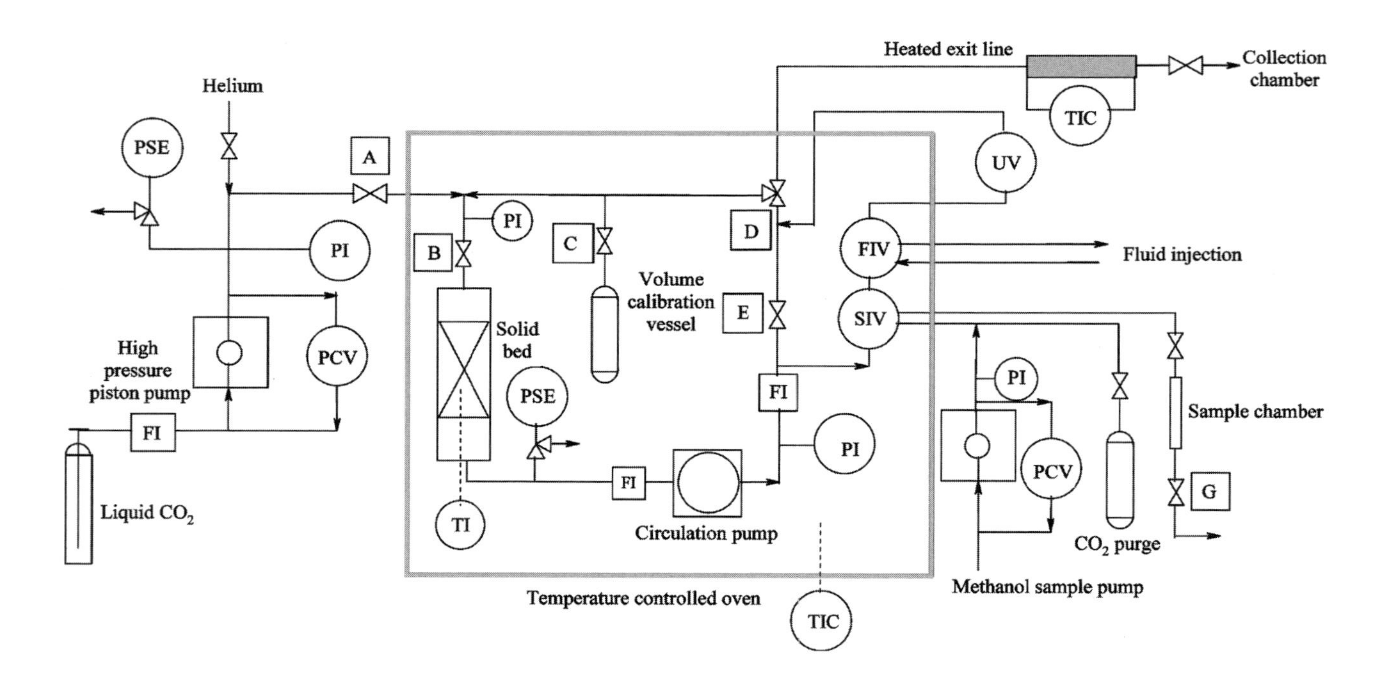

Fig. 3. A typical SCF apparatus.

sample valve. A safety rupture disk (PSE), rated at 4300 psi, protects the pump from overpressurization. The DC pump motor (24 V) is controlled by a Cole Parmer Model 2630 speed controller. The body of the pump is made with SS, and the gears are of hard polymer, RYTON, which is compatible with high-pressure CO_2. The pump flow is at 100 mL/min at 10-psi pressure drop. At the time of circulation under supercritical conditions, a flow rate of approximately 140 mL/min was achieved at a midrange pump speed.

A fraction of the main recirculation flow is delivered through a parallel sample path, which contains in series:

(1) A 6-port sample valve (SIV) (Rheodyne Model 7010, with a 142-μL loop).
(2) A 6-port injection valve (FIV) (Rheodyne Model 7010, with a 1040-μL loop).

Flow rate through this sample path, which is estimated at approx 1 mL/min, is controlled by throttling the diversion valve in the main flow circuit to achieve a ΔP of 10 psi between the pump outlet and the extractor inlet. A larger volume (142 μL) of pressurized sample is injected into a flowing stream of MeOH, which is collected in a closed chamber (2 mL). The MeOH–CO_2–analyte mixture is discharged into a tarred vial. The sample loop and the chamber are flushed with additional MeOH and purged with CO_2 from an external reservoir. On the basis of the solubility of CO_2 in MeOH, approx 80% of the gas flashes off, and the remaining MeOH solution contains less than 1% CO_2. A fluid injection valve with a 1040-μL loop is used during system calibration as well as for the injection of cosolvents.

Temperatures are measured at two points within the system: a thermocouple (TI) inside the extractor and another type-K thermocouple inside the oven near the sample valves. Output from these thermocouples is read from an Omega Model DP-462 digital thermometer. Pressures are measured by Omega Model PX425-6KGV pressure transducers at two points: at the inlet of the extractor and at the discharge of the gear pump. Output from the transducers is converted by Omega Model DP-280 digital pressure indicators (PI).

A typical discharge-collection system used during flow studies is presented in **Fig. 4**. The CO_2 flashes through a heated metering valve (Autoclave Engineers' Model 30 VRMM) into two 150 mL collection chambers (Hoke) in series. A Metheson Model 603 glass rotometer, protected from overpressurization by a 5-psi safety relief valve, is used to measure the flow rates. After passing through a 15 cm bed of activated

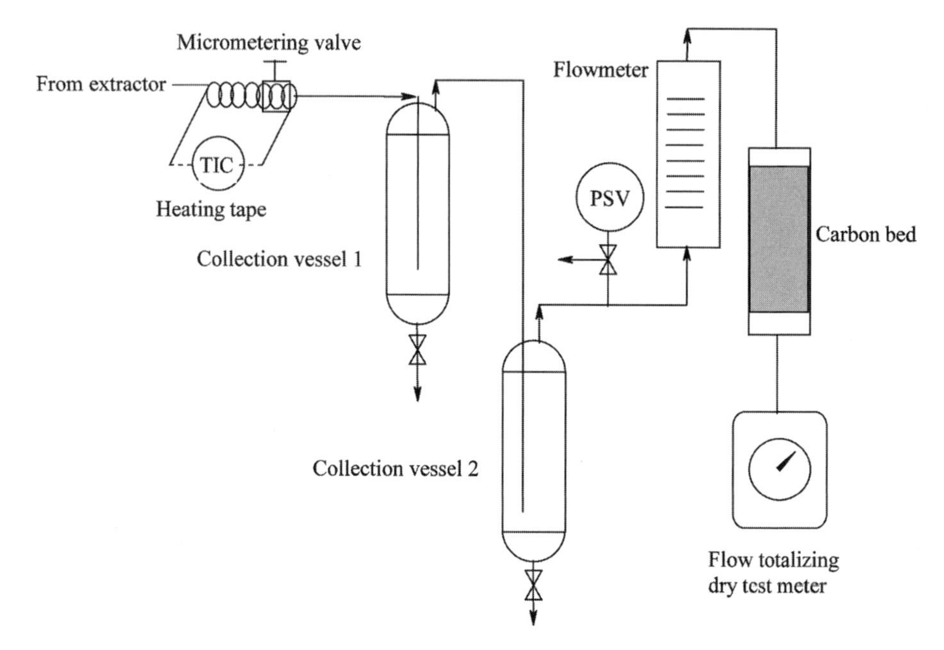

Fig. 4. A typical discharge-collection system.

carbon to remove residual organic matter, the final CO_2 gas flow is totalled in a Singer DTM 115 dry test meter.

5. Important Factors in SFE Method Development

A number of important factors are to be considered during the SFE method development of natural product extraction. Some of these are as follows:

(1) The solubility of the target compound(s) in supercritical CO_2 or other SCF has to be determined (*see* **Note 3**). It is necessary to perform solubility tests to determine the effect of temperature and pressure (which in turn control the density) on the solubility of the target compound(s) in the SCF.
(2) The effect of cosolvents on the solubility of the target compound(s) needs to be determined (*see* **Note 4**).
(3) The effect of matrix, either has the analyte lying on its surface (adsorbed), or the analyte is entrained in the matrix (absorbed), has to be considered carefully (*see* **Note 5**).
(4) The solvating power of SCF is proportional to its density, which can be affected by any temperature change for any given pressure. Therefore, strict temperature control has to be in place (*see* **Note 6**).

(5) The partition coefficient of the analyte between CO_2 and the matrix, which is often affected by the flow rate, has to be considered. Higher flow rates and longer extraction time may be necessary to sweep the analyte out of the extraction chamber. Lower flow rates may be applied if the kinetics of the system are slow.
(6) Careful consideration has to be given in choosing appropriate modifiers (*see* **Note 7**).

6. Applications of SFE to Natural Products

SFE has long been used in industries for the extraction of various commercial natural products (**Table 4**). The most cited examples are decaffeination of coffee and extraction of hops, spices, flavors, and vegetable oils. SFE has also been used in small-scale extraction of various types of natural products in the laboratory. However, the use of SFE in natural product extraction is predominantly limited to plant natural products. There are just a handful of studies carried out on the use of SFE in the extraction of microbial natural products. There are a number of reviews available on the applications of SFE, including that in natural product extraction *(7,13,14)*. Some specific examples in natural products are described below.

6.1. SFE of Essential Oils

The fragrance of certain natural product extracts is because of the presence of essential oil, which is often a complex mixture of a number of thermolabile monoterpenes and sesquiterpenes and other aromatic compounds. Correct reproduction of the natural fragrance in a concentrated extract is a complex task because of the thermolabile nature of the

Table 4
Application of SFE for the Extraction of Commercial Natural Products

Natural product	Location	Start up year	Company
Coffee	Germany	1979	HAG-General Foods
	USA	1988	General Foods
Hops	Australia	1982	Carlton and United
	Germany	1985, 1988	Barth
Hops, spices	Germany	1982	SKW Chemicals
	England	1983	English Hops
Tea	Germany	1986	SKW Chemicals

compounds, the possibility of hydrolysis and hydrosolubilization. Being Suitable for heat-sensitive compounds, SFE has been the method of choice for the extraction of natural essential oils over the past few decades. The extraction of essential oil components using solvents at high pressure or SCFs has received remarkable attention over the past several years, especially in food, pharmaceutical, and cosmetic industries, because it offers an environment-friendly and cost-effective alternative to conventional extraction processes. A typical example is the SFE of black pepper essential oil, which has been described in the literature ***(15–18)***. A semicontinuous arrangement with solvent flowing through a fixed bed of ground pepper with constant rate is generally used to yield a yellowish semisolid mass with a crystalline fraction.

A typical SFE complete setup for black pepper extraction uses a 150 mL high-pressure liquid extraction vessel immersed in a water bath of 40°C controlled by a heater. The system pressure is controlled by a back-pressure regulator or a metering valve. A syringe pump is used to deliver liquid CO_2 at 1.0 mL/min. The samples are collected by a cool trap in a 7 mL clear vial. The grinding plays a pivotal role in increasing the oil yield extracted from black pepper. When the size decreases from whole seed to 0.5 mm, the amount of pepper oil obtained increases tremendously from 2.9% to almost 7.5%. Reduction of particle size also decreases the amount of liquid CO_2 required for gaining the same level of oil yield. Various levels of pressure can be used in the SFE extraction of black pepper. For example, Ferreira et al. ***(16,17)*** carried out SFE at 200 bar for fixed-bed extraction of black pepper essential oil, whereas Sovova et al. ***(18)*** conducted their SFE at slightly higher pressure, 280 bar, to extract oleoresin from ground black pepper.

Uquiche et al. ***(19)*** described the extraction kinetics of red pepper oleoresin with supercritical CO_2 at 40°C from a pelletized substrate. The kinetics were evaluated as a function of crushed pellet particle size ($D-p = 0.273 - 3.90$ mm), superficial solvent velocity ($U-s = 0.57 - 1.25$ mm/s), and extraction pressure (320–540 bar). It was observed that the batch productivity increased with substrate pelletization, which caused a four-time increase in apparent density. It was noted that the yield of oleoresins and carotenoid pigments increased, and temperature decreased as the extraction pressure increased.

Sub- and supercritical CO_2 were used to extract two origanum samples: one commercial, and another cultivated under agronomic control ***(20)***. The experiments were performed in the temperature range of 293–313 K and from 100 to 200 bar in pressure, employing around 26 g of origanum samples. Results demonstrated that the commercial sample provided a higher yield of extract when compared to the other sample. It was also established that a rise in temperature at constant pressure led to an increase in the extraction yield despite solvent density changes. Chemical analyses were carried out in a GC-MSD, allowing the identification of around 24 compounds by the use of the library of spectra of the equipment and injection of some standard compounds for both commercial and cultivated origanum samples. It was found that the distribution of chemical components as a function of extraction time differed significantly between the origanum species. The chromatographic analysis resulted in the identification of thymol and *cis*-sabinene hydrate as the most prominent compounds present in commercial oregano sample, and carvacrol and *cis*-sabinene hydrate in the cultivated *Origanum vulgare*.

SFE of jojoba oil from *Simmondsia chinensis* seeds using CO_2 as the solvent was reported by Salgin et al. ***(21)***. The effects of process parameters, particularly pressure and temperature of extraction, particle size of jojoba seeds, flow rate of CO_2, and concentration of entrainer (*n*-hexane) on the extraction yield were examined. It was noted that an increase in the supercritical CO_2 flow rate, temperature, and pressure generally improved the SFE performance. The extraction yield increased as the particle size decreased, indicating the importance of reducing intraparticle diffusional resistance. The maximum extraction yield obtained was 50.6 wt% with a 0.23 mm particle size and a 2 mL/min CO_2 flow rate at 90°C and 600 bar. Use of an entrainer at a concentration of 5 vol% improved the yield to 52.2 wt% for the same particle size and also enabled the utilization of a relatively lower pressure and temperature, i.e., 300 bar and 70°C. There are a number of other literature citations on the application of SFE in the extraction of essential oils from various plant sources ***(16,22–24)***.

6.2. SFE of Capsaicinoids

Perva-Uzunalic et al. ***(24)*** evaluated the influence of operating parameters (pressure from 100 to 400 bar and temperatures of 40°C, 60°C, and 80°C) on the SFE extraction efficiency of capsaicinoids and color components from chilli pepper (variety Byedige). Capsaicinoid content and color value were determined in raw material and residue material after extraction. The total extraction yield and extraction efficiency of capsaicinoids increased with rise in pressure at constant temperature as well as with increasing temperature at constant pressure. The highest extraction yield for total solids of 12.8% was obtained at 400 bar and 40°C, where almost 96% of capsaicinoids and 80% of color components were removed from the raw material.

Duarte et al. ***(25)*** outlined an SFE protocol for the extraction of pepper (*Capsicum frutescens*) oleoresins. The influence of pressure and superficial velocity of supercritical CO_2 at 313 K on the *C. frutescens* oleoresins yield and capsaicinoid content was studied. The central composite, nonfactorial design was used to optimize the extraction conditions, using Statistica, version 5 software (Statsoft). The results were compared with those obtained when *n*-hexane was used for the extraction of red pepper oleoresin in a Soxhlet apparatus. At 10 min of extraction time, an optimal value of the yield was determined (5.2% w/w) for pressure of 21.5 MPa and superficial velocity of 0.071 cm/s. An optimum value for the yield in capsaicinoids (0.252% w/w) was achieved at 20.5 MPa and 0.064 cm/s. The extract of the SCF process obtained at 21.5 MPa, 0.074 cm/s of CO_2 superficial velocity at 10 min was chosen to perform a sensorial analysis. Selected extract was suspended in extra virgin olive oil (0.7% acidity) at three capsaicinoid contents (0.0006% [w/w], 0.0011% [w/w], and 0.0015% [w/w]) and analyzed in terms of pungency, aroma, taste, and aftertaste. Commercial red pepper-flavored olive oil (0.0011% [w/w] capsaicinoids) was used as reference. Panel preferred taste of the olive oil added with extract that contains 0.0006% (w/w) of capsaicinoids. At this concentration, SFE extract did not influence olive oil color. It was observed that the percentage of oleoresins recovered by SFE increased with CO_2 density and also with the superficial velocity. The optimum value for the yield in capsaicinoids occurred with the pressure and superficial velocity at 20.5 MPa and 6.4×10^{-4} m/s. The highest extraction yields in oleoresins and capsaicinoids were obtained at pressures around 20–22 MPa and about 0.06–0.07 cm/s.

6.3. SFE of Polyphenols

Polyphenols were extracted from grape marc using an alternative SFE method based on the use of a liquid trap, which allows extracted polyphenols to be retained in a saline buffer, thus avoiding the need for the organic solvent required to elute polyphenols from a solid trap ***(26)***. The major extraction variables influencing the performance of the liquid trap (viz. CO_2 modifier content, flow rate, extraction time, and trap volume) were optimized. The proposed method was applied to the SFE of 0.3-g grape marc with CO_2 modified with 3% MeOH at 350 bar at 50°C (CO_2 density 0.9 g/mL for 20 min, using a liquid flow rate of 0.9 mL/min).

6.4. SFE of Flavonoids

An SFE protocol using ethanol-modified CO_2, as solvent for the extraction of free quercetin from onion skins (red and yellow varieties), was presented by Martino and Guyer ***(27)***. In the static mode, this extraction method generated a total of 0.024 g of free quercetin per kg of onion skin for the red variety, and 0.02 g per kg for the yellow variety, at 5700 psi, 40°C, an average of 76% (molar concentration) of ethanol (EtOH), and in an extraction period of 2.5 h. The modifier enhanced the free quercetin recovery significantly. The greater the amount of EtOH collected in the trap, the greater the amount of free quercetin recovered.

The extraction yield of SFE using supercritical CO_2 and supercritical CO_2 with 10% EtOH, and activity of antioxidants, predominantly (−)-epicatechin, from the seed coat of sweet Thai tamarind (*Tamarindus indica*) was studied by Luengthanaphol et al. ***(28)***. SFE was performed on an ISCO Model SFX 3560 laboratory extractor using a temperature range 35–80°C and pressure 10–30 MPa. It was observed that the extraction of (−)-epicatechin with pure CO_2 was low (~22 μg per 100 g of seed coat). The use of 10% EtOH as a cosolvent (modifier) increased yield remarkably (~13 mg per 100 g of seed coat) under the best condition, 40°C and 10 MPa.

The feasibility of SFE using supercritical CO_2, nitrous oxide, and R134a for the extraction of ginkgo leaves (*Ginkgo biloba*) aiming at the purification of ginkgolides and flavonoids was investigated by Chiu et al. ***(29)*** and van Beek ***(30)***. It was observed that a particle size $<105\,\mu m$ produced the best extraction efficiency of terpene lactones and flavonoids. When ginkgo powder (90 g) was extracted by supercritical CO_2 at 31.2 MPa, 333 K and 10% EtOH modifier added with CO_2 in sequence, the content of ginkgolides and flavonoids in the extract was 837 and 183 g/g, respectively. The amount of flavonoids was significantly increased when two ratios of cosolvents were individually preloaded with ginkgo leaves in the extractor.

6.5. SFE of St. John's Wort

St. John's wort (*Hypericum perforatum*) was extracted on a pilot-scale batch extraction plant from SITEC Sieber Engineering AG (Maur, Switzerland) using supercritical CO_2 ***(31)***. It was concluded that this is a convenient method for the enrichment of hyperforin (**Fig. 5**) from this plant. Prior to extraction, the particle size of this herbal drug was reduced, using a cutting mill (59% between 0.355 and 1 mm). For each batch extraction, 370 g of milled St. John's wort was used, and all runs were performed

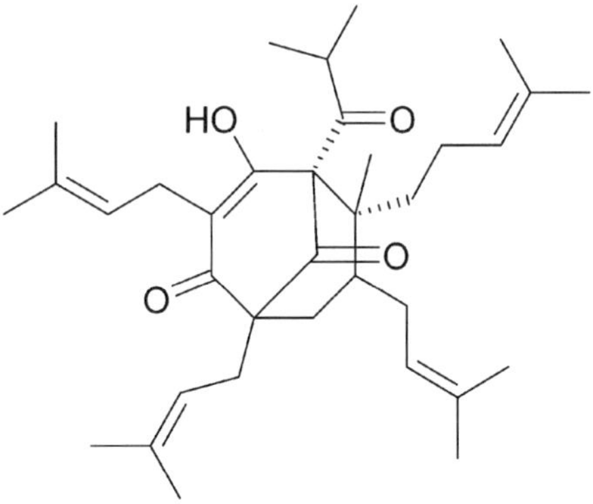

Fig. 5. Structure of hyperforin.

in the single-stage separation mode. The effects of pressure, temperature, flow rate, and extraction time were examined with respect to extraction yield and hyperforin content. Supercritical CO_2 displayed a high selectivity for phloroglucinols. Extracts were analyzed using an isocratic HPLC method with a mixture of hyperforin/adhyperforin as an external standard. Within the studied range of extraction pressure (90–150 bar) and extraction time (1–5 h), extraction at 90 bar for 3 h and 120 bar for 1 h provided the highest hyperforin content (up to 35%) in the resulting extracts. An increase in extraction temperature showed a negative effect, leading to increased degradation of hyperforin into orthoforin. When the total mass of CO_2 passing the extraction vessel was kept constant, changes in mass flow rate did not affect the extraction.

6.6. SFE of Parthenolide From Feverfew Plant

The extraction of the sesquiterpene lactone parthenolide (**Fig. 6**) from feverfew plants (*Tanacetum parthenium*) has been described in various studies *(32–34)*. SFE was compared with conventional steam distillation and solvent extraction. SFE extracted less volatile lactones and parthenolide, and steam distillation extracted volatile terpenoids. SFE resulted in incomplete extraction compared to solvent extraction. The addition of 4% MeOH increased the efficiency of SFE leading to complete extraction, but a significant loss of selectivity because of co-extraction of chlorophyll and other less volatile components. Smith and Burford and Čretnik et al. *(34,35)* exploited the relatively weak eluent strength of unmodified CO_2 in a small preparative scale. Dried and ground feverfew was extracted with CO_2 and

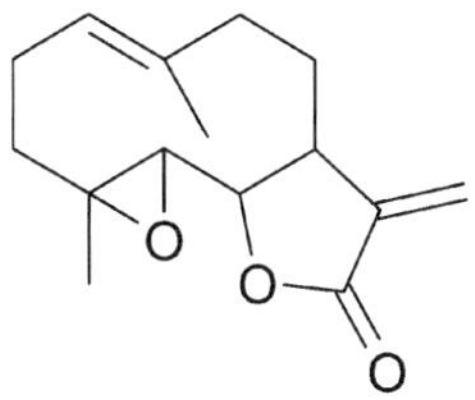

Fig. 6. Structure of parthenolide.

the extract was passed through a short silica column. The slightly polar sesquiterpene lactones were retained in the column, whereas the volatile essential oils were collected in a collection vessel. The extraction vessel was then switched off from the supercritical CO_2 flow, and 10% MeOH was added to the CO_2 flow to the column, which eluted a lactone fraction containing 80% parthenolide.

6.7. SFE of Taxol® From Pacific Yew Tree

Taxol, one of the most commercially successful and effective anticancer natural product drugs, is a complex diterpene isolated from the Pacific yew tree (*Taxus brevifolia*) (**Fig. 7**). The SFE protocol for the extraction of Taxol from the bark was introduced by Georgia Tech, Athens, GA *(35)*. About 50% of the Taxol present in the bark was selectively extracted using a CO_2–EtOH mixture as opposed to 25% extraction with supercritical CO_2 alone.

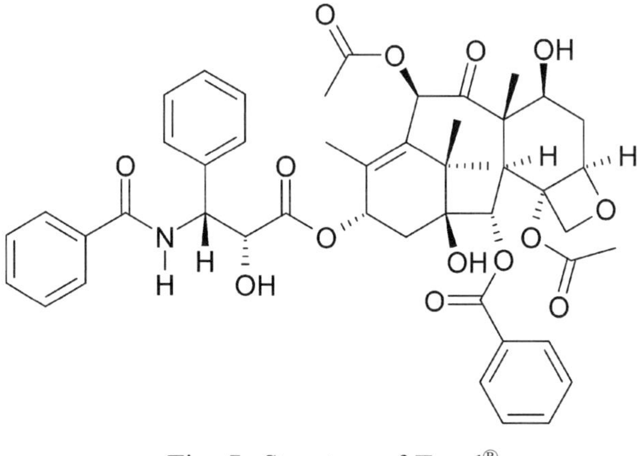

Fig. 7. Structure of Taxol®.

6.8. SFE of Resveratrol From Vitis vinifera

Resveratrol (**Fig. 8**), a well-known bioactive stilbene, was isolated selectively from grape skin (*V. vinifera*) using SFE *(36)*. An ISCO SFX3560 extractor equipped with model 260D syringe pumps was used for SFE, and a Hewlett Packard 8453 diode array spectrometer was utilized for monitoring SFE recovery of resveratrol. The overall protocol involved the following steps:

1. For the preparation of sample in spiked model, resveratrol (0.05 g) was dissolved in 10 mL of EtOH, and an aliquot (2 mL) of this solution was added to diatomaceous earth (1 g). The spiked solid was kept dry for at least 2 h at 60°C to evaporate EtOH and precipitate resveratrol over the surface of solid particles.
2. The grape skin was oven dried at 100°C for 1 h. Dry grape skin (3 g) was weighed into the extraction thimble and extracted.
3. The operating parameters for the SFE were set as follows: extraction temperature 40°C, trap temperature 40°C, restrictor temperature 40°C, pressure 150 bar, EtOH as modifier (7.5%), static extraction period 30 s, dynamic extraction time 15 min, and flow rate 2 mL/min.
4. The extract was collected in 5 mL of EtOH, diluted to 10 mL, and analyzed by UV-VIS absorption spectrometry, when the spiked model was used, and HPLC for natural sample.

It was observed that an increased EtOH percentage (up to 7.5%) caused a rise in the recovery, whereas percentages above 7.5% decreased it. A rise in pressure also increased the recovery. A minimum pressure of 110 bar with 7.5% modifier (EtOH) was found to be optimum for 100% recovery.

6.9. SFE of Dandelion Leaves

Triterpenoids, β-amyrin and β-sitosterol, were extracted from freshly ground dandelion (*Taraxacum officinale*) leaves (1 kg) in preparative scale

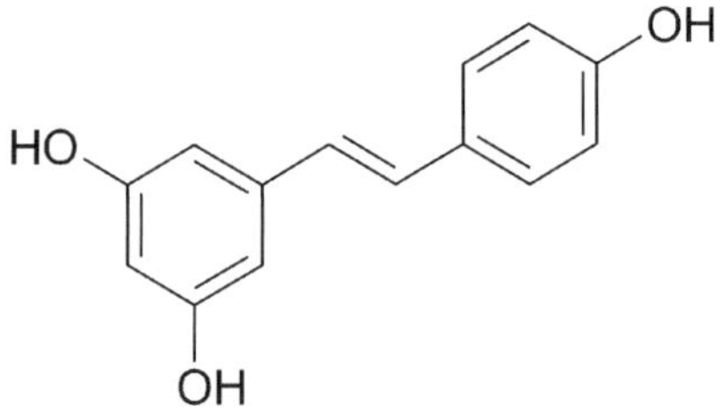

OH
HO
OH

Fig. 8. Structure of resveratrol.

by SFE on a high-pressure apparatus equipped with a 5 L NATEX extractor vessel using supercritical CO_2 *(37)*. The solvent feed was started with a flow rate of 7.4 kg CO_2/kg dried leaves per hour. The accumulated product samples were removed and weighed at regular time intervals. The extraction was stopped when the increase in yield was <0.1% while 10 kg CO_2 passed through the vessel. The extractions at pressures and temperatures near the optimum conditions gave 3.2–4.0% of extracts to the dry weight of the raw material, and TLC analysis indicated that it contained 11.0–12.7% and 3.1–3.7% β-amyrin and β-sitosterol, respectively.

6.10. SFE of Cyclosporine From Beauvaria nivea

te Bokkel reported the SFE of cyclosporine, an immunosuppressant drug from the fungus *B. nivea*, applying supercritical CO_2 *(38)*. The method of pretreatment of the fungal mycelia exhibited a profound effect on the yields and rates of cyclosporine extraction. The extraction of the drug was carried out on different samples of the fungal mycelia subjected to varying degrees of air and oven drying. The lowest yield was obtained with the mycelia that was oven-dried. The oven-dried material had a completely solid nonporous surface. However, the air-dried material, which was quite open and had many broken mycelia, was more amenable to extraction. The highest yield was obtained with the mycelia that was minimally air-dried, having a moisture content of 2.9%.

6.11. SFE of Mycotoxins

Mycotoxins are biologically active secondary metabolites produced by filamentous fungi, often found in contaminated food stuff, and are responsible for a toxic response in humans and animals, known as mycotoxicosis *(39)*. One of these mycotoxins, beauvericin (**Fig. 9**), was first reported from fungi such as *Beauvaria bassiana* and *Paecilomyces fumosoroseus*, and later it was found in various other fungi belonging to the genus *Fusarium (40)*. Ambrosino et al. *(40)* described the development and the optimization of an SFE protocol of beauvericin on spiked and natural contaminated maize samples. The following procedure was performed:

1. The maize samples were ground for 2 min in an Omni mixer and stored in sealed plastic bags at 4°C.
2. The standard beauvericin was dissolved in MeOH and used in spiking experiments. The maize was spiked at 100 ppm level with beauvericin standard solution at 2000 ppm (work solution).

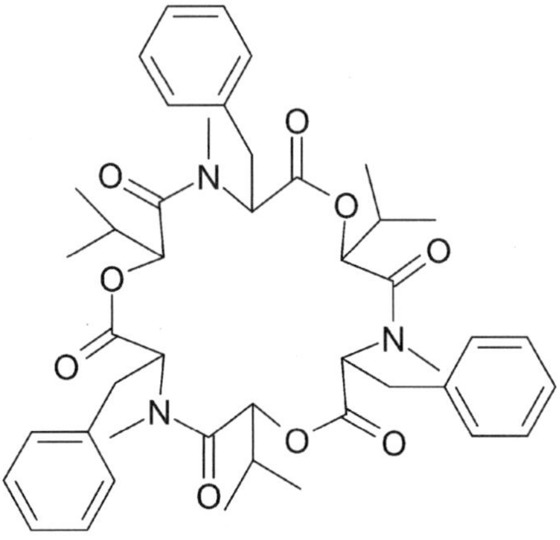

Fig. 9. Structure of beauvericin.

3. In the experiment, an aliquot (450 μL) of the work solution was added with 9 g of maize, while 900 μL to spike 18 g of blank maize.
4. To test the method, natural contaminated maize with beauvericin level of 165 mg/kg was used.
5. The SFEs were performed on a Spe-ed SSFE model 7010, which allowed precise and reproducible performance with its flexibility of both dynamic (continuous flow) or static modes.
6. The starting extraction time was set when the working pressure had been reached.
7. At regular intervals, the extracts were sampled manually using glass vials of various capacities sealed with caps and septa.
8. The modifier (10% deionized water) was added to the sample prior to the static extraction step in the extraction cell, pressurized with supercritical CO_2, and allowed to equilibrate for a period of static extraction time at a required temperature and pressure.
9. After the static extraction time, the extract-collection valve was opened, allowing the supercritical CO_2 to pass through the capillary restrictor, which was maintained at 15°C higher than the temperature of outlet valve, to the collection tube containing 2 mL of $CHCl_3$ used to trap the analyte.
10. After completion of the SFE process, the $CHCl_3$ was filtered on a 0.45-μm Millipore filter, and evaporated to dryness under nitrogen.
11. Cosolvent SFE extractions were performed using pure MeOH as a modifier, added by external HPLC pump at 3200 psi with a flow rate of 0.3 mL/min. All the extracts were reconstituted into 500 μL of pure MeOH for HPLC analysis.

12. The maize matrix used during this study was evaluated to check the presence of beauvericin content, and uncontaminated maize (9 g) was used in each experiment. For cosolvent experiments, 8.4 g of matrix was used.
13. The recovery of beauvericin was calculated by arithmetic mean of three replicate measurements carried out by HPLC.
14. Supercritical CO_2 flow rate was controlled by a micrometering valve and was fixed at 0.8 mL/min.
15. $CHCl_3$ was used to trap beauvericin extracted from maize.

The highest recovery of beauvericin was achieved at a temperature of 80°C and a pressure of 4424 psi. It was observed that the beauvericin recovery was positively correlated to operative pressures, but was not affected by CO_2 densities. For the SFE, yields of beauvericin without modifier were not better, at least from the quantitative viewpoint, than the conventional methods of extractions. However, a significant increase in the SFE yield of beauvericin was observed with the addition of modifiers, e.g., MeOH or EtOH. The main advantages of SFE of beauvericin were less use of organic solvent (more environment-friendly) and cleaner extract (more HPLC-friendly).

6.12. SFE of Astaxanthin From Red Yeast

Astaxanthin (3,3′-dihydroxy-β,β-carotene-4,4′-dione) (**Fig. 10**) is the main carotenoid pigment found in aquatic animals. This red-orange pigment is closely related to other well-known carotenoids such as β-carotene or lutein, but has stronger antioxidant activity (10 times higher than β-carotene). Astaxanthin can be more than 1000 times more effective as an antioxidant than vitamin E. In many of the aquatic animals where it can be found, astaxanthin has a number of essential biological functions, ranging from protection against oxidation of essential polyunsaturated fatty acids, protection against UV-light effects, provitamin A activity and vision,

HO O OH O

Fig. 10. Structure of astaxanthin.

immune response, pigmentation, and communication, to reproductive behavior and improved reproduction. Red yeast *Phaffia rhodozyma* is a good source of this antioxidant compound. The SFE was applied to selectively extract astaxanthin from the red yeast *P. rhodozyma* ***(41)***. The effects of extraction pressure (102–500 bar), temperature (40°C, 60°C, and 80°C), CO_2 flow rate (superficial velocities of 0.27 and 0.54 cm/min), and the use of EtOH as a modifier (1, 5, 10, 15 vol%) on the extraction efficiency were investigated. The highest yield of carotenoids and astaxanthin with equal amounts of CO_2 (50 g) was 84% and 90%, respectively, at 40°C and 500 bar. Using a two-step pressure gradient operation, on changing the pressure from 300 to 500 bar, the concentration of astaxanthin in the second fraction at 500 bar increased by about 4 and 10 times at 40°C and 60°C, respectively, and that of carotenoids by about 3.6 times at 40°C and 13 times at 60°C with the yield decreasing by about 40–50%. The commercially fermented yeast *P. rhodozyma* was used for all extractions, and the concentration of astaxanthin in this species was 1000 ppm. *P. rhodozyma* was disrupted by bead mill and dried by spray drycr at 98°C. For SFE, an ISCOSFX3560 SCF extractor equipped with two syringe pumps was used. Pure CO_2 (99.99%) and EtOH were used as an SCF and a modifier, respectively. The prepared *P. rhodozyma* was loaded in a 10 mL extraction cartridge, the remaining volume was filled with glass beads, and the restrictor was kept at operation temperature. In each SFE step, the extract was collected in acetone.

6.13. Sequential Fractionation Using SFE

The raw natural source material contains a number of compounds of various polarities and structural groups. It is possible to isolate sequentially different polarity compounds using SFE by adding suitable modifiers. A typical example of this type of fractionation is the sequential fractionation of grape seeds into oils, polyphenols, and procyanidins via a single system employing CO_2-based fluids, which was reported by Ashraf-Khorassani and Taylor ***(42)***. Pure supercritical CO_2 was used to remove >95% of the oil from the grape seeds. Subcritical CO_2 modified with MeOH was able to extract monomeric polyphenols, whereas pure MeOH was employed for the extraction of polyphenolic dimers/trimers and procyanidins from grape seed. At optimum conditions, 40% MeOH-modified CO_2 successfully removed >79% of catechin and epicatechin from the grape seeds. This extract was light yellow in color, and no higher molecular weight procyanidins were detected. Extraction of the same sample after removal of the oils and polyphenols, but under enhanced solvent extraction conditions using MeOH as a solvent, yielded a dark red solution shown via electrospray ionization HPLC-MS to contain a relatively high concentration of procyanidins.

SFE sequential extraction strategy, using EtOH as a modifier, was applied for the isolation of terpene lactones and flavonoids from the leaves of *G. biloba* ***(29)***. A semibatch extraction and absorption process was employed to collect and sequentially dissolve solutes precipitated from SCF. The extraction condition ranged from 24.2 to 31.2 MPa and from 333 to 393 K. The SCF flow rate was 5 mL/min. The overall process involved milling, sieving, loading, SFE, sampling, sample pre-treatment, and HPLC analysis. The SFE extractor was packed with 90 g of dried ginkgo powder. The absorbing system (one precipitator and two absorbent vessels) was filled with 1.4 L of 95% EtOH absorbent and maintained at 5.0 MPa and 297 K. Ginkgolides, less polar in nature than flavonoids, could be extracted efficiently at low pressure, but the extraction of more polar substances required elevated pressure and the addition of EtOH as modifier. A particle size $<105\,\mu m$ of the plant materials offered the best extraction efficiency of terpene lactones and flavonoids owing to a smaller intraparticle diffusion resistance for a smaller particle size.

7. Notes

1. SFE is generally unsuitable for liquid samples because of difficulties in handling two phases under pressure.
2. The value $<10\,g$ is a compromise between a sample mass requiring a large volume of SCF for quantitative extraction and the amount needed for a representative sample or for trace analysis.
3. If the compound is poorly soluble in SCFs, SFE should not be the preferred method of extraction.
4. Based on the information obtained from available literature, a suitable cosolvent can be chosen and used for the extraction. However, in the absence of any literature data, especially for novel molecular entities, solubility experiments have to be carried out.
5. In case of adsorption, the extraction can be accomplished with milder conditions, whereas if the analyte is absorbed into the matrix, stronger extraction conditions and prolonged extraction time may be necessary. The matrix may contain its own modifiers in the form of water, fats, or oils. If the desired analyte is of polar nature, the water content of the matrix will facilitate the extraction. If it is nonpolar, the water will inhibit the extraction. The opposite effects are evident in the case of fats and oils in the matrix.

6. The higher the density, the more the analyte to be extracted from the matrix.
7. Pure CO_2 is useful for nonpolar to slightly polar compounds. A modifier must be used to extract moderately polar compounds.

References

1. Supercritical Fluid Chromatography (SFC) (2004) IsoPro International, Menlo Park, USA. Available on-line at: http://www.isopro.net/web8.htm
2. Ramsey, E. D. (1998) *Analytical Supercritical Fluid Extraction Techniques.* Kluwer Academic Publishers, Dordrecht.
3. McHugh, M. A. and Krukonis, V. J. (1994) *Supercritical Fluid Extraction: Principles and Practice.* 2 ed., Butterworth-Heinemann.
4. Taylor, L. T. (1996) *Supercritical Fluid Extraction*, Wiley-Interscience.
5. Mukhopadhyay, M. (2000) *Natural Extracts Using Supercritical Carbon Dioxide.* 1 ed., CRC Press London.
6. Bright, F. V. and McNally, M. E. P. (1992) *Supercritical Fluid Technology: Theoretical and Applied Approaches to Analytical Chemistry.* American Chemical Society, Washington, DC.
7. Bruno, T. J. and Ely, J. F. (1991) *Supercritical Fluid Technology: Reviews in Modern Theory and Application.* CRC Press, Boca Raton, FL.
8. Johnston, K. P. and Penninger, J. M. L. (1989) *Supercritical Fluid Science and Technology.* American Chemical Society, Washington, DC.
9. Wenclawiak, B. (1992) *Analysis with Supercritical Fluids: Extraction and Chromatography.* Springer-Verlag, Berlin.
10. De Castro, M. D. L., Valcarcel, M., and Tena, M. T. (1994) *Analytical supercritical fluid extraction.* Springer-Verlag, Berlin.
11. Andrews, A. T. (1990) Supercritical carbondioxide extraction of polycyclic aromatic hydrocarbons from contaminated soil. PhD thesis, Rutgers, The State University of New Jersey, Piscataway, NJ.
12. Venkat, E. and Kothandaraman, S. (1998) Supercritical fluid methods, in *Natural Products Isolation.* 1st ed. (Cannell, R. J. P., ed.), Humana Press, New Jersey.
13. Bevan, C. D. and Marshall, P. S. (1994) The use of supercritical fluids in the isolation of natural products *Nat. Prod. Rep.* **11,** 451–466.
14. Castioni, P., Christen, P., and Veuthey, J. L. (1995) Supercritical-fluid extraction of compounds from plant origin. *Analysis* **23,** 95–106.
15. Catchpole, O. J., Grey, J. B., Perry, N. B., Burgess, E. J., Redmond, W. A., and Porter, N. G. (2003) Extraction of chill, black pepper, and ginger with near critical CO_2, propane, dimethyl ether: analysis of the extracts by quantitative nuclear magnetic resonance *J. Agric. Food Chem.* **51,** 4853–4860.

16. Ferreira, S. R. S. and Meireles, M. A. A. (2002) Modelling the supercritical fluid extraction of black pepper (*Piper nigrum* L) essential oil. *J. Food Eng.* **54,** 263–269.
17. Ferreira, S. R. S., Nikolov, Z. L., Doraiswamy, L. K., Meireles, M. A. A., and Petenate, A. J. (1999) Supercritical fluid extraction of black pepper (*Piper nigrum* L) essential oil. *J. Supercrit. Fluids* **14,** 235–245.
18. Sovova, H., Jez, J., Bartlova, M., and Stastova, J. (1995) Supercritical carbon dioxide extraction of black pepper *J. Supercrit. Fluids* **8,** 295–301.
19. Uquiche, E., del Valle, J. M., and Ortiz, J. (2004) Supercritical carbon dioxide extraction of red pepper (*Capsicum annuum* L) oleoresin. *J. Food Eng.* **65,** 55–66.
20. Rodrigues, M. R. A., Krause, L. C., Caramao, E. B., Dos Santos, J. G., Dariva, C., and De Oliveira, J. V. (2004) Chemical composition and extraction yield of the extract of *Origanum vulgare* obtained from sub- and supercritical CO_2. *J. Agric. Food Chem.* **52,** 3042–3047.
21. Salgin, U., Calimli, A., and Uysal, B. Z. (2004) Supercritical fluid extraction of jojoba oil *J. Am. Oil Chem. Soc.* **81,** 293–296.
22. Louli, V., Folas, G., Voutsas, E., and Magoulas, K. (2004) Extraction of parsley seed oil by supercritical CO_2 *J. Supercrit. Fluids* **30,** 163–174.
23. Sonsuzer, S., Sahin, S., and Yilmaz, L. (2004) Optimization of supercritical CO_2 extraction of *Thymbra spicata* oil. *J. Supercrit. Fluids* **30,** 189–199.
24. Perva-Uzunalic, A., Skerget, M., Weinreich, B., and Knez, Z. (2004) Extraction of chilli pepper (var. Byedige) with supercritical CO_2: effect of pressure and temperature on capsaicinoid and colour extraction efficiency. *Food Chem.* **87,** 51–58.
25. Duarte, C., Moldao-Martins, M., Gouveia, A. F., da Costa, S. B., Leitao, A. E., and Bernardo-Gil, M. G. (2004) Supercritical fluid extraction of red pepper (*Capsicum frutescens* L). *J. Supercrit. Fluids* **30,** 155–161.
26. Palenzuela, B., Arce, L., Macho, A., Munoz, E., Rios, A., and Valcarcel, M. (2004) Bioguided extraction of polyphenols from grape marc by using an alternative supercritical-fluid extraction method based on a liquid solvent trap. *Anal. Bioanal. Chem.* **378,** 2021–2027.
27. Martino, K. G. and Guyer, D. (2004) Supercritical fluid extraction of quercetin from onion skins *J. Food Process Eng.* **27,** 17–28.
28. Luengthanaphol, S., Mongkholkhajornsilp, D., Douglas, S., Douglas, P. L., Pengsopa, L., and Pongamphai, S. (2004) Extraction of antioxidants from sweet Thai tamarind seed coat—preliminary experiments *J. Food Eng.* **63,** 247–252.
29. Chiu, K.-L., Cheng, Y.-C., Chen, J.-H., Chang, C.J., and Yang, P.-W. (2002) Supercritical fluids extraction of *Ginkgo* ginkgolides and flavonoids. *J. Supercrit. Fluids* **24,** 77–87.

30. van Beek, T. A. (2002) Chemical analysis of *Ginkgo biloba* leaves and extracts *J Chromatography A* **967,** 21–55.
31. Rompp, H., Seger, C., Kaiser, C. S., Haslinger, E., and Schmidt, P. C. (2004) Enrichment of hyperforin from St John's Wort (*Hypericum perforatum*) by pilot-scale supercritical carbon dioxide extraction. *Eur. J. Pharm. Sci.* **21,** 443–451.
32. Smith, R. M. (1996) Supercritical fluid extraction of natural products. *LC-GC International,* 8–15.
33. Smith, R. M. and Burford, M. D. (1992) Supercritical fluid extraction and gas-chromatographic determination of the sesquiterpene lactone parthenolide in the medicinal herb feverfew (*Tanacetum parthenium*) *J. Chromatography* **627,** 255–261.
34. Čretnik, L., Škerget, M., and Knez, Z. (2005) Separation of parthenolide from feverfew: performance of conventional and high-pressure extraction techniques. *Separation Purification Technol* **41**, 13–20.
35. Jennings, D. W., Deutsch, H. M., Zalkow, L. H., and Teja, A. S. (1992) Supercritical extraction of taxol from the bark of *Taxus brevifolia J. Supercrit. Fluids* **5,** 1–6.
36. Pascual-Marti, M. C., Salvador, A., Chafer, A., and Berna, A. (2001) Supercritical fluid extraction of resveratrol from grape skin of *Vitis vinifera* and determination by HPLC *Talanta* **54,** 735–740.
37. Simandi, B., Kristo, S. T., Kery, A., Selmeczi, L. K., Kmecz, I., and Kemeny, S. (2002) Supercritical fluid extraction of dandelion leaves. *J. Supercrit. Fluids* **23,** 135–142.
38. te Bokkel (1990). Supercritical carbon dioxide extraction of cyclosporine from the fungus *Beauvaria nivea*. Ph.D. thesis, The University of Western Ontario, Canada.
39. Bottalico, A., Logrieco, A., and Visconti, A. (1989) *Taxonomy and Pathogenicity.* Elsevier, New York, pp. 85–119.
40. Ambrosino, P., Galvano, F., Fogliano, V., Logrieco, A., Fresa, R., and Ritieni, A. (2004) Supercritical fluid extraction of beauvericin from maize *Talanta* **62,** 523–530.
41. Lim, G.-B., Lee, S.-Y., Lee, E.-K., Haam, S.-J., and Kim, W.-S. (2002) Separation of astaxanthin from red yeast *Phaffia rhodozyma* by supercritical carbon dioxide extraction *Biochem. Eng. J.* **11,** 181–187.
42. Ashraf-Khorassani, M. and Taylor, L. T. (2004) Sequential fractionation of grape seeds into oils, polyphenols, and procyanidins via a single system employing CO_2 based fluids *J. Agric. Food Chem.* **52,** 2440–2444.

4

An Introduction to Planar Chromatography

Simon Gibbons

Summary

Thin-layer chromatography (TLC) is an easy, cheap, rapid, and widely used method for the analysis and isolation of natural and synthetic products. It has use also in the biological evaluation of organic compounds, particularly in the areas of antimicrobial and antioxidant metabolites, and for the determination of acetylcholine esterase inhibitors that are utilized in the treatment of Alzheimer's disease. This chapter deals with the basic principles of TLC and describes methods for the analysis and isolation of natural products. Examples of methods for isolation of several classes of natural product are detailed, and protocols for TLC bioassays are given.

Key Words: Thin-layer chromatography; TLC; bioassays; natural product isolation.

1. Introduction

Planar chromatography utilizes the separation of mixtures of organic compounds on thin layers of adsorbents that are in most cases coated on glass, plastic, or aluminum sheets. The most widely used form of planar chromatography is thin-layer chromatography (TLC), which is the easiest and cheapest technique for the isolation of natural products. TLC is one of the oldest forms of chromatography, the simplest example being the school experiment of spotting a plant extract near the bottom of thin strips of blotting paper and 'developing' in a jar with water or alcohol. As the water moves up the blotting paper, the dark extract is separated

From: *Methods in Biotechnology, Vol. 20, Natural Products Isolation, 2nd ed.*
Edited by: S. D. Sarker, Z. Latif, and A. I. Gray © Humana Press Inc., Totowa, NJ

into its component colors of light and dark greens. This is the essence of separation by TLC. This chapter describes the principles behind this technique and gives procedures for analyzing extracts and isolating natural products, so that anyone can use this technique to isolate and analyze natural products. Examples of TLC isolations of metabolites from plants are given, and the methodology behind the isolation of unknown products is detailed. Success in TLC relies on flexibility of approach and experimenting with a variety of methods.

Natural product extracts are in most cases highly complex and comprise mixtures of neutral, acidic, basic, lipophilic, hydrophilic, or amphiphilic (e.g., amino acids) compounds, and consequently there will not be one method that can be used for all eventualities. It is worthwhile to carry out ^{1}H or ^{13}C NMR spectroscopy of the extract or fraction to determine the class of compound(s) to be separated ***(1)***—deuterated NMR solvents are cheap (US $1.00 for $CDCl_3$), and one-dimensional NMR experiments are quicker to run than the extensive development of mobile and stationary phases that may be needed. The starting point should always be the simplest method first. Examples for the isolation of several classes of natural products are given in **Subheading** 7.

1.1. Basic Principles of TLC

Separation by TLC is effected by the application of a mixture or extract as a spot or thin line on to a sorbent that has been applied to a backing plate. Analytical TLC plates (thickness 0.1–0.2 mm) are commercially available from suppliers such as Merck (e.g., the commonest analytical silica gel plate is the 20×20 cm, plastic or aluminum-backed Kieselgel 60 F_{254} plate having a 0.2 mm thickness of silica sorbent [Merck No. 5554]). The plate is then placed into a tank with sufficient suitable solvent just to wet the lower edge of the plate/sorbent but not adequate to wet the part of the plate where the spots were applied (origin). The solvent front then migrates up the plate through the sorbent by capillary action, and this process is known as development (**Fig. 1**).

A factor in quantifying migration of a compound on a particular sorbent and solvent system is the R_f value. This is defined as:

$$R_f = \frac{\text{Compound distance from origin (midpoint)}}{\text{Solvent front distance from origin}}$$

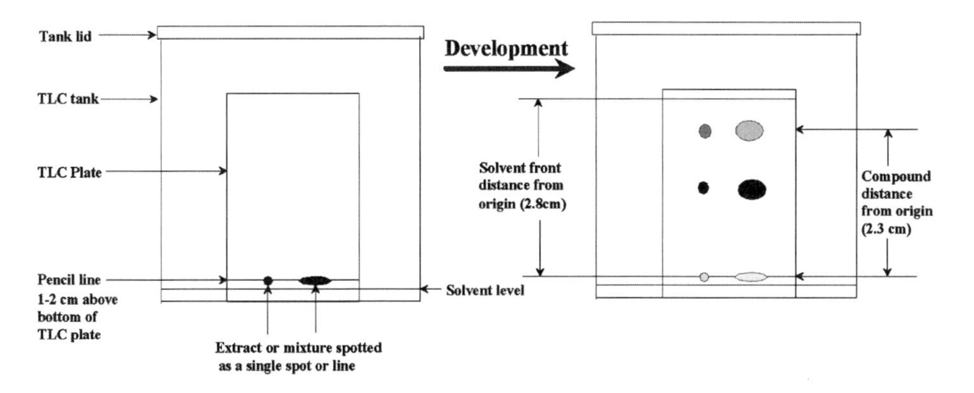

Fig. 1. TLC equipment and procedure.

In the example cited in **Fig. 1**,

$$R_f = \frac{\text{Compound distance from origin}}{\text{Solvent front distance from origin}} = \frac{2.3\,\text{cm}}{2.8\,\text{cm}}$$

$$R_f = 0.82$$

R_f values are always ratios, never greater than 1, and vary depending on sorbent and/or solvent system. These values are sometimes quoted as hR_f, i.e., relative to solvent front = 100, $hR_f = R_f \times 100$ (in our case $hR_f = 82$). In the case of adsorption chromatography (**Subheading 1.2.1.**) where the sorbent is silica, polar compounds, e.g., psilocybin (**Fig. 2**), will have a higher affinity for the sorbent (stationary phase), "stick" to the

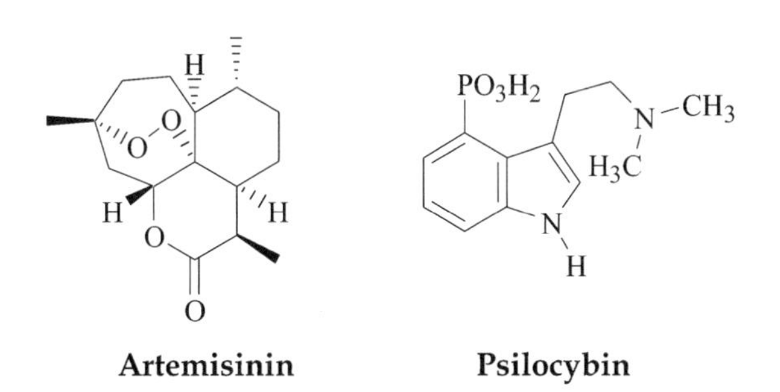

Fig. 2. Structures of artemisinin and psilocybin.

sorbent, and move slowly up the plate as the solvent (mobile phase) migrates. These compounds will have relatively small R_f values. Nonpolar compounds, e.g., artemisinin (**Fig. 2**), will have less affinity for the stationary phase, move comparatively quickly up the plate, and have relatively larger R_f values. As a consequence of development, compounds of a mixture will separate according to their relative polarities. Polarity is related to the type and number of functional groups present on a molecule capable of hydrogen bonding (c.f. 1.2.): nonpolar groups: CH_3-, CH_3O-, Ph-, CH_3CH_2; polar groups: -CO_2H, -OH, -NH_2, SO_3H, PO_3H_2.

Artemisinin, isolated from the antimalarial herb *Artemisia annua*, is a relatively nonpolar compound when compared with psilocybin (isolated from the fungus *Psilocybe mexicana*). However, it should be noted that this relative polarity will vary according to the type of stationary phase and mobile phase used. Solvent strengths are also measured in terms of polarity, and generally dielectric constants are used to quantify relative strengths (**Table 1**) *(2)*. A high dielectric constant indicates a polar solvent with a strong power of elution, whereas a low dielectric constant suggests a nonpolar solvent with a lower ability to elute a component from a sorbent. This elution strength applies to normal-phase adsorption chromatography.

1.2. Mechanisms of Separation

There are four mechanisms of chromatography by which separation can occur, and more than one may be responsible during a given separation.

Table 1
Dielectric Constant of Compounds

Solvent	Dielectric constant (20°C)	Solvent	Dielectric constant (20°C)
Pentane	1.8	Ethyl acetate[a]	6.0
Hexane	1.9	Acetic acid	6.2
Cyclohexane	2.0	Dichloromethane	9.1
Benzene[a]	2.3	Pyridine	12.3
Toluene	2.4	Acetone[a]	20.7
Diethyl ether	4.3	Methanol	32.6
Dimethyl sulfoxide	4.7	Acetonitrile	37.5
Chloroform	4.8	Water	78.5

[a]Dielectric constants were recorded at 25°C.

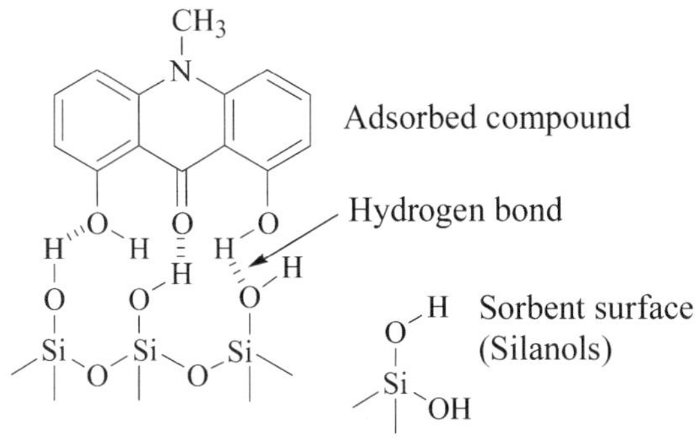

Fig. 3. Adsorption and hydrogen bonding between compound and sorbent face.

1.2.1. Adsorption Chromatography

The most commonly used sorbents of adsorption chromatography are silica and alumina. As the components move through the sorbent, their relative rates of migration are effected by their individual affinities for the sorbent. Separation occurs when one compound is more strongly adsorbed by the sorbent than the other components. When the sorbent is silica or alumina, polar natural products move slowly when compared to nonpolar ones. Adsorption takes place as a result of the interaction between the compound and groups associated with the sorbent. In the case of silica, which has silanol groups (**Fig. 3**), binding occurs between the compound and free hydroxyls on the sorbent. In this particular case, adsorption involves hydrogen bonding between compound functional groups and adsorbent surface hydroxyl groups.

1.2.2. Partition Chromatography

This mechanism involves the relative solubility of the compound between the sorbent (stationary phase) and the solvent (mobile phase). Compounds that are more soluble in the mobile phase will migrate up the plate to a greater extent than components that are more soluble in the stationary phase. Reversed-phase TLC utilizes sorbents that partition natural products between the mobile and stationary phases. These are normally fatty (lipid) stationary phases and aqueous mobile phases. The most commonly used reversed-phase sorbent is silica, which has been reacted with a straight chain 18-carbon alkyl unit to form an octadecasilyl (ODS) phase. Many alternative lipophilic phases are commercially available (**Fig. 4**).

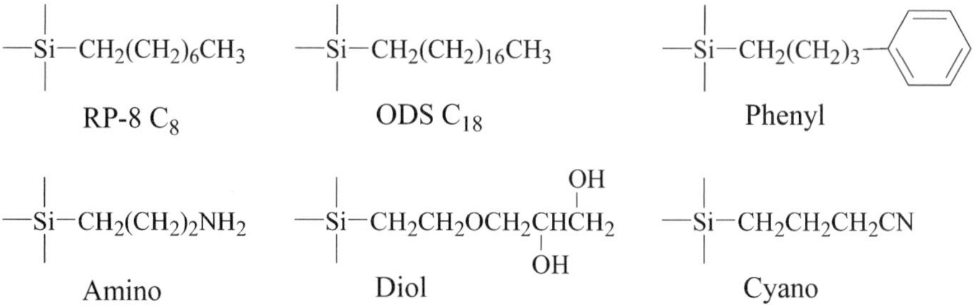

$—Si—CH_2(CH_2)_6CH_3$ RP-8 C_8

$—Si—CH_2(CH_2)_{16}CH_3$ ODS C_{18}

$—Si—CH_2(CH_2)_3—C_6H_5$ Phenyl

$—Si—CH_2(CH_2)_2NH_2$ Amino

$—Si—CH_2CH_2OCH_2CH(OH)CH_2(OH)$ Diol

$—Si—CH_2CH_2CH_2CN$ Cyano

Fig. 4. Common reverse phases for partition chromatography.

Nonpolar "fatty" compounds such as sesquiterpene artemisinin (**Fig. 2**) are readily "soluble" in stationary phases like ODS, and during solvent development a partition is set up between the two phases. Separation is effected by compounds having different rates of partition between the stationary phase and mobile phase.

1.2.3. Size-Inclusion/Exclusion Chromatography

Compounds may be separated by their relative sizes and their inclusion (or exclusion) into a sorbent (*see* Chap. 5). The most commonly used size-inclusion sorbents are the dextran gels, particularly the lipophilic versions such as Sephadex LH-20, which are mostly used for the separation of small hydrophobic natural products from their larger "contaminants," usually chlorophylls, fatty acids, and glycerides. In organic solvents such as chloroform and methanol, these gels swell to form a matrix. As compounds migrate with the solvent through the gel, small molecules become included into the gel matrix, whereas larger ones are excluded and migrate at a greater rate. It should be noted that separations on gels such as Sephadex LH-20 also involve the mechanisms of adsorption, partition, possibly ion exchange, and occasionally the trend of larger molecules eluting first and smaller ones eluting last may be reversed. This form of chromatography has found considerable use in the removal of 'interfering' plant pigments such as the chlorophylls that tend to be larger and more lipophilic than many plant natural products.

1.2.4. Ion Exchange Chromatography

This technique is limited to mixtures that contain components carrying a charge (*see* Chap. 6). In this form of chromatography, the sorbent is usually a polymeric resin, which contains charged groups and mobile counter ions that may exchange with ions of a component as the mobile phase migrates through the sorbent. Separation is achieved by differences in the affinity between ionic components and the stationary phase. In *cation exchange*, acidic groups such as -CO_2H and -SO_3H are incorporated into the resin and are able to exchange their protons with other cations of components to form -CO_2^-, H_3O^+, and -SO_3^-, H_3O^+, respectively, at particular pH ranges. In *anion exchange*, basic groups such as quaternary ammonium moieties (-N^+R_3) are incorporated into the resin and are able to exchange their anions with those of components.

1.3. Applications of TLC

Traditionally, analytical TLC has found application in the detection and monitoring of compounds through a separation process. In the case of known natural products or other compounds, e.g., pharmaceuticals, qualitative and quantitative information can be gathered concerning the presence or absence of a metabolite. An example of this is the production of the antitumor diterpene Taxol® (**Fig. 5**) from the endophytic fungus *Taxomyces andreanae*. Stierle et al. ***(3)*** isolated fungal taxol that had R_f values identical to that from the Pacific yew, *Taxus brevifolia,* on four different solvent systems (**Fig. 5**).

Use of four solvent systems gave a higher degree of confidence of the fungal taxol being authentic, although the authors did confirm this by mass spectrometry (MS). Analytical TLC has been used to chemically

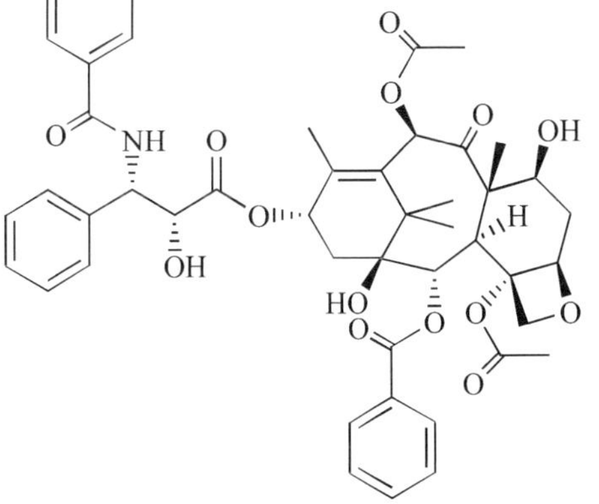

Solvent system

(1) $CHCl_3$:MeCN, (7:3)

(2) $CHCl_3$:MeOH, (7:1)

(3) CH_2Cl_2:THF, (6:2)

(4) EtOAc:iPrOH, (95:5)

Fig. 5. Various solvent systems for the isolation of Taxol®.

classify organisms by their chemical constituents, in particular, the filamentous bacteria, the actinomycetes. The important genus *Streptomyces* generally contains an LL stereoisomer of a cell wall metabolite known as diaminopimelic acid, whereas the rarer genera possess the *meso* form of this metabolite. By hydrolyzing the organism cell wall and running a TLC of the hydrolysate against the two standards, it is possible to loosely classify the actinomycete ***(4)***.

Natural products may be "tracked" by running analytical TLC after other separation processes such as column chromatography or high-pressure liquid chromatography (HPLC), and always more than one solvent system should be used for a TLC separation, as even apparently "pure" spots may consist of several compounds with identical R_f values. The similarity of different extracts from the same species can also be assessed in this way, and the decision to combine nonpolar and polar extracts can be made on identical or similar TLC chromatograms. Qualitative initial screening of extracts should be routinely performed, and the presence of ubiquitous compounds such as plant sterols and certain phenolics can be ascertained at an early stage by running the appropriate standard alongside an extract. In certain cases, classes of compounds may be determined by spraying developed plates with stains that give a colored reaction with a particular compound class (*see* **Subheading 3.**). Many natural products are still isolated by conventional preparative TLC (PTLC), and numerous examples can be found in the journals *Phytochemistry* and *Journal of Natural Products*. Although preparative HPLC is in 'vogue' and is the method of choice, PTLC is still a useful isolation method in many cases because of its simplicity, cost, speed, and ability to separate compounds in the 1 mg–1 g range.

2. System Selection

As much information as possible regarding the extract-producing organism should be gathered—this will aid in the selection of a separation system. After a full literature search, the following points will need to be addressed: Has the species been studied before? If so, what metabolites were isolated? Are there standard TLC methods available? If the species has not been studied for chemistry, is there any information at the generic level? Chemotaxonomy or the classification of an organism according to its natural products may assist in dealing with unknown genera ***(5)***—related species *may* produce related secondary metabolites.

Databases such as NAPRALERT, Berdy, *The Dictionary of Natural Products, Chemical Abstracts*, and in the case of plants, certain classical texts such as Hegnauers' *Die Chemotaxonomie der Pflanzen* ***(6)*** can give plenty of information regarding the classes of natural products present in certain taxa (*see* Chap. 12).

Information on semipurified samples, e.g., column fractions, can also be invaluable. Several workers ***(1,7)*** routinely record ^{1}H NMR spectra of column fractions prior to TLC purification. This may seem a rather expensive detection system, but much information can be gathered about the classes of compounds present and a TLC method can be "tailored" accordingly.

TLC on silica gel is still the most common method, although it suffers from some drawbacks that may be easily overcome:

1. Acidic compounds "tail" on silica because of interactions between acidic groups (e.g., -CO_2H, -OH) and silanols—this may be reduced by the addition of a small amount of an acid, e.g., 1% trifluoroacetic acid (TFA) or acetic acid, to the mobile phase, which will maintain acidic groups in a nonionized form.
2. Basic compounds may also behave poorly on silica and the addition of weak bases (e.g., 1% diethylamine or triethylamine) should eradicate poor chromatography.
3. Highly nonpolar compounds such as fatty acids, glycerides, alkanes, and some lower terpenoids require simple nonpolar solvents systems (e.g., cyclohexane, hexane, pentane, diethyl ether:hexane mixtures) and may be difficult to detect by UV (i.e., no chromophore) or by spray detection (using charring reagents, e.g., vanillin–sulfuric acid) (**Table 3**).
4. Highly polar metabolites such as sugars, glycosides, tannins, polyphenolics, and certain alkaloids require the development of polar mobile phases and in some cases such compounds may be irreversibly adsorbed onto the silica.

Choice of mobile phase should evolve through the use of a mono or binary system, i.e., 100% $CHCl_3$ or hexane:EtOAc (1:1) as a starting point and then on to the addition of acids or bases to improve chromatography, i.e., toluene:ethyl acetate:acetic acid (60:38:2), and as a last resort to use tertiary or quaternary systems, e.g., butanol:acetic acid:water (4:1:5) or hexane:ethyl acetate:formic acid:water (4:4:1:1).

2.1. Choice of Development

The decision as to whether the system is run isocratically (i.e., one solvent system of constant composition) or using a step gradient (develop once in a nonpolar solvent and increase the solvent polarity after each development) can be made by running a series of analytical plates with the sample at varying mobile phase strengths. Both systems have their merits and can be used with multiple development where the plate is developed, dried, and developed a number of times. This is especially useful in the separation of closely eluting bands. **Table 2** lists some of the more commonly used systems.

Table 2
Simple Systems for TLC

Solvent system	Sorbent	Notes
Hexane:ethyl acetate (EtOAc)	Silica gel	Universal system—can substitute hexane for petroleum spirit or pentane
Petrol:diethyl ether (Et_2O)	Silica gel	A universal system for relatively nonpolar metabolites. Excellent for terpenes and fatty acids. Care should be taken with Et_2O as explosive mixtures are formed in air
Petrol:chloroform ($CHCl_3$)	Silica gel	Considerably useful for the separation of cinnamic acid derivatives and in particular the coumarins
Toluene:ethyl acetate:acetic acid (TEA)	Silica gel	Vary the composition, i.e., 80:18:2 or 60:38:2—excellent for acidic metabolites
$CHCl_3$:acetone	Silica gel	A general system for medium polarity products
Benzene:acetone	Silica gel	Useful for the separation of aromatic products. Care should be taken as benzene is a highly carcinogenic solvent. Substitute toluene for benzene
Butanol:acetic acid:water	Silica gel	A polar system for flavonoids and glycosides
Butanol:water:pyridine: toluene	Silica gel	Sugar analysis system. Try 10:6:6:1. Development may take 4 h
Methanol:water	C_{18}	Start with 100% MeOH to determine whether metabolites will move from the origin. Increase water concentration to 'slow' down products. The addition of small amounts of acid or base may improve chromatography
Acetonitrile:water	C_{18} /C_2	A universal simple reverse-phase system
Methanol:water	Polyamide	Universal
Methanol:water	Cellulose	Used for the separation of highly polar compounds such as sugars and glycosides

3. Detection of Natural Products in TLC

At both the analytical and preparative stages of TLC, effective visualization or detection is crucial to obtain pure compounds, and poor detection will also result in a low recovery of product from the sorbent. Detection is usually either nondestructive, where the compounds may be recovered from the sorbent (ultraviolet [UV] detection), or destructive, where the compounds are contaminated by the detection reagent and are unrecoverable from the sorbent (spray detection). There are some excellent texts available on this subject, such as Wagner and Bladt's *Plant Drug Analysis* ***(8)*** or *The Merck Handbook of Dyeing Reagents for Thin Layer and Paper Chromatography* ***(9)***, covering most eventualities.

3.1. Ultraviolet Detection

UV detection involves the use of UV-active compounds (indicators) that are incorporated into the sorbent of TLC plates by the manufacturer. Typical examples of plates with these sorbents include the range of analytical plates produced by Merck: Alumina, 0.2 mm thick, 20 × 20 cm, with a 254-nm UV indicator (Merck no. 5550). Under short-wave UV light (254 nm), the indicator, which is usually a manganese-activated zinc silicate, will emit a pale green light. Under long-wave UV light (366 nm), a further indicator will emit a pale purple light. Compounds that absorb light at either 254 or 366 nm will appear as dark spots against a light background when UV light is shone onto the plate. Many compounds such as the furocoumarins will also emit a distinctive blue or yellow florescence under UV light. The major disadvantage with UV detection is that compounds that do not absorb UV light at 254 or 366 nm will be invisible and will require spray detection. The primary advantage of UV detection is, however, that it is nondestructive, and the detection of compounds can be observed readily through a separation process. UV lamps are widely commercially available from suppliers such as CAMAG (Camag Ref. 022.9230). Care should be taken not to focus light from these lamps on eyes or on skin, as UV light is mutagenic.

3.2. Spray Detection

This relies on a color reaction between the compound on the TLC plate and a spray reagent (stain) introduced onto the plate as a fine mist from a spray canister. Ten of the most common spray reagents are listed in **Table 3**. Most are universal reagents and will react with many classes of

Table 3
Ten Simple Spray Reagents for Natural Product TLC Visualization

Detection spray	Recipe	Treatment	Notes
(1) *Vanillin/ Sulfuric acid*	Dissolve vanillin (1 g) in concentrated H_2SO_4	Spray onto plate and heat at 100°C until coloration appears	A universal spray. Many terpenes give red and blue colors. Natural products with little functionality may give poor coloration—try spray (2). Spray and heat in a fume cupboard
(2) *Phosphomolybdic acid (PMA)*	Dissolve PMA in ethanol to make a 5% w/v solution	Spray onto plate and heat at 100°C until coloration appears	Useful to detect many terpenes as blue spots on a yellow background. Spray and heat in a fume cupboard
(3) *Ammonium molybdate(VI)*	Dissolve ammonium molybdate(VI) (10 g) in concentrated H_2SO_4 (100 mL)	Spray onto plate and heat at 100°C until coloration appears	A universal spray. Many diterpenes give a blue color. Spray and heat in a fume cupboard
(4) *Antimony(III) chloride*	Dissolve antimony(III) chloride in a mixture of glacial acetic acid (20 mL) and chloroform (60 mL)	Spray onto plate and heat at 100°C for 2–5 min or until coloration appears	Di- and triterpenes give a red to blue coloration. Care should be taken when handling this spray as antimony compounds are highly poisonous. Spray and heat in a fume cupboard
(5) *Tin(IV) chloride*	Add tin(IV) chloride (10 mL) to a mixture of chloroform (80 mL) and glacial acetic acid (80 mL).	Spray onto plate and heat for 5 min at 100°C or until coloration appears	Useful for the detection of flavanoids and terpenes. Tin(IV) chloride is poisonous and a lachrymator. Spray and heat in a fume cupboard

(6) *Dragendorff's reagent*	Add 10 mL of a 40% aqueous solution of KI to 10 mL of a solution of 0.85 g of basic bismuth subnitrate in acetic acid (10 mL) and distilled water (50 mL). Dilute the resulting solution with acetic acid and water in the ratio 1:2:10	Generally, no heat is required—but if reaction is not spontaneous, heat until coloration appears	This is the traditional method for alkaloid detection, although care should be taken as some nonalkaloids such as iridoids and some flavonoids give a positive reaction. Alkaloids give a dark orange to red coloration
(7) *2,4 Dinitro-phenyl-hydrazine*	Dissolve 2,4-dinitro-phenylhydrazine (0.2 g) in 2-N HCl (50 mL)	Generally, no heat is required — but if reaction is not spontaneous, heat until coloration appears	Detects aldehydes and ketones with a yellow to red coloration
(8) *Perchloric acid*	A 20% (w/v) aqueous perchloric acid solution	Heat at 100°C until coloration	A universal spray but is useful for steroids and triterpenes
(9) *Borntrager reagent*	A 10% (w/v) ethanolic solution of KOH	Heat until color detection	For the detection of coumarins and anthraquinones
(10) *Ninhydrin*	Add ninhydrin (0.3 g) to a mixture of butanol (100 mL) and acetic acid (3 mL)	Heat at 100°C until coloration	Especially useful for amino acids, amines, and as a general alkaloid spray. Alkaloids appear as a red coloration

natural products, and the most widely used sprays are 1–3. Dragendorff reagent (spray 6) is especially useful for the detection of many classes of alkaloids and is well worth the effort required to make. In some cases, heat is required to assist the color reaction, and this can be supplied in the form of a hand-held heater (hair dryer!) or a drying oven. All of the compounds required to make the spray reagents are readily available from suppliers such as Aldrich. Each of the spray reagents should be made up and used in a fume cupboard. When using spray detection in PTLC, most of the plate should be covered and only a small proportion of the edge (2 cm) sprayed with reagent. Ideally, a scalpel should be used to score a line of 2 cm from the plate edge, so that after spraying the corrosive spray reagent does not migrate into the sorbent and damage compounds.

4. Preparative Thin-Layer Chromatography (PTLC)

PTLC has long been a popular method of isolation, primarily because of its universal accessibility to students, postgraduates, and researchers working in natural product chemistry. This popularity has diminished in recent years owing to the success of high-pressure liquid chromatography (HPLC) and counter current chromatography (CCC) (*see* Chaps. 7 and 8). Unlike these two techniques, PTLC does not, however, require expensive equipment; separations can be effected rapidly and the amount of material isolated generally falls into the 1 mg to 1 g range, which is certainly sufficient for structure elucidation purposes. This section gives a breakdown of the basic steps of PTLC with emphasis on preparing, running plates, and some of the advantages and disadvantages encountered with PTLC.

4.1. When to Use PTLC

Although separations depend on the level of complexity of an extract, PTLC is nearly always used as a final purification step in an isolation procedure. A broad procedure is given here:

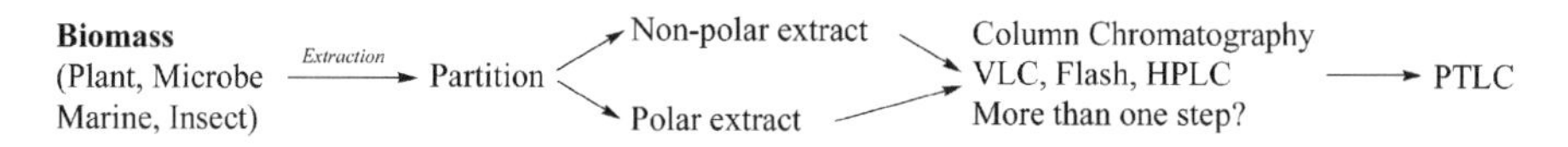

The number of compounds that can be separated on a prep plate will ultimately depend on how those compounds behave on a particular system. But as a rule, separation of no more than a mixture of three major components should be attempted. The separation of complex mixtures can be carried out on PTLC as a first stage but larger amounts of material are needed, and as the process of running many plates can be time consuming, it is more usual to separate partially purified mixtures. Complex extracts should, in the first instance, be separated via vacuum liquid chromatography (VLC), flash chromatography, or column chromatography prior to PTLC (*see* Chap. 5).

The friedelane triterpene was isolated from the bark of the cameroonian rain forest tree *Phyllobotryon spathulatum* ***(10)*** using the following separation process:

Species	Extract	Step (1)	Step (2)	Structure
P. spathulatum	Petrol and $CHCl_3$ (bark)	VLC on silica gel. Fraction eluted with 25% EtOAc in hexane	PTLC Toluene 80 EtOAc 18 AcOH 2	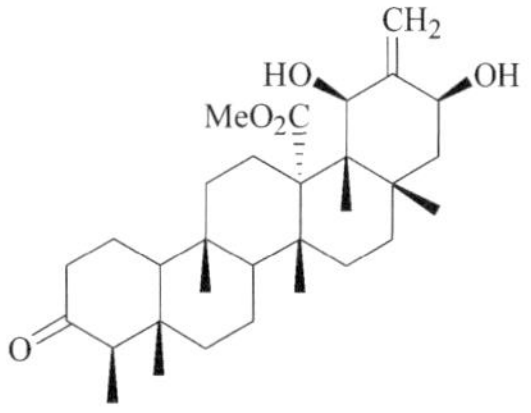 friedelane triterpene

This triterpene needed only one purification step prior to PTLC, and because of its lack of distinctive chromophore, required visualization by spray detection (vanillin reagent, **Table 3**).

4.2. Scale Up From Analytical to PTLC

The scale up procedure from analytical (0.1–0.2 mm sorbent thickness) to PTLC (0.5–4 mm sorbent thickness) is of paramount importance, as changing the size of a separation can drastically affect the chromatography of natural products. The chromatography of a compound separated on analytical plates where milligrams of material are involved can alter significantly when milligrams or 10s of milligrams are employed. On normal-phase silica, when moving from analytical to prep scale, reducing the polarity of the solvent markedly is required. Often, this is a trial and error procedure, but as an example the separation of a mixture of two components achieved using hexane:EtOAc (60:40) as a mobile phase on an analytical plate would possibly require the less polar system hexane:EtOAc (90:10) on a prep plate to give comparable R_f values. This is a general rule and will vary according to compound class and the types of stationary phases used—the best method being to sacrifice a small portion of mixture and experiment.

4.3. Commercially Available PTLC Plates

These plates are usually limited to the sorbents silica, alumina, C_{18}, and cellulose and are usually of thicknesses 0.5, 1.0, and 2.0 mm. The glass-backed silica gel 60 plates from Merck have a particle size distribution of 5–40 mm when compared to the corresponding analytical plate of 5–20 mm. These silica plates have a high specific surface area, are homogeneous, and give excellent results.

The use of commercially available prep plates with a concentration zone will enhance separation. This zone is a layer of inert large pore silica at the bottom of the plate onto which the sample is applied. As the solvent migrates through this zone, the mixture is unretained and focuses at the interface between the zone and "normal" sorbent. Uneven applications of mixtures are focused as discrete lines and this will greatly improve separation.

Normal-phase plates such as the 2 mm Merck Kieselgel $60F_{254}$ 20×20 cm plate (Merck No. 5717) require pre-elution with a nonpolar solvent such as dichloromethane to "clean" and remove contaminants. These impurities will be carried with the solvent to the top of the plate, and then the plates should be dried prior to use. When a new box is opened, unused plates should be stored in a dessicator, as moisture from the air will affect the "activity" of the sorbent (especially in the case of silica) resulting in poor resolution and poor separation.

4.4. Home-Made Preparative Plates

Making one's own plates allows greater flexibility of choice of sorbent, whereas commercial ones are restricted to three or four sorbents. With the correct recipe, home-made plates offer much wider scope for experimentation. They also allow the variation of thickness to accommodate the separation of large amounts of material. Binders such as calcium sulfate (gypsum) are required to bind the sorbent to the plate, but some silica sorbents, e.g., Merck 7749, contain sufficient binder for the purpose. Preparing one's own plates will also give the choice of selecting a sorbent with or without a UV indicator and will also enable the incorporation of additives into the sorbent that enhance separation. An example of this (recipe shown later) is the addition of a small quantity of silver nitrate to a silica sorbent that will aid in the resolution of olefinic compounds.

If cost is an issue, then making plates is cheaper than the commercial alternative, and the removal of sorbent from the plate backing during the desorption process is easier than on commercial plates—a point to consider if compounds are poorly resolved. The following example is a method for making silica prep plates of 0.5 mm thickness with optional silver nitrate additive. For plates of 1 or 2-mm thickness, two or four times the amount of water and sorbent are required and all of the equipment are readily available from suppliers such as CAMAG or Merck:

Equipment:		
	Five glass-backing plates 20 × 20 cm	Adjustable gate applicator
	Two glass spacer plates 5 × 20 cm	Kieselgel 60 (45 g) Merck 7749
	Plate holder	Silver nitrate (1 g) (optional)
	TLC plate coater	Distilled water (90 mL) 200-mL conical flask and stopper

Procedure:

1. The glass-backing plates should be cleaned with 1N aqueous KOH and then acetone, and finally dried prior to spreading.
2. The plates should be put into the TLC plate coater with spacer plates at either end. The applicator should be adjusted to the correct plate size (0.5 mm in this case) and placed at one end on the spacer plate.
3. Silica (45 g) or alternative sorbent and silver nitrate (if required) should be put into the conical flask and water (90 mL) added. The stoppered conical flask is then shaken vigorously for 30 s to ensure that a homogeneous slurry is produced.
4. The slurry should be poured immediately into the applicator, and in one steady movement the applicator should be pulled across the plate faces to rest on the far plate spacer.
5. The plates should be left to air dry for 1 h, put into a plate holder, and activated in an oven at 115°C for 4 h prior to use.

This general method may be applied to other sorbents, although additional binder may be necessary. When incorporating silver nitrate into the sorbent, the plates should be stored and developed in the dark to avoid discoloration and degeneration of the sorbent.

4.5. Sample Application

Prep plates such as those mentioned earlier should be removed from the oven and allowed to cool to room temperature before use. The sample to be separated should be dissolved in a minimum volume of solvent as possible (usually in the concentration range 10–20 mg/mL). This sample is then applied to the bottom of the plate (1.5 cm from the bottom) as a thin line (2–4 mm thick) using either a capillary or a pasteur pipet that has been extruded over a Bunsen burner. Thinner capillaries (5–10 mL) give greater control over sample application and result in a finer, more concentrated line. To apply the sample in a straight fashion, it is preferable to lightly draw a pencil line (without scoring the sorbent!) approximately 1.5 cm above the plate edge or use a piece of A4 paper placed on the plate as a guide. Application of the sample as a straight line is necessary as this forms the origin, and assuming that the plate is homogeneous, during development, the sample will separate into compounds with even bands. If the sample is applied in an irregular fashion (i.e., a wavy line), then during development the sample will separate into compounds with irregular bands that are difficult to remove in a pure form from the plate during the desorption process. The sample should not also be applied right up to both edges of the plate, as edging effects (the rapid movement of solvent up the plate sides or poor sorbent homogeneity) will result in the uneven movement of solvent up the plate during development and as a consequence in an irregular band shape.

4.6. Development and Detection

A suitable solvent system and sorbent phase are chosen (some of the simpler cases are given in **Subheading 2.** and in the examples [*see* **Subheading 7.**]). Mobile phases for prep systems should be made up freshly, and usually 100 mL volumes are suitable to run one or two plates in the same tank. A solvent-saturated atmosphere in the tank is favored to improve chromatography, and this can be produced by adding some filter paper (15 × 15 cm) in the tank. Silica gel is quite a "reactive" sorbent and some natural products are unstable on such a phase. It should be noted that during development, plates should be kept out of sunlight as degradation may occur. This development is achieved by allowing the solvent front to reach the plate top or be within a few millimetres of it, and then removing from the tank and air drying it in a fume cupboard. Hand-held dryers should be avoided to remove excess solvent from plates because of the risk of degradation.

Most semipurified samples have some residual color or natural products of interest may be colored. Hence, it might be possible to gauge how far the compounds have migrated up the plate. In the case of plant extracts, nonpolar pigments such as chlorophylls or carotenes can give a visual aid to separation and an idea of how far the compounds of interest have migrated relative to the pigments. If the natural products of interest absorb long or short UV light, then the success of the separation can be readily observed—this is especially useful in multiple development and it is rewarding to see compounds resolved through the isolation process! With poor UV-absorbing products, use of a spray reagent is required (*see* **Subheading 3.**), and spraying only the plate edge will give an idea of how far the natural products of interest have migrated. If after spray detection the compounds have not migrated far enough up the plate to effect separation, then the contaminated sprayed silica (or other sorbent) must be cut away to avoid contamination before further redevelopment.

4.7. Desorption and Recovery of Natural Products

Once the decision has been made that the separation of compounds has been satisfactorily achieved, natural products need to be effectively recovered from the sorbent, dried, and stored for structure elucidation. On the silica example mentioned earlier, where a UV fluorescent indicator is incorporated into the sorbent, UV-absorbing bands (compounds) may be marked out by a pencil or scalpel, and scrapped off the backing plate onto tin foil or paper. With spray detection, bands may be cut from the spray colored edge (but not incorporating it!) along the plate using a ruler. These bands may be scrapped from the plate onto foil. Compounds may be desorbed from the sorbent in three simple ways:

1. The compound-rich sorbent can be put in a conical flask and solvent added. The suspension should be left for 30 min to facilitate the leaching of compound into the solvent and then filtered. This process should be repeated two or three times to ensure good recovery. The type of solvent used should be slightly more polar than is normally required to dissolve the sample, and as an example if the sample dissolves readily in chloroform then desorption should be carried out using chloroform:methanol (9:1) or (8:2). This should ensure maximum recovery from the sorbent and minimize the possibility of a product being strongly bound to the solid phase.
2. The compound-rich sorbent should be put into a sintered glass funnel (3-porosity frit) attached to a glass buchner flask to which vacuum is applied. The sorbent is then washed with solvent, and the resulting solution can be recovered in the flask and evaporated to yield the product. Repeated washings with solvent will recover compounds effectively and this is the method of choice for recovery from PTLC plates.

3. A pasteur pipet blocked with a small amount of defatted cotton wool or a microcolumn with a 3-porosity frit can be packed with compound-rich sorbent. These "mini" column can then be eluted with a solvent to recover compounds. Care should be taken not to overpack these columns, as compound elution time may be considerable. However, one benefit of these desorption methods, assuming that the natural products of interest are sufficiently stable, is that they may be set up with sufficient solvent and left to desorb for a long period of time.

In all three cases where silica is the sorbent, methanol can be used as a final wash stage to ensure full compound recovery—many natural products such as glycosides of flavonoids and triterpenes are highly polar and may require the addition of 1% or 2% acetic acid in this final methanol elution. Products should be dried quickly after elution, preferably using a high-purity N_2 blow down apparatus and stored in a freezer. Rotary evaporators using heat and turbovaps using air should be avoided because of the risk of heat decomposition and oxidation. It should be noted that the procedure of removing sorbents (especially silica and alumina) from the plate backings should always be performed in a fume cupboard, and a dust mask should be worn to avoid breathing in dangerous fine particulate material.

4.8. Assessing Purity by TLC

After desorption, analytical TLC should be performed on recovered products to ascertain purity. The smaller particle size of analytical plates compared to PTLC enables better resolution and a greater ability to measure purity. At least two different solvent systems should be used to distinguish between compounds that have similar (or identical!) R_f values on a particular system. It should be noted that if a recovered compound appears to be impure after what promised to be a successful purification by PTLC, then it is possible that the natural product was unstable on the sorbent or in solution, and an alternative stationary phase or separation process should be sought.

4.9. Advantages and Disadvantages of PTLC

In the last 10 years, there has been quite a considerable movement away from "wet" techniques such as PTLC and conventional column chromatography toward instrumental techniques such as HPLC and CCC. Many natural product chemists prefer these instrumental methods because of the greater control over a separation process and reproducibility they afford, although the high cost and need for routine servicing of these machines will play a significant role for PTLC in an isolation procedure. The following lists detail some pros and cons of PTLC:

Advantages:

1. Cost effective compared to the instrumentation required, for example, HPLC or CCC.
2. A simple technique that requires little training or knowledge of chromatography to be used.
3. An analytical method may be easily scaled up to a preparative method.
4. Ability to isolate natural products quickly in the milligram to gram range.
5. Flexibility of solvent and stationary phase choice, i.e., the solvent system can be changed quickly during a run.
6. The separation can be optimized readily for one component, i.e., it is relatively easy to "zero in" on a particular product.
7. Methods are quickly developed.
8. Almost any separation can be achieved with the correct stationary phase and mobile phase.
9. A large number of samples can be analyzed or separated simultaneously.

Disadvantages:

1. Poor control of detection when compared to HPLC.
2. Poor control of elution compared to HPLC.
3. Loading and speed are poor compared to VLC.
4. Multiple development methods to isolate grams of material may be time consuming.
5. Restricted to simple sorbents, such as silica, alumina, cellulose, and RP-2.

5. Centrifugal Preparative Thin-Layer Chromatography (CPTLC)

This excellent and underexploited technique can be utilized as a primary "clean up" process of natural product extracts or as a final purification step. CPTLC makes use of a rotor that is coated with a sorbent to form a circular plate, which is then attached to a spindle and rotated using a motor. A solvent is then introduced into the middle of the circular plate by a pump to equilibrate the sorbent. Plates should be saturated with the solvent and allowed to equilibrate at a given flow rate for 10 min. The sample mixture can then be introduced to the plate in the same fashion. As the plate rotates and solvent migrates through the sorbent, the sample is separated into circular compound bands that may be collected readily (**Fig. 6**).

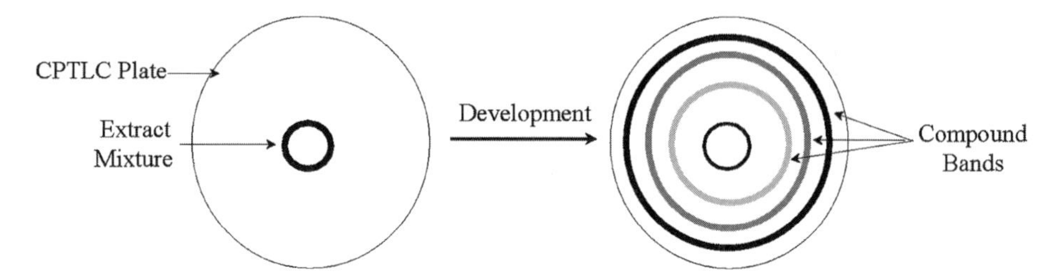

Fig. 6. Circular band separation through CPTLC development.

The plate and motor are housed in an apparatus where a nitrogen atmosphere can be applied. A good example of CPTLC apparatus is the Chromatotron (Harrison Research Model 7924), which has a duct to collect eluting bands and a quartz window that fits over the plate allowing visualization of the plate with UV light.

CPTLC has a number of advantages over PTLC. First, because development is centrifugally accelerated by plate rotation, separation is rapid. Second, solvent changes can be made quickly, and one can operate in a gradient or isocratic mode. Larger amounts of material (1–2 g) may be loaded onto the plate in one go than with PTLC. As with homemade prep plates, CPTLC plates allow the choice of sorbent, additives, and binders.

Recipes for making a variety of plates with different sorbents accompany CPTLC apparatus manuals; however, 2- or 4-mm thickness silica gel plates may be made in the following way.

Silica (Kieselgel 60 PF_{254} Merck Art 7749) (65 or 100 g; for 2- or 4-mm thickness) and binder ($CaSO_4$; 4 or 6 g) are added to distilled water (100 or 190 mL) and shaken thoroughly. The resulting slurry should be poured

Fig. 7. Structure of the clerodane diterpene isolated from *Z. guidonia*.

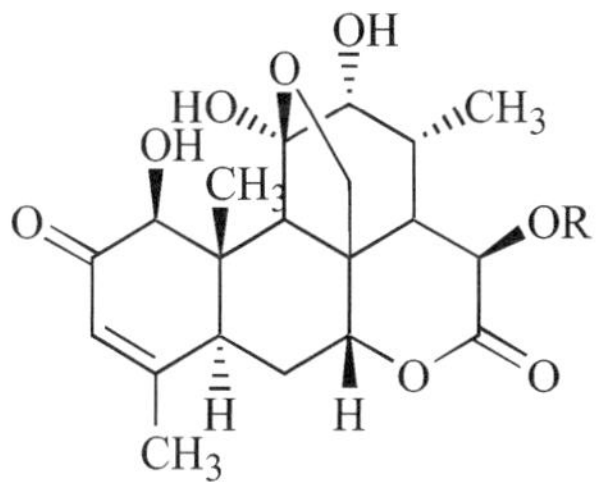

Ailanthinone R = $COCH(Me)CH_2CH_3$
2'-Acetylglaucarubinone R = $COC(OAc)(Me)CH_2CH_3$
[Silica gel; CH_2Cl_2:isopropanol (49:1)]

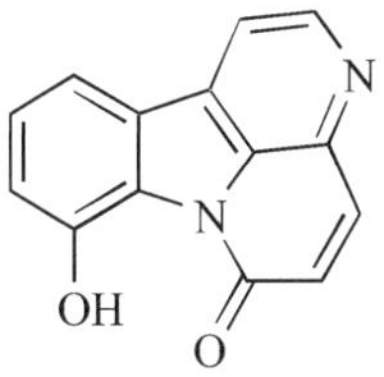

8-Hydroxycanthin-6-one
[Silica gel; toluene:EtOAc:AcOH (5:4:1)]

Fig. 8. Compounds isolated by CPTLC from the rain forest tree *O. gabonensis*.

onto the rotor at the edge and the plate tapped gently to remove air bubbles and to ensure a homogeneous layer. The plate should be air dried for 30 min and oven dried at 50°C for 12 h. The resulting plate should then be scraped to the required thickness and stored in an oven at 50°C prior to use.

The correct choice of solvent system should be ascertained by using a series of analytical plates with increasingly polar solvent to determine R_f values. When using an isocratic system, a solvent system where the R_f of the least polar compound is 0.3 can be used. This will result in a steady separation in which fractions (and compound bands) can be collected and analyzed by TLC. A more polar system (e.g., R_f of the least polar component is 0.8) can be utilized in which concentrated compounds will appear as discrete bands that move quickly through the plate—smaller volume fractions should be collected and analyzed. As with PTLC, use of a sorbent incorporating a UV indicator will aid in the monitoring of UV-active compounds.

Khan et al. *(11)* used a Chromatotron CPTLC apparatus in the separation of some unusual clerodane diterpenes (**Fig. 7**) in a chemotaxonomic study of the Flacourtiaceous species *Zuelania guidonia*. Silica gel sorbent and a mobile phase of petrol/ethyl acetate (49:1) were used and the compound was visualized under UV light.

Ampofo and Waterman *(12)* used CPTLC to isolate the cytotoxic quassinoids (**Fig. 8**), ailanthinone, 2′-acetylglaucarubinone, and the alkaloid 8-hydroxycanthin-6-one from the rain forest tree *Odyendyea gabonensis* (Simaroubaceae).

6. Other TLC Techniques

6.1. Overpressure (OPTLC)

OPTLC was introduced by Tyihak et al. in 1979 *(13)* in an attempt to combine the advantages of conventional TLC and HPLC. This technique employs the use of a pressurized circular ultramicrochamber (PUM chamber), which houses a TLC plate and inlets for the introduction of sample and solvent onto the sorbent. The thin sorbent layer is covered by a membrane kept under external pressure, so that the vapor phase above the sorbent is nearly eliminated. A substantially shorter time is required for separation than in conventional TLC and classical column (CC) chromatography, and greater resolution and separation efficiency are achieved. The rate at which a solvent migrates is as stable as for HPLC, and consequently the technique can be used to model CC methods. Separations can be carried out 5–20 times faster than conventional TLC; hence, this method may be applicable to a large number of mixtures. Tyihak et al. validated this technique using the separation of the synthetic dyes indophenol, Sudan G, and Butter Yellow. Natural products such as capsaicin from *Capsicum anuum* and furocoumarins from *Heracleum sphondylium* have been separated by Nyiredy et al. *(14)*.

6.2. Automated Multiple Development (AMD)

This method utilizes a fully automated developing chamber, which consists of a sensor to optically detect the solvent front position, a mechanism to lift the plate out of the developing chamber, multiple solvent reservoirs, a solvent pump, and an integrated fan to dry the plate and remove solvent vapor. Modern systems contain microprocessor-controlled programming to vary solvent composition after each run. Multiple development dramatically increases separation power, improves reproducibility and precision,

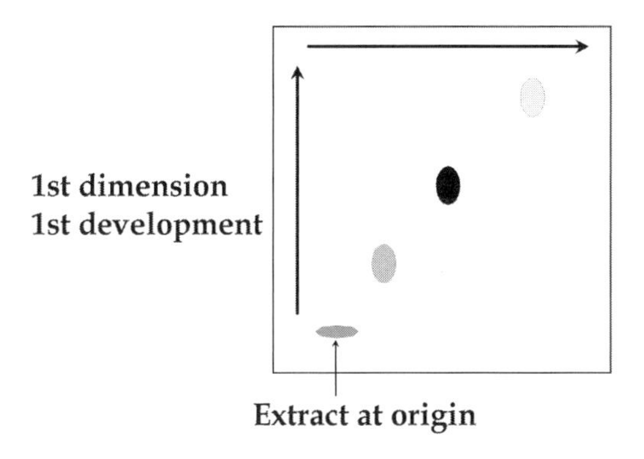

Fig. 9. Two-dimensional TLC plate after two developments.

and can be run without supervision. This apparatus can also be used in conjunction with a TLC plate scanner, which will detect UV-active bands. This can be interfaced with a PC and linked to a printer for hard copy. An excellent example of an AMD device is the CAMAG AMD system available through Merck.

6.3. Two-Dimensional TLC

Two-dimensional TLC is frequently used for the screening of complex mixtures. If the object is to find known compounds and standards available, then this is a powerful form of TLC. The extract is spotted onto the plate in the normal fashion. The plate is developed, dried, and then turned through 90° and developed a second time (**Fig. 9**). This has the advantage of resolving compounds into the second dimension, which gives further resolution. Also, different solvent systems may be used for the second elution, which further enhances the resolving power of this technique. The resulting chromatogram may then be observed under UV light or stained for detection purposes.

7. Analytical and PTLC: Some Natural Product Examples

The majority of examples cover classes of compound from plant sources (**Fig. 10**). Only the final PTLC stage is given in the isolation and further details can be obtained from the references.

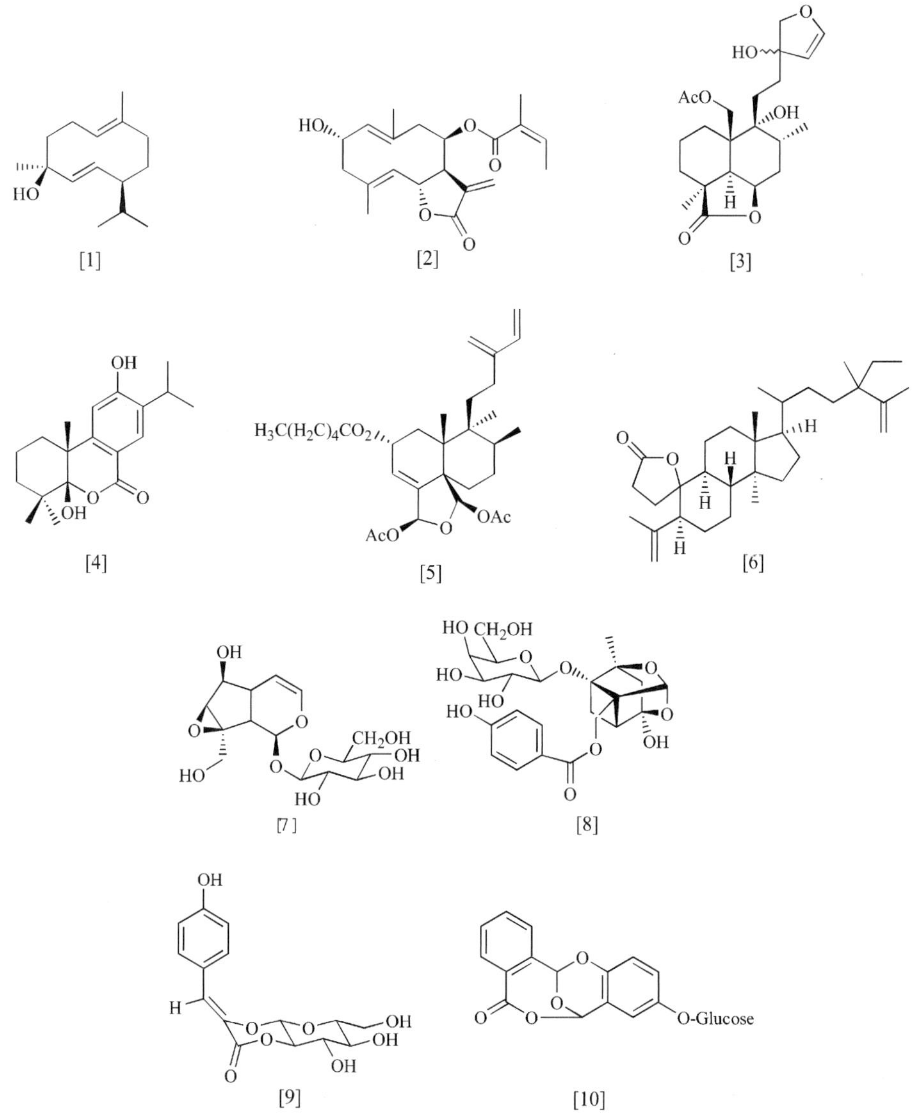

Fig. 10. Examples of some natural products isolated by planar chromatography.

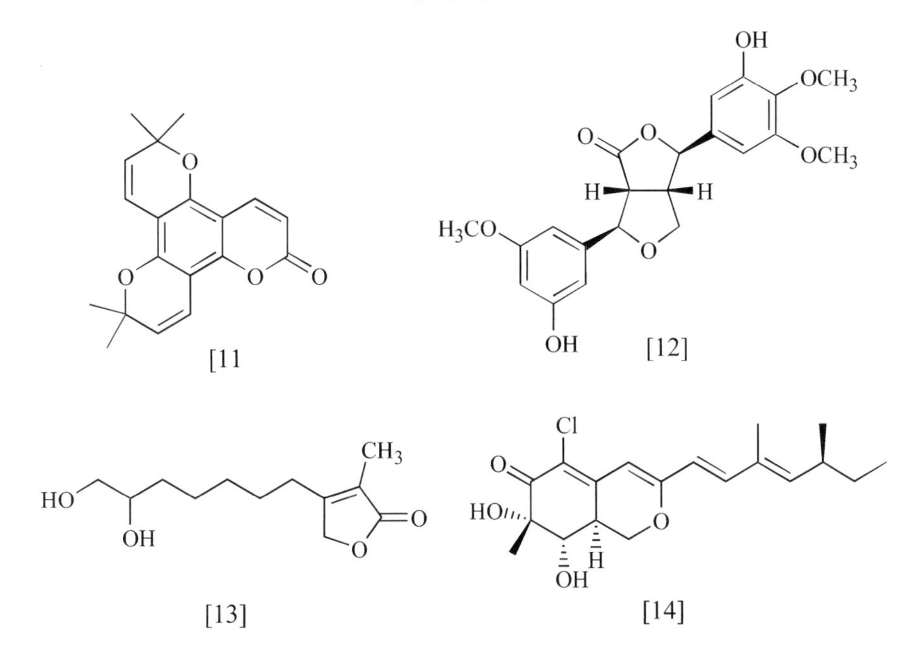

Fig. 10. (*Continued*)

1. Ex. 1. Sesquiterpene [**1**] from *Boronia inornata* ***(15)***. Silica gel, toluene:EtOAc (96:4) then toluene:EtOAc (92:8).
2. Ex. 2. Sesquiterpene [**2**] from *Calea divaricata* ***(16)***. Silica gel, Me_2CO:hexane (4:1).
3. Ex. 3. Diterpene [**3**] from *Leonitis ocymifolia* ***(17)***. Silica gel, hexane:$CHCl_3$:EtOAc (2:3:2).
4. Ex. 4. Diterpene [**4**] from *Cupressus goveniana* ***(18)***. Silica gel, CH_2Cl_2:EtOAc (94:6).
5. Ex. 5. Diterpene [**5**] from *Casearia tremula* ***(19)***. Silica gel, toluene:EtOAc:AcOH (88:10:2)(two developments).
6. Ex. 6. Steroidal triterpene [**6**] from *Sabal blackburniana* ***(20)***. Silica gel, petroleum ether:EtOAc (4:1), two developments.
7. Ex. 7. Glycoside [**7**] from *Castileja rhexifolia* ***(21)***. Alumina, *n*-BuOH:H_2O:MeOH (7:3:1).
8. Ex. 8. Glycoside [**8**] from *Paeonia emodi* ***(22)***. Silica gel, $CHCl_3$:MeOH:H_2O; (80:19.5:0.5).
9. Ex. 9. Glycoside [**9**] from *Gunnera perpensa* ***(23)***. Silica gel, MeOH:$CHCl_3$ (3:17).

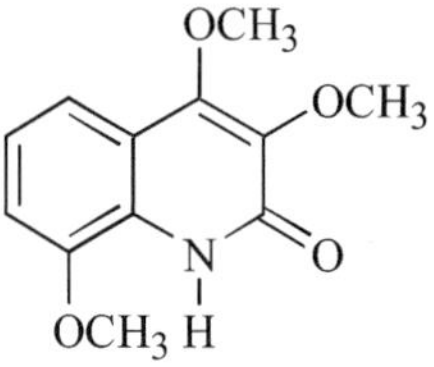

OCH_3 OCH_3 N O OCH_3 H

[15]

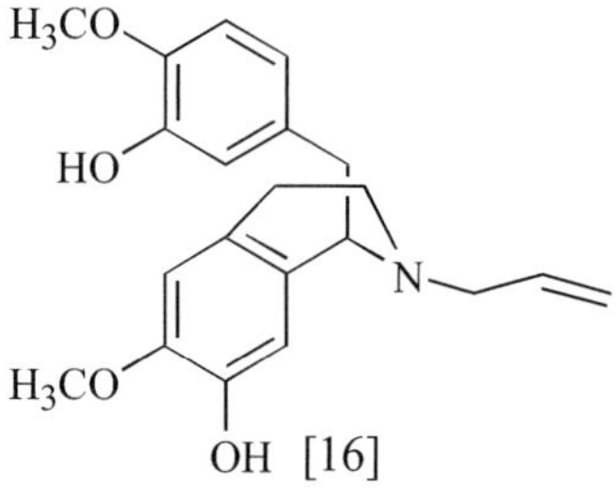

H_3CO HO N H_3CO OH [16]

O O N OH O OCH_3 OCH_3

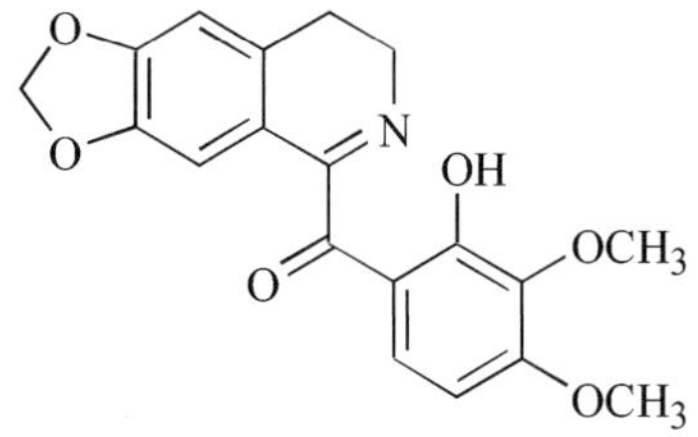

[17]

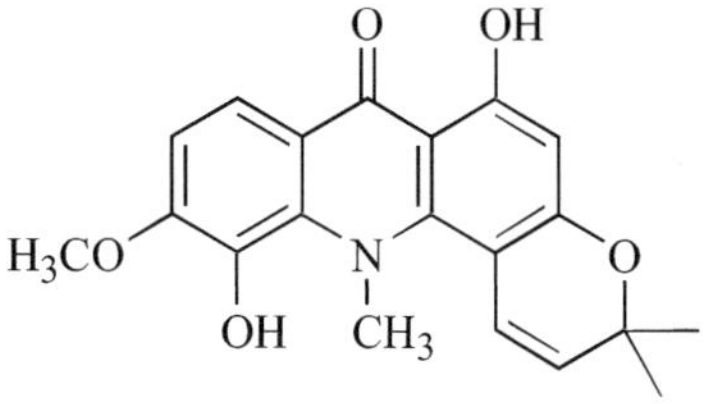

O OH H_3CO N O OH CH_3

[18]

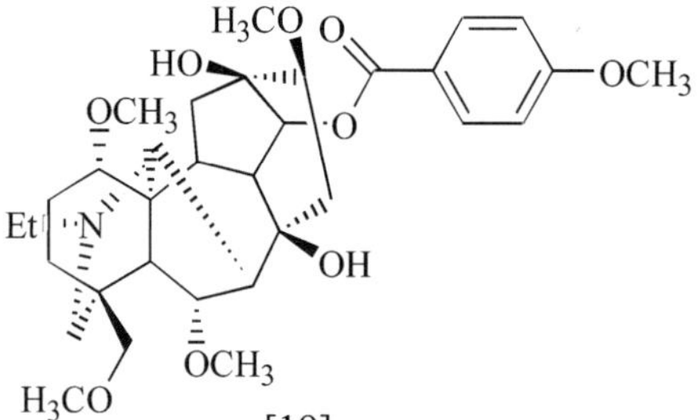

H_3CO O HO OCH_3 O OCH_3 Et N OH OCH_3 H_3CO

[19]

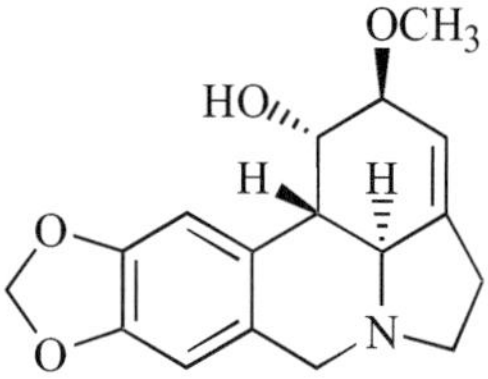

OCH_3 HO H H O O N

[20]

OCH_3 OH H_3CO N H_3C H OCH_3 H_3CO N CH_3 H O H_3CO OH

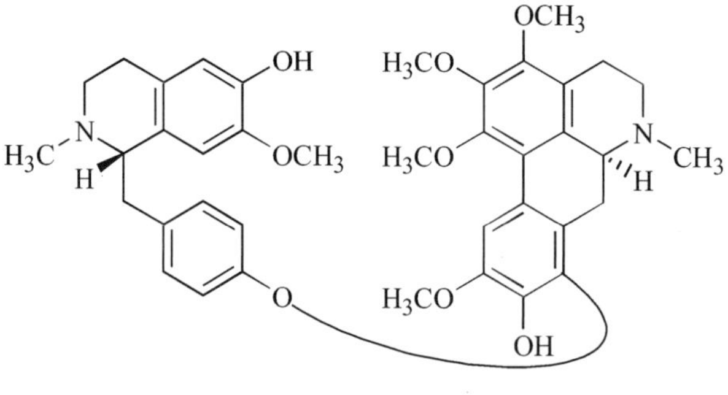

[21]

Fig. 10. (*Continued*)

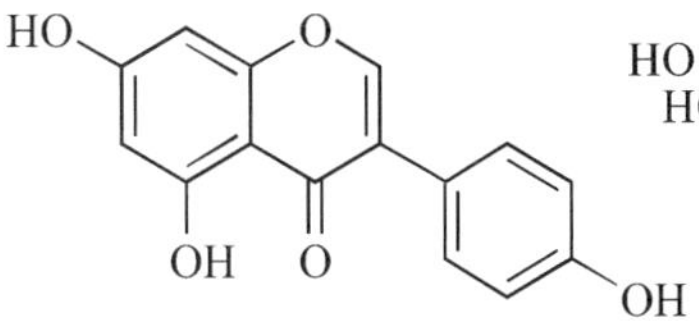

[22]

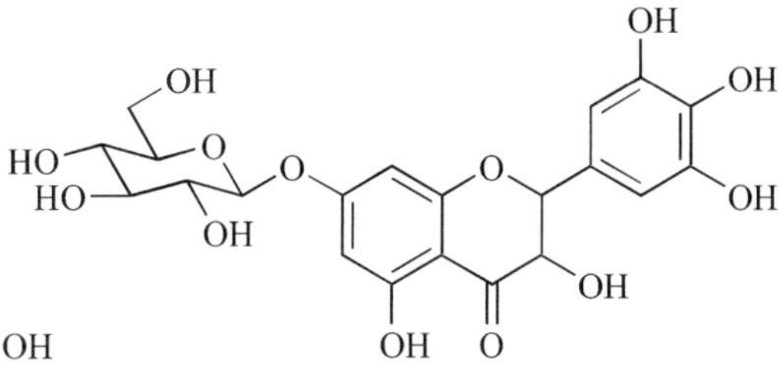

[23]

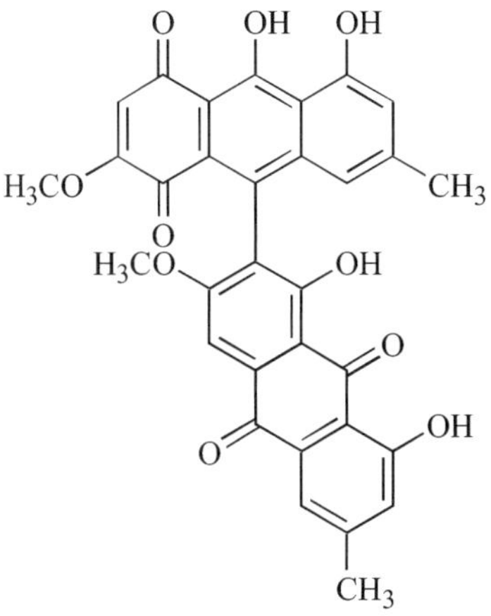

[24]

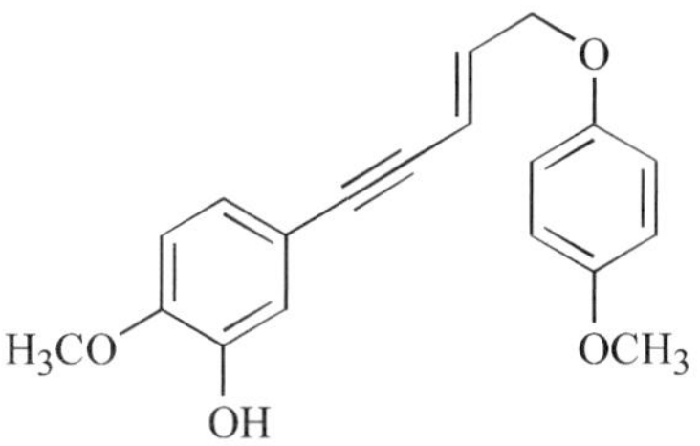

[25]

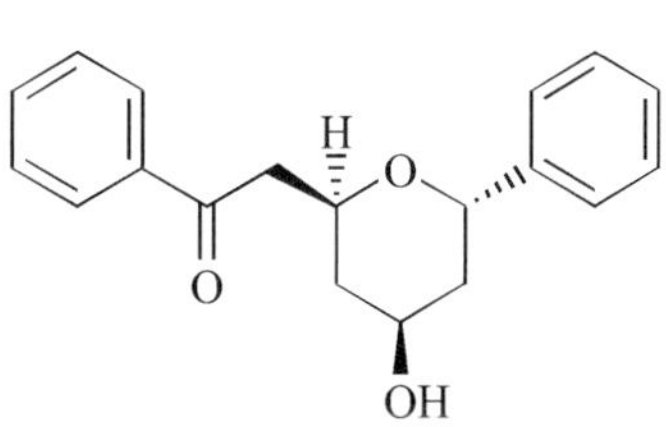

[26]

Fig. 10. (*Continued*)

10. Ex. 10. Glycoside [**10**] from *Homalium longifolium* ***(24)***. Silica gel, EtOAc: MeOH (1:1).
11. Ex. 11. Coumarin [**11**] from *Asterolasia drummondita* ***(25)***. Silica gel, $CHCl_3$:EtOAc (9:1) (two developments.)
12. Ex. 12. Lignan [**12**] from *Imperata cylindrica* ***(26)***. Silica gel, C_6H_6:EtOAc (1:1).
13. Ex. 13. Polyketide [**13**] from *Seiridium* sp. ***(27)***. Silica gel, (1) petroleum ether:acetone, (6:4), (2) $CHCl_3$:iso-propanol (9:1).
14. Ex. 14. Azaphilone [**14**] from *Penicilliun sclerotiorum* ***(28)***. Silica gel, CH_2Cl_2:MeOH (19:1).
15. Ex. 15. Alkaloid [**15**] from *Eriostemon gardneri* ***(25)***. Silica gel, 1. hexane: EtOAc, 8:2. 2. $CHCl_3$:EtOAc, 8:2.
16. Ex. 16. Alkaloid [**16**] from *Papaver somniferum* ***(29)***. Silica gel, $CHCl_3$: MeOH, 8:2.
17. Ex. 17. Alkaloid [**17**] from *Berberis* sp. ***(30)***. Silica gel, 1. $CHCl_3$:MeOH: NH_4OH (90:10:1). 2. C_6H_6:Me_2CO:MeOH:NH_4OH (45:45:10:1).
18. Ex. 18. Alkaloid [**18**] from *Citrus decumana* ***(31)***. Silica gel, C_6H_6:EtOAc, (19:1).
19. Ex. 19. Alkaloid [**19**] from *Aconitum forrestii* ***(32)***. Alumina, Me_2CO: Hexane (45:55).
20. Ex. 20. Alkaloid [**20**] from *Sternbergia lutea* ***(33)***. Silica gel, $CHCl_3$:MeOH (9:1).
21. Ex. 21. Alkaloid [**21**] from *Thalictrum faberi* ***(34)***. Silica gel, cyclohexane:E-tOAc:diethylamine (6:4:1 to 8:4:1).
22. Ex. 22. Isoflavone [**22**] from *Derris scandens* ***(35)***. Silica gel, Me_2CO:$CHCl_3$ (2:98).
23. Ex. 23. Flavonoid glycoside [**23**] from *Picea abies* ***(36)***. Silica gel, *n*-BuOH: AcOH:H_2O (4:1:5).
24. Ex. 24. Anthraquinone dimer [**24**] from *Senna multiglandulosa* ***(37)***. Silica gel, $CHCl_3$:MeOH (100:2).
25. Ex. 25. Norlignan [**25**] from *Asparagus gobicus* ***(38)***. Silica gel, $CHCl_3$.
26. Ex. 26. Diarylheptanoid [**26**] from *Dioscorea spongiosa* ***(39)***. ODS (reverse phase), MeOH:H_2O (5:2).

8. TLC Bioassays

The simplicity and the ability of TLC to separate mixtures quickly with little expense means that it can be readily used to detect biological activity of separated components. A number of these assays are described in the literature, and the majority rely on the separated compounds giving a color reaction with a sprayed reagent either as a final step or as a consequence of

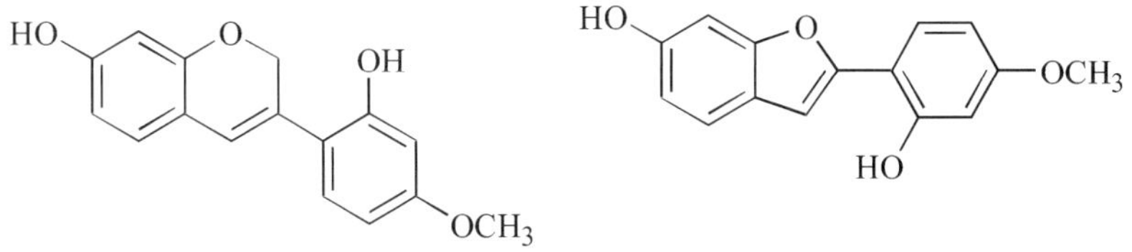

Fig. 11. Antioxidant compounds detected by DPPH assay.

enzyme activity, chemical reaction, or organism activity. These assays are easily developed and performed. Examples are given here.

8.1. Antioxidant TLC Assay

Antioxidant substances are important components of diet and it is widely recognized that they contribute to a healthy state in humans. There is much evidence to suggest that antioxidants are important in retarding cancer cell occurrence and proliferation, and consequently a TLC assay that detects this activity is useful. Erasto et al. ***(40)*** investigated the antioxidant properties of flavonoids using 2,2-diphenyl-picrylhydrazyl (DPPH) radical. This compound is a stable radical, and in the presence of radical scavengers (antioxidants) it is converted from a purple to a yellow color. This contrasting difference in color is distinct and enables recognition of antioxidant substances (**Fig. 11**). Their antioxidant activity was discovered by running a TLC plate with these samples, with concentration ranging from 0.1 to 100 μg. The plates were then dried, sprayed with a DPPH solution (2 mg/mL in methanol), and left for half an hour. Antioxidant compounds appeared as yellow spots against a purple background ***(40)***.

8.2. Acetylcholine Esterase TLC Assay

Inhibitors of the enzyme acetylcholine esterase increase the levels of this neurotransmitter at synapses in the cerebral cortex, and this is beneficial for patients suffering from Alzheimer's disease ***(41)***. Galanthamine (**Fig. 12**),

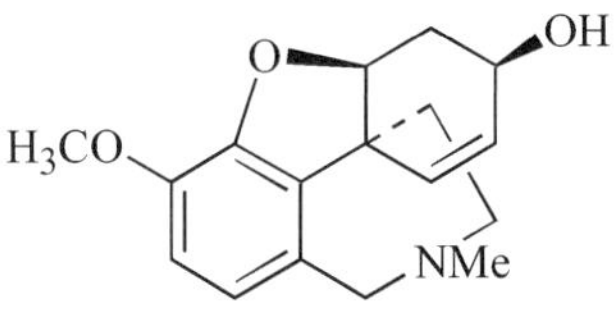

Fig. 12. Structure of galanthamine.

a natural product from snowdrop (*Galanthus nivalis*), is used as an inhibitor of this enzyme in the treatment of this disease. In this assay, the TLC plate is run with prospective inhibitors (e.g., galanthamine, physostigmine, or a plant extract), and after development the plate is dried with a hair dryer to ensure complete removal of the developing solvent. The plate is then sprayed with a stock solution of acetylcholine esterase (1000 U dissolved in 150 mL of 0.05-M Tris–hydrochloric acid buffer at pH 7.8 with bovine serum albumin (150 mg), which was added to stabilize the enzyme during the assay). The TLC plate is then placed in a water-humidified chamber for 20 min while ensuring that water does not touch the plate.

To detect the enzyme, two solutions are prepared: (1) 1-naphthyl acetate (250 mg) in ethanol (100 mL) and (2) Fast Blue B salt (400 mg) in distilled water (160 mL). Following the 20-min incubation of the plate, 10 mL of the naphthyl acetate solution and 40 mL of the Fast Blue B salt solution were mixed together and sprayed onto the TLC to give a purple coloration. Where inhibitors of acetylcholine esterase are present, the white background of the TLC spot is evident.

This assay works by conversion of naphthyl acetate to alpha-naphthol by acetyl choline esterase. Alpha-naphthol then reacts with Fast Blue salt B to give a purple azo dye ***(41)***. Where acetylcholine esterase is inhibited, production of α-naphthol is stopped and therefore azo dye production is inhibited. Marston et al. ***(41)*** have shown that physostigmine inhibited the enzyme at 0.001 μg and galanthamine at 0.01 μg, showing that this assay is sensitive. The authors state that this assay is rapid, simple, and easy to apply to several samples at one time. They also postulate that this bioassay could be extended to other enzymes as long as the enzymes are stable under the test conditions.

8.3. Antimicrobial TLC Bioassays

TLC bioassays against fungi and bacteria have proved exceptionally popular owing to their ease of use, cost, rapidity, and ability to be scaled up to assess antimicrobial activity of a large number of samples. Generally, TLC plates are run and then the microorganism is introduced to the plate as a spray (in the case of direct bioautography) or the plate is covered with a growth medium containing the microorganism in a dish or tray (overlay assay). With the occurrence of multiple drug-resistant bacteria, such as methicillin-resistant *Staphylococcus aureus* (MRSA), and the need for new antimycotic drugs, these simple bioassays will continue to prove useful in the assessment of antimicrobial activity of natural product extracts. A review of key antifungal and antibacterial assays has been made by Cole ***(42)***, and the reader is referred to a number of authors, namely Spooner and Sykes ***(43)***, Holt ***(44)***, Rios et al. ***(45)***, Homans and Fuchs ***(46)***, Betina ***(47)***, Ieven et al. ***(48)***, and Begue and Kline ***(49)***.

8.3.1. TLC Direct Bioautography

This technique may be utilized with either spore-forming fungi or bacteria, and can be used to track activity through a separation process. It is a sensitive assay and gives accurate localization of active compounds ***(50)***. For the assessment of antifungal activity, the plant pathogen *Cladosporium cucumerinum* (IMI-299104) can be used as it is nonpathogenic to humans, readily forms spores, and can be easily grown on TLC plates with the correct medium. A simple method is outlined as follows:

1. Extracts or pure compounds may be spotted onto analytical TLC plates in duplicate (plastic-backed, Kieselgel 60 PF254, Merck Art 5735), developed with the appropriate mobile phase and dried.
2. A slope of *Cladosporium cucumerinum* (IMI-299104) is prepared from a culture and allowed to sporulate for two days.
3. A TLC growth medium is prepared as follows: NaCl (1 g), KH_2PO_4 (7 g), $Na_2HPO_4 \cdot 2H_2O$ (3 g), KNO_3 (4 g), $MgSO_4$ (1 g) and Tween-80 (20 drops added to water (100 mL). A volume of 60 mL of this solution should be added to 10 mL of aqueous glucose (30% w/v).
4. A fungal suspension is prepared by adding the above solution to the fungal slope and shaking it.
5. The suspension is sprayed onto one of the TLC plates and incubated at 25°C for 2 d in an assay tray with wet cotton wool to ensure a moist atmosphere.
6. The inoculated TLC plate is observed at regular intervals, and the presence of antifungal compounds is indicated by inhibition or reduced lack of mycelial growth. This is frequently observed as light spots against a dark green background. The spraying is performed in a lamina flow cabinet.
7. The remaining TLC plate is visualized using a spray reagent and/or UV detection and compared with the incubated plate.

Aspergillus niger, a more easily sporulating fungus, may be used in the place of *Cladosporium* sp., but care must be taken with this organism because of the risk of aspergillosis, and all microbes should be handled aseptically in a lamina flow cabinet. Controls of antifungal compounds such as amphotericin B should be used each time this assay is performed. This assay does not distinguish between fungicidal and fungistatic metabolites, and

Aristolen-2-one Prostantherol

Fig. 13. Antifungal sesquiterpenes aristolen-2-one and prostantherol from two species of *Prostanthera*.

further assays such as a liquid broth assay will need to be performed to measure minimum inhibitory concentration (MIC) ***(42)***. It should be noted that amphotericin B is highly toxic and care must be exercised in its use.

Dellar et al. ***(51)*** isolated the antifungal sesquiterpenes aristolen-2-one ***(38)*** and prostantherol ***(39)*** from two species of *Prostanthera* (Labiatae) (**Fig. 13**). Activity was assessed and tracked through the separation procedure by the use of direct bioautography with *Cladosporium cucumerinum* as the target fungus. Compound ***(38)*** inhibited the growth of *C. cucumerinum* for 70 h at a dose of 1 µg, whereas ***(39)*** caused inhibition at 10 µg for the same duration.

The antifungal activity of many plant phenolic compounds can be readily assessed using this simple procedure. Hostettmann and Marston ***(52)*** have investigated a series of xanthones (**Fig. 14**) from *Hypericum brasiliense* (Guttiferae) for activity against *C. cucumerinum*. One of these compounds

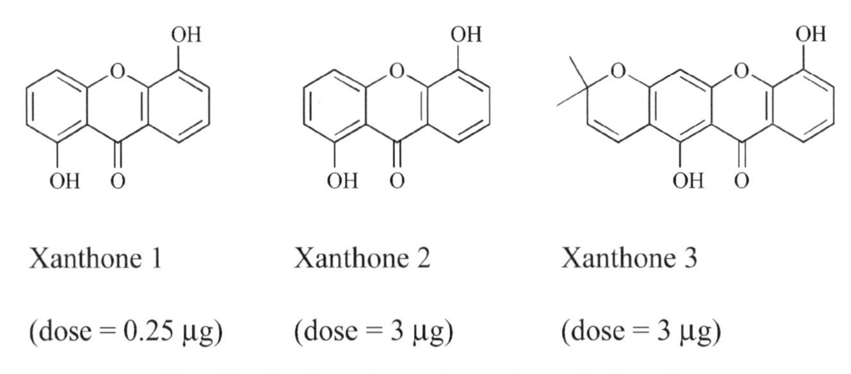

Fig. 14. Antifungal xanthones from *H. brasiliense*.

(xanthone 1) exhibited a low inhibitory dose (25-μg), which may warrant further investigation of the antimycotic function of these interesting compounds.

8.3.2. TLC Bioautographic Overlay Assay

In this form of assay, the extract or pure compound is run on a TLC plate, which is then covered by a medium seeded with the appropriate micro-organism. As with direct bioautographic assays, both fungi and bacteria may be investigated. Rahalison et al. ***(50)*** have applied this technique for the evaluation of antimicrobial extracts against the yeast *Candida albicans* and the bacterium *Bacillus subtilis.* A simple overlay assay against *S. aureus* (Manohar, R. 1996, personal communication) may be carried out as follows:

1. A base of nutrient agar (NA)(oxoid) should be poured into an assay dish and allowed to set.
2. The extract, fraction, or pure compound should be run on a TLC plate (in duplicate) with the appropriate developing solvent. One of these plates should be visualized under UV light and then stained to observe developed compounds. R_f values should be accurately measured.
3. An innoculum of *S. aureus* at a titer of 10^9/ mL in Mueller Hinton Broth is prepared and NA is added at 7.5 g/L to thicken the medium. This is then diluted out to give a final titer of 10^5 CFU /mL.
4. The remaining TLC plate should be placed on the NA base and then the medium containing the test organism is poured over the plate and incubated at 37°C for 24 h.
5. Antibacterial zones appear as clear spots against a background of bacterial colonies. The zones may be visualized more clearly by the use of tetrazolium salts (such as *p*-iodonitrotetrazolium chloride [INT] or methylthiazoyltetrazolium chloride [MTT]), which indicate bacterial lactate dehydrogenase activity. These solutions may sprayed onto the face of the medium. Zones of inhibition (and therefore antimicrobial compounds) appear as clear zones against a purple background.
6. Zones of inhibition should be compared with the previously developed TLC plate, so that active metabolites may be readily isolated.
7. An appropriate control substance such as ampicillin or chloramphenicol should be used.

Drug-resistant bacteria such as methicillin resistant *S. aureus* will need to be cultured in the presence of methicillin (1 mg/mL) to minimize the risk of loss of resistance.

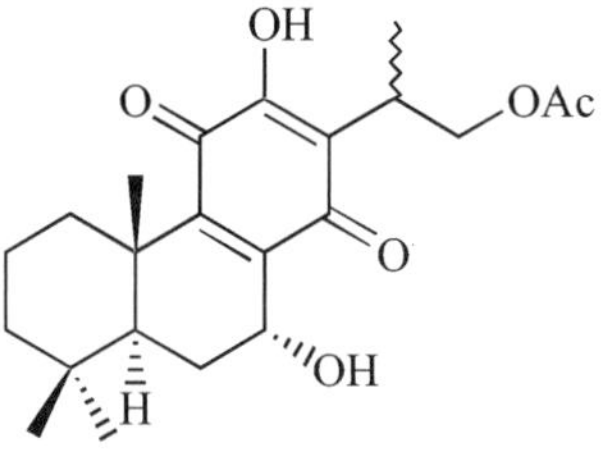

Fig. 15. Antibacterial diterpene from *Plectranthus hereroensis*.

Batista et al. ***(53)*** used an overlay method in the bioassay-guided fractionation of an acetone extract of the roots of *Plectranthus hereroensis* (Labiatae) to isolate the antibacterial diterpene (**Fig. 15**). *S. aureus* was used as the test organism. This compound was then assessed in a broth dilution assay and found to have a MIC of 31.2 µg /mL. Hamburger and Cordell ***(54)*** used a variant of this assay to investigate the activity of plant sterols and phenolic compounds. An overlay of nutrient broth containing the test organism was spread over the TLC plate and then incubated. Interestingly, this assay was insensitive to some cytotoxic compounds, including camptothecin, glaucarubolone, and β-peltatin when tested at 5 µg.

9. Conclusion

The ease of use, speed, and low cost of TLC make it a widely used and versatile technique that can be readily learnt. While HPLC is becoming increasingly popular as the method of choice for a final clean up of extract fractions to get to a purified natural product, analysis of papers from journals *Planta Medica, Phytochemistry*, and *Journal of Natural Products* in 2004 shows that this technique still has a central place in natural product isolation and analysis. TLC is also easily interfaced with bioassays, and it is likely that many new enzyme inhibition assays that use a colorimetric change on a TLC plate will be developed as a first simple test for biological activity. The author encourages any scientist making the first foray into natural product research to experiment with the plethora of sorbents and solvents that TLC has to offer, because slight changes in a method can dramatically effect a separation.

References

1. Gray, A. I. (1993) Quinoline alkaloids related to anthranilic acid. *Methods in Plant Biochemistry*. vol. 8. Chapter 8, 288. Academic, London.

2. *CRC Handbook of Chemistry and Physics*. 72 ed. CRC Press.
3. Stierle, A., Strobel, G., Stierle, D., Grothaus, P., and Bignami, T. (1995) The search for a Taxol® producing microorganism among the endophytic fungi of the pacific yew *Taxus brevifolia*. *J. Nat. Prod.* **58,** 1315–1324.
4. Staneck, J. L. and Roberts, G. D. (1974). *Simplified approach to identification of aerobic actinomycetes by thin-layer chromatography. Appl. Microbiol.* **28,** 226–231.
5. Waterman, P. G. and Grundon, M. F., eds. (1983) *Chemistry and Chemical Taxonomy of the Rutales*. Academic, London.
6. Hegnauer, R. (1989) *Chemotaxonomie de Pflanzen.* vol. 1–10. Birkhäuser Verlag, Berlin.
7. Gibbons, S. (1994) Phytochemical studies on the Flacourtiaceae and Simaroubaceae. PhD Thesis, University of Strathclyde.
8. Wagner, H. and Bladt, S. (1996). *Plant Drug Analysis—A Thin Layer Chromatography Atlas.* Springer-Verlag, Berlin.
9. Merck Handbook (1980) *Dyeing Reagents for Thin Layer and Paper Chromatography*. E. Merck, Darmstadt, Germany.
10. Gibbons, S., Gray, A. I., Hockless, D. C. R. et al. (1993) Novel D: A *friedo*-oleanane triterpenes from the stem bark of *Phyllobotryon spathulatum. Phytochemistry* **34**, 273–277.
11. Khan, M. R., Gray, A. I., and Waterman, P. G. (1990) Clerodane diterpenes from *Zuelania guidonia* stem bark. *Phytochemistry* **29,** 2939–2942.
12. Ampofo, S. and Waterman, P. G. (1984) Cytotoxic quassinoids from *Odyendyea gabonensis* stem bark: isolation and high field NMR. *Planta. Med.* **50,** 261–263.
13. Tyihak, E., Mincsovics, E., and Kalasz, H. (1979) New planar liquid chromatographic technique: overpressured thin layer chromatography. *J. Chromatogr.* **174,** 75–81.
14. Nyiredy, S., Dallenbach-Tölke, K., Erdelmeier, C. A. J., Meier, B. and Sticher, O. (1985) Abstracts, 33rd Annual Congress of the Society for Medicinal Plant Research, Regensburg.
15. Ahsan, M. (1993) PhD Thesis, University of Strathclyde.
16. Ober, A. G., Fronczek, F. R., and Fischer, N. H. (1985) Sesquiterpene lactones of *Calea divaricata* and the molecular structure of leptocarpin acetate. *J. Nat. Prod.* **48,** 302.
17. Habtemariam, S., Gray, A. I., and Waterman, P. G. (1994) Diterpenes from the leaves of *Leonotis ocymifolia* var. *raineriana*. *J. Nat. Prod.* **57,** 1570–1574.
18. Jolad, S. D., Hoffmann, J. J., Schram, K. H., Cole, J. R., Bates, R. B., and Tempesta, M. S. (1984) A new diterpene from *Cupressus govenia* var. *abramasiana*: 5β-hydroxy-6-oxasugiol (Cupresol). *J. Nat. Prod.* **47,** 983–987.
19. Gibbons, S., Gray, A. I., and Waterman, P. G. (1996) Clerodane diterpenes from the bark of *Casearia tremula*. *Phytochemistry* **41,** 565–570.

20. El-Dib, R., Kaloga, M., Mahmoud, I., Soliman, H. S. M., Moharram, F. A., and Kolodziej, H. (2004) Sablacaurin A and B, two 19-*nor*-3,4-*seco*-lanostane-type triterpenoids from *Sabal causiarum* and *Sabal blackburniana*, respectively. *Phytochemistry* **65,** 1153–1157.
21. Roby, M. R. and Stermitz, F. R. (1984) Penstemonoside and other iridoids from *Castilleja rhexifolia.* Conversion of penstemonoside to the pyridine monoterpene alkaloid rhexifoline. *J. Nat. Prod.* **47,** 854–857.
22. Riaz, N., Malik, A., Rehman, A. et al. (2004) Lipoxygenase inhibiting and antioxidant oligostilbene and monoterpene galactoside from *Paeonia emodi. Phytochemistry* **65**, 1129–1135.
23. Khan, F., Peter, X. K., Mackenzie, R. M. et al. (2004) Venusol from *Gunnera perpensa*: structural and activity studies. *Phytochemistry* **65**, 1117–1121.
24. Shaari, K. and Waterman, P. G. (1995) Further glucosides and simple isocoumarins from *Homalium longifolium. Nat. Prod. Lett.* **7,** 243–250.
25. Sarker, S. (1994) Phytochemical and chemotaxonomic studies in the tribe Boronieae (Rutaceae). PhD Thesis, University of Strathclyde.
26. Matsunaga, K., Shibuya, M., and Ohizumi, Y. (1994) Graminone B, a novel lignan with vasodilative activity from *Imperata cylindrica. J. Nat. Prod.* **57,** 1734–1736.
27. Evidente, A. and Sparapano, L. (1994) 7′-Hydroxyseiridin and the 7′-hydroxyisoseiridin, two new phytotoxic $\Delta^{\alpha,\beta}$-butenolids from three species of *Seiridium* pathogenic to cypresses. *J. Nat. Prod.* **57,** 1720–1725.
28. Pairet, L., Wrigley, S. K., Chetland, I. et al. (1995) Azaphilones with endothelin receptor binding activity produced by *Penicillium sclerotiorum:* taxonomy, fermentation, isolation, structure elucidation and biological activity. *J. Antibiot.* **48**, 913–923.
29. Brochmann-Hanssen, E. and Cheng, C. Y. (1984) Biosynthesis of a narcotic antagonist: conversion of *N*-allylnorreticuline to *N*-allylnormorphine in *Papaver somniferum. J. Nat. Prod.* **47,** 175–176.
30. Valencia, E., Weiss, I., Shamma, M. et al. (1984) Dihydrorugosine, a pseudobenzylisoquinoline alkaloid from *Berberis darwinii* and *Berberis actinacantha*. J. Nat. Prod. **47**, 1050–1051.
31. Basa, S. C. and Tripathy, R. N. (1984) A new acridone alkaloid from *Citrus decumana. J. Nat. Prod.* **47,** 325–330.
32. Pelletier, S. W., Ying, C. S., Joshi, B. S., and Desai, H.K. (1984) The structures of Forestine and Foresticine, two new C_{19} -diterpenoid alkaloids from *Aconitum forrestii* stapf. J. Nat. Prod. **47**, 474–477.
33. Evidente, A., Iasiello, I., and Randazzo, G. (1984) Isolation of Sternbergine, a new alkaloid from the bulbs of *Sternbergia lutea. J. Nat. Prod.* **47,** 1003–1008.

34. Lin, L.-Z., Hu, S.-H., Zaw, K. et al. (1994) Thalfaberidine, a cytotoxic aporphine-benzylisoquinoline alkaloid from *Thalictrum faberi*. *J. Nat. Prod.* **57**, 1430–1436.
35. Mahabusarakam, W., Deachathai, S., Phongpaichit, S., Jansakul, C., and Taylor, W. C. (2004) A benzil and isoflavone derivatives from *Derris scandens* Benth. *Phytochemistry* **65,** 1185–1191.
36. Slimestad, R., Andersen, O. M., and Francis, G. W. (1994) Ampelopsin 7-glucoside and other dihydroflavonol 7-glucosides from needles of *Picea abies*. *Phytochemistry* **35,** 550–552.
37. Abegaz, B. M., Bezabeh, M., Alemayehu, G., and Duddeck, H. (1994) Anthraquinones from *Senna multigladulosa*. *Phytochemistry* **35,** 465–468.
38. Yang, C. X., Huang, S. S., Yang, X. P., and Jia, Z. J. (2004) Non-lignans and steroidal saponins from *Asparagus gobicus*. *Planta Med.* **70,** 446–451.
39. Yin, J., Kouda, K., Tezuka, Y. et al. (2004) New diarylheptanoids from the rhizomes of *Dioscorea spongiosa* and their antiosteoporotic activity. *Planta Med.* **70**, 54–58.
40. Erasto, P., Bojase-Moleta, G., and Majinda, R. R. T. (2004) Antimicrobial and antioxidant flavonoids from the root wood of *Bolusanthus speciosus*. *Phytochemistry* **65,** 875–880.
41. Marston, A., Kissling, J., and Hostettmann, K. (2002) A rapid TLC bioautographic method for the detection of acetylcholineesterase and butyrlcholine esterase inhibitors in plants. *Phytochem. Anal.* **13,** 51–54.
42. Cole, M. D. (1994) Key antifungal and antibacterial assays—a critical review. *Biochem. Syst. Ecol.* **22,** 837–856.
43. Spooner, D. F. and Sykes, G. (1972) *Laboratory Assessment of Antibacterial activity*. In: Norris J. R., Ribbons D.W. (Eds.), Methods in Microbiology, Vol. 7B. Academic Press, London, pp. 216–217.
44. Holt, R. J. (1975) Laboratory tests of antifungal drugs. *J.Clin. Pathol.* **28,** 767–774.
45. Rios, J. L., Recio, M. C., and Villar, A. (1988) Screening methods for natural products with antimicrobial activity: a review of the literature. *J. Ethnopharmacol.* **23,** 127–149.
46. Homans, A. L. and Fuchs, A. (1970) Direct bioautography on thin-layer chromatograms as a method for detecting fungitoxic substances. *J.Chromatogr.* **51,** 327–329.
47. Betina, V. (1973) Bioautography in paper and thin layer chromatography and its scope in the antibiotic field. *J. Chromatogr.* **78,** 41–51.
48. Ieven, M., Vanden Berghe, D. A., Mertens, F., Vlietinck, A., and Lammens, E. (1979) Screening of higher plants for biological activity I. Antimicrobial activity. *Planta Med.* **36,** 311–321.
49. Begue, W. J. and Kline, R. M. (1972) The use of tetrazolium salts in bioautographic procedures. *J. Chromatogr.* **64,** 182–184.

50. Rahalison, L., Hamburger, M., Hostettmann, K., Monod, M., and Frenk, E. (1991) A bioautographic agar overlay method for the detection of antifungal compounds from higher plants. *Phytochem. Anal.* **2,** 199–203.
51. Dellar, J. E., Cole, M. D., Gray, A. I., Gibbons, S., and Waterman, P. G. (1994) Antimicrobial sesquiterpenes from *Prostanthera* aff. *melissifilia* and *P. rotundifolia*. *Phytochemistry* **36,** 957–960.
52. Hostettmann, K. and Marston, A. (1994) Search for new antifungal compounds from higher plants. *Pure Appl. Chem.* **66,** 2231–2234.
53. Batista, O., Simoes, M. F., Duarte, A., Valdeira, M. L., De la Torre, M. C., and Rodriguez, B. (1995) An antimicrobial abietane from the roots of *Plectranthus hereroensis*. *Phytochemistry* **38,** 167–169.
54. Hamburger, M. O. and Cordell, G. A. (1987) A direct bioautographic TLC assay for compounds possessing antibacterial activity. *J. Nat. Prod.* **50,** 19–22.

Suggested Readings

1. Grinberg, N. ed. (1990) *Modern Thin Layer Chromatography. Chromatographic Science Series*. vol. 52. Marcel Dekker, Inc.
2. Hostettmann, K., Hostettmann, M., and Marston, A. (1986). *Preparative Chromatography Techniques—Applications in Natural Product Isolation.* Springer Verlag, Berlin.
3. *Merck Handbook—Dyeing Reagents for Thin Layer and Paper Chromatography* (1980) E. Merck, Darmstadt, Germany. (A comprehensive set of spray reagents.).
4. Touchstone, J. C. and Dobbins, M. F. (1982) *Practice of Thin Layer Chromatography*. John Wiley and Sons Publishers.
5. Wagner, H. and Bladt, S. (1996) *Plant Drug Analysis—A Thin Layer Chromatography Atlas*. Springer-Verlag, Berlin. (The first point of call for anyone interested in TLC of natural products. There are many excellent examples of systems and detection sprays).

5

Isolation of Natural Products by Low-Pressure Column Chromatography

Raymond G. Reid and Satyajit D. Sarker

Summary

This chapter deals with the isolation of natural products using low-pressure liquid column chromatography (LPLC). A brief summary of the adsorption and size-exclusion processes involved in the LPLC is presented. Different types of stationary phases used in both adsorption and size-exclusion LPLC are described along with examples of each type used in LPLC. Operational parameters are also discussed in detail in relation to column packing (both wet and dry), column equilibration, sample application, and column development. An outline of generic procedures for adsorption and size-exclusion LPLC is provided. Sixty-one specific examples of the use of LPLC for the isolation of various natural products, including flavones, coumarins, alkaloids, and triterpenoid saponins from various plants are given. Many of these examples incorporate a combination of adsorption and size-exclusion chromatography to obtain specific compounds.

Key Words: Natural products; low-pressure column chromatography; adsorption; size exclusion; packing materials.

1. Introduction

Since the invention of chromatography by M. S. Tswett over a century ago, remarkable advances have been accomplished in this area. From the first use of filter paper to examine plant pigments to the development of the most modern high-performance columns and affinity gels, chromatography continues to play an important role in the separation of compounds from

From: *Methods in Biotechnology, Vol. 20, Natural Products Isolation, 2nd ed.*
Edited by: S. D. Sarker, Z. Latif, and A. I. Gray © Humana Press Inc., Totowa, NJ

complex mixtures, such as natural product extracts. Prior to the 1970s, there were only a few reliable chromatographic methods available commercially *(1)*. During 1970s when most works on chemical separation were performed by a variety of techniques, including open column chromatography (CC), paper chromatography, and thin layer chromatography (TLC), the concept of pressure liquid chromatography to decrease the flow through time, and to reduce purification time of compounds being isolated by CC, began to turn to reality *(1,2)*. In low-pressure liquid column chromatography (LPLC), a mobile phase is allowed to flow through a densely packed adsorbent. Different separation mechanisms can be applied depending on the choice of packing material and mobile phase selected. There are a number of books that summarize the advances in the field of chromatography, the history and the fundamental theoretical aspects, available to date *(3–11)*. Hence, the aim of this chapter is not to describe the history or theory of chromatography, but to discuss the application of and recent advances in LPLC in relation to the isolation of natural products. However, a brief description of the theoretical basis and various separation mechanisms and techniques are presented with specific examples of their applications. This chapter deals with LPLC in the context of a liquid mobile phase used in conjunction with some solid packing material. Different packing materials and their uses in the isolation of natural products are also covered.

2. Separation Processes

One of the fundamental principles behind the separation of components in liquid chromatography (LC) is illustrated in **Fig. 1**. The separation takes place through selective distribution of the components between a mobile phase and a stationary phase. However, there are a number of factors, related to the physical and chemical nature of the mobile and stationary phases, as well as the solutes controlling the various interactions between the solutes and two phases, involved in the separation process. The number of possible interactions between the solutes and the stationary phase (adsorbent) depends on the particle size of the stationary phase—the greater the surface area of the stationary phase, the more the number of interactions. A stationary phase with a high surface area tends to give enhanced separations. The equilibrium between the solute and the stationary phase is termed *distribution constant*, which is dependent on the chemical nature of the system. Different types of separation processes involved in LPLC are discussed here.

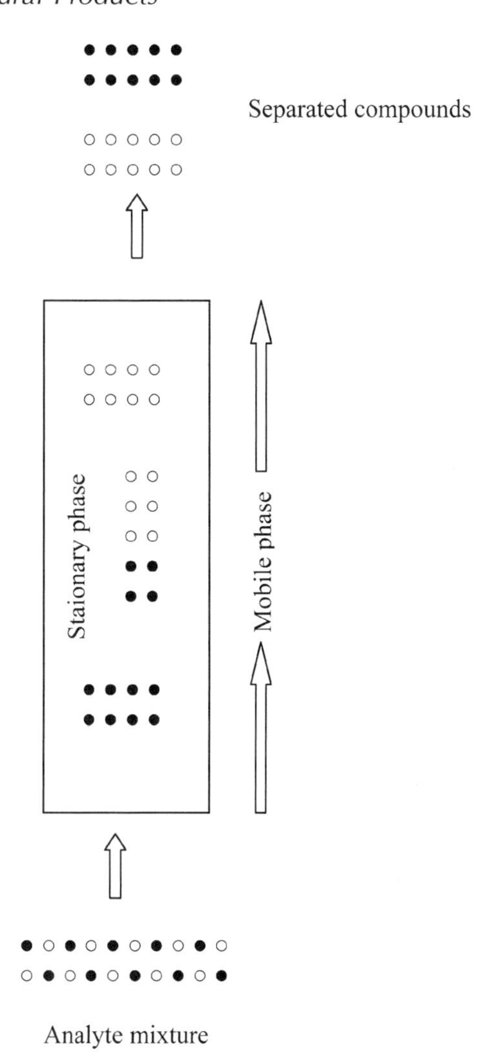

• = Strongly adsorbed to the stationary phase and less soluble in the mobile phase

○ = Loosely bound to the stationary phase and more soluble in the mobile phase

Fig. 1. Illustrated basic principle of adsorption chromatography.

2.1. Adsorption

Adsorption is one of the basic liquid chromatographic modes that rely on the adsorption process to effect the separation. The analyte molecules are retained by their interaction with the surface of the stationary phase (**Fig. 1**). Therefore, the separation is based mainly on the differences between the adsorption affinities of the analyte molecules for the surface of the stationary phase. The extent of adsorption of analyte molecules onto the stationary phase is governed by a number of factors, e.g., hydrogen bonding, van der Waal forces, dipole–dipole interactions, acid–base properties, complexation, charge transfer, etc. The detailed criteria that control the adsorption process in LC have been outlined by Kazakevich and McNair *(12)* in their web-based publication. The observed retention of a solute is most commonly a result of a combination of these interactions, which are, however, reversible. The choice of appropriate stationary phases as well as the mobile phase is crucial for obtaining optimum separation of components, maximizing the recovery of solutes, and avoiding irreversible adsorption of solutes onto the packing material.

2.2. Size Exclusion

This technique is often referred to as size-exclusion chromatography (SEC), or gel-permeation chromatography (GPC) when organic solvents are used. If aqueous solvents are used, it is termed as gel filtration chromatography (GFC). The fundamental principle of SEC or GPC is the separation of molecules according to their hydrodynamic volume. The stationary phases used are nonadsorbing porous particles with pores of approximately the same size, as the effective dimensions in solution, of the molecules to be separated. Various types of materials are used to make beads with various pore sizes. Large molecules cannot enter the matrix, intermediate size ones can enter part of the matrix, and those that are small freely enter the matrix. The function of the matrix is to provide a continuous decrease in accessibility for molecules of increasing size. There is no interaction between the solute and the stationary phase, and separation is on the basis of molecular size and shape of the analyte molecules. As a result, molecules elute in order of decreasing size; the largest ones are eluted from the column first and the smallest last. Because packing materials are inert, sample recovery from this type of column is high. **Figure 2** illustrates the fundamental process involved in this chromatography.

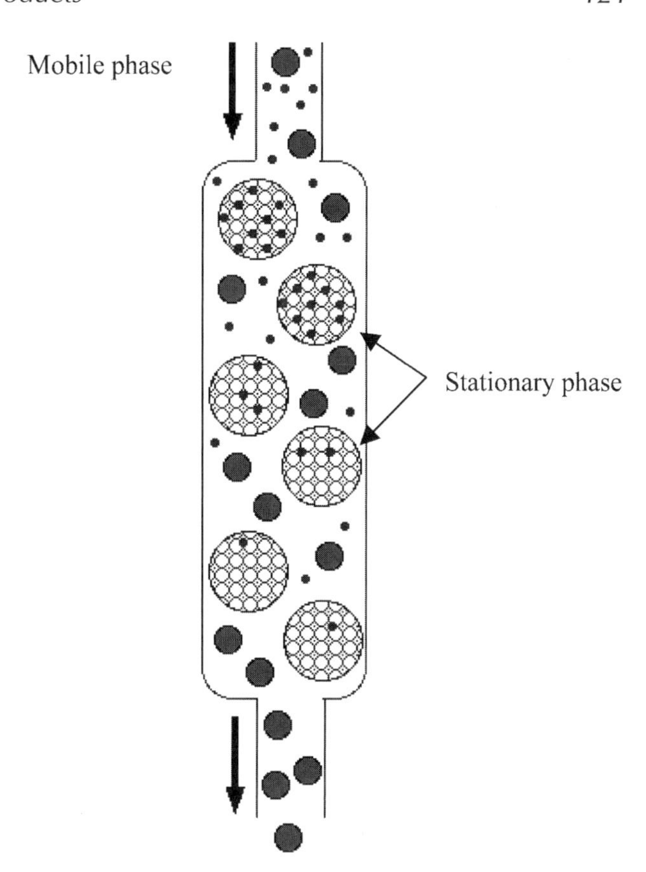

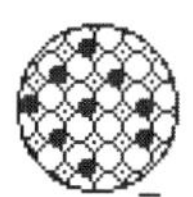 = Semisolid beads of a polymeric gel (stationary phase)

• = small molecules; ● = large molecules

Fig. 2. Illustrated basic mechanism of SEC.

3. Types of Stationary Phases

There are many types of stationary phases available for LPLC ***(13)***. A description of the physical characteristics of the packing materials usually consists of the particle size, shape, porosity, and surface area. For LPLC, particle sizes are normally between 10 and 200 μm, while for HPLC, they are between 2 and 10 μm. Particles can be irregularly shaped or completely spherical. The porosity of the particle represents the ratio of the volume of surface pores to the total volume of the particle. In commercial packing materials, size, shape, and porosity can be spread over a wide range or controlled within a very narrow range. Stationary phases of the same type can also vary from manufacturer to manufacturer.

3.1. Adsorption Stationary Phases

3.1.1. Silica Gel

Ever since the birth of modern LC, silica gel has been the adsorbent of choice for chromatographers. It is the most commonly used stationary phase in LPLC for the separation of natural products. Silica gel is available commercially in a wide range of forms. Its unsurpassed capacity for both linear and nonlinear isothermal separations, and above all, its almost complete inertness toward labile compounds make it a logical first choice as a general purpose adsorbent for LPLC. Silica gel may be regarded as a typical polar sorbent. The relative adsorptivity of different molecular groups on silica gel is similar to that observed for other polar adsorbents. The silica gel surface is weakly acidic and there is a tendency toward preferential adsorption of strongly basic substances relative to adsorption on neutral or basic adsorbent. Silica gel is a three-dimensional polymer of tetrahedral units of silicon oxide, chemically represented by $SiO_2 \cdot H_2O$. It is a porous material and offers a large surface area in the range of 5–800 m^2/g. For LPLC, the particle size of the silica gel is normally in the range of 40–200 μm with pore sizes between 40 and 300 Å. Smaller particle sizes are usually considered for HPLC work.

The silica gel surface consists of exposed silanol groups, and these hydroxyl groups form the active centers. The hydroxyl groups can potentially form strong hydrogen bonds with various compounds. In chromatographic terms, the stronger the hydrogen bond, the longer the compound will be retained on the silica gel. Polar compounds containing carboxylic acid, amines, or amides are strongly absorbed on silica gel. Nonpolar compounds, that lack in polar functional groups, i.e., a few or no hydrogen-bonding sites, are weakly absorbed with little or no retention. The retention of a specific compound also depends on the polarity of the mobile phase. The stronger the hydrogen-bonding capability of a mobile phase, the better it elutes polar compounds on silica gel columns. Alternatively, nonpolar solvents can be used to elute nonpolar compounds.

Nonpolar solvents used for chromatography with silica gel include pentane, hexane, and dichloromethane (DCM), while ethyl acetate (EtOAc), methanol (MeOH), and acetonitrile (ACN) are examples of polar solvents. Sometimes, small concentrations of water in the mobile phase can be used to elute strongly absorbed components. However, care must be taken when using water or MeOH, as the silica gel stationary phase will dissolve and could possibly contaminate the compounds of interest. When using water with silica gel, the pH of the mobile phase must be less than 7.0. At any pH above 7.0, silica gel dissolves readily.

3.1.2. Bonded-Phase Silica Gel

Silica gel can be chemically modified in a variety of ways to alter both its physical properties and chromatographic behavior. The main purpose of surface modification is shielding of the active silanol groups and attachment to the accessible adsorbent surface organic ligands that are responsible for specific surface interactions. The silanol groups can be blocked with a variety of silylchlorides to produce either a nonpolar (reversed phase) or an intermediate polarity (bonded normal phase) chromatography support. A broad variation of functionality of bonded species could be attached to the silica surface (**Table 1**). The most popular is an organofunctional group R, which may carry substituents of various functionalities such as alcoholic or phenolic hydroxyl, amine, phenyl, carbonyl, nitrile, etc. The functional group R can be linked to the surface silicon atoms in the following ways:

1. Si–R: R is directly bonded to the surface silicon atoms. The elimination of the original hydroxyl groups can be achieved by chlorination of the surface with subsequent treatment of the chlorinated surface with organometallic compounds. This type of surface modification is usually laborious and does not offer a stable reproducible surface modification.
2. Si–O–R: This bond is known as an ester bond and is easily formed by the reaction between an alcohol and surface hydroxyl groups.
3. Si–O–Si–R: This structure is obtained by treatment of a hydroxylated silica surface with organosilanes of R_nSiX_{4-n}, where X is the reactive groups such as halogen, ethoxy, and methoxy. This type of bonded ligands is the most widely used adsorbent in HPLC.

The nonpolar (reversed phase) stationary phase is prepared by treating silica gel with chlorodimethylalkylsilanes or chloroalkoxysilanes of

Table 1
Commonly Used Bonded-Phase Silica

Phase	Description	Structure	Application
C1	TMS, trimethyl Reversed-phase material	$-Si-CH_3$	Ideal for polar and multifunctional compound separation
C2	RP-2, dimethyl Reversed-phase material	$-Si-C_2H_5$	Ideal for polar and multifunctional compound separation
C4	Butyl Reversed-phase material	$-Si-C_4H_9$	Ideal for HIC and ion-pairing chromatography of large proteins and macromolecules
C5	Pentyl Reversed-phase material	$-Si-C_5H_{11}$	Useful for hydrophobic proteins and oligonucleotide
C6	Hexyl Reversed-phase material	$-Si-C_6H_{13}$	Useful for ion-pairing chromatography
C8	MOS, RP-8, LC8, Octyl Reversed-phase material	$-Si-C_8H_{17}$	Useful for various types of nonpolar and moderately polar natural product isolation

C12	Dodecyl Reversed-phase material	$-Si-C_{12}H_{25}$	Useful for various types of nonpolar and moderately polar natural product isolation
C18	ODS, RP-18, LC18, octadecyl Reversed-phase material	$-Si-C_{18}H_{37}$	Most retentive for nonpolar compounds. Ideal for various types of moderately polar and polar natural product isolation
CN	CPS, PCN, cyano, cyanopropyl, Nitrile Reversed-phase or normal-phase material	$-Si-(CH_2)_3CN$	Unique selectivity for polar natural products in both reversed and normal-phase modes.
NH_2	Amino, amino propyl silyl (APS) Reversed-phase, normal-phase or weak ion-exchange material	$-Si-(CH_2)_3NH_2$	Ideal for carbohydrate separation
OH	Diol, Glycerol Reversed-phase or normal-phase material	$-Si-(CH_2)_3OCH_2CH(OH)CH_2OH$	Used for various medium polar and polar natural product separation
Phenyl	C_6H_5 Reversed-phase material	$-Si-(CH_2)_3Ph$	Useful for analyzing aromatic natural products
Phenyl ether	$C_6H_5(C_3H_6O$ linker) Reversed-phase material	$-Si-(CH_2)_3Ph$	Ideal for extremely polar aromatic compound separation

different chain lengths. The incorporated alkyl chain length could contain C_2, C_4, C_6, C_8, or C_{18}, although C_8 and C_{18} are prefered by most chromatographers. The use of monochlorosilanes ensures monomolecular surface coverage, which results in much greater reproducibility of packing material performance.

For the intermediate polarity (bonded normal phase), short chain or functionalized silanes are attached to the silica gel support. Currently available stationary phases include cyanopropyl (CN), nitro (NO_2), aminopropyl (NH_2), and diol. Both types of bonded phases are commonly available, although the reverse-phase class is much more prevalent in chromatography today. Despite both these phases being polar in nature, their applicability in systems where highly polar mobile phases are required quickly became evident. These intermediate polarity stationary phases have been used in areas incompatible with standard normal-phase chromatography. These include analyses of peptides, proteins, a wide range of pharmaceuticals, and natural products.

The uses of these types of stationary phases in LPLC are very limited because of the higher costs associated with bonded-phase silica phases. These phases are used more extensively in HPLC, although the high resolution and selectivity of the various types of stationary phases make them appropriate for use in LPLC. They are now available in short cartridge-type prepacked columns for sample isolation, using 5–20 g of packing material at reasonable prices from various manufacturers. Lots of information regarding various aspects of bonded-phase silica, e.g., properties, preparation, and applications, has been incorporated in a number of studies ***(14)***.

3.1.3. Alumina

Alumina is a porous polymer of aluminum oxide (Al_2O_3) and can be produced with an acidic, basic, or neutral surface based on the pH of the final wash of the synthetic absorbent. The reactivity of alumina can be optimized by varying the moisture content of the sample in a manner similar to the control of silica gel activity. The acidic alumina ($pH \approx 4.0$) is useful for the separation of carboxylic acids; the basic form ($pH \approx 10.0$) is used for basic compounds such as alkaloids, while the neutral one ($pH \approx 7.0$) is appropriate for the separation of nonpolar compounds such as steroids. Alumina is quite sensitive to the amount of water that is bound to it—the higher its water content, the less polar sites it has to bind organic compounds, and thus the less "sticky" it is. This stickiness or activity is designated as I, II, or III, with I being the most active. Alumina is usually purchased as activity I and deactivated with water before use according to specific procedures. The neutral form of activity II or III, 150 mesh (a surface area of 155 m^2/g), is most commonly employed. However, the use of alumina has decreased significantly in recent years owing to its ability to catalyze a variety of different reactions. Compounds that are susceptible to base are likely to be degraded upon chromatographic separation on alumina. For this reason, it is essential that small trial chromatographic separations are carried out to check for degradation of the samples.

3.1.4. Polystyrene

Styrene–divinylbenzene polymers are commonly used as the backbone for ion-exchange resins. However, in the absence of ionizing groups, the polymer can form a gel that can be used as adsorbent in low-pressure reversed–phase chromatography. The resin is a relatively large particle size bead (250–600 μm), which does not provide high-resolution chromatographic separations, but is useful for desalting and adsorption–elution of natural products from fermentation broths. The main advantages of polystyrene-based resins are: they are much less expensive than bonded-phase silica gel stationary phases, the potential problems caused by exposed silanols in silica-based materials can be avoided, and natural products, e.g., tannins, which give poor separation and recovery on silica-based columns, can be easily purified on polystyrene adsorbent. Chromatographic grades of polystyrene resins, e.g., MCI Gel CHP20p and SEPABEADS SP20ms, are available in particle sizes suitable for LPLC as well as for HPLC. Polystyrene resin or gel can also be used as the stationary phase in SEC for the purification of macromolecules. For example, Polyspher PST10 is a highly crosslinked, macroporous polystyrene–DVB copolymer with a mean particle diameter of 10 μm, and the pore size of 300 Å allows peptides, proteins, and other macromolecules to penetrate into the pore system.

3.2. Size-Exclusion Stationary Phases

3.2.1. Polyacrylamide

Copolymerization of acrylamide and *N,N′*-methylene-*bis*-acrylamide leads to the formation of porous polyacrylamide beads, Bio-Gel P gels, that can be used as a stationary phase in LPLC to carry out purification of macromolecular natural products, e.g., carbohydrates, peptides, and tannins. The particle sizes of the gels range from 45 to 180 μm. The gels are hydrophilic and essentially free of charge, and provide efficient gel filtration of labile natural products. They swell in water and are almost exclusively used with water as the mobile phase, although up to 20% alcohol can be used to improve the solubility of the sample. The type of Bio-Gel gels can be chosen on the basis of the molecular weight range of the molecules to be separated. For example, P-10 gel allows separation of molecules within the molecular weight range from 1500 to 20,000.

3.2.2. Carbohydrates

For the chromatography of labile natural products, one of the most commonly used materials is an inert polymer of carbohydrates (**Table 2**). Crosslinking of polysaccharides produces three-dimensional networks that can be converted to beads ideal for SEC. These highly specialized gel filtration and chromatographic media are composed of macroscopic beads synthetically derived from the polysaccharide, dextran. The organic chains are crosslinked to give a three-dimensional network having functional ionic groups attached by ether linkages to glucose units of the

Table 2
Commonly Used Sephadex G-series

Sephadex type	Bead size (μm)	Fractionation range (molecular weight)
G-10	40–120	≤700
G-15	40–120	≤1500
G-25	20–50	100–5000
	20–80	100–5000
	50–150	100–5000
	100–300	100–5000
G-50	20–50	500–10,000
	20–80	500–10,000
	50–150	500–10,000
	100–300	500–10,000
G-75	20–50	1000–50,000
	40–120	1000–50,000
G-100	20–50	1000–100,000
	40–120	1000–100,000
G-150	40–120	>100,000
G-200	40–120	>100,000

polysaccharide chains. Available forms include anion and cation exchangers, as well as gel filtration resins, with varying degrees of porosity; bead sizes fall in discrete ranges between 20 and 300 μm.

Sephadex® is prepared by crosslinking water-soluble dextran with epichlorohydrin. These gels swell in water. There are a variety of different carbohydrate gels available that are usually defined by the amount of solvent picked up by dry beads upon swelling. For example, Sephadex G-15 and Sephadex G-100 pick up 1.5 and 10 mL/g, respectively, of dry beads. There are gels available to separate compounds with masses ranging from 10 to 100,000 amu. For natural product purification, the most commonly used Sephadex media are highly crosslinked G-10 and G-15. There are also gels available that swell in organic solvents such as dimethylformamide, dimethylsulfoxide, ethylene glycol, and aqueous MeOH.

One of the most extensively used gels in natural product separation, especially nonpolar or intermediate polarity compounds, is Sephadex LH-20, a hydroxypropylated form of Sephadex G-25. The derivatization offers lipophilicity to the gel, at the same time preserving its hydrophilicity. As a result of the added lipophilicity, LH-20 gel swells sufficiently in organic solvents and allows handling of natural products that are soluble in organic solvents. The useful fractionation range of LH-20 is approx 100–4000 amu, and is particularly ideal for removal of chlorophyll from plant extracts. The gel filtration mode is operational when a single eluent is used. The partition mechanism comes into play when the eluent is a mixture of solvents (usually a mixture of polar and nonpolar solvents); the more polar of the solvents is taken up by the gel, resulting in a two-phase system with stationary and mobile phases of different compositions.

While the major mode of separation is associated with gel filtration, additional adsorption mechanisms, e.g., hydrogen bonding, also aid the separation process. Sephadex gels may also show unusual affinity toward certain types of compounds, e.g., phenolic and heteroaromatic compounds are more strongly retained than would otherwise be expected based on their molecular weights, especially when the eluent is a lower alcohol. Most of these polysaccharide gels are stable in eluting solvents, except for strong acids. However, polysaccharide columns are prone to microbial attack. Two most important advantages of these gels are: (1) they are fairly inert and do not usually adsorb compounds irreversibly, and (2) the column can be used for several experiments without the need for regeneration.

4. Column Operation

This section deals mainly with the operation of an LPLC system with consideration given to the choice of stationary phase and mobile phase, and a general guide for packing the column is also provided. A classical LPLC system is presented in **Fig. 3**. The column is packed with the stationary phase and the sample is applied to the top of the column. A solvent reservoir is located above the stationary phase and flowed through the column under gravity. Fractions are collected in the fraction collector after separation. The separation and purification of compounds from a crude extract becomes easier if the identity of the compounds is known. The properties of the compounds can also be helpful in the selection of a suitable stationary phase. In most cases of isolation of compounds from natural sources, the specific identity of components is rarely known. However, the polarity criteria and the results from the preliminary qualitative tests for various types of compounds present in the extract, e.g., alkaloids, flavonoids, and steroids, could be helpful.

4.1. Selection of Stationary Phase

The choice of stationary phase depends on the polarity of the sample. For highly polar compounds, ion exchange (*see* Chap. 6) or GPC is the preferred option. Where the expected compounds are related to certain known compound classes, published compound separation protocols available for their purification can be employed as a starting point. For samples where the polarity of compounds is not known, TLC can be used to determine suitable stationary phases, because TLC plates coated with most of the stationary phases, e.g., silica gel, and alumina, used in CC are now readily available. By using several types of TLC plates and running the samples with different solvent systems, an idea about the system required can be developed (**Table 3**). This can provide useful information regarding starting conditions for the separation stage and the required gradient to elute as many compounds as possible. Once the sample has been fractionated, other separation techniques can be used.

4.2. Column Packing and Equilibration

The column is usually packed by the chromatographer prior to use. However, it is now possible to purchase prepacked columns of different stationary phases and sizes. Column packing materials are usually discarded after use to avoid contamination of future samples. However,

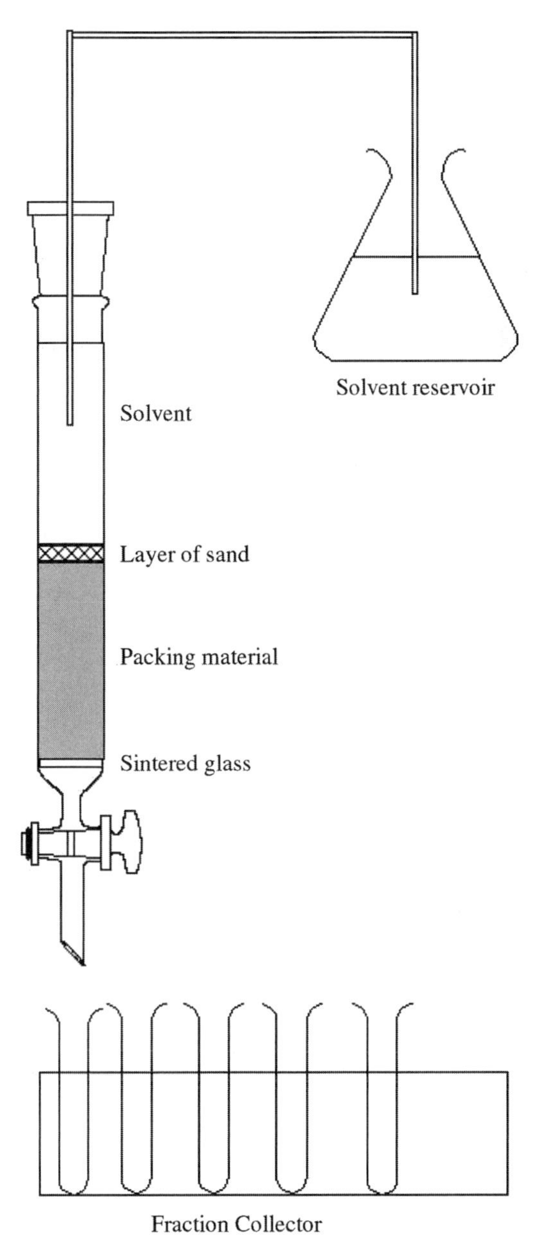

Fig. 3. Classical LC chromatography.

Table 3
Solvent System Development Using TLC

Step	Procedure
1	Prepare a solution of the crude extract or compound mixture, in a low boiling point organic solvent, at a concentration of at least 10 mg/mL.
2	Apply this solution (2–5 μL) to different TLC plates (2.5×10 cm), let the applied spots dry completely.
3	Use various mobile phases to develop each TLC plate.
4	Once developed, visualize the plates under UV lamp or by spraying with appropriate TLC reagents.
5	Compare the TLC plates and choose the solvent system that retains the compound(s) of interest at $R_f = 0.2$–0.3. In the case of crude extract, the mobile phase that puts the most mobile component (highest spot) at about $R_f = 0.5$ should be the initial solvent composition for the CC. The solvent system that keeps the least mobile (lowest spot) at about $R_f = 0.2$ can be used as the final eluent for CC.

for some gel filtrations, the packing material, e.g., Sephadex LH-20, can be washed thoroughly and used again. The stationary phase is normally introduced into the column, dry or in slurry, using a suitable solvent. The LPLC columns are normally made of thick-walled glass, which is resistant to most solvents, and can withstand the low-to-medium pressures used during column development. A glass frit is normally used to support the stationary phase. The alternative is to use a plug of glass wool covered with a layer of sand for this purpose (**Fig. 3**). The amount of stationary phase required depends on the amount of sample to be fractionated. The general guideline is to use 100–500 g of packing material per gram of crude sample.

4.2.1. Slurry Packing

This is the easiest and the most commonly used method for column packing. It is the only method used to pack columns that swell in the mobile phase such as carbohydrate packings (e.g., Sephadex G-10). To prepare the slurry, the required amount of stationary phase is taken in a beaker, a solvent is added, and the mixture is stirred. If necessary, more solvent can be added to achieve a pourable consistency of the slurry, which should neither be so thick that air bubbles are trapped in the column, nor so thin that it requires more than one pour to pack the column. For stationary phases that swell, e.g., Sephadex, sufficient time has to be allowed for the phase to become completely solvated. The slurry is poured into the column, which is kept partially open during pouring, and the solvent is allowed to flow through the column, leaving a packed bed of stationary phase. Once the packed bed is settled, the flow of solvent should be stopped leaving sufficient amounts of solvent on top of the packed bed, which is necessary, in most cases, to avoid cracking of the packed bed.

4.2.2. Dry Packing

Dry packing, if performed properly, is an efficient way to pack a column and is commonly used for regular or bonded silica gel. The dry stationary phase is poured into the column. It is essential at this stage that the column be vibrated in some way so that the packing is allowed to settle. Alternatively, the column can be tapped with a cork ring during the fill operation. The stationary phase is then "wetted" using an appropriate solvent by allowing solvent to flow through the column. The column can then be equilibrated with the mobile phase required for the sample. Dry packing is particularly useful for vacuum liquid chromatography (VLC) of nonpolar or intermediate polarity natural products where Silica Gel 60H is normally used as the adsorbent (**Fig. 4**). In VLC, however, vacuum is used to achieve a compact packing of the column.

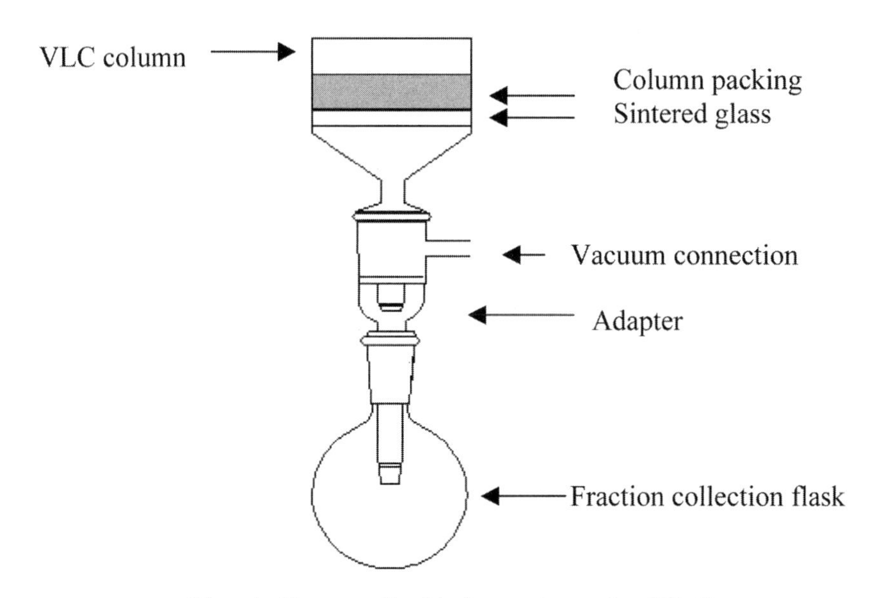

Fig. 4. Vacuum liquid chromatography (VLC).

4.3. Sample Application

The sample can be applied to the top of the column in a variety of ways depending on the stationary phase and the development method used. The sample is usually dissolved in a small amount of the initial mobile phase or a noneluting solvent and gently applied to the top of the column bed. The sample is allowed to flow by opening the exit valve. When the sample has been loaded, the mobile phase is carefully applied to the column to prevent disturbing the stationary phase bed. Sometimes, to avoid disintegration of the top of the column bed where the sample is adsorbed, a layer of sand (5–10 mm thick), a filter paper, or glass wool can be applied. If the sample is not soluble in the initial or noneluting solvent, then a sample can be loaded as dry free flowing powder. This is an option used for silica gel columns, where the starting mobile phase is fairly nonpolar. In this technique, the sample is dissolved in a small amount of appropriate solvent (DCM, EtOAc, or MeOH), and a weight of silica gel (factor of 10) is added to this solution. The solvent is removed under vacuum in a rotary evaporator leaving the sample adsorbed onto the silica gel. This dry silica gel containing the sample can be transferred to the top of the column bed and wetted with a little of the initial mobile phase to remove air bubbles.

4.4. Column Development

The column can be developed by elution of samples using various methods. The mobile phase can flow under gravity, by applying a nitrogen pressure at the inlet, a vacuum at the outlet (e.g., VLC), or pumping the mobile phase through the column at varying pressures (e.g., flash chromatography, FC). In all these cases, a solvent gradient needs to be applied, although isocratic solvents are often the preferred option.

4.4.1. Gradient Formation

A step gradient is often the method of choice in LPLC because of the simplicity and the quality of separation. If the composition of solvents in a step gradient is chosen properly according to the need of changing polarity, excellent fractionation of natural compounds can be achieved. This is unlike ion-exchange chromatography, where step gradients are not desirable. With modern HPLC gradient elution systems, complex gradients can be programmed. In general, for LPLC, one to three column volumes of each solvent step are required. Step gradients are generated by simply preparing a range of mobile phases composed of polar/nonpolar solvents of varying ratios. During the column operation, the column inlet reservoir is refilled with the new solvent. For any finer gradient elution, a gradient maker can be used.

4.4.2. Gravity

Generally, good results can be achieved with gravity elution where the particle size is greater than 60 μm. Smaller particle sizes result in back pressure, which does not allow the eluent to pass through the column at a desired flow rate. Gravity elution is easy to run where the mobile phase is poured on top of the open column and allowed to flow naturally under gravity. A solvent reservoir can be used to increase capacity, and the flow rate can be controlled by adjusting the outlet valve (**Fig. 3**).

4.4.3. Pressure

Positive pressure can be applied to the top of the column to accelerate the flow rate and achieve better resolution in LPLC (**Fig. 5**). This technique is called FC and it uses particle sizes in the 40–60 μm range. An accurate flow rate can be achieved by using a needle release valve. Glass columns used in FC must be of appropriate wall thickness and strong enough to withstand the pressure. It is advisable to use plastic mesh netting as a column jacket or simply to tape the outside of the column to avoid any danger associated with column explosion. Nowadays, metallic columns, empty or prepacked, especially designed for FC, can be purchased commercially. Biotage® flash chromatographic systems of various sizes have been found to be useful for initial fractionation of nonpolar and medium polarity natural products ***(15–22)***.

4.4.4. Vacuum

An alternative to applying pressure at the top of the column is to apply vacuum at the end of the column. This technique is called VLC (**Fig. 4**). The operation is similar but it is more difficult to control the mobile phase flow. However, this technique is safer than FC. A common use of this technique is the rapid purification of a specific compound from a sample, especially a reaction mixture. In natural product isolation, this technique is applied for initial fractionation of crude nonpolar or intermediate polarity extracts ***(23–33)***. The sample is applied to the adsorbent in a sintered glass and the mixture eluted with a mobile phase directly into a vacuum flask. Generally, TLC grade silica without any binder in it (Silica Gel 60H) is used to dry pack the column.

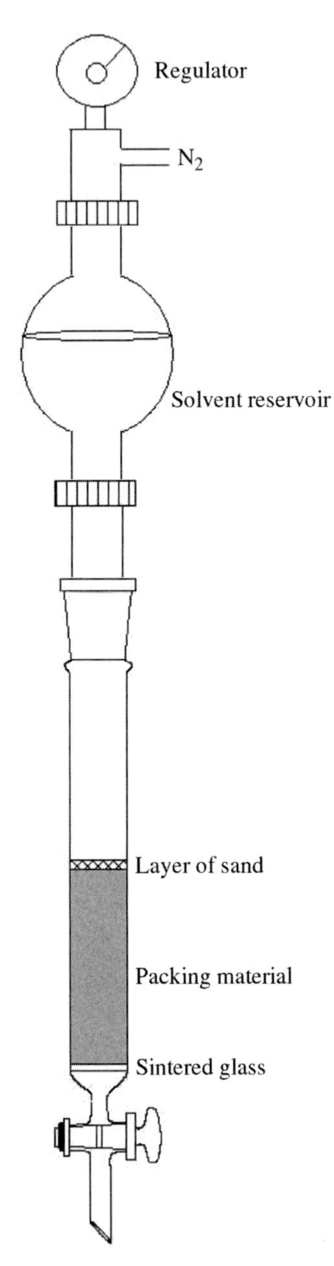

Fig. 5. Flash chromatography.

For the pre-HPLC fractionation of polar extracts, e.g., MeOH extract, solid-phase extraction cartridges (10 g) prepacked with reversed-phase silica C_{18} (ODS) have been found to be useful ***(34–54)***. These cartridges can be placed on the ports of a vacuum manifold or connected to a vacuum flask, similar to VLC, with appropriate adapter (**Fig. 6**). Cartridges are available in different sizes and also in various packing materials, e.g., diol, C_8, C_6, or ion-exchange resins. As the particle sizes are fine and the cartridges are packed mechanically, the column bed is compact and offers excellent separation. Both isocratic and step gradients can be used.

4.4.5. Pumped

A pump is a more controllable solvent delivery system. It can deliver a smooth and constant flow of solvent (**Fig. 7**). The pump must be inert and designed for use with flammable solvents. Differences in low-, medium-, and high-pressure chromatography can be made essentially on the basis of the particle size of the stationary phase and resulting operating pressure of the packed column. The LPLC is run with 40–200-μm particles at a flow rate that generates no pressure significantly greater than the atmospheric pressure. The medium pressure liquid chromatography (MPLC) uses 25–40-μm particles with pressures between 75 and 600 psi, and the HPLC (*see* Chap. 8) is used with 3–12-μm particles with pressures between 500 and 3000 psi. The resolution of separation achieved by these chromatographic techniques follows the order: HPLC > MPLC > LPLC. The run time is also reduced considerably in the following order: LPLC > MPLC > HPLC. With the greater access and availability of various HPLC systems, the use of MPLC has reduced significantly.

4.5. Detection

During the isolation of natural products, individual fractions can be analyzed for chemical profiling by either TLC or HPLC (using an autosampler). Small samples can be spotted on a TLC plate, and after development a guideline regarding the performance of the separation can be obtained. A more advanced method involves the use of a UV or refractive index detector prior to fraction collection. A variable wavelength UV detector allows different wavelengths to be monitored for the duration of the chromatography. It should be noted that for such large columns, the cell path length must be reduced to 0.5 mm, to decrease the absorbance values expected from the analysis. For compounds with little or no UV,

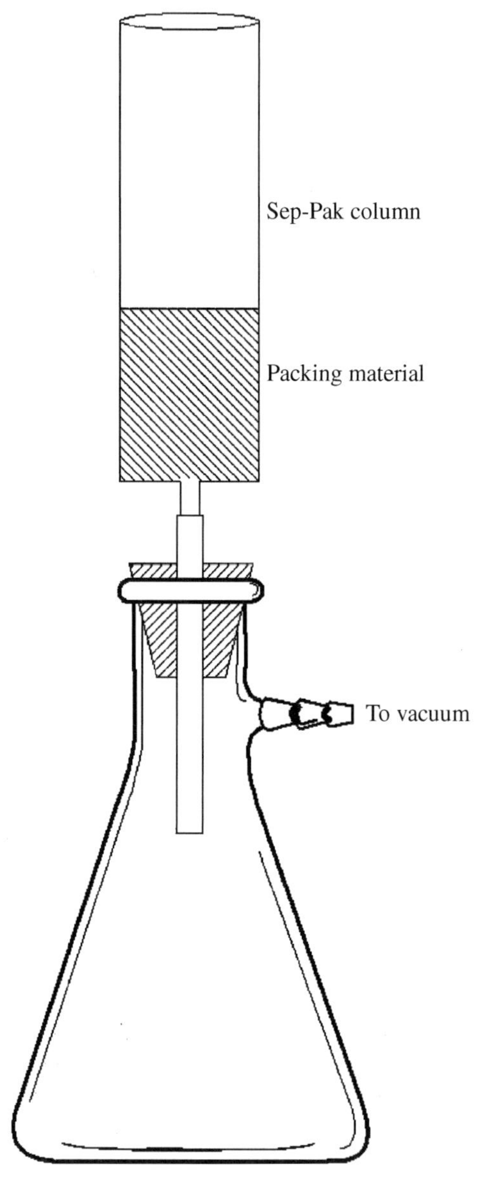

Fig. 6. Solid-phase extraction (Sep-Pak).

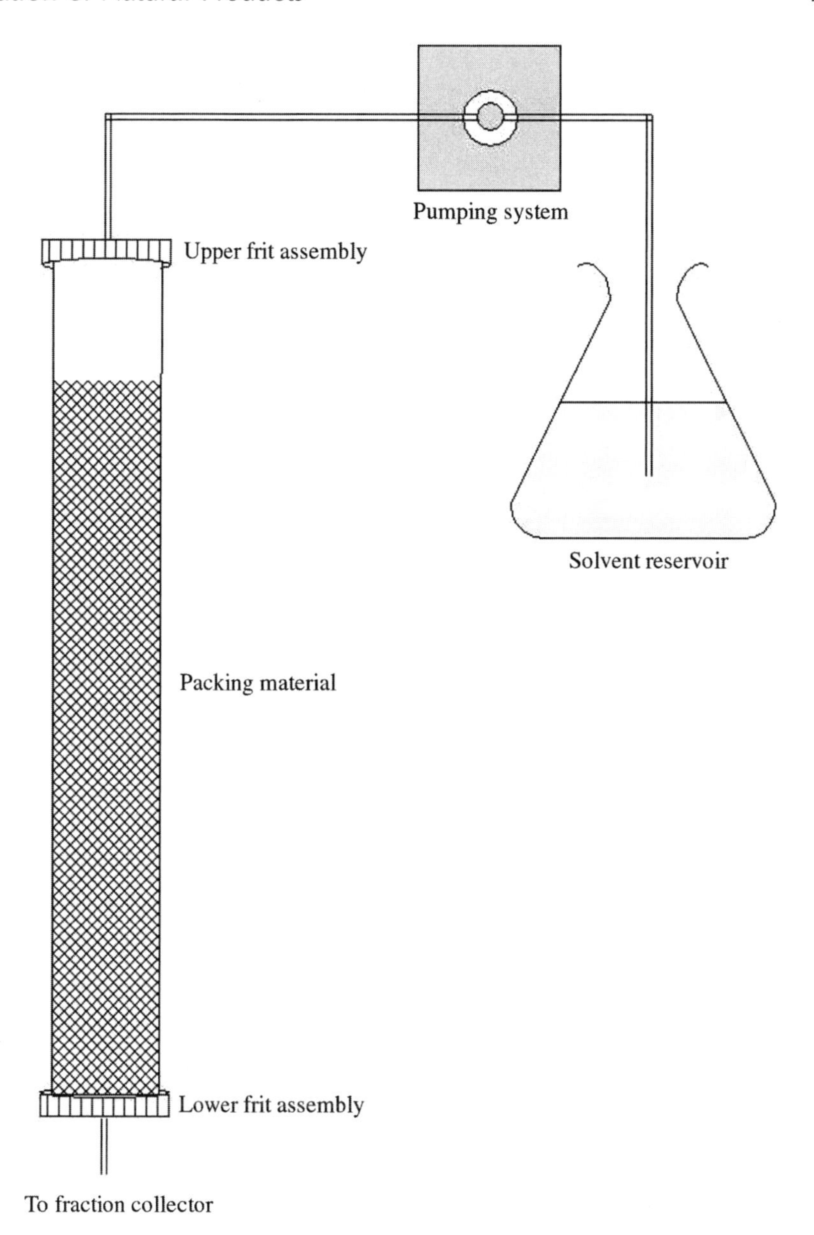

Fig. 7. Application of pump in LPLC.

absorption refractive index detectors can be useful. However, for bioassay-guided isolation of bioactive natural products, a robust and rapid high-throughput assay is the assay of choice for identifying the fraction containing bioactive compounds. Depending on the assay capabilities, optimization of the number of fractions to be submitted for bioassay is necessary. If the LPLC of the extract produces a large number of fractions, in the first round, representative fractions, for example, one in each five or ten fractions, can be subjected to bioassay. In the second round, all five or ten fractions from the active composite need to be assayed separately to pinpoint the active fraction(s). In this way, the number of bioassays required can be reduced significantly.

5. Outline of Generic Procedures for Adsorption and SEC

5.1. Adsorption

5.1.1. Silica Gel Chromatography

Column Preparation

1. Select a suitable heavy wall glass column, which is about three times the volume of the silica gel required for the separation.
2. Use a ratio of 100 g of silica gel/g of crude sample for relatively easy separations. For more difficult separations, a ratio of up to 500 g of silica gel/g crude material is required. A bed height of 20–30 cm is suitable with 40–60 cm head space available to hold the mobile phase.
3. The bottom of the column should have either a sintered glass frit or a plug of glass wool to support the stationary phase (**Fig. 3**). While vibrating the column (or tapping), carefully pour the stationary phase in and open the outlet. Allow the stationary phase to settle while the liquid is flowing. Continue adding stationary phase until the bed is complete.

Column Equilibration

1. Fill the column with the initial mobile phase, taking care to ensure that the bed is not disturbed.
2. Open the column outlet to allow the solvent to flow naturally under gravity.
3. Continue the equilibration stage until the stationary phase bed has a uniform appearance, i.e., no visible dry areas or air pockets.
4. Reduce the solvent level to just above the stationary phase. Stop the flow by closing the outlet valve.

Sample Application

1. Dissolve the sample in a minimum of the mobile phase. The volume (in milliliters) should not exceed $0.4 \times D^2$ (where D is the column diameter in centimeters).
2. Using a Pasteur pipet carefully apply the sample solution, taking care not to disturb the bed.
3. Open the outlet valve, and allow the sample to flow into the bed. When the sample has been adsorbed completely, the glass column sides can be washed carefully with a little amount of the mobile phase and then allowed to flow again. Care must be taken to ensure the bed is not disturbed and is not allowed to become dry.
4. This stage can be repeated to ensure complete adsorption of the sample on the bed.
5. Fill the column with the initial mobile phase required for the separation.

Sample Elution and Fraction Collection

1. Allow the initial mobile phase to flow.
2. Collect the fractions. An automated sample fraction collector can be used or fractions can be collected manually. For complex separations, small fractions can be collected (for a 100-mL column, fractions of 5–10 mL are ideal). For crude separations, fractions of larger volumes can be collected for a particular solvent polarity.
3. A different polarity mobile phase is now loaded and samples collected as above.
4. Repeat as required for solvents of different polarity.
5. For a nonpolar extract, e.g., *n*-hexane extract, a typical step gradient sequence might be as follows: initial composition of 100% *n*-hexane, followed by at least 1 column volume each of 5% EtOAc in *n*-hexane, 10–100% EtOAc in *n*-hexane (increment of 10% in each step), and 5–20% MeOH in EtOAc (increment of 5% in each step).

5.1.2. Reversed-Phase Silica Gel Chromatography (Solid-Phase Extraction)

Column Preparation

1. Solid-phase extraction cartridges prepacked with suitable stationary phase can be purchased commercially. The most extensively used cartridge in natural product separation is 10 g reversed-phase C_{18} silica cartridge. Connect the cartridge with the port of a vacuum manifold or a vacuum flask (similar to VLC), and start the pump (water pump or mechanical pump).
2. To clean and wet the cartridge, run 50 mL of MeOH or ACN, as appropriate, through the column under vacuum.

Column Equilibration

1. Run at least 100 mL of water through the column under vacuum.

Sample Application

1. Dissolve the sample in the initial mobile phase (e.g., 20% or 30% MeOH or ACN in water). The sample volume should not exceed 10 mL.
2. Place a small filter paper on top of the column to avoid any clogging.
3. Apply sample solution from the top end of the column and allow the solution to flow under vacuum.

Sample Elution and Fraction Collection

1. Use a step gradient for fractionation of crude polar extract, e.g., MeOH. The step gradient is generally composed of MeOH–water or ACN–water of various ratios. The volume of eluent for each fraction depends on the amount of sample mixture to be fractionated. Usually, the volume of each step is 200–250 mL for the fractionation of maximum 2 g of crude sample. Apply the eluent from the top end of the column and let it run under vacuum.
2. Once the total volume of eluent from one step has been run through the column, the next step of a different polarity mobile phase composition is loaded.
3. Collect the fractions (200–250 mL) obtained from running each step. If necessary, fractions of smaller volume can also be collected.
4. For a MeOH extract, a typical step gradient sequence might be as follows: 10%, 30%, 60%, 80%, and 100% MeOH in water, 200–250 mL.

5.2. Size-Exclusion Chromatography (SEC)

Column Preparation

1. Choose a glass column (preferably with a sintered glass frit) with a diameter between 15 and 25 mm and length between 40 and 100 cm. The size of the column depends on the required bed volume. In practice, 1 mL of swollen gel is required for the separation of 1 mg of extract.
2. Swell the gel (Sephadex LH-20) overnight by suspending the stationary phase in methanol. Use sufficient MeOH to get slurry with a pourable consistency. Allow enough time to ensure the gel is completely swollen.
3. The packing process should be completed in one complete pouring session. Partially open the outlet valve on the column and slowly fill the column with the gel. It is important to have some form of mechanical vibration of the column during the packing to ensure a good column.

4. More nonpolar solvents like chloroform ($CHCl_3$) or DCM can also be used if necessary, especially when dealing with nonpolar or medium polarity extracts.

Column Equilibration

1. Adjust the flow rate to between 1 and 5 mL/min to allow the packing bed to settle. This may take a few hours.
2. On completion, allow the MeOH level to drop to a level just above the packing material.

Sample Application

1. Dissolve the sample in MeOH, ensuring that the sample volume is not more than 5% of the column volume. The sample solution viscosity should be similar to that of the mobile phase.
2. Using a Pasteur pipet, carefully apply the sample solution taking care not to disturb the bed.
3. Open the outlet valve and allow the sample to flow into the bed. Care must be taken to ensure the bed is not disturbed and is not allowed to become dry.
4. Add another 10 mL of MeOH to the bed and flow through so that the sample solution has been completely washed into the bed.

Sample Elution and Collection

1. Fill the column with MeOH and attach a solvent reservoir or pumping system.
2. Open the column outlet and adjust the flow rate to 2–4 mL/min. The flow rate should be set to give a contact time of approx 50–100 min.
3. Collect the fractions, normally about 5 mL per fraction.

6. Practical Examples

Practical examples of the use of various types of LPLC are available in the chapters dedicated to the isolation of natural products from plant, microbes, and marine sources (*see* Chaps. 13–15). However, a few typical examples are presented here for better understanding of this chapter.

6.1. Adsorption Chromatography

To date, literally thousands of examples of the application of LPLC for the isolation of various types of natural products are available in the literature. The uses of some of the different stationary phases are summarized here. However, isolation of natural products, almost in all cases, requires a combination of various separation techniques.

6.1.1. Alumina

This stationary phase has long been used for the isolation of alkaloids from various plant sources. One of the typical examples of the application of alumina is the isolation of diterpenoid alkaloids from the aerial parts of *Delphinium staphisagria* (*55*). The plant material was dried (2.9 kg) and Soxhlet extracted with 80% ethanol (EtOH). After solvent removal under vacuum, the extract was acidified and filtered. The acidic solution was extracted with $CHCl_3$ and the solvent was removed. The $CHCl_3$ extract was adsorbed on neutral alumina and subjected to FC using the same adsorbent. Elution with *n*-hexane, *n*-hexane–EtOAc (1:1), and MeOH afforded two alkaloids, delphinine (2.6 g) and atisinium chloride (1.3 g) (**Fig. 8**). The acidic solution was neutralized and extracted with $CHCl_3$. The sample was chromatographed on alumina using gradient elution with *n*-hexane–EtOAc mixtures of various proportions, resulting in the isolation of delphinine (17 mg), and four more alkaloids, bullatine C (1.3 g), chasmanine (300 mg), 14-acetylchasmanine (450 mg), and neoline (525 mg). The neutral solution was basified using sodium hydroxide and extracted with $CHCl_3$. Alumina CC of this extract, followed by purification using Sephadex LH-20, afforded three more alkaloids, 19-oxodihydroatisine (93 mg), 22-*O*-acetyl-19-oxodihydroatisine (5 mg), and dihydroatisine (76 mg).

Similar isolation protocols using alumina columns were applied for the isolation of alkaloids from *Consolida oliveriana* (*56*). After extraction of 3.9 kg of dried plants with 80% EtOH and solvent removal, the acidified extract was subjected to a pH gradient, and the aqueous phases were extracted with DCM to obtain a neutral residue and a basic residue. The neutral one was subjected to alumina column eluting with EtOAc and EtOAc–MeOH (9:1), resulting in 57 fractions. The EtOAc fractions were individually subjected to further CC over alumina, yielding six diterpenoid alkaloids, olividine (1.6 mg), olivimine (12 mg), consolidine (60 mg), delphatine (7 mg), delsoline (10 mg), and pubescenine (5 mg). The EtOAc–MeOH fractions were pooled together and chromatographed on alumina using a similar mobile phase resulting in the isolation of a further 11 alkaloids (56 mg in total). The basic residue was analyzed by Sephadex LH-20 using an eluent of *n*-hexane–DCM–MeOH mixture (4:3:3). The residue obtained from this was subjected to CC (alumina) yielding a further five alkaloids (323 mg). These two typical examples demonstrate that alumina alone, or in conjunction with other separation modes, is still the method of choice for the isolation of plant-derived alkaloids.

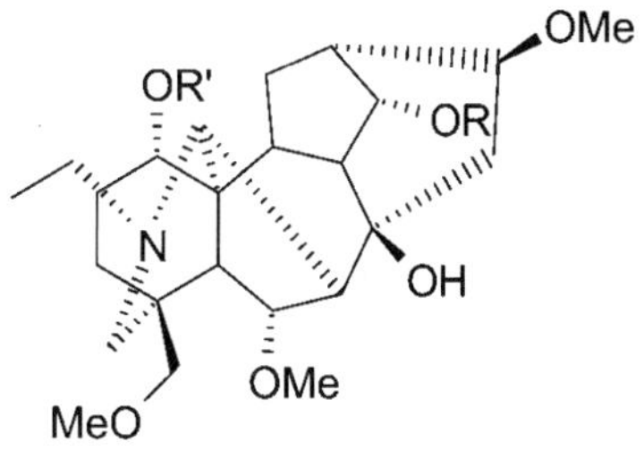

Neolin	R = H	R' = H
Bullatine C	R = Ac	R' = H
Chasmanine	R = H	R' = Me
14-Acetylchasmanine	R =Ac	R' = Me

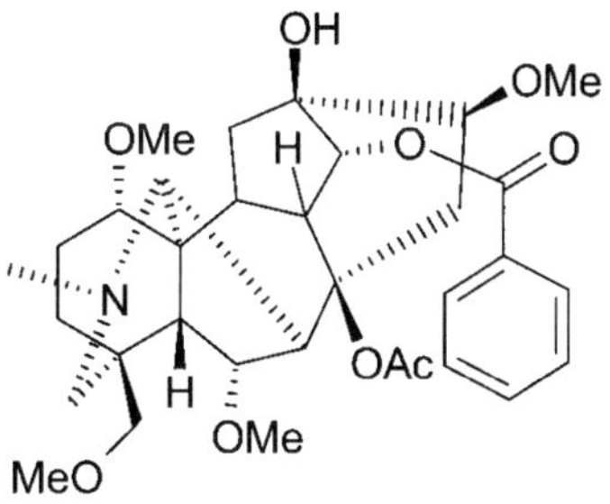

Delphinine

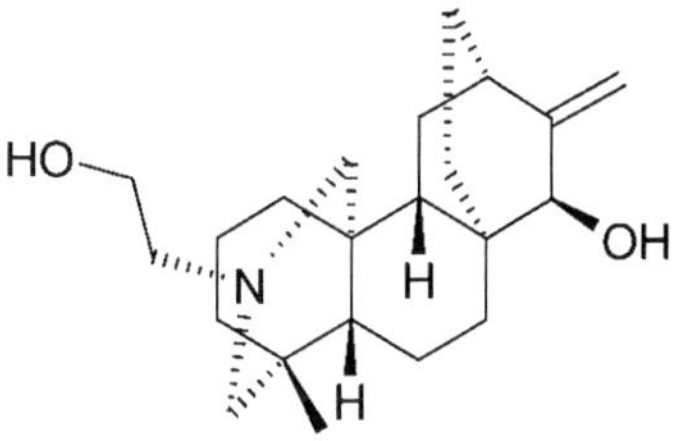

Dihydroatisine

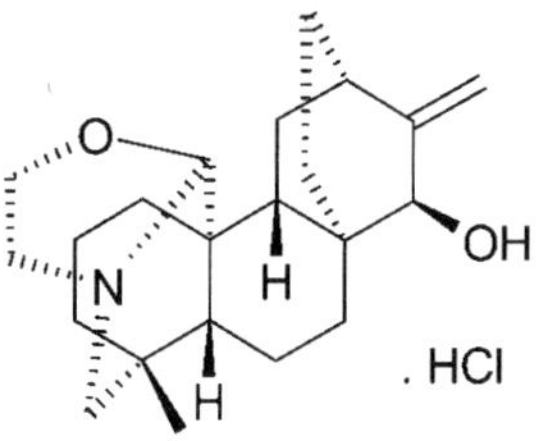

Atisinium chloride

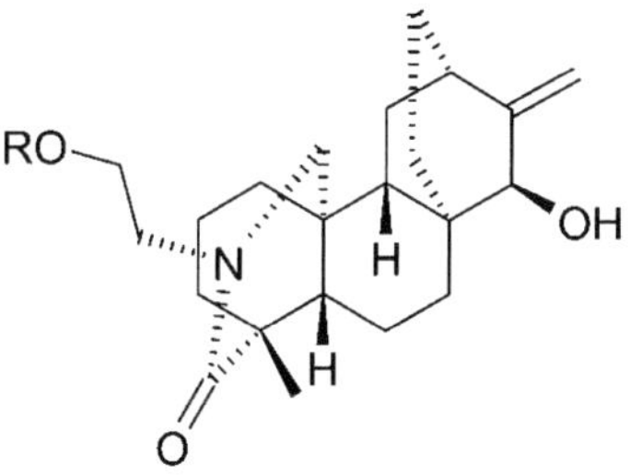

19-Oxodihydroatisine	R = H
22-*O*-Acetyl-19-oxodihydroatisine	R = Ac

Fig. 8. Structures of diterpene alkaloids isolated from *Delphinium staphisagria* using alumina as adsorbent.

6.1.2. Silica Gel

In LPLC, silica gel is the most extensively used adsorbent for the separation of various types of natural products, particularly nonpolar and intermediate polarity compounds. Avicennin (40 mg), a prenylated angular pyranocoumarin, was isolated from a portion (6 g) of the *n*-hexane extract of the aerial parts of *Eriostemon apiculatus* (0.7 kg) by a combination of CC (silica gel, mobile phase: *n*-hexane containing increasing amounts of EtOAc) and preparative thin layer chromatography (PTLC) (**Fig. 9**). ***(33)***. Guaiene-type sesquiterpenes, (–)-guaiol and (–)-1,12-oxaguai-10(15)-ene, were isolated from *E. fitzgeraldii* ***(24)*** (**Fig. 10**) using silica column. Powdered aerial parts (470 g) were Soxhlet extracted with, successively, *n*-hexane, $CHCl_3$, and MeOH. The concentrated *n*-hexane extract (10.2 g) was subjected to VLC (Silica gel 60H) eluting with solvents with increasing polarity starting from 100% gasoline via 100% EtOAc to 100% MeOH. The VLC fraction eluted with 20% EtOAc in gasoline was subjected to CC (silica gel, mobile phase: 100% $CHCl_3$, 4-mL fractions were collected) followed by PTLC on silica gel (solvent system: $CHCl_3$: EtOAc = 9:1) to obtain (–)-guaiol (183 mg). The concentrated VLC fraction eluted with 50% EtOAc in gasoline was subjected to Sephadex LH-20 filtration, eluting with 0.5% MeOH in $CHCl_3$, and fractions of 4 mL were collected. (–)-1,12-Oxaguai-10(15)-ene was isolated from the

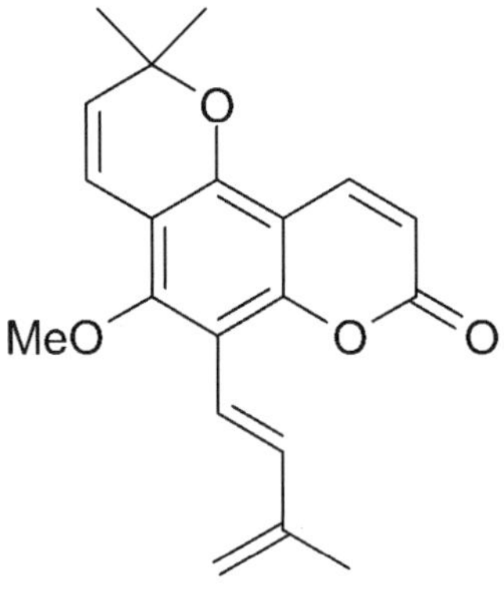

MeO O O O

Avicenin

Fig. 9. Structure of avicenin isolated from *E. apiculatus* by silica gel column chromatography.

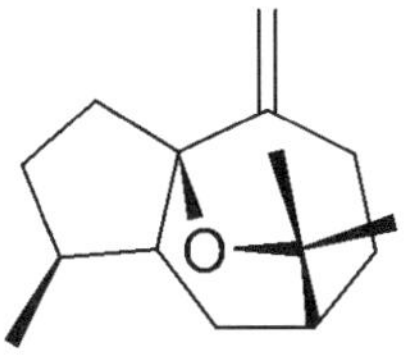

Fig. 10. Structure of (–)-1,12-oxaguai-10(15)-ene isolated from the aerial parts of *E. fitzgeraldii*.

combined Sephadex fractions ***(10–16)*** by CC followed by PTLC. The column eluting conditions were similar to those used for (–)-guaiol.

LPLC has been applied successfully for the isolation of sapogenin as well as saponin-type polar compounds from plants. The dry roots of *Cimicifuga foetida* (4.5 kg) ***(57)*** were extracted with 90% EtOH at low temperatures (<65°C). The EtOH was evaporated and the residue was mixed with water and filtered. The water-insoluble material (50 g) was dissolved in DCM and chromatographed using silica gel. The sample was eluted using DCM and increasing concentrations of MeOH in DCM, which yielded seven fractions. Fraction 3 was subjected to CC (silica gel) and eluted with a mixture of cyclohexane–acetone giving two further fractions. Further CC (silica gel) of subfraction 1 using $CHCl_3$–MeOH as eluent, provided 2′-*O*-acetylactein (10 mg) (**Fig. 11**) Subfraction 2 was also subjected to the same chromatography and produced 2′-*O*-acetyl-27-deoxyactein (12 mg). Fraction 4, after chromatography on silica gel with cyclohexane–acetone as mobile phase, yielded actein (1.2 g) and 27-deoxyactein (1.5 g). Fraction 5, after silica gel CC with $CHCl_3$–MeOH as mobile phase, yielded cimicifugoside (1 g) and 15α-hydroxycimicidol 3-*O*-β-D-xyloside (160 mg). Numerous examples of the application of silica gel LPLC in the isolation of saponins can be found in many published literature. Silica gel LPLC was employed to isolate oleanene triterpenoid saponins from *Madhuca longifolia* ***(58)*** and *Albizia myriophylla* ***(59)***. Application of this adsorbent in the isolation of other classes of compounds is also evident in the isolation of phenolics from *Polygala fallax* ***(60)*** and sesquiterpene pyridine alkaloids from *Maytenus chiapensis* ***(61)***.

Epicoccamide (**Fig. 12**), an antibacterial agent, has been isolated from a jellyfish-derived culture of *Epicoccum purpurascens* by VLC ***(62)***. The

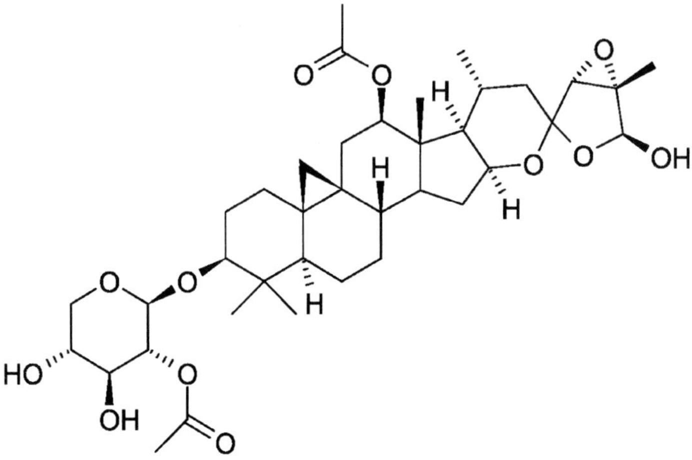

2′-*O*-Acetylactein

Fig. 11. Structure of the saponin, 2′-*O*-acetylactein from the roots of *C. foetida*.

solid medium and fungal mycelium were diluted with water, and blended and extracted with 13.5 L of EtOAc. The resultant EtOAc extract (3.9 g) was subjected to VLC eluting with a gradient of DCM–EtOAc–MeOH to yield 13 fractions of 250 mL each. The reversed-phase HPLC (Eurospher 100 C_{18} semiprep column, eluted with 85% MeOH in water)

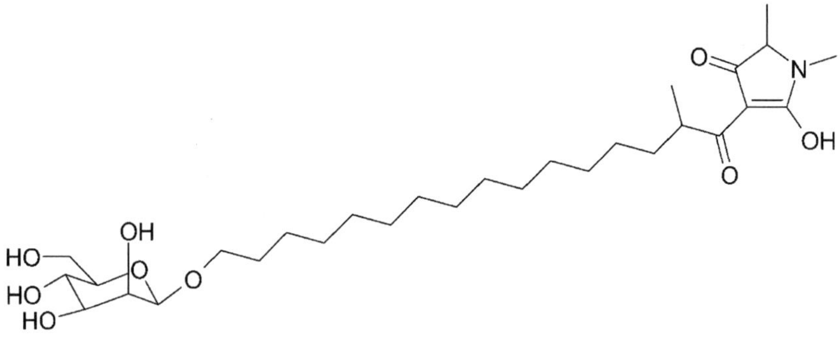

Epicoccamide

Fig. 12. Structure of epicoccamide isolated from *E. purpurascens* by VLC and HPLC.

of the VLC fraction 11 (2.2 g, eluted with EtOAc:MeOH = 40:60) afforded epicoccamide. These are just a few examples of the application of silica gel LPLC in combination with other forms of chromatography and stationary phases in the isolation of natural products of many classes.

6.1.3. Bonded-Phase Silica

With the increasing number of stationary phases available, the use of bonded-phase silica in LPLC for the isolation of natural products has increased remarkably. Solid-phase extraction offers initial fractionation of the crude extract. Cartridges packed with C_{18} reversed-phase silica are useful for pre-HPLC fractionation of various polar extracts, generally using a MeOH–water of ACN–water step gradient as eluent.

The isolation of two new alkaloids from the MeOH extract of the aerial parts of *Glechoma hederaceae* using a combination of solid-phase extraction on a reversed-phase C_{18} silica cartridge, and preparative HPLC on a reversed-phase C_{18} column has recently been reported ***(37)***. The dried plant material (135 g) was Soxhlet extracted successively with *n*-hexane, DCM, and MeOH. The MeOH extract was concentrated using a rotary evaporator. The extract was fractionated on Waters Sep-Pak Vac (10 g) C_{18} cartridge, using step gradients of MeOH–water mixtures (10–100% MeOH). The preparative HPLC of the Sep-Pak fraction eluted with 40% MeOH in water yielded two new alkaloids, hederacine A (3.8 mg) and hederacine B (3.2 mg) (**Fig. 13**). The use of C_{18} bonded-phase silica in LPLC for the pre-HPLC fractionation of MeOH extract is evident in a number of other recent publications. The isolation protocols for dibenzylbutyrolactone lignans from the seeds of *Centaurea scabiosa* ***(63)***, dibenzylbutyrolactone lignans ***(64)*** and serotonin conjugates ***(65)*** from *C. nigra*, and flavonol glycosides from seeds of *Agrimonia eupatoria* ***(66)*** and *Alliaria petiolata* ***(36)*** demonstrate the successful application of solid-phase extraction method in natural product isolation.

The use of other bonded stationary phases in LPLC has also been demonstrated in various publications. A C_8 adsorbent was employed in the isolation of spirostanol and furostanol glycosides from tubers of *Polianthes tuberosa* ***(67)***. After chromatography with the C_8 stationary phase, five components were isolated. A cyano-bonded stationary phase was employed in the isolation of two xanthones from *Maclura tinctoria* ***(68)***. The use of ion-exchange stationary phases has also been reported, although their use is not as widespread as C_{18} bonded phase. In the isolation of alkaloids from *Thalictrum wangii* ***(69)***, an anion-exchange stationary phase was employed.

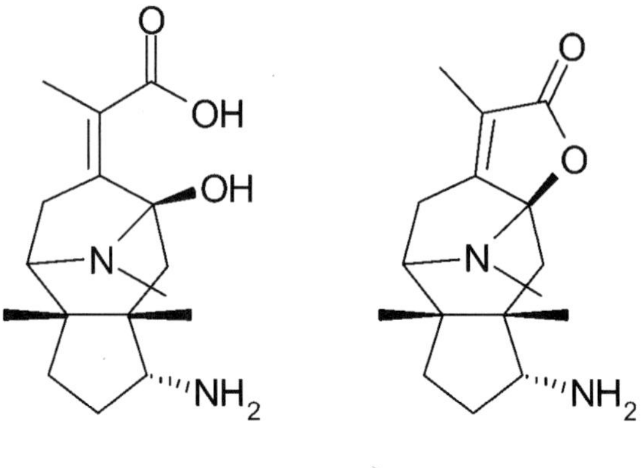

Hederacine A Hederacine B

Fig. 13. Structure of hedaracine A and hederacine B isolated from *G. hederaceae* by solid-phase extraction and HPLC.

The use of ion-exchange LPLC for large-scale separation of proteins has also been reported recently ***(70)***.

6.2. Size-Exclusion Chromatography

Many reports on the use of SEC and GPC in the isolation of natural products are available in the literature. Most of them are used in conjunction with adsorption chromatography to isolate or purify compounds of interest, or to remove high molecular weight unwanted compounds, e.g., chlorophyll. However, the use of SEC is mainly restricted to the isolation and purification of proteins, peptides, and tannins. In the published protocols for the isolation of small natural compounds, Sephadex LH-20 is the most extensively used stationary phase. It has also been used routinely for the removal of chlorophyll from nonpolar or medium polarity natural product extracts.

In the recovery of β-phycoerythrin, a light-harvesting pigment and cyanobacteria used as a fluorescent probe, from the red microalga *Porphyridium cruentum,* GPC with a cellulose column was used ***(71)***. After complex sample preparation, the phycobiliproteins were applied to a 15.0 × 9.0-cm column containing DEAE-cellulose, previously conditioned with 50-mM acetate buffer (pH 5.5). The sample was washed on the column with 800 mL of the acetate buffer. The sample elution was achieved by, first, applying 800 mL of 0.25 M acetic acid–sodium acetate buffer (pH 5.5), and finally, 1000 mL of 0.35-M acetic acid–sodium acetate buffer (pH 5.5). The eluate was collected in 25 mL fractions. The colored fractions were saturated with 65% ammonium sulfate and allowed to stand overnight. Samples were centrifuged, and the pellets resuspended in pH 7.0 sodium phosphate buffer and freeze-dried. Another example of SEC is the fractionation and characterization of the gum exuded from *Prosopis laevigata* trees ***(72)***. The gums were separated on a phenyl-Sepharose CL-4B and eluted with different concentrations of sodium chloride and water. Fractions were collected, dialyzed against de-ionized water, and freeze-dried.

The use of Sephadex LH-20 in the removal of chlorophyll is evident in the protocol described for the isolation of umbelliferone from the $CHCl_3$ extract of *E. apiculatus* ***(33)***. A portion of the $CHCl_3$ extract (3.6 g) was subjected to Sephadex LH-20 column eluted with 1% MeOH in $CHCl_3$ to remove chlorophyll, and the chlorophyll-free extract was analyzed by chromatotron and PTLC to obtain umbelliferone (6.3 mg). Sephadex LH-20 LPLC was used successfully for the isolation of anthocyanins from blackcurrant fruits ***(73)***, and the quantification of procyanidins in cocoa and chocolate samples ***(74)***. These separations used MeOH as the eluting solvent. The isolation of the antibiotic rachelmycin from the EtOAc crude extract of a fermentation broth was achieved by Sephadex LH-20 CC

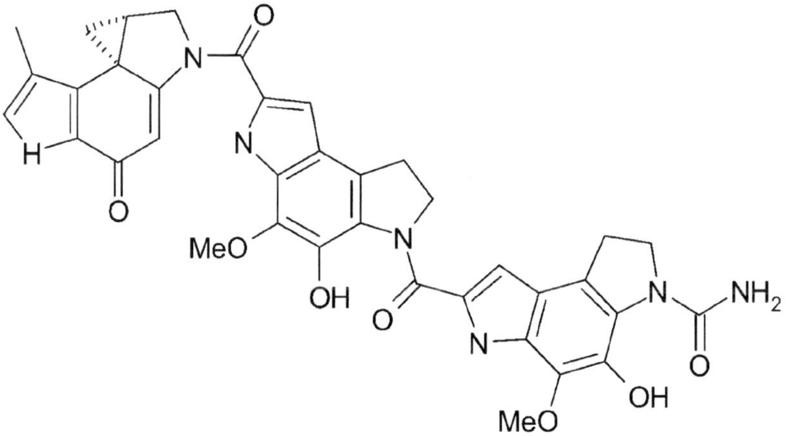

Fig. 14. Structure of rachelmycin, an antibiotic, isolated from a fermentation broth, using Sephadex LH20 chromatography.

(75). The EtOAc extract was dried over anhydrous sodium sulfate and evaporated to dryness to obtain an oily residue (380 mg), which was dissolved in MeOH (10 mL) and applied to a Sephadex LH-20 column (400 mL bed volume) equilibrated with MeOH. The elution was carried out with MeOH, at a flow rate of 20 mL/min, and a total of 160 fractions of 20 mL were collected. The fractions were subjected to bioassay, which revealed that fractions 93–135 contained the active compound. These fractions were combined and evaporated to dryness to yield 3.2 mg residue, which was mainly composed of the known antibiotic rachelmycin (**Fig. 14**) with a few other small impurities.

References

1. Brown, P. R. (1990) High performance liquid chromatography—past developments, present status, and future trends. *Anal. Chem.* **62,** A995–A1014.
2. Warner, M. (1990) Pioneers in gas chromatography. *Anal. Chem.* **62,** A1015–A1017.
3. Issaq, H. J. (2001) (ed.) *A Century of Separation Science*. Marcel Dekker, New York.
4. Ettre, L. S. (2001) (ed.) *Milestones in the Evolution of Chromatography*. ChromSource Inc., Portland.
5. Gehrke, C. W. (2001) (ed.) Chromatography—a century of discovery 1900–2000, in *Journal of Chromatography Series*. vol. 64, Elsevier, Amsterdam.
6. Brown, P. and Hartwick, R. A. (1989) *High Performance Liquid Chromatography* Wiley, New York.
7. de Neue, U. (1997) *HPLC Columns: Theory, Technology and Practice*. Wiley, New York.
8. Robards, K., Haddad, P., and Jackson, P. (1994). *Principles and Practice of Modern Chromatographic Methods.* Academic, Elsevier, Amsterdam.
9. Snyder, L. R., Kirkland, J. J., and Glajch, J. L. (1997). *Practical HPLC Method Development.* Wiley, New York.
10. Wu, C.-S. (2003) *Handbook of Size Exclusion Chromatography and Related Techniques.* Marcel Dekker, New York.
11. Braithwaite, A. and Smith, F. J. (1995) *Chromatographic Methods.* 5 ed., Kluwer Academic Publishers.
12. Kazakevich, Y. and McNair, H. (1996) *Basic Liquid Chromatography*. Available on-line at http://hplc.chem.shu.edu/NEW/HPLC_Book/index.html.
13. Unger, K. K. (1990) (ed.) Packings and stationary phases in chromatographic techniques, in *Chromatographic Science Series*. vol. 47, Dekker, New York.

14. Scott, R. P. W. (1993). *Silica Gel and Bonded Phases: Their Production, Properties and use in LC. Wiley, New York.*
15. Sarker, S. D., Bartholomew, B., Nash, R. J., and Simmonds, M. (2001) Sideroxylin and 8-demethylsideroxylin from *Eucalyptus saligna* (Myrtaceae) *Biochem. Syst. Ecol.* **29,** 759–762.
16. Sharp, H., Latif, Z., Bartholomew, B. et al. (2001) Emodin and syringaldehyde from *Rhamnus pubescens* (Rhamnaceae). *Biochem. Syst. Ecol.* **29**, 113–115.
17. Sharp, H., Bartholomew, B., Bright, C., Latif, Z., Sarker, S. D., and Nash, R. J. (2001) 6-Oxygenated flavones from *Baccharis trinervis. Biochem. Syst. Ecol.* **29,** 105–107.
18. Sharp, H., Latif, Z., Bright, C., Bartholomew, B., Sarker, S. D., and Nash, R. J. (2001) Totarol, totaradiol and ferruginol: three diterpenes from *Thuja plicata* (Cupressaceae). *Biochem. Syst. Ecol.* **29,** 215–217.
19. Sharp, H., Thomas, D., Currie, F. et al. (2001) Pinoresinol and syringaresinol: two lignans from *Avicennia germinans* (Avicenniaceae). *Biochem. Syst. Ecol.* **29**, 325–327.
20. Cowan, S., Bartholomew, B., Bright, C. et al. (2001) Lignans from *Cupressus lusitanica* (Cupressaceae). *Biochem. Syst. Ecol.* **29**, 109–111.
21. Turnock, J., Cowan, S., Watson, A. A. et al. (2001) *N-trans*-feruloyltyramine from two species of the Solanaceae. *Biochem. Syst. Ecol.* **29,** 209–211.
22. Cowan, S., Stewart, M., Latif, Z., Sarker, S. D., and Nash, R. J. (2001) Lignans from *Strophanthus gratus. Fitoterapia* **72,** 80–82.
23. Murphy, E. M., Nahar, L., Byres, M. et al. (2004) Coumarins from the seeds of *Angelica sylvestris* (Apiaceae) and their distribution within the genus Angelica. *Biochem. Syst. Ecol.* **32,** 203–207.
24. Sarker, S. D., Armstrong, J. A., and Waterman, P. G. (1995) (–)-1,12-Oxaguai-10(15)-ene: a sesquiterpene from *Eriostemon fitzgeraldii. Phytochemistry* **40,** 1159–1162.
25. Sarker, S. D., Waterman, P. G., and Armstrong, J. A. (1995) Coumarin glycosides from 2 species of *Eriostemon. J. Nat. Prod.-Lloydia* **58,** 1109–1115.
26. Sarker, S. D., Armstrong, J. A., and Waterman, P. G. (1995) An alkaloid, coumarins and a triterpene from *Boronia algida. Phytochemistry* **39,** 801–804.
27. Sarker, S. D., Waterman, P. G., and Armstrong, J. A. (1995) 3,4,8-Trimethoxy-2-quinolone—a new alkaloid from *Eriostemon gardneri. J. Nat. Prod.-Lloydia* **58,** 574–576.
28. Sarker, S. D., Gray, A. I., Waterman, P. G., and Armstrong, J. A. (1994) Coumarins from Asterolasia trymalioides. *J. Nat. Prod.-Lloydia* **57,** 1549–1551.
29. Sarker, S. D., Gray, A. I., Waterman, P. G., and Armstrong, J. A. (1994) Coumarins from 2 *Asterolasia species. J. Nat. Prod.-Lloydia* **57,** 324–327.

30. Sarker, S. D., Armstrong, J. A., and Waterman, P. G. (1994) Sesquiterpenyl coumarins and geranyl benzaldehyde derivatives from the aerial parts of *Eriostemon myoporoides*. *Phytochemistry* **37,** 1287–1294.
31. Sarker, S. D., Armstrong, J. A., and Waterman, P. G. (1994) Angular pyranocoumarins from *Eriostemon thryptomenoides* (Rutaceae). *Biochem. Syst. Ecol.* **22,** 863–864.
32. Sarker, S. D., Armstrong, J. A., Gray, A. I., and Waterman, P. G. (1994) Coumarins from *Asterolasia phebalioides* (Rutaceae). *Biochem. Syst. Ecol.* **22,** 433.
33. Sarker, S. D., Armstrong, J. A., Gray, A. I., and Waterman, P. G. (1994) Pyrano coumarins from *Eriostemon apiculatus*. *Biochem Syst. Ecol.* **22,** 641–644.
34. Shoeb, M., Jaspars, M., MacManus, S. M., Majinda, R. R. T., and Sarker, S. D. (2004) Epoxylignans from the seeds of *Centaurea cyanus. Biochem. Syst. Ecol.* 32, 1201–1204.
35. Egan, P., Middleton, P., Shoeb, M. et al. (2004) GI5, a dimer of oleoside, from *Fraxinus excelsior* (Oleaceae). *Biochem. Syst. Ecol.* 32, 1069–1071.
36. Kumarasamy, Y., Byres, M., Cox, P. J. et al. (2004) Isolation, structure elucidation and biological activity of flavone C-glycosides from the seeds of *Alliaria petiolata*. *Chem. Nat. Compounds* **40**, 106–110.
37. Kumarasamy, Y., Cox, P. J., Jaspars, M., Nahar, L., and Sarkar, S. D. (2003) Isolation, structure elucidation and biological activity of two unique alkaloids, hederacine A and B, from *Glechoma hederaceae*. *Tetrahedron* **59,** 6403–6407.
38. Kumarasamy, Y., Cox, P. J., Jaspars, M., Nahar, L., and Sarker, S. D. (2003) Bioactivity of secoiridoid glycosides from *Centaurium erythraea*. *Phytomedicine* **10,** 344–347.
39. Kumarasamy, Y., Nahar, L., Cox, P. J. et al. (2003) Biological activities of lignans from *Centaurea scabiosa*. *Pharm. Biol.* **41**, 203–206.
40. Kumarasamy, Y., Cox, P. J., Jaspars, M., Rashid, M. A., and Sarker, S. D. (2003) Bioactive flavonoid glycosides from the seeds of *Rosa canina*. *Pharm. Biol.* **41,** 237–242.
41. Kumarasamy, Y., Cox, P. J., Jaspars, M., Nahar, L., and Sarker, S. D. (2003) Cyanogenic glycosides from *Prunus spinosa* (Rosaceae). *Biochem. Syst. Ecol.* **31,** 1063–1065.
42. Sarker, S. D., Whiting, P., Dinan, L., Sik, V., and Rees, H. H. (1999) Identification and ecdysteroid antagonist activity of three resveratrol trimers (suffruticosol A, B and C) from *Paeonia suffruticosa*. *Tetrahedron* **55,** 513–524.
43. Sarker, S. D., Whiting, P., Sik, V., and Dinan, L. (1999) Ecdysteroid antagonists (Cucurbitacins) from *Physocarpus opulifolius* (Rosaceae). *Phytochemistry* **50,** 1123–1128.

44. Dinan, L., Sarker, S. D., Bourne, P., Whiting, P., Sik, V., and Rees, H. H. (1999) Phytoecdysteroids in seeds and plants of *Rhagodia baccata* (Labill.) Moq. (Chenopodiaceae). *Arch. Insect Biochem. Physiol.* **41,** 18–23.
45. Sarker, S. D., Dinan, L., Sik, V., Underwood, E., and Waterman, P. G. (1998) Moschamide: an unusual alkaloid from the seeds of *Centaurea moschata. Tetrahedron Lett.* **39,** 1421–1424.
46. Sarker, S. D., Sik, V., Rees, H. H., and Dinan, L. (1998) 2-Dehydro-3-*epi*-20-hydroxy ecdysone from *Froelichia floridana. Phytochemistry* **49,** 2311–2314.
47. Sarker, S. D., Sik, V., Rees, H. H., and Dinan, L. (1998) (20*R*)-1α,20-Dihydroxy ecdysone from *Axyris amaranthoides. Phytochemistry* **49,** 2305–2310.
48. Sarker, S. D., Sik, V., Dinan, L., and Rees, H. H. (1998) Moschatine: an unusual steroidal glycoside from *Centaurea moschata. Phytochemistry* **48,** 1039–1043.
49. Sarker, S. D., Girault, J. P., Lafont, R., and Dinan, L (1998) (20*R*) 15α-Hydroxy-8β,9α,14α,17α-pregn-4-en-3-one 20-*O*-β-D-glucopyranoside from *Centaurea moschata. Pharm. Biol.* **36,** 202–206.
50. Sarker, S. D., Savchenko, T., Whiting, P., Sik, V., and Dinan, L. N. (1997) Two limonoids from *Turraea obtusifolia* (Meliaceae), prieurianin and rohitukin, antagonise 20-hydroxyecdysone action in a *Drosophila* cell line. *Arch. Insect Biochem. Physiol.* **35,** 211–217.
51. Sarker, S. D., Girault, J. P., Lafont, R., and Dinan, L. (1997) Ecdysteroid xylosides from *Limnanthes douglasii. Phytochemistry* **44,** 513–521.
52. Sarker, S. D., Dinan, L., Sik, V., and Rees, H. H. (1997) 9ξ-*O*-β-D-Glucopyranosyloxy-5-megastigmen-4-one from *Lamium album. Phytochemistry* **45,** 1435–1439.
53. Sarker, S. D., Savchenko, T., Whiting, P., Sik, V., and Dinan, L. N. (1997) Moschamine, *cis*-moschamine, moschamindole and moschamindolol: four novel indole alkaloids from *Centaurea moschata. Nat. Prod. Lett.* **9,** 189–199.
54. Sarker, S. D., Dinan, L., Girault, J. P., Lafont, R., and Waterman, P. G. (1996) Punisterone [(20*R*, 24*S*)-25-deoxy-11α,20,24-trihydroxyecdysone]: a new phytoecdysteroid from *Blandfordia punicea. J. Nat. Prod. Lloydia* **59,** 789–793.
55. Diaz, J. G., Ruiz, J. G., and de la Fuente, G. (2000) Alkaloids from *Delphinium staphisagria. J. Nat. Prod.* **63,** 1136–1139.
56. Grandez, M., Madinaveitia, A., Gavin, J. A., Alva, A., and de la Fuente, G. (2002) Alkaloids from *Consolida oliverianna. J. Nat. Prod.* **65,** 513–516.
57. Zhu, N., Jiang, Y., Wang, M., and Ho, C-T. (2001) Cycloartane triterpene saponins from the roots of *Cimicfuga foetida. J. Nat. Prod.* **64,** 627–629.
58. Yoshikawa, K., Tanaka, M., Arihara, S. et al. (2000) New oleanene triterpenoid saponins from *Madhuca longifolia*. J. Nat. Prod. **63,** 1679–1681.

59. Yoshikawa, K., Morikawa, T., Nakano, K., Pongpiriyadacha, Y., Murakami, T., and Matsuda, H. (2002) Characterisation of new sweet triterpene saponins from *Albizia myriophylla*. *J. Nat. Prod.* **65,** 1638–1642.
60. Ma, W., Wei, X., Ling, T., Xie, H., and Zhou, W. (2003) New phenolics from *Polygala fallax*. *J. Nat. Prod.* **66,** 441–443.
61. Nunez, M. J., Guadano, A., Jiminez, I. A., Ravelo, A. G., Gonzalez-Coloma, A., and Bazzocchi, I. L. (2004) Insecticidal sesquiterpene pyridine alkaloids from *Maytenus chiapensis*. *J. Nat. Prod.* **67,** 14–18.
62. Wright, A. D., Osterhage, C., and König, G. M. (2003) Epicoccamide, a novel secondary metabolite from a jellyfish-derived culture of *Epicoccum purpurascens*. *Org. Biomol. Chem.* **1,** 507–510.
63. Ferguson, C. A., Nahar, L., Finnie, D. et al. (2003) *Centaurea scabiosa*: a source of dibenzylbutyrolactone lignans. Biochem. Syst. Ecol. **31,** 303–305.
64. Middleton, M., Cox, PJ., Jaspars, M. et al. (2003) Dibenzylbutyrolactone lignans and indole alkaloids from the seeds of *Centaurea nigra* (Asteraceae). Biochem. Syst. Ecol. **31,** 653–656.
65. Kumarasamy, Y., Middleton, M., Reid, R. G., Nahar, L., and Sarker, S. D. (2003) Biological activity of serotonin conjugates from the seeds of *Centaurea nigra*. *Fitoterapia* **74,** 609–612.
66. Tomlinson, C. T. M., Nahar, L., Copland, A. et al. (2003) Flavonol glycosides from the seeds of *Agrimonia eupatoria* (Rosaceae). Biochem. Syst. Ecol. **31,** 439–441.
67. Jin, J-M., Zhang, Y. J., and Yang, C.-R. (2004) Spirostanol and furostanol glycosides from the fresh tubers of *Polianthes tuberosa*. *J. Nat. Prod.* **67,** 5–9.
68. Groweiss, A., Cardellina, J. H., and Boyd, M. R. (2000) HIV-inhibitory prenylated xanthones and flavones from *Maclura tinctoria*. *J. Nat. Prod.* **63,** 1537–1539.
69. Al-Howiriny, T. A., Zemaitis, M. A., Gao, C.-Y., et al. (2001) Thalibealine, a novel tetrahydroprotoberberine-aporphine dimeric alkaloid from *Thalictrum wangii*. J. Nat. Prod. **64,** 819–822.
70. Levison, P. R. (2003) Large-scale ion-exchange column chromatography of proteins: comparison of different formats. *J. Chromatogr. B* **790,** 17–33.
71. Roman, R. B., Alvarez-Pez, J. M., Fernandez, F. G. A., and Grima, E. M. (2002) Recovery of pure β-phycoerythrin from the microalga *Porphyridium cruentum*. *J. Biotechnol.* **93,** 73–85.
72. Orozco-Villafuerte, J., Cruz-Sosa, F., Ponce-Alquicira, E., and Vernon-Carter, E. J. (2003) Mesquite gum: fractionation and characterisation of the gum exuded from *Prosopis laevigata* obtained from plant tissue culture and from wild trees. *Carbohydr. Polymers* **54,** 327–333.
73. Froytlog, C., Slimestad, R., and Andersen, O. M. (1998) Combination of chromatographic techniques for the preparative isolation of anthocyanins-applied on blackcurrant (*Ribes nigrum*) fruits. *J. Chromatogr. A* **825,** 89–95.

74. Adamson G. E., Lazarus, S. A., Mitchell A. E., et al. (1999) HPCL method for the qualification of procyanidins in cocoa and chocolate samples and correlation to total antioxidant capacity. *J. Agric. Food Chem.* **47**, 4184–4188.
75. Salituro, G. M. and Dufresne, C. (1998) Isolation by low-pressure column chromatography, in *Natural Products Isolation*, 1 ed. (Cannell, R. J. P., ed.), Humana Press, New Jersey.

6

Isolation by Ion-Exchange Methods

David G. Durham

Summary

Ion-exchange chromatography techniques have been applied for the isolation of charged or ionizable natural products. The basis of the ion-exchange process is the reversible binding of either a cationic or an anionic molecule to an oppositely charged insoluble resin matrix by displacement of a counterion. Resins may be characterized by their matrix structure, and the "strength" of the exchanger is dependent upon the pK_a of the ionized group. Choice of the type of resin and manipulation of pH conditions can be made to selectively bind ionized molecules. Elution of bound solute from the resin may be achieved by either adjustment of ion strength or changes in pH. The efficiency of the process may be enhanced by the use of gradient elution and inclusion of organic cosolvents. Application of the versatility of the technique is illustrated by a number of examples drawn from the literature.

Key Words: Ion-exchange chromatography; ion-exchange resins; solid phase extraction; pH gradient elution; ion-gradient elution.

1. Introduction

This revised chapter is based on the original version authored by Dufresne in the first edition of this book *(1)*. In the past, ion-exchange chromatography (IEC) techniques had been applied for the isolation of natural products. The method is used for the separation of charged or ionizable compounds, a property exhibited by large numbers of natural products. IEC was first reported some 70 years ago, with its application as a technique for the separation of natural products being described within a few years for the isolation and

From: *Methods in Biotechnology, Vol. 20, Natural Products Isolation, 2nd ed.*
Edited by: S. D. Sarker, Z. Latif, and A. I. Gray © Humana Press Inc., Totowa, NJ

separation of amino acids from protein hydrolysis. It continues to be widely used even today in the context of the analysis and separation of proteins and nucleotides. The adjunct of solid phase extraction (SPE) as "clean-up" procedures for liquid chromatography (LC) and hybrid liquid chromatography–mass spectrometry has further extended the applications of the principle.

The application of IEC as a separation and analytical tool has been extensively addressed in the chemical literature *(2,3)*. With this in mind the chapter focuses on the more practical aspects of the choice of conditions likely to result in the separation and isolation of a natural product rather than chromatography *per se*.

A major problem that confronts the natural product chemist is the challenge of the isolation of compounds from a heterogeneous mixture of compounds, present in low concentrations, usually because of their limited solubilities in their environmental matrix. A direct objective in the application of the ion-exchange method will be to achieve both a concentrating and selective extraction of the desired target molecule. The presence of an ionizable or charged grouping within the molecular structure serves as the basis of the manipulative process. Ion-exchange purification steps have proved particularly fruitful in the isolation of compounds from crude extracts and fermentation broths. In principle, the method allows ready development of scale up to manufacturing production, a process achieved efficiently and with economic advantage. The latter is clearly demonstrated by the use of ion-exchange-based processes in the industrial production of many antibiotics.

2. Theory of Ion-Exchange

The basis of the ion-exchange process is the reversible binding of charged molecules to an oppositely charged insoluble matrix. A charged molecule (M) interacts with an oppositely charged group (G) attached to a supporting resin matrix (R) and is retained by displacement of a counterion (C). Because the functional group on the support matrix may carry either a negative or positive charge, exchange of the counterion gives either cationic or anionic resins. The process is represented by the following equations:

Cationic exchange

$$R\text{–}G^- \; C^+ + M^+ \rightleftharpoons R\text{–}G^- \; M^+ + C^+$$

Anionic exchange

$$R\text{–}G^+ \; C^- + M^- \rightleftharpoons R\text{–}G^+ \; M^- + C^-$$

The process is usually reversible, which when manipulated leads to adsorption and elution cycles, with regeneration of the adsorbent resin.

Most of the charged organic molecules are the result of the presence of ionizable weak acid or basic groups. The charge on the molecule can be adjusted by alteration of the pH. The Henderson–Hasselbach equation predicts that organic acids will be fully dissociated to form anions at two pH units above the p*K*a value of the acid, while amines are fully protonated to yield cations up to two pH units below the p*K*a value of their conjugate acid. Consequently, the majority of carboxylic acids will carry a negative charge at pH values above 6.0, leading to retention on anionic exchange resins carrying a positively charged group. Amines are generally protonated at pH values below 8.0 and would be retained on cationic exchange resins carrying a negatively charged functional group.

For a charged molecule to be retained on the resin, it must displace the counterion associated with the resin. The equilibrium involved may be expressed in terms of a partition constant (K_p), which is the ratio of the dissociation constants for the binding of the resin charged group to the counterion (K_c) or charged molecule (K_m). This relationship is expressed in the following equations irrespective of charge:

$$K_p = \frac{K_c}{K_m} = \frac{[RGM][C]}{[RGC][M]}$$

$$\text{where} \quad K_c = \frac{[RG][C]}{[RGC]} \quad \text{and} \quad K_m = \frac{[RG][M]}{[RGM]}$$

2.1. Charge Attraction

The effectiveness of an ion-exchange process will be dependent upon the affinity of the charged resin group toward the ions (expressed as K_c and K_m) and their relative concentrations ([C] and [M]). More highly polarized ions will be more strongly attracted to the exchange resin. Polyvalent ions usually have greater affinity for a charged resin than do monovalent ions, as a result of multiple binding at more than one charged group on the resin.

2.2. Role of the Counterions

The presence of the counterion maintains the charge neutrality of the resin. The reversibility of the binding process may be manipulated by changing the concentration of counterion or its substitution by a counterion of greater affinity (increased value of K_c). Although the affinity of the counterion for the resin may be affected by the nature of the resin itself, the affinity of anions for resins usually follows the increasing sequence: hydroxide < acetate < bicarbonate, formate < chloride < phosphate, citrate. This sequence illustrates why chloride ions (greater affinity) are effective in eluting carboxylate anions (represented by acetate of lower affinity) from an anion exchange resin.

An increasing affinity of cations toward cation exchangers is given by the following sequence: lithium < hydrogen < sodium < ammonium < calcium. In this case, it is apparent that sodium ions will not be very effective in eluting amines (at a pH in which they exist as positively charged ammonium ions) from a cationic exchange resin.

2.3. Hydrophobic Interactions

Although interaction between ions and resin is primarily influenced by interacting charges, significant effects may occur because of additional hydrophobic interactions between the residue of the resin matrix structure and uncharged functionalities of the adsorbed molecule. These nonionic effects contribute to the overall adsorption and retention process, but may be largely reduced by utilizing the effects of modifying organic solvents to minimize these partitioning effects.

2.4. Exchange Capacity and Rate of Process

The ability of the exchange resin to adsorb molecules will depend upon the density or number of charged groups per unit mass of resin. This may be practically expressed as the number of milliequivalents of ion adsorbed per gram of resin (meq/g). Hence, a resin of capacity of 2 meq/g has a theoretical capacity to adsorb 2 mmol of monovalent or 1 mmol of divalent charged molecule per gram of resin.

The rate of the ion-exchange process will depend upon the "availability" of the interacting groups. This will be governed by two steps: the diffusion of the exchanging ion into the resin as the counterion diffuses out, and the chemical ion exchange to maintain electrical neutrality. This latter effect is a fast step, with the overall rate being controlled by the rate limiting diffusion process. This process will be influenced by the surface area of the resin phase. Increasing the surface area of the resin, in association with decreasing the resin particle size, will reduce diffusion time and raise the efficiency of the ion-exchange process.

3. Materials for Ion-Exchange

Ion-exchange resins are used in the form of water-insoluble particles or beads. The insoluble matrix material is covalently chemically bonded to negatively or positively charged groups. Exchangeable counterions are associated with these groups and maintain the electrical neutrality of the phase. The support matrices may be based on a number of polymer materials, which include synthetic polymers, silica, or polysaccharides. The following sections outline the types of materials used for resin matrices and the chemical nature of the attached ion-exchanging groups.

3.1. Support Matrices

The physical properties of the ion-exchange resin are determined by the chemical nature of the matrix material. These include rigidity and resistance to mechanical shock, porosity of particles, and the effect of shape on solvent flow over the particle. In addition, polymers need to be both thermally and chemically stable. Three major groups of materials have been used as support matrices for natural product isolation: polystyrene or acrylate resins, carbohydrate polymers, and silica gel. Both porous and nonporous support materials have been used. The term "macroreticular" has been used to describe highly porous materials that have rigid resin pores. For porous materials, the rate of the ion-exchange process is limited by the ability of ions to diffuse into the porous matrix. In contrast, nonporous material, in which charged groups are only exposed externally on the particle, will offer the advantage of appreciably more rapid exchange times. Diagrammatic representation of resin types is shown in (**Fig. 1**). When porous resins are placed in water, the solvent molecules diffuse into the resin pores and hydrate the charged groups and their counterions, resulting in a

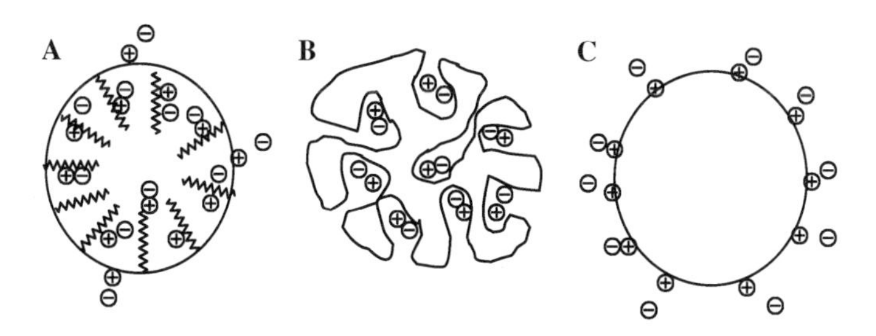

Fig. 1. Schematic representation of cationic resin matrix types (with counterions) **A**, porous gel; **B**, macroreticular resin; **C**, bonded silica.

swelling of the resin. The amount of swelling will depend upon the nature of the polymer. Polymers, that are more rigid by virtue of a greater crosslinked matrix structure will show less capacity for swelling. The terms "hydrophobic" and "hydrophilic" have been used to classify types of matrices on the basis of their propensities to interact with nonpolar and polar groups, respectively.

Because of volume changes on hydration, bead dimensions of resin polymers are usually expressed by mesh size, whereas rigid silica-based supports are measured in microns. The wide applications of resins indicate that larger beads (50–100 mesh) could be used for direct resin extraction techniques or employed chromatographically in large preparative columns. Smaller beads (200–400 mesh) or silica supports (5–10 μm) would be more appropriate for analytical or rapid small-scale separations.

3.1.1. Polystyrene Resins

Hydrophobic resins based on polystyrene or methacrylate polymers have been extensively used as ion-exchange matrix materials. These polymer types allow a high degree of substitution by charged functional groups to give good resin capacity, and offer both thermal and chemically stable systems. Particularly, they may be used over wide pH ranges (polystyrene: 1–14 and methacrylate: 2–10 pH units).

Polystyrene resins, manufactured by the reaction of styrene with divinylbenzene containing variable amounts of divinylbenzene (2–8%), result in a range of polymers with different degrees of crosslinking. Owing to this polymer crosslinking, the matrices offer good mechanical strength and rigidity. The higher degrees of crosslinking increase the capacity and selectivity of the resins, but at the expense of equilibration times as diffusion to the interior of the matrices becomes increasingly restricted. One of the most widely used range of such polymers is marketed under the name "Dowex," with the percentage of incorporated divinylbenzene and hence degree of crosslinking indicated by the terms X-2, X-4, and so on.

A disadvantage of polystyrene supports is the probability of hydrophobic interactions between molecules and the matrix backbone, which may be sufficient to significantly affect recovery of adsorbed molecules from the resin. However, the adsorption onto unsubstituted hydrophobic resins may be used advantageously as a technique for desalting extracts before and after ion-exchange purification. Adsorbed materials are recovered by washing with solvents of increasing organic content.

3.1.2. Carbohydrate Polymers

Polysaccharide polymers, in contrast, are hydrophilic. Crosslinked dextran matrices in particular have been used in the separation and purification of biomolecules. These polymers have proved useful for protein purification, because denaturation of proteins occurs less readily in these hydrophilic matrices, when hydrophobic interactions with the polymer backbone are infrequent.

A widely available polymer of this crosslinked dextran type is marketed as "Sephadex" (*see* Chap. 5). The ion-exchangeable functionalities are bonded to the glucose units of the core matrix by ether bonds. The polymer, supplied as beads, hydrates readily in water, swelling to a gel, which shows good flow properties for bed packing. This gel swelling property of the matrix produces changes in bed volume with variations in ionic strength of the surrounding solvent. Two ranges of resin are available based on the degree of polymer crosslinking and resulting porosity. The more highly crosslinked (25-type) are less porous, swell less, and are more mechanically robust. These properties make them better suited to the isolation of lower molecular weight natural products, while the more porous (50-type) resins are more applicable for larger biomolecules. Polymers are stable across a wide range of pH (2.0–12.0). Other available carbohydrate resin supports are based on functionalized crosslinked agarose or cellulose.

3.1.3. Silica Gel

Silica-based particulate matrices offer an alternative to the organic polymers. The nonporous nature of the material and its more limited chemical stability result in advantages and disadvantages. Silica is available as small sized nonporous particles (5–50 μm). The ion-exchange process is rapid, because the functional groups are distributed over a large external surface area and ions do not need to diffuse into the resin matrix. The matrix volume is unaffected by solvent composition as the silica does not shrink or swell. The small particle sizes give rise to resistance to solvent flow in chromatographic columns. Silica is stable over a restricted pH range (2.0–7.5), which limits extended operating times with solvents at high pH values. However, the efficiency of the ion-exchange process makes possible the use of a wider pH range, provided the exposure time is not unduly prolonged.

3.2. Functional Groups

The ion-exchange process is dependent upon the nature of the functional group anchored to the matrix phase and defines the type of resin. The pK_a of this group is used to define the "strength" of the exchanger and is dependent upon the ionized state of the group and its ability to effect a separation. The apparent pK_a values for types of functional groups normally used in ion-exchange resins are shown in **Table 1**. The number of groups per unit volume of resin will define the capacity of the resin. The functional ionized group will exchange its counterion for a complementary charged molecule. The charge on the exchanging molecule will in most cases be that of a weakly ionizing species.

3.2.1. Anion Exchangers

A large number of natural products are carboxylic acids and will therefore be ionized and negatively charged at pH >6.0. These may be purified by the use of cither strong or weak anionic exchangers. Examples of commercially available types of strong anionic exchangers (SAX) are given in **Table 2**. These include two types of quaternary ammonium groups (QAE-resins, quaternary amino ethyl) and are positively charged over the full range of pH. Weak anion exchangers that are commercially available are provided in **Table 3**. Weak anion exchangers arise from protonation of primary, secondary, or tertiary amines at pHs below their pK_a values. Adsorbed carboxylic acid group containing molecules may be eluted from resins by adjustment of the solvent pH, either to suppress

Table 1
Approximate pK_a Values for Various Types of Functional Groups

Functional group	Approximate pK_a value	Ionized state (90% ionized)	Exchanger type
$-SO_3H$ Sulfonic acid	<1	Anion >1	Strong cation
$-PO_3H$ Phosphoric acid	pK_{a1} = 2.5, pK_{a2} = 7.5	Anion >3.5	Strong cation
$-CO_2H$ Carboxylic acid	5	Anion >6	Weak cation
$-NR_2$ Tertiary amine	8	Cation <7	Weak anion
Quanternary ammonium $-NR_3^+$ Type I Type II	>13	Cation <13	Strong anion

Table 2
Selection of Commercially Available Strong Anion Exchangers to Illustrate Products by Major Brand Names and Manufacturers

Resin	Manufacturer	Functional group	Counterion	Matrix support	Capacity (meq/dL)	pH range	Bead mesh range size (μm)
		(Type I)					
Dowex-1X2-100	D	Trimethyl benzyl ammonium	Cl^-	Polystyrene gel 2% Xlink	70	0–14.0	50–100 150–250
Dowex-1X8-400	D	Trimethyl benzyl ammonium	Cl^-	Polystyrene gel 8% Xlink	120	0–14.0	200–400 35–75
Amberlite IRA-400	R&H	Quaternary ammonium	Cl^-	Polystyrene gel 8% Xlink	140	0–14.0	16–50 500–1000
Dianion PA 312	M	Trimethyl benzyl ammonium	Cl^-	Polystyrene porous 6% Xlink	120	0–14.0	16–50 500–1000
QAE Sephadex A25	P	Quaternary amino ethyl	Cl^-	Dextran	50	2.0–12.0	100–400 40–120
TSK gel SAX	T	Quaternary ammonium	Cl^-	Polystyrene-DVB	>370	1.0–14.0	>400
		(Type II)					
Dianion SA20A	M	Trimethyl benzyl hydroxyethyl ammonium	Cl^-	Polystyrene gel	130	0–14.0	16–50

Manufacturers key: D, Dow Chemicals; R&H, Rohm and Haas; M, Mitsubishi; P, Pharmacia; T, Tosoh.

Table 3
Selection of Commercially Available Weak Anion Exchangers to Illustrate Products by Major Brand Names and Manufacturers

Resin	Manufacturer	Functional group	Counterion	Matrix support	Capacity (meq/dL)	pH range	Bead mesh range size (µm)
Dowex-M-43	D	Dimethyl amine	Free base	Polystyrene macropore		0–14.0	
Amberlite IRA-67	R&H	Diethyl amino ethyl	Free base	Acrylic gel	160	0–7.0	16–50 500–1000
Dianion WA 10	M	Dimethyl amino ethyl	Free base	Polyacrylate gel	120	0–9.0	16–50 500–1000
DEAE Sephadex A25	P	Diethyl amino ethyl	Cl^-	Dextran	50	2.0–12.0	100–400 40–120
TSK gel DEAE-2SW	T	Diethyl amino ethyl	Cl^-	Silica	> 30	2.0–7.5	> 400 5

Manufacturers key: D, Dow Chemicals; R&H, Rohm and Haas; M, Mitsubishi; P, Pharmacia; T, Tosoh.

ionization of the carboxylic acid group (pH lowered to <4.0), or in the case of a weak exchanger by suppression of the resin charge by raising the pH above 9.0. For more comprehensive details about the range of commercially available resins, the reader is referred to suppliers' catalogs or manufacturers' web sites.

3.2.2. Cation Exchangers

Amine groups in natural products will be protonated and positively charged at pH<7.0. These may be purified by the application of strong or weak cationic exchangers. Examples of commercially available types of strong cationic exchangers (SCX) are given in **Table 4**. Sulfonic acids typify strong cation exchangers, being dissociated and negatively charged over the full pH range. Even though they are weaker acids, losing their charge below pH 2.0, phosphoric acids are classified as strong cation exchangers and above pH 8.0 they will act as divalent cation exchangers.

Examples of commercially available types of weak cation exchangers are given in **Table 5**. Carboxylic acids are the only negatively charged groups above pH 6.0 and are therefore considered as weak cationic exchangers. Adsorbed amine group containing molecules may be eluted from resins by adjustment of the solvent pH, either to deprotonate the amine group (pH raised to >9.0), or in the case of a weak exchanger by the suppression of the resin charge by lowering the pH below 4.0.

4. Column Operation

This section addresses the steps involved in the setting up and carrying out an ion-exchange separation. These may be summarized as selecting the most appropriate exchange resin, setting up the column, loading the sample, followed by the separation processes of washing and elution. The pH and buffering of solvents require particular attention, because the manipulation of the charge of the resin or on the sample molecule is essential for the process. Examples of useful buffers are shown in **Table 6**.

4.1. Selection of Packing Material

There will probably be no single ideal resin to achieve a separation. The initial choice of a resin may largely be empirical, being most strongly influenced by the availability of the resin rather than rational design. A number of suppliers' web sites have simple decision "trees" to guide the selection of a suitable resin. If the charge on a solute is known, then the selection of ion-exchange resin type is self-evident. For example, isolation of a basic

Table 4
Selection of Commercially Available Strong Cation Exchangers to Illustrate Products by Major Brand Names and Manufacturers

Resin	Manufacturer	Functional group	Counterion	Matrix support	Capacity (meq dL^{-1})	pH range	Bead mesh range size (μm)
Dowex-50WX2-100	D	Sulfonic acid	H^+	Polystyrene gel 2%Xlink	60	0–14.0	50–100
Amberlite IR-122	R&H	Sulfonic acid	Na^+	Polystyrene gel 10%Xlink	210	0–14.0	16–50 500–1000
Dianion SK1B	M	Sulfonic acid	Na^+	Polystyrene gel 8%Xlink	190	0–14.0	40–60 400–600
SP-Sephadex A25	P	Sulfopropyl	Na^+	Dextran	30	2.0–12.0	100–400 40–120
TSK gel SP-5PW	T	Sulfonic acid	Na^+	Methacrylate gel	>10	2.0–12.0	>400 10

Manufacturers key: See footnote of Table 3.

Table 5
Selection of Commercially Available Weak Cation Exchangers to Illustrate Products by Major Brand Names and Manufacturers

Resin	Manufacturer	Functional group	Counterion	Matrix support	Capacity (meq dL^{-1})	pH range	Bead mesh range size (μm)
Dowex-MAC-3	D	Carboxylic acid	H^+	Acrylic macroreticular	380	4.0–14.0	16–50
Amberlite IRC-50	R&H	Carboxylic acid	H^+	Acrylic mr 4% Xlink	350	5.0–14.0	16–50 500–1000
Dianion WK10	M	Carboxylic acid	H^+	Methacrylic porous	250	5.0–14.0	16–50 500–1000
CM-Sephadex C25	P	Carboxymethyl	Na^+	Dextran	56	2.0–12.0	100–400 40–120
TSK gel SP-5PW	T	Sulfonic acid	Na^+	Methacrylate gel	>10	2.0–12.0	>400 10

Manufacturers key: See footnote of Table 3.

Table 6
Useful Buffers for Ion-Exchange Separations

Buffer	pH range
Citrate	2.0–6.0
Phosphate (pK_{a1}, pK_{a2})	2.0–3.0, 7.1–8.0
Formate	3.0–4.5
Acetate	4.2–5.4
Triethanolamine (Tris)	6.7–8.7
Borate	8.0–9.8
Ammonia	8.2–10.2

compound containing an amine group would prompt the initial use of a strong cationic exchange resin (SCX).

The starting point of a natural product separation will often require the isolation of compounds of unknown structure. A strategy to achieve a suitable resin phase for separation should be based on preliminary experiments using a series of probe columns, eluted under different conditions. The development of commercially available solid phase extraction cartridges based on ion-exchangers facilitates this process, although a series of small-scale "self packed" columns may substitute if materials are available. A disadvantage of commercial cartridges is that many are silica based and hence unstable at pH above 8.0, somewhat limiting their use. However, this may be compensated in comparison to other resins by a more rapid kinetic exchange resulting from small particle size and lack of porosity. Owing to the presence of competing salt ions at varying concentrations in product extracts, it is advantageous to desalt the samples before carrying out probe testing.

In some cases, the focus of the natural product isolation may be driven by a bioassay screening for biologically active compounds. In these situations, it will be necessary to first carry out preliminary stability studies to ensure that activity is not lost because of changes in manipulative conditions rather than deficiencies in the separation process. A pH stability study is carried out subjecting the material to a range of pH values (e.g., 2.0, 4.0, 9.0, 11.0) for different periods of time (e.g., 0.5, 4, 24 h). After the time interval, the pH is adjusted to pH 7.0 and the sample bioassayed. The results will indicate the range of pH to which the sample may be exposed and hence the most appropriate resin to use for the recovery of biologically active material.

Monitoring of sample breakthrough in spent fractions during the loading procedure will give an indication of the capacity of the resin. Efficient loading of extract per unit volume will be important when a scaling up of processes is envisaged. A probe scheme for resin selection is outlined in **Fig. 2**, together with cartridge columns and elution conditions mentioned in **Table 7**, to illustrate changing pH and counterion strength. When recovery of a sample from a resin is poor, replacement by a weaker exchanging resin or with a lower capacity is indicated. A carefully chosen resin can result in a very selective concentration and purification of a desired compound in a highly efficient process.

Sample extract

↓ 1. pH-stability study

2. Preliminary desalting of extract

↓ 3. Equilibrate probe column bed (2 mL) by washing with water (5 mL)

Sample loading

↓ 4. Adjust sample pH for retention (Table 7)

5. Load sample onto column bed (20 mL, collecting 2 x 5 mL spent fractions to monitor for sample breakthrough)

↓ 6. Wash column with water (5 mL)

Sample elution

↓ 7. Elute with eluate solvent 1.(Table 7)

8. Elute with eluate solvent 2. (Table 7)

↓ 9. Adjust eluate to pH 7 and bioassay.

Fig. 2. Sequential scheme for resin selection.

Table 7
Probe Cartridge Column Examples and Elution Conditions

	Anion exchanger		Cation exchanger	
Conditions	Silica-based	Polymer-based	Silica-based	Polymer-based
Probe column	Waters Accell QMA Sep-Pak Discovery DSC-SAX	Self-packed Dowex 1 (Cl^- cycle)	Waters Accell CM Sep-Pak Discovery DSC-SCX	Self-packed Dowex 50 (Na^+ cycle)
Loading pH	$6.5 < pH < 8.0$	$6.5 < pH < 8.0$	$4.0 < pH < 6.0$	$4.0 < pH < 6.0$
Load time	1–2 min	5–10 min	1–2 min	5–10 min
Elution 1	1% $NaCl/H_2O$	0.1M aq $NaHCO_3$	5% NaCl	0.1M NH_4OH
Elution 2	3% NH_4Cl in 90% MeOH aq.	0.5M aq HCl	2% aq pyridine	1M NH_4OH

4.2. Resin Preparation

Careful preparation of ion-exchange resins is an essential prerequisite for a successful separation. Commercially available resins usually need a degree of preconditioning.

4.2.1. Washing

Initial washing of resins removes impurities and "fines" that result from the manufacturing process or storage. One should consult manufacturers' literature for the recommended procedures for washing their resins. Washing resin polymers with methanol or solvents that simulate the elution process will minimize the likelihood of unwanted contaminants in the final products.

4.2.2. Swelling

The second important process in resin preparation is swelling the resin. The swelling process exposes those functional groups in the interior of the resin to the mobile phase, hydrating groups, and counterions by penetrating pores. In particular, suspending carbohydrate polymers in water may not produce complete swelling. Charged functional groups in the hydrophilic matrix may remain hydrogen bonded, be unavailable for the exchange process, and limit the capacity of the resin. Pretreatment of the resin with dilute

HCl and NaOH solutions (0.5 M) breaks the hydrogen bonds, thereby ensuring full swelling. The resin can then be neutralized and equilibrated with the required buffer solution.

4.2.3. Resin Cycle

The cycle of the resin is very important. The resin may be commercially supplied in a counterion form, which is not directly suitable for use. Conversion of the resin from one cycle to another may be achieved by treatment with large volumes of solution of the desired counterion, followed by large volumes of water to remove excess ion. The most usual cycle conversions for cation exchangers are from a hydrogen to sodium cycle, and for anion exchangers from a chloride to acetate cycle.

4.3. Column-Size Selection

Important criteria in selecting a column are the volume and shape. A resin bed volume should be selected that is several times the expected exchange capacity. If the amount of sample is known, this volume may be calculated from the manufacturers' resin data. For situations where the sample size is uncertain, the resin bed volume should be 0.1–0.05 of the applied loading volume of solution. The resin volume may be adjusted in subsequent scaling up procedures. Long columns are not desirable, because long bed heights lead to back pressure in the column and allow band broadening by diffusion. A short wide diameter column with a bed height of 10–20 cm is usually preferred.

4.4. Sample Loading

Loading samples in isolation processes are often derived from cellular cultures. The samples should be clarified before loading, as the residual biomass will coat the resin and seriously impede solute penetration into resin pores. Centrifugation is the preferred method for clarifying, because filtration may result in nonspecific adsorbance on either filter paper (cellulose) or filter aid (silica).

The samples may be loaded onto resin by one of the two methods: "batch" or "column." In the former, the resin and clarified sample are premixed together for adequate time (15–30 min) to allow ion-exchange adsorption. The resin can be recovered by filtration through a coarse sieve or decanting the spent solvent and then slurry packed for elution. Alternatively in the column method, the prepared resin is slurry packed into a column, allowed to settle in a bed, and then the sample applied. A flow time of 10–20 min (5–10 mL/min for a 100 mL column) is required for application of the sample to ensure that molecules diffuse into resin pores and are adsorbed. Times for nonporous silica matrix resins may be somewhat less.

4.5. Elution

4.5.1. Ionic Strength

Elution of bound solute from the resin may be achieved by either adjustment of ion strength or changes in pH. The use of a high salt concentration causes displacement of solute by a shift in the equilibrium in favor of the bound counterion. Sodium chloride can be used to elute solutes from both cationic and anionic-exchange columns by exchange of the appropriate ion with the complementary charged solute.

4.5.2. pH Adjustment

Alternatively, adjustment of pH may modify the charge on the solute or in the case of weak exchange resins on the resin itself. A change in pH may achieve a more selective elution of solute by removing the charge on the solute so that it is no longer bound to the resin. A carboxylic acid containing solute may be adsorbed on an anionic resin at pH 6.0 and eluted at pH 4.0 when the carboxylate anions become protonated and uncharged. This may, however, significantly alter the solubility of the solute leading to precipitation or hydrophobic interaction with the resin polymer backbone resulting in retention of solute within the resin.

4.5.3. Cosolvents

The use of water-miscible cosolvents such as methanol or acetonitrile with either strategy for elution, ion strength or pH adjustment, is useful to disrupt the hydrophobic interactions between the resin and the neutral molecule. A particularly useful solvent for the elution of anions from an anion-exchange resin such as Dowex-1 is a solution of 3% ammonium chloride (approx 0.5 M) in 90% methanol in water. Organic salt mixtures of pyridine and acetic acid, whose compositions may be adjusted to give a wide range of pH values (**Table 8**), may also be useful and offer the advantage of being readily removed by evaporation owing to their volatility.

Table 8
Composition of Pyridine and Acetic Acid Buffers

Composition	pH 3.0	pH 5.0	pH 8.0
Water	970	970	980
Acetic acid	28	14	0.2
Pyridine	2	16	20

4.5.4. Solvent Gradients

Gradient elution may be carried out "stepwise" or by a "smooth" incremental gradient. While for probe experiments an adequate result may be obtained using a stepwise technique, a smooth gradient is usually more effective and can be readily produced. A simple smooth gradient may be generated by constant addition of a second solvent component to a mixing reservoir containing the initial elution solvent. Gradients with variable slope characteristics (linear, concave, or convex) may be generated by control of flow rates. Large gradient volumes give rise to better resolution; but this is achieved at the expense of band broadening and the need to collect larger fractions. A total elution volume of 5–10 times the column volume capacity will be necessary to ensure recovery of exchangeable analyte.

The elution flow rate may be more rapid than the flow rate for column loading, with the contact time between resin and eluant solvent being halved. However, the volume of fractions collected should be such to exploit the resolution achieved on the column. For example, to elute a 100 mL resin bed column volume, a total gradient volume of 600 mL is collected over a period of 30 min, in 10–20 mL fractions (i.e, 5–10 fractions per column volume).

5. Applications

Ion-exchange extraction methods may be applied to the isolation of strong and weak acids and bases as well as to amphoteric compounds. Because the process, especially when involving salt elution, will usually also involve adjustments of pH, it will be necessary to determine an initial pH-stability profile for the potential isolate, to select an appropriate ion-exchange process. A number of illustrative isolation processes are outlined to indicate the versatility of the method to effect separations of natural products using different strategies.

5.1. Anionic Compounds

Selectivity in the isolation of acidic compounds on anionic exchangers may be achieved by adsorption under low pH conditions. At pH 4.0 and below, carboxylic acids will be unionized and unretained on the resin, giving only selective adsorption of stronger acids. Where a compound is unstable under these conditions, it may be less selectively adsorbed on a strong anion-exchange resin at higher pH values along with carboxylic acids. Elution from a strong anion resin is achieved with a salt concentration gradient.

If a weak anion resin is employed, adjustment of mobile phase pH to above the pK_a of the resin group will suppress charge on the resin resulting in elution of the adsorbate. Again, stability of the isolate to higher pH conditions may be a limiting factor.

5.1.1. Cephamycins A and B

The cephalosporin antibiotics (**Fig. 3**) were isolated some 30 yr ago from fermentation cultures. The procedure outlined *(4)* illustrates both the multiple use of anion-exchange resins under different pH conditions to achieve selectivity, as well as the application of adsorption resins in desalting steps.

5.1.2. Zaragozic Acids

Zaragozic acids (**Fig. 4**) were discovered while screening for squalene synthase inhibitors *(5)*. The tricarboxylic acid core flanked by lipophilic side chains provided a particular challenge for ion-exchange processes. The amphipathic nature of the molecules resulted in significant backbone interactions with the polystyrene resin matrices, resulting in limited recovery of isolate. Adaptation of the ion-exchange process to the features of the compounds resulted in a useful isolation process.

5.2. Cationic Compounds

Strongly basic compounds are selectively separated by the use of either strong or weak cation-exchange resins at high pH values. At these values, weak bases are not protonated and will not be retained on the resin. When a strong cation exchanger is used, elution is carried out using a salt gradient. If a weak cation exchanger is used, lowering the pH to values that suppress resin protonation will result in sample elution. When alkaline instability is a problem, it may be difficult to separate weak from strong

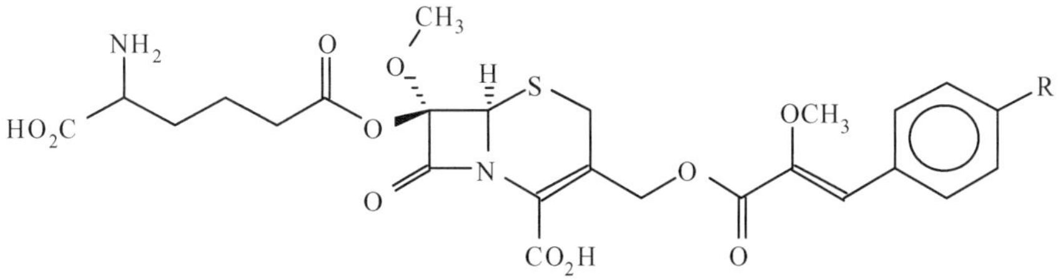

1.	Acidify the fermentation broth and filter.	Removal of biomass.
2.	Pass filtrate through XAD-2 resin and elute with 60% aqueous MeOH.	Initial desalting of fermentation broth with adsorption resin.
3.	Concentrate eluate and adjust pH to 3.5 with aqueous NH_4OH.	
4.	Dilute with H_20 and pass through Amberlite IRA-68 (Cl) resin column.	Adsorbed on weak anionic exchange resin: low pH.
5.	Elute with 1M $NaNO_3$ in 0.1M NaOAc (pH 7.5), collecting fractions.	Elute with nitrate ion.
6.	Bioassay and combine active fractions; adjust to pH 3.0 and desalt on XAD-2 as under 2 above.	Removal of nitrate ions.
7.	Concentrate eluate, adjust to pH 4.0; lyophilize.	
8.	Dissolve in 0.5M NH_4Br-0.05M AcOH buffer.	Small volume solution.
9.	Chromatograph on DEAE-Sephadex A-25.	Weak anion exchange resin; Elute with same buffer.
10.	Bioassay fractions; pool active fractions.	Resolves cephamycin A from B.
11.	Pass each pool through XAD-2 resin and elute with 90% aqueous MeOH.	

Fig. 3. The structures of cephamycin A ($R=OSO_3H$) and cephamycin B (R=OH) with an extraction scheme.

bases. A solution may be to adsorb the base onto a weak cation exchanger at neutral pH, followed by elution with a salt gradient or strong acid to suppress resin ionization if stability permits.

Weak bases may be adsorbed on strong cation exchangers. Separation of weak bases from strong bases is achieved by selective elution with a buffer at a pH 1.0–2.0 units above the p*K*a of the required base. These conditions may often be achieved by the use of the weak base pyridine in aqueous solution (**Table 8**). If the weak base is unstable to alkaline

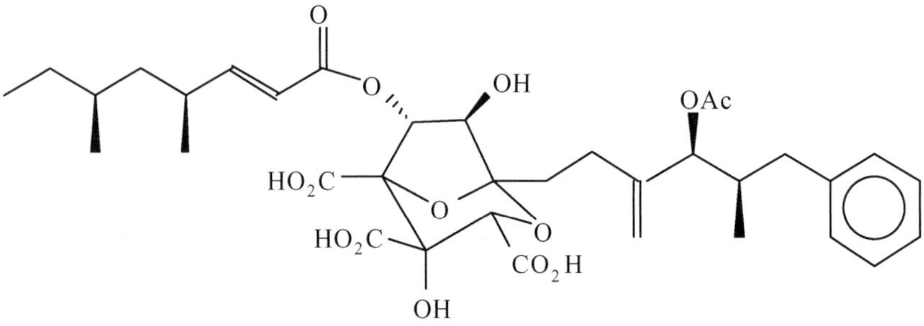

Step	Comment
1. Extract whole fermentation broth with EtOAc at pH 2.	Solvent extraction removed acids from broth and biomass.
2. Adsorbed onto Amberlyst A-21 (acetate cycle)	Solvent resistant macroreticular resin allows adsorption from crude non-aqueous extract. Weak anion exchanger with acetate ions (pK_a 5) sufficiently basic to deprotonate acids for selective adsorption of monocarboxylate ions (pK_{a1} 3.5).
3. Elute with 3% NH_4Cl in 90% aqueous MeOH	
4. Desalt eluate with Diaion HP-20	Desalting with adsorption resin.

Fig. 4. The structure of zaragozic acid A with an extraction scheme.

conditions, it may not be possible to separate it from stronger bases by ion-exchange processes.

5.2.1. Palau'amine

Palau'amine (**Fig. 5**), which shows antibiotic properties, has been isolated from a marine sponge native to the South Pacific *(6)*. The compound, with a basic guanidine group, may be effectively isolated from aqueous extracts using a weak cation-exchange resin. At neutral pH values, the cationic exchanger will carry a negative charge while the guanidine group will be protonated, resulting in retention of the natural material on the resin. The antibiotic is recovered by salt gradient elution with sodium ions.

5.2.2. Gualamycin

Gualamycin (**Fig. 6**) was isolated from the culture broth of a *Streptomyces* species during a screening for novel acaricides *(7)*. The isolation

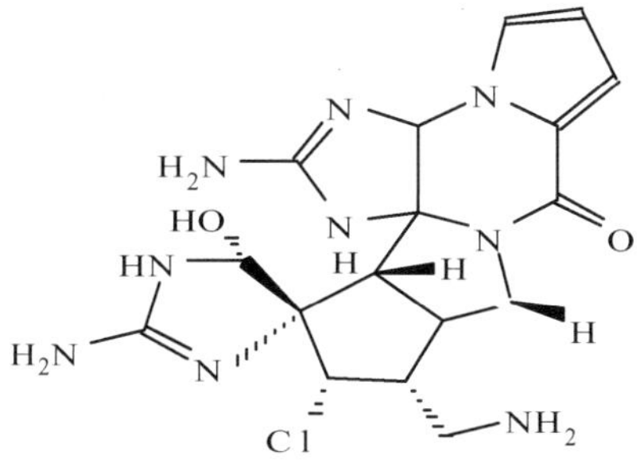

	Step	Note
1.	Lypophilize sponge and extract with MeOH.	
2.	Evaporate solvent and triturate with water.	
3.	Pass through Cellex CM (Na cycle) resin column.	Weak cation exchanger.
4.	Elute with step gradient NaCl; bioassay fractions.	Activity in 0.5M NaCl fraction.
5.	Lypophilize and desalt by trituration with EtOH.	
6.	Chromatograph ethanol soluble material on Sephadex LH-20, eluting with MeOH.	

Fig. 5. The structure of Palau'amine with an extraction scheme.

process typifies the difficulties encountered in the isolation of such types of molecules. Small amphoteric water-soluble compounds are both difficult to solvent-extract and chromatograph on silica or by RP-HPLC. The lack of an identifying chromophore in the structure means that detection is often dependent on the biological screen employed. The presence of an amine group allows manipulation by ion-exchange methods.

5.2.3. Paromomycins

Paromomycin (**Fig. 7**) represents an example of a class of water-soluble aminoglycoside antibiotics isolated some 40 yr ago ***(8)***. As with most of this class of compounds, the presence of a basic amino functionality allows purification by ion-exchange methods.

6. Concluding Remarks

The miniaturization of coupled analytical techniques (e.g., HPLC-MS) applied to natural product chemistry has led to the development and use of solid phase extraction media based on ion-exchange phenomena. The use of ion-exchange chromatography remains an important technique in the isolation of natural products from complex mixtures. The multiple

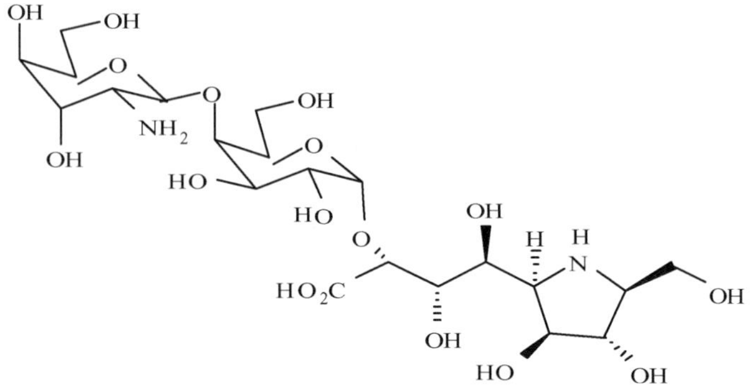

1. Filter broth; pass through charcoal column.
2. Wash with water; elute with step gradient of MeOH.
3. Bioassay fractions; pool active fractions.
4. Pass through Dowex-50W(H^+ cycle) resin column.
5. Wash with water; elute with 2.8% aqueous ammonia.
6. Bioassay fractions; lypophilize and redissolve in water.
7. Pass through CM-Sephadex (Na^+) resin column.
8. Elute with step gradient of aqueous NaCl.
9. Bioassay: desalt.

Fig. 6. The structure of gualamycin with an extraction scheme.

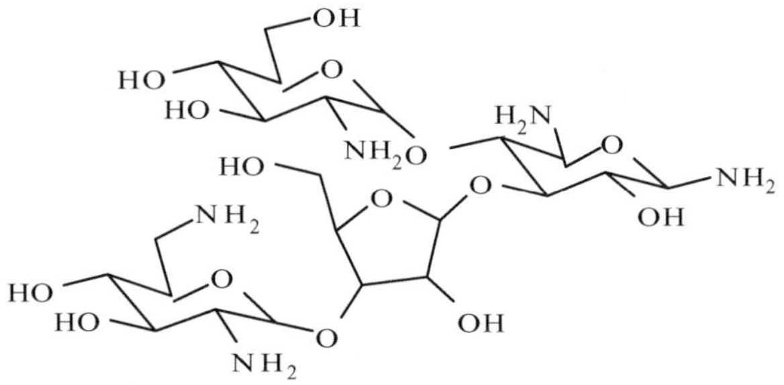

1. Adjust broth to pH 3.0; filter.
2. Adjust filtrate to pH 7.0 with NaOH.
3. Pass through Amberlite IRC-50 (NH_4^+) resin column.
4. Wash with water; elute with 0.5M aqueous NH_4OH.
5. Bioassay eluate fractions; concentrate active fractions *in vacuo*; adjust to pH 7.0.
6. Pass through Amberlite CG-50 (NH_3) resin column.
7. Wash with water; elute with step gradient aqueous NH_4OH (0.05–0.3M).
8. Bioassay fractions; evaporate.

Fig. 7. The structure of paromomycin with an extraction scheme.

applications of resins in the isolation of pyrrolizine alkaloids from phytochemical sources ***(9)*** are a good example. In addition, the increasing demands of biotechnological techniques have led to developments in the use of hydrophilic matrix ion-exchange resins in the isolation of proteins and their applications for the isolation of enzyme natural products from plants. The use of ion exchange resins remains a major tool in the development of large-scale natural product isolation.

References

1. Dufresne, C. (1998) Isolation by ion-exchange methods, in *Natural Products Isolation*, 1 ed. (Cannell, R. J. P., ed.), The Humana Press Inc, New Jersey.
2. Poole, C. F. and Poole, S. K. (1991) *Chromatography Today*. Elsevier, Amsterdam, pp. 422–439.
3. Harland, C. E. (1994) *Ion Exchange: Theory and Practice*. 2nd ed., Royal Society of Chemistry, London.
4. Hamill, R. L. and Crandall, L. W. (1978) Cephalosporin antibiotics, in *Antibiotics; Isolation, Separation and Purification*, vol. 15 (Weinstein, M. J. and Wagman, G.H., eds.), *Journal of Chromatography Library*, pp. 87–91.
5. Bergstrom, J., Dufresne, C., Bills, G., Nallin-Omstead, M., and Byrne, K. (1995) Discovery, biosynthesis, and mechanism of action of the zaragozic acids: potent inhibitors of squalene synthase. *Ann. Rev. Microbiol.* **49,** 607–639.
6. Kinnel, R. B., Gehrken, H.-P., and Scheuer, P. J. (1993) Palau'amine: a cytotoxic and immunosuppressive hexacyclic bisguanidine antibiotic from the sponge *Stylotella agminata*. *J. Am. Chem. Soc.* **115,** 3376, 3377.
7. Tsuchiya, K. Kobayashi, S., Harada, T., Takashi, N. T., Nakagawa, T., and Shimada, N. (1995) Gualamycin, a novel acaricide produced from *Streptomyces* sp. NK11687. I. Taxonomy, production, isolation and preliminary characterisation *J. Antibiot.* **48,** 626–629.
8. Marquez, J. A. and Kershner, A. (1978) Deoxystreptamine-containing antibiotics, in, *Antibiotics; Isolation, Separation and Purification*, vol. 15 (Weinstein, M. J. and Wagman, G. H., eds.), *Journal of Chromatography Library*, pp. 202–207.
9. Kato, A., Kano, E., Adachi, I. et al. (2003) Australine and related alkaloids: easy structural confirmation by ^{13}C NMR spectral data and biological activities. *Tetrahedron: Asymmetry* **14**, 325–331.

7

Separation by High-Speed Countercurrent Chromatography

James B. McAlpine and Patrick Morris

Summary

High-speed countercurrent chromatography provides the natural product chemist with a high-resolution, separatory method that is uniquely applicable to sensitive (unstable) compounds and virtually allows quantitative recovery of the load sample. Different instruments use several means of retaining a stationary liquid phase. The solvent system can be chosen to optimize the separatory power and the number of systems available is limitless. Several examples are provided to illustrate the power of the method and to guide the chemist in the choice of an appropriate system.

Key Words: High-speed countercurrent chromatography (HSCC); centrifugal partition; Ito coil; pH-zone refining; coil planet centrifuge; pristinamycin; Taxol®, cephalomannine; niddamycin; tirandamycin; arizonin; concanamycin; squalestatin; auxin; australifungin; phomopsolide; trichoverroid; 5-*N*-acetylardeemin; oxysporidone; coloradocin; 2-norerythromycin; pentalenolactone; aselacin; siderochelin; 1,3-dimethylisoguanine; michellamine; 6-*O*-methylerythromycin; dorrigocin; halishigamide; tunichrome; gibberellin; tetracycline.

1. Introduction

Modern high-speed countercurrent chromatography (HSCC) has developed only over the last 20 yr or so and offers the natural product chemist a further separation tool with many unique advantages. It is inherently

From: *Methods in Biotechnology, Vol. 20, Natural Products Isolation, 2nd ed.*
Edited by: S. D. Sarker, Z. Latif, and A. I. Gray © Humana Press Inc., Totowa, NJ

the mildest form of chromatography with no solid support and hence no chance of loss of substrate by binding to the column. The only media encountered by the sample are solvent and Teflon® tubing. The former is common to all forms of chromatography and the latter to most. It is true that many of the solvent systems have more components than most other forms of chromatography, but these can be chosen from the most nonreactive and innocuous solvents. Hence, the chromatographer is virtually assured of near 100% recovery of sample from a chromatography. The number of two-phase systems, which can be employed, is limited only by the imagination of the chromatographer, and the systems can be explored by any of several simple tests, prior to a preparative separation, to ensure success. Two similar compounds of almost identical polarity can have surprisingly different partition coefficients in a specific two-phase system resulting in baseline separation by countercurrent chromatography.

Countercurrent methodology had its beginning in the 1950s with the Craig machine *(1)*—a mechanical system of sequential separating cells in which one phase of a two-phase solvent system could be equilibrated with the other phase in successive cells, thereby carrying a solute along according to its partition coefficient between the two phases. Solutes would be washed from the train in the order that their partition coefficients favored the mobile phase. These instruments were cumbersome, delicate, and required a major air-handling system as they invariably leaked organic solvents to some extent. A typical system would involve 200–400 cells, and a separation using such an instrument would take a week to accomplish. They were used because they could achieve separations, which could not be otherwise effected. They were both displaced and replaced—displaced by the high-pressure liquid chromatography (HPLC) and the large number of possible solid supports available for this methodology, and replaced by droplet countercurrent (DCC) instruments that could be used to effect the same separation method in a fraction of the laboratory space. DCC instruments are still available. They are composed of vertical tubes of a diameter, which will allow droplets of one phase of a two-phase solvent system to rise (or fall) through the other phase. These mixing tubes are then connected top to bottom with fine tubing, such that the droplets would completely fill the tubing and exclude the stationary phase from these interconnecting tubes. Hence, the mobile phase is added slowly and allowed to percolate through the mixing chambers under the force of gravity, to achieve a distribution of solute according to its partition coefficient. DCC systems, while sharing the high separatory power of the Craig machine and the high overall recovery of load sample, still suffer from the problem of being slow, and this is further complicated by the need to maintain constant temperature during the course of a chromatography.

HSCC instruments became commercially available around 1980 and have overcome all of their earlier drawbacks. They make countercurrent chromatography a useful means of achieving delicate separations on the milligram-to-gram scale in a few hours. This chapter offers the reader a primer in the use of this technique for separations in natural product isolations.

2. Current Instruments

Although several minor variants are available, instruments are basically of two types: The centrifugal partition chromatography instrument as sold for many years by Sanki Instruments (Kyoto, Japan) and more recently by SEAB (Villejuif, France); the coil planet centrifuge as designed by Yochiro Ito and sold currently by Conway Centri Chrom Inc (Williamsville, NY, USA), PharmaTech (Baltimore, MD, USA), Dynamic Extraction Ltd. (Uxbridge, UK) and AECS (Bridgend, UK). The centrifugal partition instruments effectively replace gravity as the driving force of a DCC procedure with centrifugal force and thus achieve a remarkable increase in the speed of the process. This does necessitate the use of two rotating seals. The Teflon tubing is replaced by small solid blocks or sheets of Tefzel®, honeycombed with a channel system analogous to the tubing of a droplet system. The blocks (cartridges) in the Sanki instrument are connected by Teflon tubing and can be individually replaced should one become clogged or spring a leak. A typical instrument would contain 12 such cartridges. A distinct advantage of this system is the ability to use relatively viscous solvents such as *n*-butanol at room temperature. Although *n*-butanol containing solvent systems can be used in the coil planet centrifuge, they usually result in poor retention of stationary phase at ambient temperatures, and hence suboptimal performance. An obvious weak point in the design of the centrifugal partition chromatograph is the rotating seal, which must remain solvent resistant and leak-proof to a wide variety of solvents under speeds as high as 2000 rpm. The Sanki instrument has a ceramic-graphite spring-loaded seal with a specified 1500 h life expectancy. The authors' limited experience with one of these instruments would suggest that this seal is surprisingly robust. One distinct advantage of this system over the coil planet centrifuge is that the apparatus is inherently symmetrical and there is no need for a counterbalance.

The coil planet centrifuge is just one of a large number of instruments that have been the life work of Yochiro Ito. His study of the movement of one phase of a two-phase solvent system with respect to the other under a variety of imposed vectors, and the use of this behavior as a separatory tool is without equal. The coil planet centrifuge is available both in horizontal and vertical configurations, and although the forces imposed on the solvents are slightly different, the practical effects are essentially equivalent and the two systems can be used interchangeably to effect a separation. The coil planet centrifuge consists of Teflon tubing wound in a spiral around a central cylinder. When the coil is filled with liquid and spun around its axis, an Archimedean screw force is exerted on the liquid, tending to drive it toward the center (head) of the spiral or toward the outside (tail) depending on the direction of spin. This Archimedean screw force provides the counter to the flow, and the means by which one phase is held stationary, while the other is pumped through it. Moreover, if the coil is spun in a synchronous planetary motion such that the period of orbital rotation is the same as that for spin around the axis of the coil, it is possible to thread the feed and exit lines through the center spindle of the coil and out of the center spindle of the orbit. These two lines do not entwine as the instrument spins and hence the need for a rotating seal is removed. In practice, the orbital axis is the drive axis and the coil spin is driven from this axis by two identical cogs. Given that the usual operating speed is around 800 rpm and that a typical chromatography takes 2–3 h and therefore involves an excess of 100,000 rotations, it is not surprising that the inlet and outlet do become twisted with constant use. It is necessary to inspect and occasionally untwist them. For an instrument in steady use, this needs to be done only about once a month or even less frequently and should never be done while the instrument is in motion.

The two-phase system undergoes some interesting dynamics during operation as explained theoretically by Ito *(2)* and demonstrated by strobe light photography by Conway and Ito *(3)*. Within each orbital rotation, the two phases undergo a mixing and a separation step. This is reasonably postulated to increase the partition efficiency, and hence the separatory resolution of the method. Typical instruments of both types hold approximately 300 mL in a column, although analytical coils are available for the coil planet centrifuge with a capacity of 90 mL and are common as are larger coils with up to 1 L capacity. The system must be counterbalanced, and the counterbalance has to be tailored to the solvent system. Newer instruments have as many as four identical coils connected in series and symmetrically placed around the center axis. This provides an internal counterbalance; however, the system is only in balance to the extent that each coil maintains the same amount of stationary and mobile phases. The difficulty with increasing the scale lies in the large centrifugal forces generated and the need to keep such a system in balance as the solvent system changes in composition while filling the columns or if any significant bleeding of stationary phase should occur during a chromatography.

2.1. Vendors

HSCCs have not caught the attention of large instrument makers, and their manufacture tends to be a "cottage industry." Currently, they are available from:

1. Conway Centri Chrom Inc., 52 MacArthur Drive, Williamsville, NY 14221, USA. Tel.: 716-634-3825.
2. AECS, PO Box 80 Bridgend, South Wales CF31 4LH, United Kingdom. Tel.: 1656-649-073.
3. S.E.A.B. (Société d' Etudes et d' Application industrielle de Brevets) 64 Rue Pasteur 94807 Villejuif Cedex, France. Tel.: 1-4678-9111.
4. Pharma-Tech Research Corporation, 6807 York Road, Baltimore, MD 21218, USA. Tel.: 301-377-7018.
5. Sanki Engineering, Ltd., Imazato 2-16-17, Nagaoka-cho, Kyoto 617, Japan.
6. Dynamic Extraction Ltd c/- Brunel University, Uxbridge, UK. Tel.: 44 -1661 854734. E-mail: Ian.Sutherland@brunel.ac.uk.

3. Operation

The use of HSCC as a separation tool in natural product chemistry can have various aspects. Scientists at PanLabs in Bothell, Washington, chose to use it as a dereplication tool *(4)*, by choosing a single solvent system, and building a database of the elution times of known bioactive microbial metabolites. All bioactive extracts, while still at the crude extract stage, were subjected to HSCC on an Ito coil. The retention time of the bioactive peaks eluting from the column can then be correlated with those of similar activities in the database, and the presumptive identity of the bioactive component can be checked by spectroscopic methods. A much more common usage will be as a preparative chromatographic method, either for the purification of crude or semipurified mixtures, or alternatively to separate two closely related congeners that have already been the subject of several other separation steps. The initial approach may be tailored to the particular problem.

3.1. Separation of Crude Mixtures

One can liken the course of an HSCC chromatography to a thin layer chromatography (TLC). Analytes, which strongly favor the stationary phase, tend to behave as would those in a TLC, which have an R_f of zero, while those that strongly favor the mobile phase behave like those with an R_f of 1.0. When running a TLC, the highest resolving power is usually obtained for those analytes with R_f in the vicinity of 0.4. Similarly, there are optimal partition coefficients (*see* Chap. 10) to effect the highest chance of separation. Hence, if the researcher knows some analytical aspect of the metabolite or chemical entity he or she wishes to isolate, a solvent system can be selected to maximize the chances of a successful purification. In many cases, this will be a particular bioactivity. Here, a crude partition coefficient can be determined by distributing the mixture in the two-phase system and bioassaying both phases. Optimal partition coefficients for the Ito coil planet centrifuge are between 1.0 and 2.0 favoring the stationary phase, whereas for centrifugal partition chromatograph, they are between 2.0 and 5.0, again favoring the stationary phase. Partition coefficients based on bioactivity of crude mixtures have the inherent problem that several congeners of a natural product extract may be bioactive, and the determined value is a weighted mean based on potencies, quantities, and partition coefficients of individual components. Although this is theoretically disturbing, in practice it seldom seems to present a problem as one of the two situations will prevail. Either of the congeners will have similar partition coefficients, and one will have chosen a system in which they fall in the area of maximum resolution leading to separation from other components in the mixture, and at worst case no separation from one another. Alternatively, the bioactive components will have different partition coefficients, and the system will be useful in separating them from one another, even if less effective in separating them from other components. Other analytical determinants can be used to estimate the partition coefficients. If the desired compounds are known to be colored, this feature can be used to rapidly assess solvent systems either by eye or with a spectrometer. If the desired isolate can be detected on TLC plates, comparison of the spot intensity from TLC chromatography of equal aliquots of the two phases can give an adequate assessment of the system. TLCs of the two phases can be used with bioautography detection to overcome all of the disadvantages of simple bioassay of the phases.

3.2. Separation of Two Closely Related Congeners

In the course of a natural product isolation, the chemist is often presented with mixtures containing very closely related biosynthetic relatives that may differ only by one or two methylenes, the placement of an olefin, or the stereochemistry of a nonpolar substituent. If the molecules have strongly polar groups common to their structure, the difference in polarity associated with these structural differences can be insignificant and render an adsorption-type chromatographic method useless. In this situation, HSCC is often the separation method of choice, especially if the mixture has already been the subject of multiple chromatographic steps. Although the methods given here for choice of a solvent system may well work, this may be the time to employ more sophisticated analytical techniques to ensure success. TLC or analytical HPLC of each phase in the two-phase system will work if the congeners are sufficiently separated in such a system to assay them. However if it does not, it is worth examining the two phases by ^{1}H-NMR or an HPLC-MS assay to ascertain the partition coefficients of the congeners with a high level of confidence.

3.3. Choosing and Tailoring the Solvent System

A cursory glance at the solvent systems used in **Table 1** will reveal that most of the two-phase systems are multicomponent, and many different systems can arise from the same three or four components by differing the ratio of those components. In the early literature, carbon tetrachloride (CCl_4) and chloroform ($CHCl_3$) were common components in solvent systems. They have several desirable properties, including low viscosity and high density. But with the discovery of their carcinogenicity, they have

Table 1
Application of HSCC in Natural Product Isolation

Solvent system	Compounds	Reference
n-Hexane–EtOAc–MeOH–H_2O 70:30:15:6	Tirandamycins A and B (**Fig. 5**)	*(10)*
n-Hexane–EtOAc–MeOH–H_2O 1:1:1:1	Arizonins and concanamycins (**Fig. 6**)	*(11,12)*
n-Hexane–EtOAc–MeOH–H_2O (0.01 N H_2SO_4) 5:6:5:6	Squalestatins	*(13)*
n-Hexane–EtOAc–MeOH–H_2O 2:3:3:2	Arizonins (**Fig. 6**)	*(11)*
n-Hexane–EtOAc–MeOH–H_2O 3:7:5:5	Auxins (**Fig. 7**)	*(14)*
n-Hexane–EtOAc–MeOH–H_2O (25 mM PO_4^{3-} buffer pH 6.9) 7:3:5:5	Australifungins (**Fig. 8**)	*(15)*
n-Hexane–EtOAc–MeOH–H_2O 2:2:2:1	Phomopsolides (**Fig. 8**)	*(16)*
n-Hexane–EtOAc–MeOH–EtOH–H_2O 10:14:10:2:13	Taxol® and cephalomannine (**Fig. 3**)	*(8)*
n-Hexane–$CHCl_3$–MeOH–H_2O 1:1:1:1	Trichoverroids (**Fig. 9**)	*(17)*
n-Hexane–DCM–MeOH–H_2O 5:1:1:1	Bu2313 B (A tetramic acid) and 5-*N*-acetylardeemin (**Fig. 10**)	*(10,18)*
n-Hexane–DCM–MeOH–H_2O 10:40:17:8	Steroids	*(19)*
Heptane–EtOAc–MeOH–H_2O 1:1:1:1	Oxysporidinone (**Fig. 19**)	*(20)*
$CHCl_3$–EtOAc–MeOH–H_2O 12:8:15:10	Pristinamycins (**Fig. 2**)	*(8)*

$CHCl_3$–MeOH–H_2O 1:1:1	Coloradocin, 2-norerythromycins, pentalenolactone and ascelacins (**Figs. 11** and **12**)	*(22–24)*
$CHCl_3$–MeOH–H_2O 7:13:8	Siderochelin (**Fig. 19**)	*(10)*
$CHCl_3$–MeOH–H_2O 4:3:3	1,3-Dimethylisoguanine (**Fig. 19**)	*(25)*
$CHCl_3$–MeOH–H_2O (0.5% HBr) 5:5:3	Michellamines (**Fig. 13**)	*(26)*
n-Hexane–EtOAc–CH_3CN-MeOH 5:4:5:2	Triterpene acetates (**Fig. 20**)	*(27)*
n-Heptane–C_6H_6–IPA–acetone–H_2O 5:10:3:2:5	2-Norerythromycins (**Fig. 11**), and 6-***O***-methylerythromycin metabolites (**Fig. 14**)	*(21,28)*
n-Hexane–EtOAc–H_2O 3:7:5	2-Norerythromycins (**Fig. 11**)	*(21)*
EtOAc–EtOH–H_2O 3:1:2	Dorrigocins (**Fig. 15**)	*(29)*
EtOAc–MeOH–H_2O	Halishigamides (**Fig. 16**)	*(30)*
i-AmOH–*n*-BuOH–*n*-PrOH–H_2O–HOAc- *t*-Bu_2S 8:12:10:30:25:1	Tunichromes (**Fig. 17**)	*(31)*
Et_2O–MeOH–H_2O (PO_4^{3-}) 3:1:2	Gibberellins (**Fig. 17**)	*(14)*
n-BuOH–H_2O (0.01 N HCl)	Tetracyclines (**Fig. 18**)	*(32)*

become less used. Both dichloromethane (DCM) and diethyl ether can be used, but the researcher should be aware that a vapor lock will force the stationary phase from the system and abort the chromatography. Hence, these solvents need to be used only if the instrument comes with temperature control or if ambient temperature permits. A common approach to a four-component system such as the hexane–ethyl acetate (EtOAc)–methanol (MeOH)–water system is to assume that for organics of medium polarity, hexane and water will be poor solvents, while EtOAc and MeOH are good solvents. The lower phase will consist mainly of MeOH and water, and the upper phase will comprise mainly of hexane and EtOAc. Hence if in the 1:1:1:1 system, the desired compound favors the upper phase, it can be displaced toward the lower, by increasing the proportion of hexane or MeOH. Increasing the proportion of MeOH has its limits, as at some level the system will become monophasic. In choosing the solvent system, it is necessary to avoid those that form an emulsion. A useful practical test is to shake well together a milliliter of each of the two phases and allow the mixture to separate under gravity. The separation should be complete in 5 s.

When working with ionizable compounds, it is advisable to be sure that these are maintained in the same ionization state throughout the chromatography, and that this has been taken into account when the solvent system is being chosen. This can be effected by including small amounts of an acid or a base in the solvent system, or by including low concentration buffers as the aqueous component. A good alternative for ionizable compounds is to use the pH-zone-refining technique, a partition ion-exchange displacement method, developed by Ito over the last decade ***(5)*** (*see* later). In any of these approaches, it is important to take into account the effect on the solubility of the load sample. The amount of material, which can be successfully chromatographed, is determined by this solubility (and of course by the differences in partition coefficients of the components). Typical loads for a 300 mL coil are in the order of 200 mg, but this can vary by almost an order of magnitude in either direction dependent on solubility.

3.4. Physical Aspects of Operation

All of the HSCC instruments are effectively closed systems. It is not necessary to locate the actual instrument in an exhaust hood, however, the solvent will be pumped from reservoirs, and the eluent is usually collected in a fraction collector. Because almost all systems involve volatile organic solvents, it is advisable to locate these peripherals in a hood. For smaller systems (~ 250 mL capacity), the pump must be capable of delivering between 2 and 5 mL/min and should not produce large pulses. A typical 3-way injection valve is required, and the sample can be loaded in any volume from 1 to 10 mL. It is of paramount importance that the sample be completely dissolved. To avoid any possibility of salting out, it is common to load the sample in a mixture of the two phases. The machine should be fully loaded and at least the initial parts of the instrument should be equilibrated and rotated before the sample is loaded.

3.5. pH-Zone Refining Chromatography

Over the last decade, Ito has demonstrated the use of HSCC in a semi-displacement mode for ionizable compounds. pH-zone refining chromatography uses a two-phase aqueous/organic system, in which each phase is modified with an "ion-pairing" reagent. Ito's nomenclature is to regard the normal displacement mode as that with the aqueous phase stationary, and the reverse displacement mode as that with this phase as mobile. He refers to the modifier in the stationary phase as a "retainer" and that in the mobile phase as a "displacer." The optimal concentration of each modifier is different for the two modes. For example, in the separation of *Crinum moorei* alkaloids, in a *t*-butyl methyl ether/water system, in the displacement mode the aqueous stationary phase was modified with 10 mM HCl and the mobile organic phase with 10 mM triethylamine. However, in the reverse displacement mode the same modifiers were used at half of this concentration. In each case, a 3 g crude mixture of the three alkaloids, crinine, powelline, and crinamidine (**Fig. 1**), were completely separated on a 300 mL capacity Ito coil instrument. However, the reversed-phase mode was complete in 150 min, whereas the normal phase required almost 400 min *(6)*. Unlike typical countercurrent chromatography, pH-zone-refining results in elution of the analytes as rectangular peaks, with impurities concentrated at the interfaces between peaks. Ito states that the two modifiers, HCl and triethylamine, have proved successful for all of the separations of organic bases that he has undertaken, but that in the separation of acidics the most versatile modifiers are trifluoroacetic acid and ammonium hydroxide. However, other acids have sometimes been more desirable *(5)*. It is noticeable that the loading capacity of most

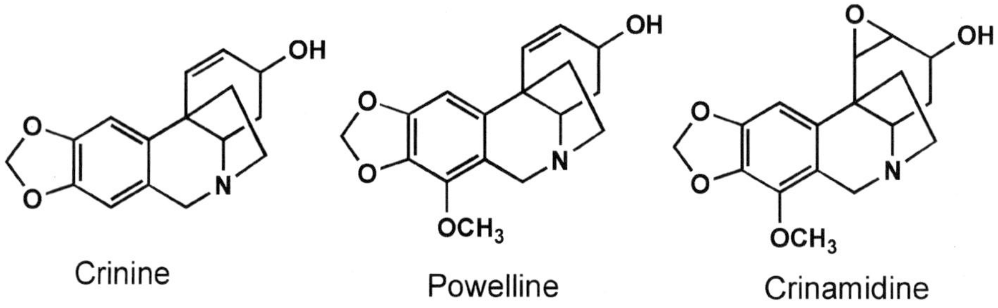

Crinine Powelline Crinamidine

Fig. 1. Structures of *C. moorei* alkaloids.

pH-zone refining chromatographies is greatly enhanced over typical partition chromatography.

3.6. Use of the Ito Coil Planet Centrifuge

When using the Ito coil, the researcher is presented with choosing from three twofold variables:

1. The question of which phase to select as the stationary phase, i.e., which phase to fill the column with.
2. The choice of the inlet tube, either the "head" or the "tail" of the column.
3. The question of which direction to spin the colmn, i.e., to have the Archimedean force directed to the inside or the outside of the spiral.

Two of the eight possibilities will usually work well, two poorly and the other four will result in no retention of the stationary phase.

The tubings to the columns are labeled "head" and "tail," and spin directions as "forward" and "reverse". It is advisable to fill the column with the stationary phase while it spins with an Archimedean force against the fill. This ensures that the column is filled without any vapor blocks. The column does not have to be spun at normal running speed as this will create balancing problems. When the column is filled with stationary phase, it should be spun at 800–1000 rpm while pumping the mobile phase. It is possible to introduce the sample load with the solvent front of the mobile phase, but in a new system this is usually unwise. The system can be tested to ascertain the displacement of stationary phase before introducing the sample. In the better systems, only about 10% of the stationary phase is displaced before breakthrough of the mobile phase. After that, only the mobile phase is eluted from the column.

It is useful to keep in mind the theoretical shape of the elution curves. Components with partition coefficients strongly favoring the mobile phase will be eluted very early in chromatography and will be in a sharp peak. As the chromatography continues, the peaks eluted with the mobile phase broaden. After two to three column volumes of mobile phase have been eluted, it is possible to reverse the direction of spin and displace the stationary phase either by continuing to pump mobile phase or preferably by a stream of nitrogen. The stationary phase should also be collected in fractions, as separations could have been effected but without the compounds eluting. Components eluted with the stationary phase will also come as sharp peaks. If the stationary phase is displaced with nitrogen, the researcher should be wary of the increasing flow rate as the column empties.

3.7. Use of the Centrifugal Partition Chromatograph

With this mode of HSCC, the choices are more straightforward for the novice. The only choice that needs to be made in running the instrument is to identify which is the mobile phase. The S.E.A.B. instrument, the Kromaton, comes with a conveniently labeled switch, "ascending"–"descending," and this can be mentally placed in the old droplet format, in which the upper phase mobile will be "ascending" and the converse. One does not have to carefully estimate counterbalances. The Kromaton has neoprene seals that are relatively inexpensive and easy to change, compared to the ceramic once in the Sanki instrument. They do need to be changed more frequently, however, and the cartridge in the 1 L instrument is quite heavy. The instrument should be loaded with a spin speed of 200 rpm for the 200 mL instrument and 900 rpm for the 1 L instrument. A flow rate of between 1 and 5 mL/min is appropriate for the 200 mL instrument, dependent on the relative densities of the two phases. However, the 1 L instrument is run with flow rates between 20 and 30 mL/min, provided that the pressure on the column does not exceed 800 psi. This instrument is quiet and stable.

3.8. Detection

In principle, any of the detection systems commonly used in chromatography can be used to monitor the analytes in the eluate. In practice, however, UV detectors give esthetically displeasing trace, as the trace leakage of the stationary phase gives a spiked curve. Evaporative light scattering detectors have proven satisfactory *(7)* and have the advantage of not requiring a chromophore. In practice, with biologically active compounds, it is usually convenient to simply collect fractions and subsequently assay these for activity, and/or by TLC.

4. Examples of the Use of HSCC for the Separation of Natural Products

The following few examples from the literature have been chosen to represent the power of the method. In each case, baseline or near baseline separation of two close structurally related congeners has been obtained. Each represents, by any separatory method, a considerable challenge.

4.1. Separation of Pristinamycins

The pristinamycins (**Fig. 2**) are an unusual complex of antibiotics in that they consist of two pairs of peptolide antibiotics very closely related within the two pairs but with virtually no structural relationship from one pair to

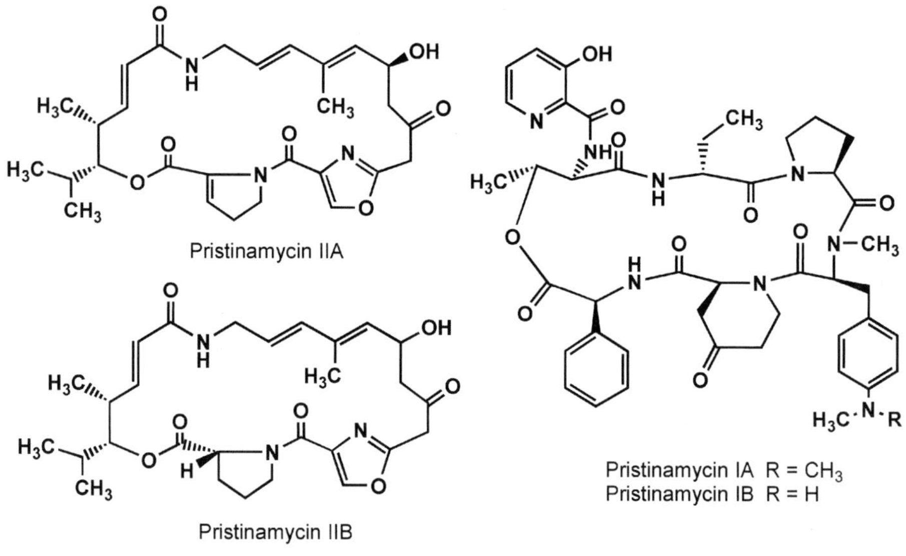

Pristinamycin IIA

Pristinamycin IIB

Pristinamycin IA R = CH_3
Pristinamycin IB R = H

Fig. 2. Structures of pristinamycins.

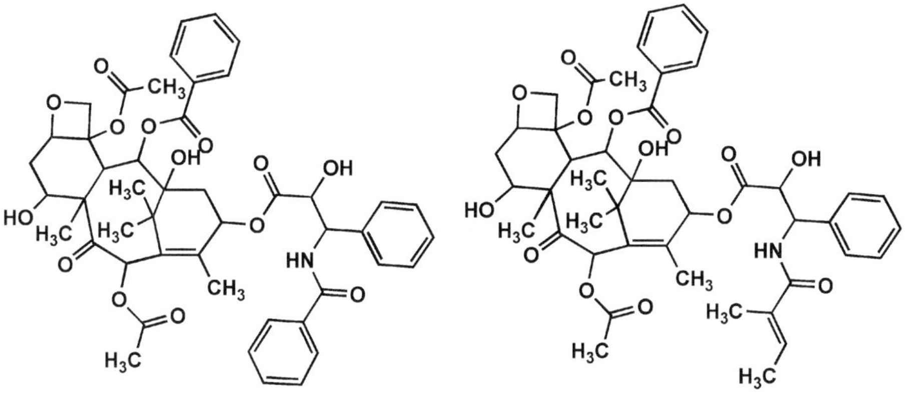

Fig. 3. Structures of Taxol® and cephalomannine.

the other. Pristinamycins IA and IB differ only in the degree of *N*-methylation of a 4′-aminophenylalanine moiety, while pristinamycins IIA and IIB differ only in that IIA has a 2,3-dehydroproline moiety where IIB has a proline. Thiébaut and his group *(8)* were able to achieve baseline separations between IIA and IIB with a system comprising $CHCl_3$–EtOAc–MeOH–water (12:8:15:10) on a triple planetary coil instrument at 1400 rpm with the upper phase mobile. Pristinamycins IA and IB were best separated with a system in which the same components were in the ratio 6:4:8:1, where the last component was formic acid "to control the pH" but of otherwise unspecified strength.

4.2. Separation of Taxol and Cephalomannine

The anticancer agent, *Taxol*®*, can be obtained from a number of *Taxus* species, but invariably it occurs with sizeable amounts of the congener, cephalomannine (**Fig. 3**). These complex diterpenes differ only in the nature of the amide carboxylic acid attached to the amine of the phenylisoserine side chain. In the case of Taxol®, this is a benzoic acid moiety whereas in cephalomannine it is a tiglic acid group. These two impart very

*In an aberration, the US Department of Patents and Trademarks allowed Bristol Myers Squibb to trademark the name taxol despite the fact that this name had been given to the natural product many years earlier by its discoverer and had been used in the literature consistently. The current approved generic name for the compound is paclitaxel.

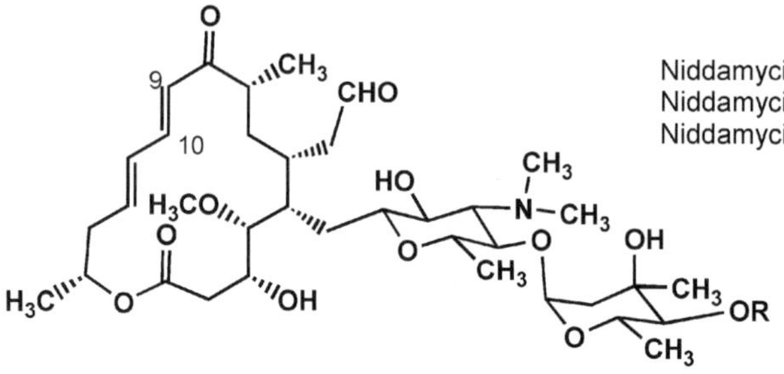

Niddamycin A1 R = $COCH_2CH(CH_3)_2$
Niddamycin B R = $COCH_2CH_2CH_3$
Niddamycin F = 9,10-Dihydroniddamycin A1

Fig. 4. Structures of niddamycins.

little selective polarity to the two natural products, and their separation is notoriously difficult. Almost baseline separation of a small sample (6.1mg) was achieved however by Chiou et al. ***(9)*** using a system of hexane-EtOAc–MeOH–ethanol–water (10:14:10:2:13) with the aqueous phase mobile. In this system, Taxol® had a partition coefficient of 1.8 and cephalomannine 1.42.

4.3. Separation of Niddamycins

The 16-membered antibacterial macrolide niddamycin complex (**Fig. 4**) is produced by *Streptomyces djakartensis* as a mixture of aliphatic esters of the 3″-hydroxyl, the secondary alcohol on the neutral sugar mycarose.

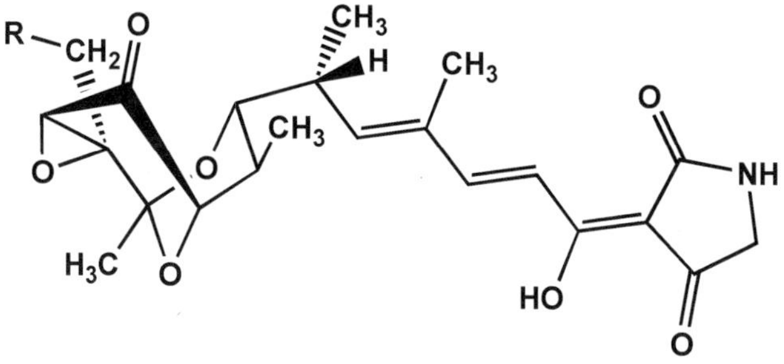

Tirandamycin A , R=H
Tirandamycin B , R=OH

Fig. 5. Structures of tirandamycins.

Niddamycin A1 has a butyryl ester, whereas niddamycin B has the isovaleryl ester at this position. In addition, another congener, niddamycin F is similar to niddamycin A1, except that the 9,10 olefinic bond in the macrolide ring of A1 is fully reduced in niddamycin F. Chen and coworkers achieved baseline separation of a 200 mg sample of all three niddamycins on an Ito coil in a system of carbon tetrachloride–methanol–0.01 M aqueous phosphate buffer at pH 7.0 in the ratio 2:3:2. With the aqueous phase mobile, niddamycin A1 was eluted first followed by niddamycin B, while niddamycin F was retained and was recovered from fractions of the stationary phase when it was pumped from the column (**Table 1**).

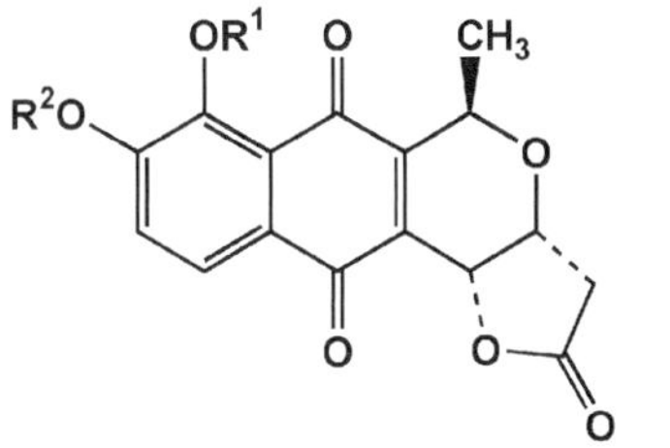

Arizonin A1 $R^1 = CH_3$, $R^2 = H$
Arizonin B1 $R^1 = H$, $R^2 = CH_3$
Arizonin C1 $R^1 = R^2 = CH_3$

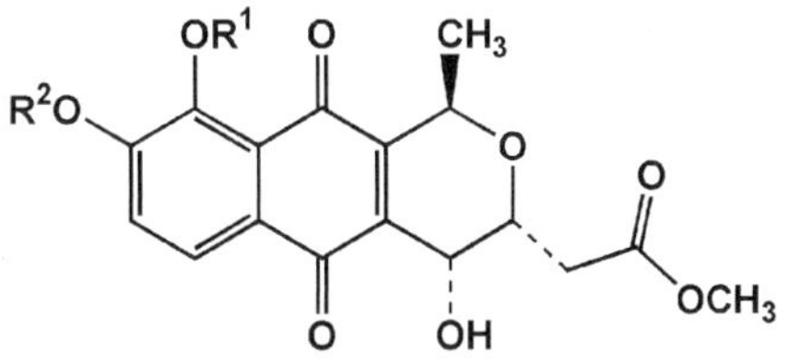

Arizonin A2 $R^1 = CH_3$, $R^2 = H$
Arizonin B2 $R^1 = H$, $R^2 = CH_3$
Arizonin C2 $R^1 = R^2 = CH_3$

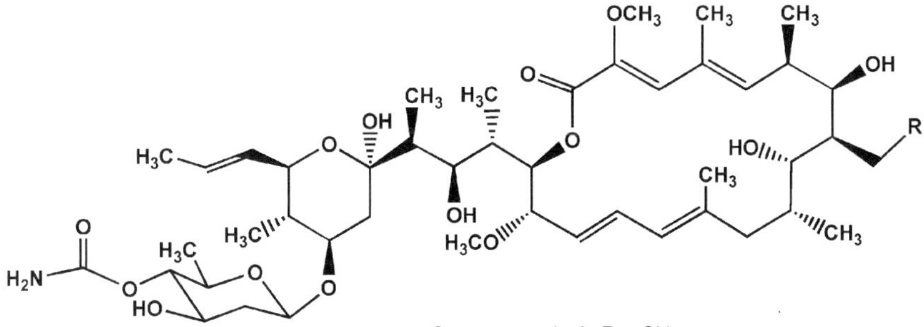

Concanamycin A $R = CH_3$
Concanamycin B $R = H$

Fig. 6. Structures of arizonins and concanamycins.

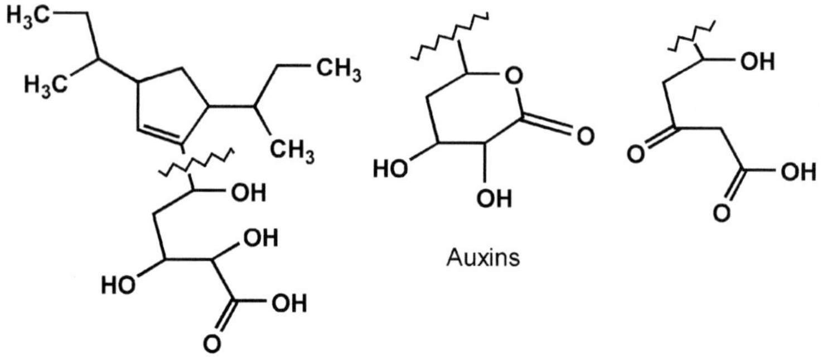

Auxins

Fig. 7. Structures of auxins.

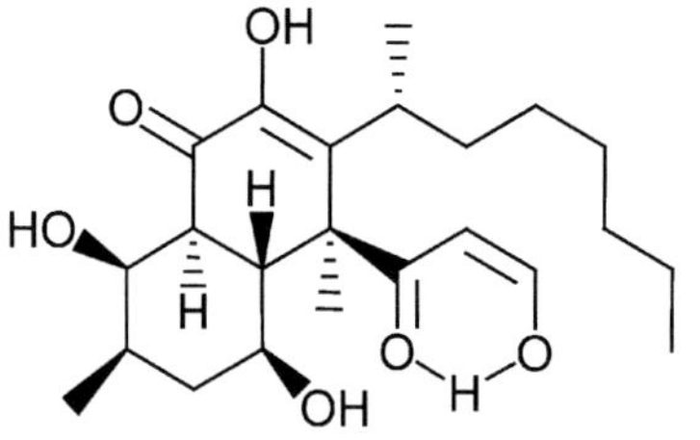

Australifungin

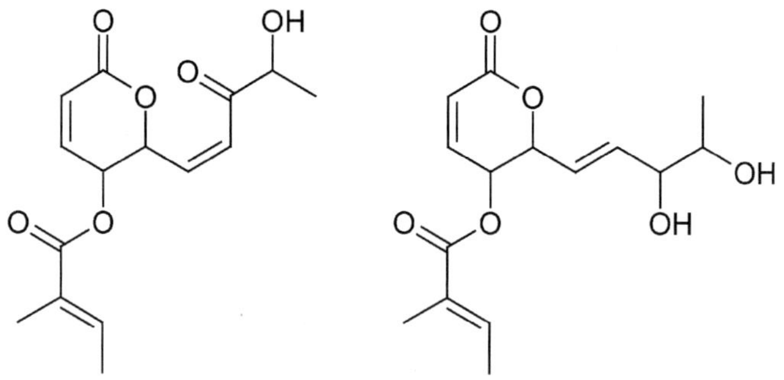

Phomopsolide A

Phomopsolide B

Fig. 8. Structures of australifungin and phomopsolides.

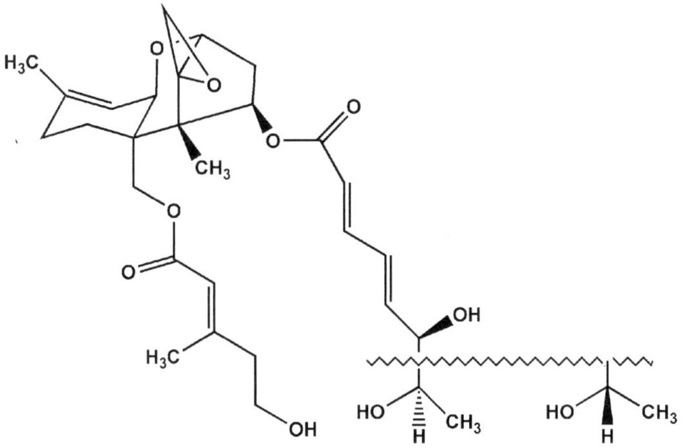

Isotrichoverrin A Isotrichoverrin B

Fig. 9. Structures of trichoverroids.

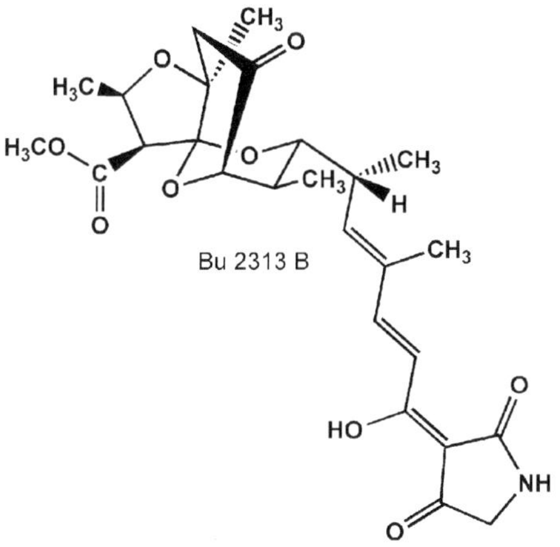

Bu 2313 B

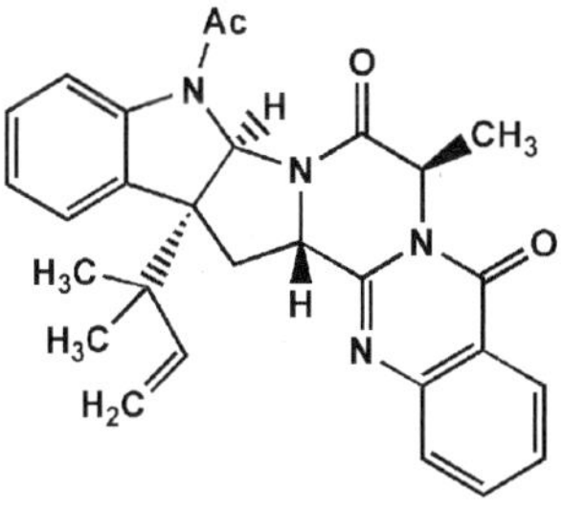

5-*N*-Acetylardeemin

Fig. 10. Structures of Bu2313 B and 5-*N*-acetylardeemin.

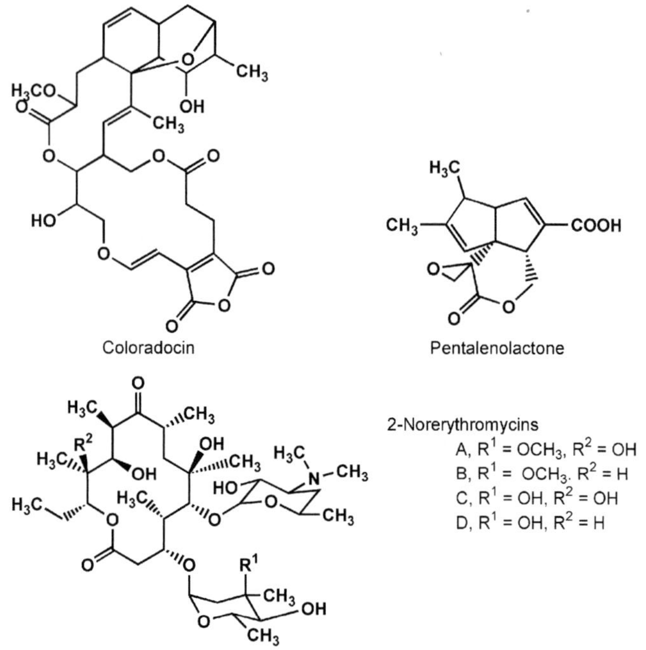

H_3CO O O OH CH_3 CH_3 O O HO O O O O

Coloradocin

H_3C CH_3 COOH O O O

Pentalenolactone

O H_3C CH_3 R^2 H_3C OH OH CH_3 H_3C H_3C N CH_3 HO H_3C O O O CH_3 O O R^1 CH_3 OH O CH_3

2-Norerythromycins
A, R^1 = OCH_3, R^2 = OH
B, R^1 = OCH_3, R^2 = H
C, R^1 = OH, R^2 = OH
D, R^1 = OH, R^2 = H

Fig. 11. Structures of coloradocin, pentalenolactone, and 2-norerythromycins.

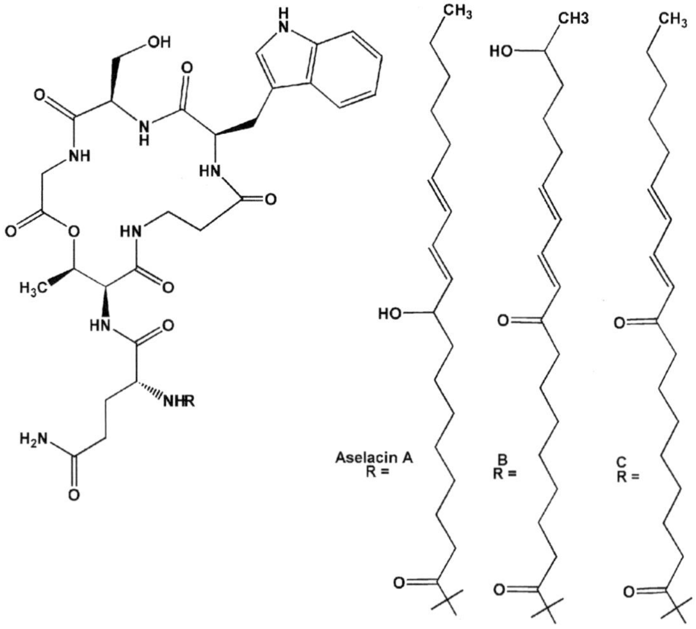

OH O O H N N H NH HN O O O HN H_3C O HN O NHR H_2N O

CH_3 CH3 CH_3 HO HO O O

Aselacin A R =

B R =

C R =

O O O

Fig. 12. Structures of aselacins.

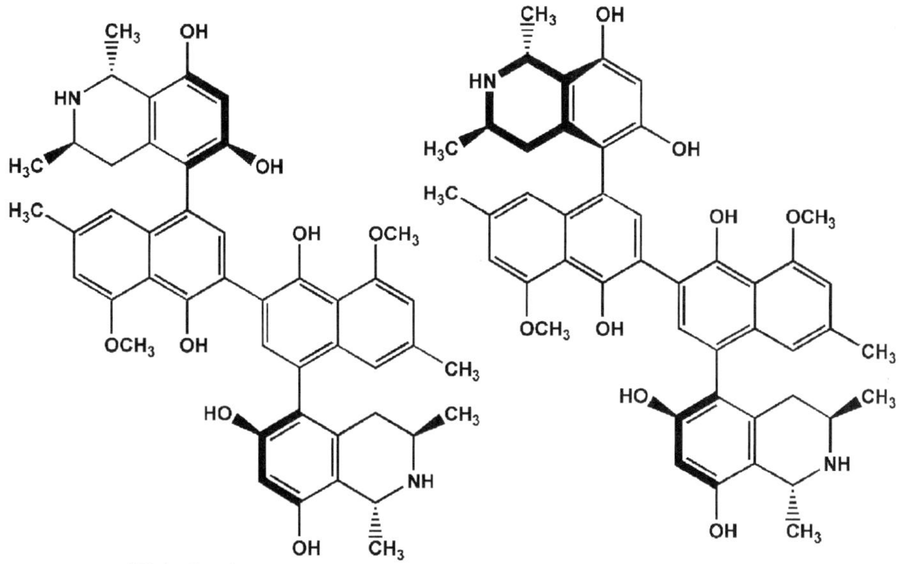

CH_3 OH
HN
H_3C OH
H_3C
OH OCH_3
OCH_3 OH
CH_3
HO CH_3
NH
OH CH_3

Michellamine A

CH_3 OH
HN
H_3C OH
H_3C
OH OCH_3
OCH_3 OH
CH_3
HO CH_3
NH
OH CH_3

Michellamine B

Fig. 13. Structures of michellamines.

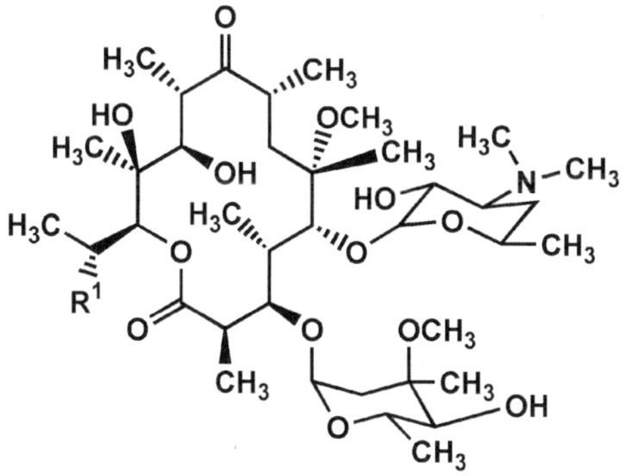

O
H_3C CH_3
HO OCH_3
H_3C OH CH_3 H_3C
N CH_3
HO
H_3C H_3C O O
O CH_3
R^1
O O OCH_3
CH_3 CH_3
OH
O
CH_3

6-O-Methylerythromycin
(Clarithromycin) $R^1 = H$
(14R)-14-Hydroxyclarithromycin $R^1 = OH$

Fig. 14. Structures of 6-*O*-methylerythromycin metabolites.

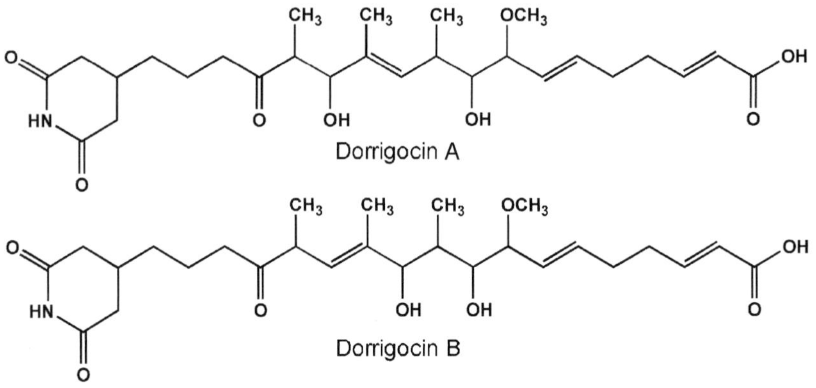

Dorrigocin A

Dorrigocin B

Fig. 15. Structures of dorrigocins.

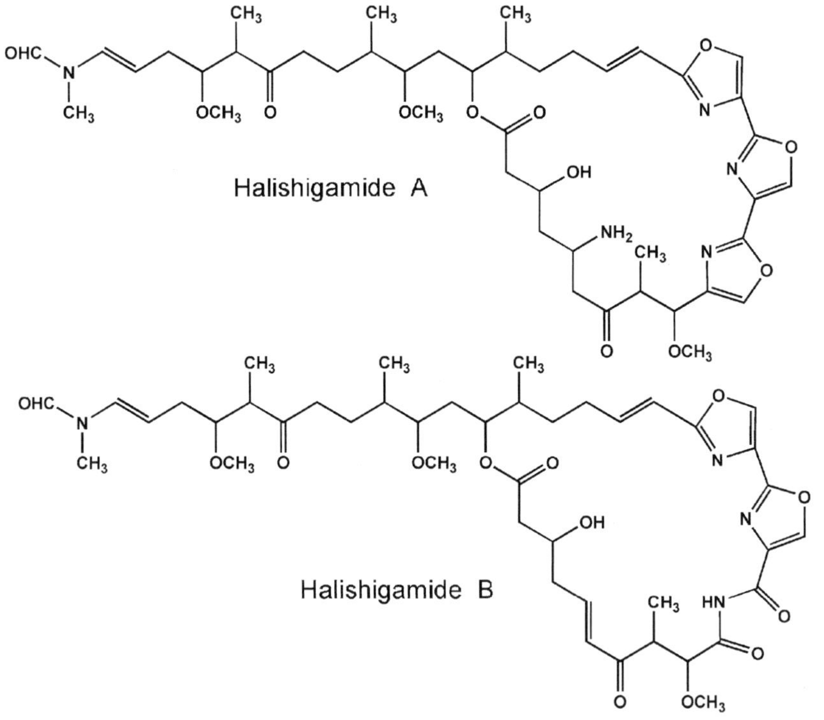

Halishigamide A

Halishigamide B

Fig. 16. Structures of halishigamides.

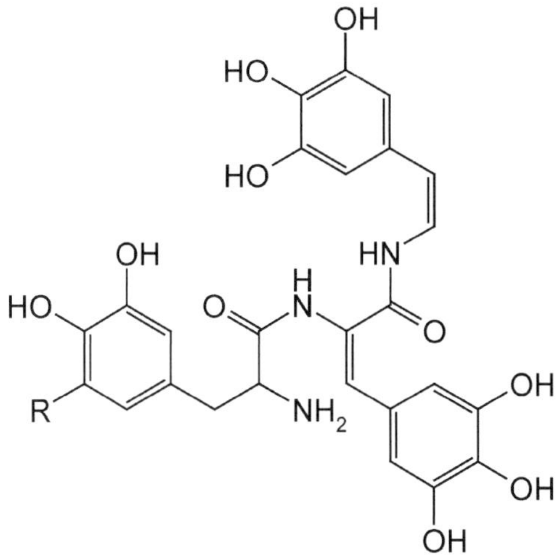

Tunichrome A1 (R = H) and tunichrome B1 (R = OH)

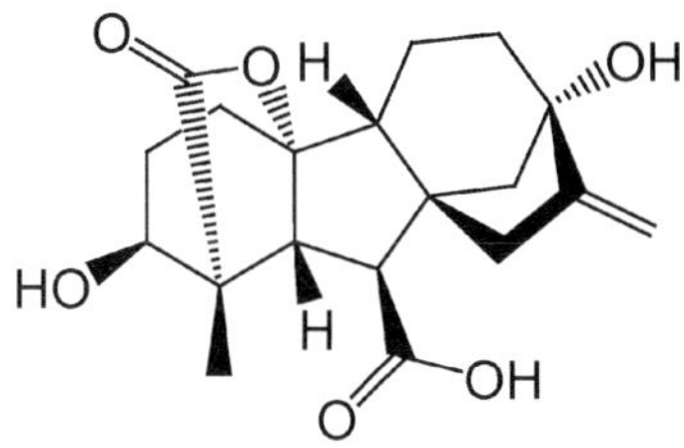

Gibberellic acid

Fig. 17. Structures of tunichromes and gibberellic acid.

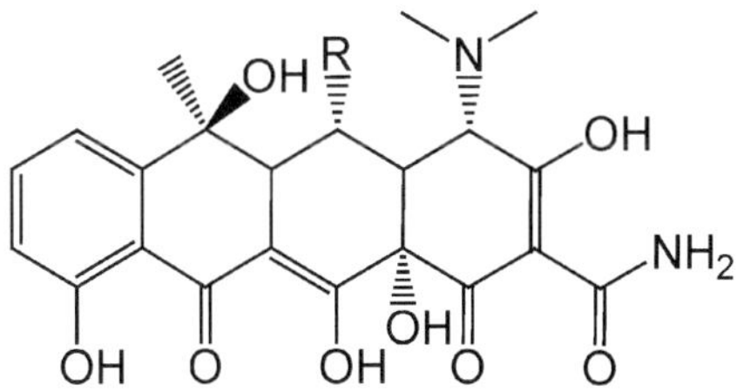

Tetracycline (R = H) and oxytetracycline (R = OH)

Fig. 18. Structures of tetracyclines.

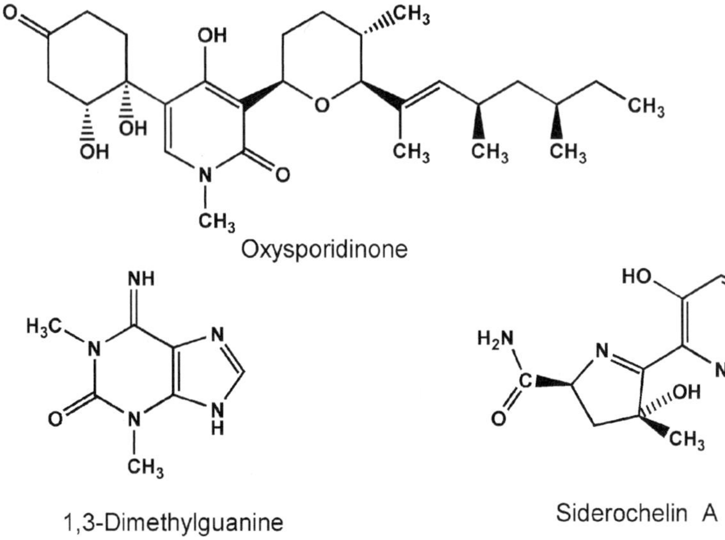

Fig. 19. Structures of oxysporidinone, siderochelin, and 1,3-dimethylisoguanine.

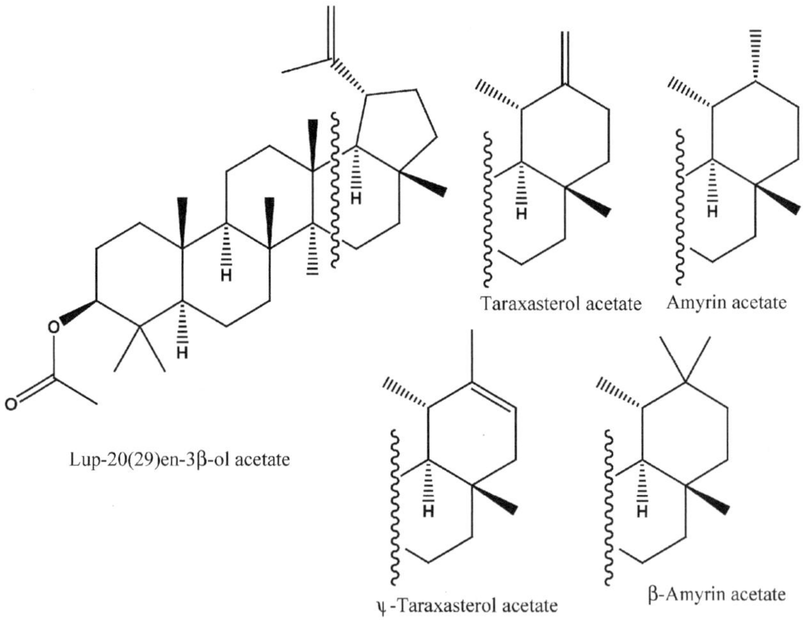

Fig. 20. Structures of triterpene acetates.

References

1. Craig, L. C. and Craig, D. (1956) *Techniques in Organic Chemistry, vol. III, Separation and Purification* (Weissberger, A. ed.), Interscience Publishers Inc. New York, Part I pp. 247–254.
2. Ito, Y. (1986) High-speed countercurrent chromatography. *CRC Crit. Rev. Anal. Chem.* **17,** 65–143.
3. Conway, W. and Ito, Y. (1984) *Analytical Chemistry—Applied Spectroscopy Section.* Pittsburgh Conference and Exposition. Atlantic City, Abstract 472.
4. Baker, D. (1997) Optimizing microbial fermentation diversity for natural product discovery. *IBC Conference on Natural Product Discovery—New Technologies to Increase Efficiency and Speed.* March 17, 18, Coronado, California.
5. Ito, Y (1996) pH-peak-focusing and pH-zone-refining countercurrent chromatography, in *High-Speed Countercurrent Chromatography* (Ito, Y. and Conway, W.D., eds), John Wiley and Sons, New York, Chichester, Brisbane, Toronto, Singapore, pp. 121–175.
6. Ito, Y. and Ma, Y (1995) pH-zone-refining countercurrent chromatography (review). *J. Chromatogr.* **753,** 1–36.
7. Schaufelberger, D. E. and McCloud, T. G. (1991) Laser-light-scattering detection for high-speed countercurrent chromatography. *Chromatography* **726,** 87–90.
8. Drogue, S., Rolet, M.-C., Thiébaut, D., and Rosset, R. (1992) Separation of pristinamycins by high-speed countercurrent chromatography, I. Selection of solvent system and preliminary preparative studies. *J. Chromatogr.* **593,** 363–371.
9. Chiou, F. Y., Kan, P., Chu, I.-M., and Lee, C.-J. (1997) Separation of taxol and cephalomannine by countercurrent chromatography. *J. Liquid Chromatogr. Rel. Technol.* **20,** 57–61.
10. Brill, G. M., McAlpine, J. B., and Hochlowski, J. E. (1985) Use of coil planet centrifuge in the isolation of antibiotics. *J. Liquid Chromatogr.* **8,** 2259–2280.
11. Hochlowski, J. E., Brill, G. M., Andres, W. W., Spanton, S. G., and McAlpine, J. B. (1987) Arizonins, a new complex of antibiotics related to Kalafungin II. Isolation and characterization. *J. Antibiot.* **40,** 401–407.
12. Martin, D. G., Biles, C., and Peltonen, R. E. (1986) Countercurrent chromatography in the fractionation of natural products. *Am. Lab.* **18,** 21–26.
13. Dawson, M. J., Farthing, J. E., Marshall, P. S., et al. (1992) The Squalestatins, novel inhibitors of squalene synthase produced by a species of *Phoma*. I. Taxonomy, fermentation, isolation, physico-chemical properties and biological activity. *J Antibiot.* **45**, 639–647.
14. Mandava, N. B. and Ito, Y. (1982) Separation of plant hormones by countercurrent chromatography. *J. Chromatogr.* **247,** 315–325.

15. Mandala, S. M., Thorton, R. A., Frommer, B. R., et al. (1995) The discovery of Australifungin, a novel inhibitor of sphinganine *N*-acyltransferase from *Sporormiella australis.* Producing organism, fermentation, isolation, and biological activity. *J. Antibiot.* **48**, 349–356.
16. Stierle, D. B., Stierle, A. A., and Ganser, B. (1987) New phomopsolides from a *Penicillium* sp. *J. Nat Prod.* **60,** 1207–1209.
17. Jarvis, B. B., DeSilva, T., McAlpine, J. B., Swanson, S. J., and Whittern, D. N. (1992) New trichoverroids from *Myrothecium verrucaria* isolated by high speed countercurrent chromatography. *J. Nat Prod.* **55,** 1441–1446.
18. Hochlowski, J. E., Mullally, M. M., Spanton, S. G., Whittern, D. N., Hill, P., and McAlpine, J. B. (1993) 5-*N*-Acetylardeemin, a novel heterocyclic compound which reverses multiple drug resistance in tumor cells II. Isolation and elucidation of the structure of 5-*N*-acetylardeemin and two congeners. *J. Antibiot.* **46,** 380–386.
19. Williams R. G. (1985) *Analytical Chemistry—Applied Spectroscopy Section.* Pittsburgh Conference and Exposition. New Orleans, LA, Abstract 300.
20. Breinholt, J., Ludvigsen, S., Rassing, B. R., Rosendahl, C. N., Nielsen, S. E., and Olsen, C. E. (1997) Oxysporidinone: a novel antifungal, *N*-methyl-4-hydroxy-2-pyridinone from *Fusarium oxysporum. J. Nat. Prod.* **60,** 33–35.
21. Ōmura, S., Iwata, R., Iwai, Y., Taga, S., Tanaka, Y., and Tomoda, H. (1985) Luminamicin, a new antibiotic. Production, isolation and physico-chemical and biological properties. *J. Antibiot.* **38,** 1322–1326.
22. Chen, R. H., Hochlowski, J. E., McAlpine, J. B., and Rasmussen, R. R. (1988) Separation and purification of macrolides using the Ito multi-layer horizontal coil planet centrifuge. *J. Liquid Chromatogr.* **8,** 2259–2280.
23. McAlpine, J. B., Tuan, J. S., Brown, D. P., et al. (1987) New antibiotics from genetically engineered actinomycetes I. 2-Norerythromycins, isolation and structural determinations. *J. Antibiot.* **40**, 1115–1122.
24. Hochlowski, J. E., Hill, P., Whittern, D. N., et al. (1994) Ascelacins, novel compounds that inhibit binding of endothelin to its receptor II. Isolation and elucidation of structures. *J. Antibiot.* **47**, 528–535.
25. Mitchell, S. S., Whitehall, A. B., Trapido-Rosenthal, H. G., and Ireland, C. M. (1997) Isolation and characterization of 1,3-dimethylisoguanine from the Bermudian sponge *Amphimedon viridis. J. Nat. Prod.* **60,** 727–728.
26. Hallock, Y. F., Manfredi, K. P., Dai, J.-R., et al. (1997) Michellamines D-F, new HIV- inhibitory dimeric naphthylisoquinoline alkaloids, and Korupensamine E, a new antimalarial monomer, from *Ancistrocladus korupensis. J. Nat Prod.* **60**, 677–683.
27. Abbott, T., Peterson, R., McAlpine, J., Tjarks, L., and Bagby, M. (1987) Comparing centrifugal countercurrent chromatography, nonaqueous reversed phase HPLC and Ag ion exchange HPLC for the separation and characterization of triterpene acetates. *J. Liquid Chromatogr.* **12,** 2281–2301.

28. McAlpine, J. B., Theriault, R. J., Grebner, K. D., Hardy, D. J., and Fernandes, P. B. (1987) Minor products from the microbial transformation of 6-*O*-methylerythromicin A by *Mucor circinelloides. 27th Interscience Conference on Antimicrobial Agents and Chemotherapy*, New York, Abstract.
29. Hochlowski, J. E., Whittern, D. N., Hill, P., and McAlpine, J. B. (1994) Dorrigocins: novel antifungal antibiotics that change the morphology of *ras*-transformed NIH/3T3 cells to that of normal cells II. Isolation and elucidation of structures. *J. Antibiot.* **47,** 870–874.
30. Kobayashi, J., Tsuda, M., Fuse, H., Sasaki, T., and Mikami, Y. (1997) Halishigamides A–D, new cytotoxic oxazole-containing metabolites from Okinawan sponge Halichondria sp. *J. Nat. Prod.* **60,** 150–154.
31. Bruening, R. C., Oltz, E. M., Furukawa, J., and Nakanishi, K. (1985) Isolation and Structure of Tunichrome B-1, a reducing blood pigment from the tunicate *Ascidia nigra* L. *J. Am. Chem. Soc.* **107,** 5298–5300.
32. Zhang, T. (1984) Horizontal flow-through coil planet centrifuge: some practical applications of countercurrent chromatography. *J. Chromatogr.* **315,** 287–297.

8

Isolation by Preparative High-Performance Liquid Chromatography

Zahid Latif

Summary

Preparative high-performance liquid chromatography (HPLC) has become a mainstay of natural product isolation and purification. The various modes available (e.g., normal-phase, reversed-phase, size exclusion, and ion-exchange) to date can be used to purify most classes of natural products. This chapter presents an overview of the different modes along with a practical guide as to how to purify a natural product using the most robust and widely used of the modes, namely reversed-phase preparative HPLC. Instrumentation setup and detection methods, sample preparation, method development, and sample work up are also discussed.

Key Words: Natural products; preparative HPLC; method development; column; solvent; fraction; purification.

1. Introduction

The use of preparative high-performance/pressure liquid chromatography (prep HPLC) has become a mainstay in the isolation of most classes of natural products over the last 10 yr. The relative cost of prep HPLC systems has fallen because of increased competition, with the arrival of numerous column and equipment manufacturers. In addition, the constant innovation and new applications within the area of HPLC have meant that systems, which were "state of the art" some 10 yr ago, are now within the reach of most research groups.

From: *Methods in Biotechnology, Vol. 20, Natural Products Isolation, 2nd ed.*
Edited by: S. D. Sarker, Z. Latif, and A. I. Gray © Humana Press Inc., Totowa, NJ

Prep HPLC is a robust, versatile, and usually rapid technique by which compounds can be purified from complex mixtures. The main difference between prep HPLC and other "lower pressure" column chromatographic system is the consistency and size of the particles in the stationary phase (*see* Chap. 5). Particle size distribution is critical when trying to separate a mixture of two compounds, because the smaller the particle size, better the separating power (or resolution) between the two compounds. The "average" particle size of prep HPLC stationary phases is typically between 3 and 10 μm, substantially smaller than other stationary phases. The particles are synthesized to be spherical and the size distribution to be narrow, which allows the stationary phase to be tightly packed in a highly uniform and reproducible manner. In addition, this minimizes the occurrence of voids or channels, which would disrupt the mobile phase traveling uniformly through the stationary phase and lead to inefficient separation. The small particle size results in having to use high pressures (up to 3–4000 psi) to push the mobile phase through the system. However, the high surface area available for the solutes to interact with the stationary phase results in a chromatography with high powers of resolution that are necessary for purifying complex natural product mixtures.

Crude natural product extracts and mixtures can sometimes consist of hundreds of compounds, and the isolation of particular components presents its own unique problems. Invariably, a fast and efficient technique is required to purify out the compounds of interest. However, the end requirement of the pure compounds is what drives the size and scale of the isolation technique used. If microgram quantities of compound are needed for initial bioassay screening, then purification can sometimes be carried out using analytical-scale HPLC systems, where the column internal diameter (i.d.) is usually around 4.6 mm. Greater quantities are usually needed for structure elucidation purposes, and a laboratory-scale prep HPLC system is required to isolate the milligram quantities needed for NMR or X-ray crystallography. Column internal diameters usually range from 10 to 100 mm. If gram quantities are called for, then typically pilot plant-scale prep HPLC systems are needed (column i.d. $>$ 100 mm) that present their own unique issues, though the theory behind the isolation process is essentially the same.

The materials and methods used for isolating natural products by prep HPLC also depend upon the type of compound that is encountered in the extract, which in turn is dependent upon the extraction procedure. A polar extract of a plant carried out using aqueous ethanol will differ substantially in compounds encountered than if the same plant was extracted with *n*-hexane. Therefore, polarity of the compound mixture is a major deciding factor as to which prep HPLC method is to be used.

This chapter concentrates on the practical aspects of carrying out a lab-scale prep HPLC separation to purify natural products. It covers the various modes of prep HPLC and selecting the right mode to achieve separation. Instrumentation setup and detection methods, sample preparation, method development, and sample work up are also presented. Discussion of chromatographic theory is kept to a minimum, and further information can be found in the excellent reviews listed under the Suggested Readings section at the end of this chapter.

2. Materials

When considering practical aspects of prep HPLC, we also cover stationary phases, instrumentation, and solvents used.

2.1. Modes of Separation and Stationary Phases

Prep HPLC purification of natural products typically uses one of the following four chromatographic modes: normal-phase, reversed-phase, gel permeation chromatography. (GPC), and ion exchange chromatography. The modes are determined by the stationary phase and the preparative column used, and the solvents utilized for elution. The mode to be used depends on the compatibility of the extract or mixture with the different column modes. **Table 1** illustrates the different stationary phases available, and the separation modes they utilize. The brand of stationary phase also plays a significant role in the purification process. Not all C_{18} silicas, for example, are the same; a separation achieved using a Waters brand column may look completely different from a Merck brand. This column selectivity therefore has to be taken into account when considering a separation strategy ***(1)***.

2.1.1. Normal-Phase HPLC

Normal-phase chromatography uses a polar stationary phase (usually silica) and less polar (nonaqueous) eluting solvents. Compounds are separated by adsorption onto the surface of the polar stationary phase as they elute down the column and the affinity they have to the eluting nonpolar solvent. In general, the more polar the compound, the more likely it is to be adsorbed onto the stationary phase, and less polar compounds will be eluted first from the column. Increasing the polarity of the eluting solvent reduces elution time. Normal-phase HPLC is best suited to lipophilic compounds, long chain alkane derivatives, or where the mixture of interest is

Table 1
Stationary Phases Commonly Used in HPLC

Stationary phase	Structure	Modes
C_{18}	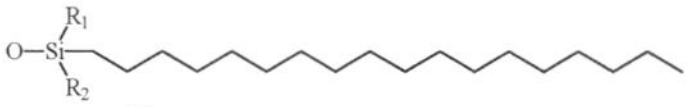 $O-Si(R_1)(R_2)-$ (C18 alkyl chain)	Reversed phase
C_8	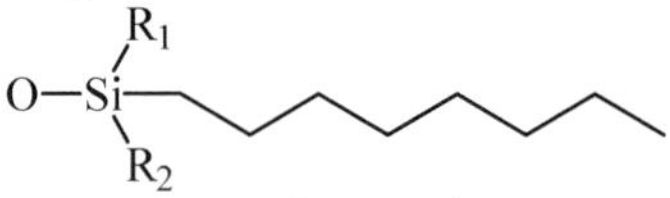 $O-Si(R_1)(R_2)-$ (C8 alkyl chain)	Reversed phase
Polymeric polystyrene divinyl benzene	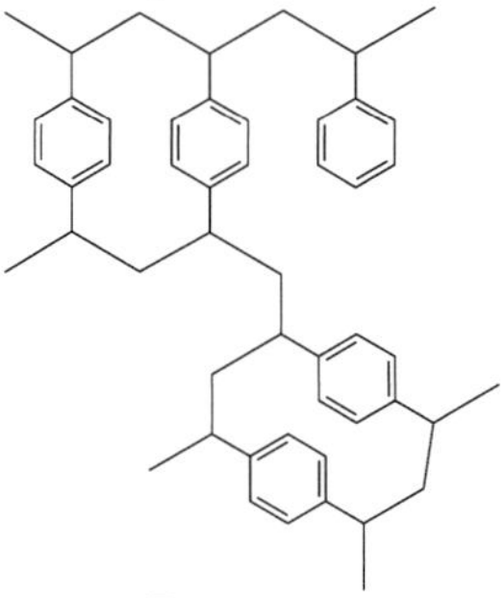	Reversed phase
CN (cyano)	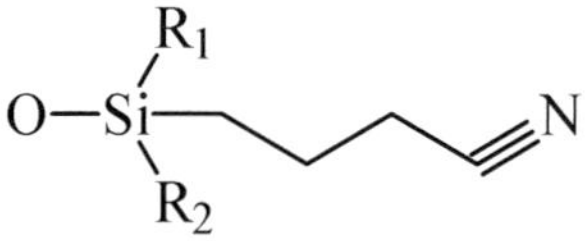 $O-Si(R_1)(R_2)-$ ... N	Normal and reversed phase
Diol	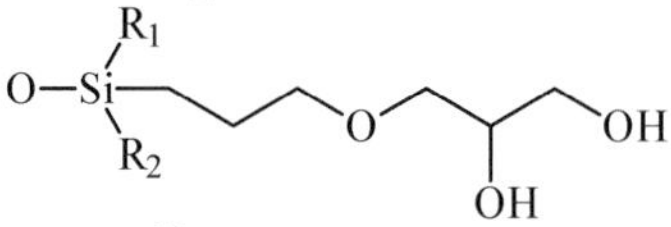 $O-Si(R_1)(R_2)-$ ... O ... OH, OH	Normal and reversed phase
Silica	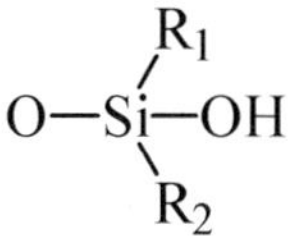 $O-Si(R_1)(R_2)-OH$	Normal phase
Benzenesulfonic acid	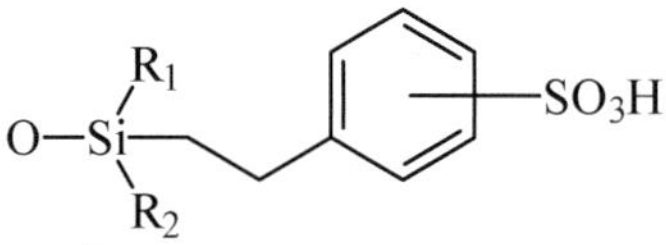 $O-Si(R_1)(R_2)-$ ... SO_3H	Strong cation exchange
Quaternary ammonium	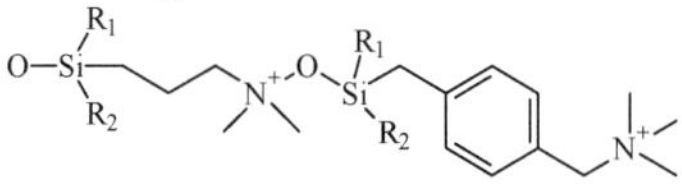 $O-Si(R_1)(R_2)-$ ... N^+ ... O ... $Si(R_1)(R_2)$ ... N^+	Strong cation exchange

sparingly soluble in aqueous conditions. It is often successful in separating geometric and positional isomers though not quite so in separating compounds differing only by alkyl groups. In most cases, normal-phase HPLC has been superseded by reversed-phase HPLC. The eluants used in normal-phase HPLC are usually mixtures of aliphatic hydrocarbons (*n*-hexane, *n*-heptane), halogenated hydrocarbons (chloroform, dichloromethane), more polar oxygenated hydrocarbons (diethyl ether, ethyl acetate, acetone), or hydroxylated solvents such as isopropanol and methanol (*see* **Note 1**). Care must be taken to control the aqueous content of the solvents as water deactivates silica causing a breakdown in the separation. This problem is seen particularly when using the hydroxylated solvents, and they should be avoided or another separation mode used to maintain the robustness of the separation system. In addition, the toxic and flammable nature of the solvents must be taken into account and the prep HPLC system should be positioned in a fume cupboard. Efforts must made to make sure the system is "earthed" sufficiently to prevent the possibility of a spark being created by static electricity causing an explosion.

2.1.2. Reversed-Phase HPLC

As the name indicates, this technique is the reverse of normal-phase HPLC, whereby the stationary phase is more nonpolar than the eluting solvent. Examples of reversed stationary phases are given in **Table 1** (including a nonsilica-based reversed-phase HPLC sorbent). Silica-based reversed-phase sorbents are also called "bonded-phase" materials, whereby the silica particles are derivatized with alkylsilyl reagents. The degree of silanization (or carbon loading) can result in columns from different manufacturers having substantially varying chromatographic characteristics, and in some cases several columns may be used for separating different mixtures *(1)*. The cost of columns can make it prohibitive to have more than one or two different brands of prep HPLC column. Therefore, a compromise may have to be struck between price and optimal separation. The eluant used in reversed-phase HPLC commonly comprises a mixture of water and miscible organic solvents, usually acetonitrile (MeCN), methanol (MeOH), or tetrahydrofuran (THF). In addition, buffers, acids, or bases may be added to suppress compound ionization or to control the degree of ionization of free unreacted silanol groups to reduce peak tailing and improve chromatography. The issues of free silanol groups have been addressed in numerous other ways to improve chromatography, such as the use of inert nonsilica supports to remove the silanol issue, or that of end-capping in an attempt to mop up the free silanol groups. Each of these innovations adds to the cost of the column and in some ways may not be necessary for the particular compounds being examined. Reversed-phase

HPLC lends itself well to the purification of most classes of natural products *(2)*. Because of this, it is usually the first technique used when analyzing and attempting to purify compounds from a complex mixture, especially when the identity of the compounds of interest is unknown.

2.1.3. Other Modes of Chromatography

Gel permeation chromatography (also called size exclusion chromatography) is predominantly used for fractionating and purifying proteins and oligosaccharides but has been used in some cases for separating lower molecular weight molecules (*see* Chap. 5). The stationary phase is typically made of rigid spherical particles of macroporous polystyrene/divinylbenzene copolymers. The stationary phase is inherently hydrophobic (similar to reversed-phase packing materials), and is essentially chemically and physically inert. The pore size in the particles is strictly controlled. Compounds are separated by their ability to enter the pores—the smaller molecules are "trapped" temporarily in the pores, while larger molecules are not held up and pass through the column relatively unhindered. The extent of retardation of the molecules is a function of their molecular size, and as such this type of chromatography has found a use in purifying biomolecules. While natural product mixtures invariably contain many compounds of similar molecular weights, gel permeation chromatography has become a useful adjunct to the other modes of HPLC separation of natural products where some prior knowledge of the molecular weight of the various components may be known.

Ion exchange chromatography (*see* Chap. 6) uses an anionic or cationic stationary phase for the separation of acids and amines. Compounds with a net charge bind reversibly to the ionizable groups on the stationary phase and are eluted through displacement of a stronger ionized species in the eluent. The support in the stationary phase may be of a silica or a styryl-divinylbenzene origin. Again, the use of ion exchange columns assumes that there is some prior knowledge of the chemical content of the sample mixture and as such is not used as a first-line separation method.

2.2. Solvents

Solvents used in HPLC (*see* **Table 2**) typically have to be:

1. Of high purity to maintain the integrity of the system and sample.
2. Compatible with the detector and not interfere with the observation of one's target compounds, i.e., "transparent."

Table 2
Properties of Solvents Used in Normal and Reversed-Phase HPLC

Solvent	Molecular weight	b.p. (°C)	Refractive index at 20°C	Eluotropic value (silica)	UV cutoff (nm)
Acetone	58.1	56	1.359	0.43	330
Acetonitrile	41.1	81	1.344	0.5	190
t-Butylmethyl ether	88.2	53–56	1.369	0.29	—
Chloroform	119.4	60.5–81	1.426	0.03	200
Heptane	100.2	98	1.387	0	200
Hexane	86.2	69	1.375	0	200
Methanol	32	64.6	1.329	0.73	205
Tetrahydrofuran	72.1	67	1.407	0.35	215
Water	18	100	—	0.73	—

3. Compatible with the sample (solubility and nonreactive).
4. Low viscosity to keep system back pressure low.
5. Reasonably priced (a typical prep HPLC run may use a liter or more of solvent each time).

Furthermore, the solvents need to be "degassed" to remove dissolved oxygen, which comes out of solution to form microscopic bubbles under the high pressures seen in the system. These bubbles interfere with the detector causing sharp spikes to be seen. There are numerous ways to degas solvents, including applying a vacuum to the solvent or placing the container of solvent into an ultrasonic bath before use. Most prep HPLCs however come fitted with in-line degassers and helium "sparge" systems that purge the solvents with helium gas, initially and periodically, during the use of the instrument and maintain the solvents in a degassed state.

2.3. Buffers and Ionization Control

As the ionic state of the compounds and the stationary phases are critical for producing efficient and reproducible chromatography, the pH of the eluting solvents must be controlled. Keeping compounds in the unionized form means that they are more likely to interact with the stationary phase in reversed-phase chromatography. Buffers have been used extensively in reversed-phase HPLC to do this, and a few of them are listed in **Table 3**. Care must be taken with their use to ensure that they

Table 3
Properties of Some Buffers Used in Reversed-Phase HPLC

Name	Molecular weight	Usual concentration range (mM)
Ammonium acetate	77	5–20
Ammonium dihydrogen phosphate	115	10–100
Sodium phosphate	120	10–100
Sodium phosphate (dibasic)	142	10–100
Potassium phosphate (monobasic)	136	10–100
Potassium phosphate (dibasic)	174	10–100

do not precipitate out in the presence of organic solvents and that the salts are removed from the final purified product (*see* **Note 2**). In addition, one must be careful not to use excessive amount of base as most silica-based columns are unable to operate at pHs greater than 8.

The use of buffers can be bypassed to some extent by the utilization of straight acids or bases. This is particularly useful when the compounds are unknown, and the use of a small amount of acid or base during the method development phase can help greatly in achieving good chromatography. Ion suppression of carboxylic acids in samples can be brought about by the addition of either mineral or organic acids to the mobile phase (**Table 4**). Peak tailing, caused by free silanol groups, can lead to poor chromatography and may be overcome to some extent by adding triethylamine (0.05–0.1% v/v) to the mobile phase. Care should be taken to ensure that the acid and/or base is removed quickly after purification takes place to avoid compound breakdown or unwanted reactions.

Table 4
Properties of Some Acids Used in Reversed-Phase HPLC

Name	Molecular weight	Usual concentration range (% v/v)	Removal from HLC fraction
Acetic acid	60	0.01–0.1	Rotary evaporation
Formic acid	46	0.01–0.1	Rotary evaporation
Trifluoroacetic acid	114	0.01–0.2	Freeze drying or desalting

3. Method

3.1. Instrumentation Setup

Prep HPLC systems are made up of a number of components as shown in **Fig. 1**. It should be noted that "method development" is usually carried out on analytical-scale HPLC systems, which require much less solvent and sample. Once a suitable solvent system and method have been established, they are then scaled up to the prep HPLC system. The level of sophistication depends on the age and the cost of the system, but they are also made up of a number of essentials.

System controller: a small computer, which controls the pump(s), flow rate, and solvent composition in binary, ternary, and quaternary systems in both isocratic and gradient modes. In some cases, the computer may also record the detector and fraction collector outputs.

Pumps: designed to pump solvent at high pressure with minimal pulsing. Flow rates may vary from 5 to 100 mL/min depending on the size of the pump heads and the system.

Injection loop: usually a Rheodyne type injector system. The sample is dissolved and injected as a solution in (or as close to) the mobile phase starting conditions. Loop capacity tends to range from 1to 30 mL in semi-

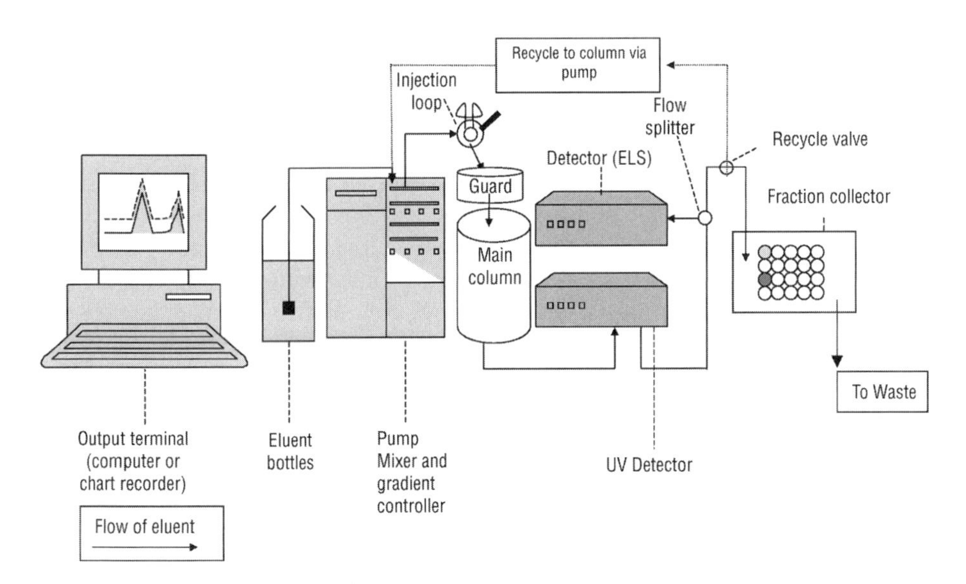

Fig. 1. A prep HPLC system featuring an ELS detector and a recycle valve.

prep and prep HPLC systems. The sample should be dissolved in the smallest volume of solvent possible, and therefore the loop should be changed to match this volume. Injecting a 1 mL sample into a 20 mL loop will lead to the sample "diluting" in the loop and cause band broadening, and poor peak shape and chromatography.

Guard column: designed to protect the main column from particulate matter that may be in the sample. The use of a guard may be negated by prefiltering the sample and passing it through a solid-phase extraction (SPE) cartridge (*see* **Note 2**).

Column: usually of stainless steel construct. The column may be a radially compressed column (manufactured by Waters) in which case it is important to ensure that the column is at the correct pressure before pumping solvent through the system.

Detectors: the three main detector types used in prep HPLC are ultraviolet/visible (UV/Vis), refractive index (RI), and evaporative light scattering (ELS) detectors.

1. UV/Vis detectors: used for compounds that absorb electromagnetic radiation in the UV/visible wavelength range (200–600 nm). This covers organic compounds that possess a degree of unsaturation or chromophore. Detectors may detect at a single wavelength that can be set by the user, or the instrument may be a photodiode array (PDA) detector, which can detect UV/Vis absorbance over a wavelength range. The PDA tends to be fitted to analytical systems predominantly, and hence during the method development phase, a suitable wavelength may be selected for use on the single wavelength prep HPLC system. Many solvents used in reversed-phase HPLC are transparent at the lower wavelengths (down to 190 nm for "far UV" acetonitrile), but the normal-phase solvents such as ethyl acetate can mask a substantial UV range at which compounds would be detected.
2. RI detectors: all compounds in solution have the ability to refract light to a greater or lesser degree. The RI detector exploits this by detecting differences in the refractive index of the eluate from the column to determine the presence or absence of a compound. As such, the RI detector represents a "universal" detector, not reliant on the compound containing a chromophore necessary for detection with UV/Vis detectors. However, the major disadvantage of the RI detector is that it can only be used in isocratic conditions (*see* **Subheading 3.6.**), where the eluting solvent conditions are constant as any background change in solvent conditions results in considerable baseline drift against which the detection of compounds cannot be made.
3. ELS detectors: quite recently, ELS detectors have come to prominence as an alternative to both UV/Vis and RI detectors *(3)*. ELS detectors work by

passing the eluate through a heated nebulizer to volatilize the eluate and evaporate the solvent. The solvent is carried away as a gas but the solute forms a stream of fine particles, which passes between a light source and detector and scatters the light. The detector measures this scattering effect. The main advantages of ELS detectors are that they will detect compounds not possessing a chromophore and can be used in isocratic and gradient elution methods, as such superceding RI detectors. ELS detectors are promoted as "mass detectors" as they are independent of the chemical characteristics such as UV absorption coefficients or refractivity that compounds may possess, and the response is that of mass alone. The only problem compounds that may evade detection are those with a boiling point close to the eluting solvent mixture as they may volatilize and not be detected. In addition, ELS detectors are destructive unlike UV/Vis and RI detectors. As such, the detector cannot be put "in-line" similar to the UV/Vis or RI detectors; hence, the eluate must be split so that a small proportion (typically 0.5–1 mL/min) flows to the ELS detector while the rest goes to the fraction collector (*see* **Fig. 1** for configuring the ELS detector into the system and **Fig. 2** for an example of use).

Fraction collectors: in many cases, the simplest fraction collector tends to be the instrument operator who will decide when and how to collect fractions coming off the column depending on the chromatography seen. However, the reliability and control systems of modern fraction collectors are such that manual intervention is not needed and the fraction collector can be programmed and operated in an automated fashion (*see* **Note 3**).

3.2. Carrying Out a Prep HPLC Isolation

As reversed-phase HPLC using C_8 or C_{18} is the "first-line method" for the isolation of most classes of natural products, we concentrate on this technique. Deciding which type of column to be used comes from experience with working on different types of compounds. It is difficult to decide where to start when dealing with molecules of unknown structure. As mentioned previously, many labs do not have the luxury of having many different types of prep HPLC column, thereby restricting the choice. However, it may be useful to try two or three different types of reversed-phase analytical column to get a feel for the compound mixture and the separations achievable.

3.3. Method Development

Finding the correct solvent system by which to achieve separation is the key to purifying natural compounds from complex mixtures. The vast

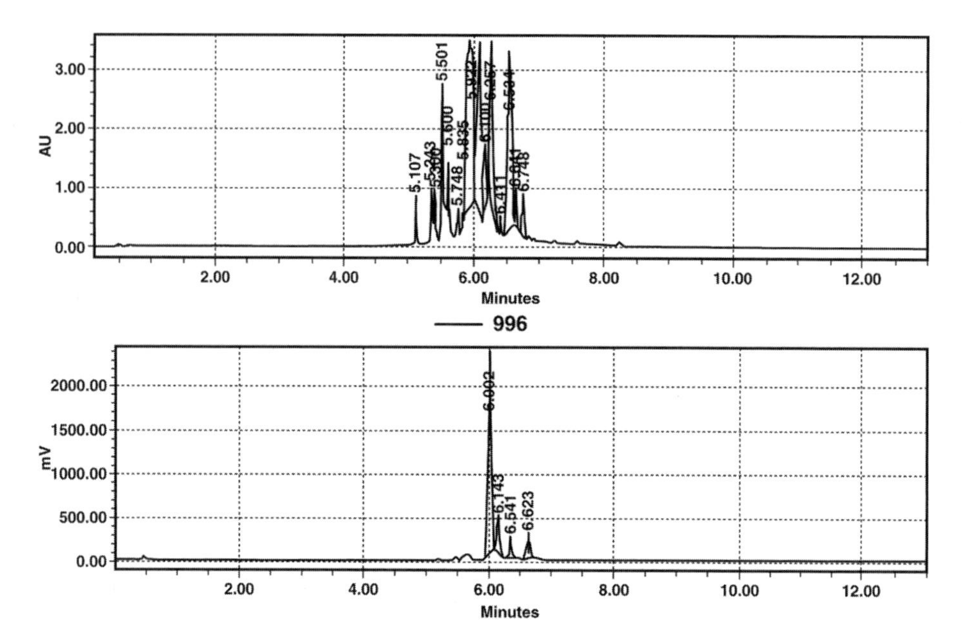

Fig. 2. HPLC analysis of a fraction of licorice (*Glycyrrhiza uralensis*) using a PDA detector (Waters 996) with UV set at maxplot (above) vs an ELS detector (Polymer Labs, below). As can be seen, the UV detector shows many peaks of similar height, whereas the ELS detector indicates that there is only one major compound in the fraction.

majority of the work is involved in developing the method, which will give optimal separation and lead to a successful scale up. It should be noted that not all separations can be achieved in one prep HPLC step. A particular complex mixture may need "prefractionation" using another technique such as flash chromatography to reduce the number of components to make method development simpler.

The amount injected and collected is dependent upon the end requirements. The amount loaded onto the column is limited by sample solubility and how well the column can achieve separation before the stationary phase becomes overloaded. If large quantities of pure compounds are required, aliquots may have to be injected repeatedly to achieve the final weights required. It should be noted that the stationary phases used in the analytical HPLC column must be the same brand and make (or carry similar specifications) to that used in the prep HPLC column. Ideally, the length of the analytical and prep columns should be the same as well to make the scale-up easier and more predictable.

3.4. Solvent Selection

For reversed-phase HPLC, water is used as the weak solvent against which a stronger organic solvent is used to elute compounds from the column. The three most commonly used solvents are acetonitrile (MeCN), MeOH, and THF, all offering good UV "transparency" and differing chemical characteristics for separation selectivity *(4,5)*. This means that if one solvent fails to achieve the required separation, then one of the others may be used. In some cases, binary mixture may be enough to elicit separation but ternary or quaternary systems may be required (*see* **Note 4**).

3.5. Gradient Analysis

An analysis has to be carried out of the mixture to assess the number of components and identify the compounds of interest. This is done initially by carrying out an analytical-scale "scouting" run *(6)*. Essentially, the sample is injected onto the column in high aqueous conditions (e.g., 5% MeCN in H_2O), and the organic proportion is increased over time to elute all the compounds off the column (at 100% MeCN). For a standard analytical column (4.6 mm i.d. × 150 mm), the typical flow will be 1 mL/min with the gradient time taking 30 min with a hold at the end of the gradient to ensure that all compounds have been eluted. Gradient elution can be used to achieve separation of complex mixtures over a range of polarities but can be time-consuming. This is because the column must be re-equilibrated back to the starting conditions at the end of the run, using around 10-column volumes of solvent *(7)*, thus reducing sample throughput and increasing solvent usage.

3.6. Gradient to Isocratic Conditions

If possible, it is desirable to achieve separation in isocratic conditions where the solvent mixture is kept constant throughout, and re-equilibration is not needed. The gradient analysis of the sample is used as a tool to find the best isocratic elution method. As a general rule, and using our typical gradient scout above, an isocratic system can be achieved if all the compounds of interest elute within a 25–30% change in organic solvent in the gradient analysis *(6)*. To estimate an isocratic solvent system from the gradient scout, one must measure the solvent conditions to elute the first peak having taken off an amount to take into account the column dead volume. Therefore, if a compound elutes at 10 min, and the time taken away to elute the column dead volume (e.g., 1.5 mL for our typical column = 1.5 min), then the solvent system estimated for the isocratic system is that seen at 8.5 min for our gradient (= approx 32% MeCN:H_2O).

If the compounds elute across a broader range, the isocratic run will not be suitable as the time taken to elute all the compounds would be excessively long and wasteful on solvent. In such cases, a truncated gradient can be used, where the starting conditions will be the solvent system calculated as above for the isocratic system, and the end system will be the concentration required to elute the last peak from the column as seen in the scouting gradient. These calculations represent starting points from which the solvent system can be adjusted in smaller steps (e.g., 5–10%) to find the optimal solvent conditions for the separation. The solvent system might elute all the compounds within a reasonable time frame, but the degree of separation of peaks and peak shape may not be optimal. Slight changes in the solvent proportions might be needed to optimize the chromatography. If this is not enough to give the desired separation, it may be that different columns or solvent systems need to be employed. Consideration may be given to adding a small amount of acid or base to the solvent system to sharpen up chromatography or it may become apparent that buffers will be required. However, by changing one variable at a time, it soon becomes apparent to the chromatographer what may need to be adjusted to achieve the separation required. Once a solvent system has been found to achieve adequate separation, the amount of sample loaded onto the analytical column can be increased and the solvent system adjusted until loading and separation limits have been reached. The simplest and quickest way to estimate loading is to increase the sample quantity injected by factors of two until the limits have been reached ***(8)***. Once this has been achieved, then the system can be scaled up onto the prep HPLC system.

3.7. Scale-Up to Prep HPLC

Scaling up to prep HPLC can be relatively straightforward if the only variables being increased are the diameter of the column/and or length of the column (*see* **Notes 5** and **6**). A direct linear scale-up can be achieved using the following equation ***(9)***:

$$\text{Direct scale-up factor} = \frac{L_{(\mathrm{P})}\, A_{c(P)}}{L_{(\mathrm{A})}\, A_{c(A)}}$$

where L is the length and A_c the cross-sectional area of the (P) preparative and (A) analytical columns. The direct scale factor allows the calculation of the scaled up flow rate and an estimation of the amount that can be injected onto the prep column. Again, the differences between the analytical and prep systems mean that certain adjustments may have to be made at the preparative stage to achieve the optimal separation. Optimization is to be carried out if the flow rates indicated by the scaling equation produce high column backpressure. This may necessitate a reduction of the flow rate and/or a change in the solvent composition to keep the backpressure down (**Fig. 3**).

3.8. Fraction Collection

Fraction collectors can be programmed in three ways: collect by time, collect by peak threshold, or by peak gradient. To collect "time fractions," the fraction automatically switches tube after a set time period (e.g., 30 s). This allows the collection of all components and is useful if wishing to create many fractions to form a library of fractions or for bioassay-guided fraction when the identity of the active is unknown. The fractions collected may not be pure, but this method helps narrow the field down to a particular region of the active mixture. Collecting by peak threshold involves the collection of peaks over a set threshold (e.g., 10% of total detector response). This allows the collection of the major peaks within a mixture but results in the loss of the minor ones. Collecting by peak gradient results in the fraction collector measuring the upslope of peaks, and when the upslope is high enough the fraction collector begins collecting. When the peak has been reached, the fraction collector measures the downslope and stops collecting when the gradient of the peak becomes shallow. However, broad shallow peaks may not be detected by this manner. Collecting by peak threshold or by peak gradient is also reliant on achieving baseline separation of the peaks. Where two peaks elute closely or overlap, then the fraction collector tends to get "confused" and cannot be relied on to do the job. In general, it is useful to watch the fraction collector during its first automated run to ensure no problems occur and also, if the sample is particularly valuable, to collect the waste or uncollected fractions (*see* **Note 3**).

In natural product mixtures and where systems are overloaded, achieving baseline separation can be near impossible. If the peaks are sufficiently close together with minimal peak broadening and the prep system is fitted with a recycle valve, the peaks can be passed through the column again and separated further. If two peaks are involved, and there is some broadening, the chromatographer may wish to use peak shaving and recycling where the leading and tail ends of two merged peaks are collected as pure components, and the center merged portion is passed through the column to be recycled and re-separated ***(2,10)***.

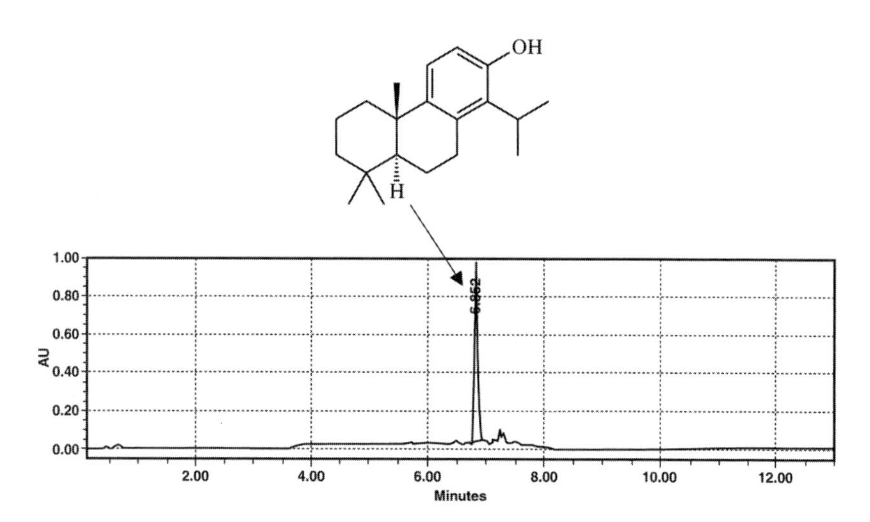

Fig. 3. An illustration of isocratic method development and scale-up to isolate totarol from *Thuja plicata*. The gradient analysis (top) was carried out using a C_8 (6 µm) Waters Symmetry (4.6 mm i.d. × 50 mm length, flow rate 1 mL/min). The gradient runs from 90:0:10 (H_2O:MeCN: 0.1% TFA in MeCN) to 0:90:10 (H_2O:MeCN: 0.1% TFA in MeCN). Over 10 min with a 2 min wash at the end. Totarol elutes at 6.8 min and the column dead volume approx 1 mL (equivalent to 1 min). The solvent conditions at 5.8 min therefore are approx 36:54:10 H_2O:MeCN: 0.1% TFA, in MeCN (equivalent to 36:64 H_2O:MeCN + 0.1% TFA), which can be used as a starting point for isocratic method development. The semiprep, scale-up column used was a C_{18} (6 µm) NovaPak Waters Rad-Pak column, which shows some similarities in stationary-phase selectivity to the analytical column used. The dimensions for the scale-up column were 8 mm i.d. × 100 mm length. This gave a scale-up factor of 6× (i.e., flow rate for column would be 6 mL/min and loading would be 6× greater). The lower chromatogram shows the actual separation carried out. The backpressure generated at 6 mL/min was too high for the system and the separation was carried out at 3 mL/min. Hence, it took longer for the compounds to elute from the column (as can be seen by the retention times). The solvent system used was 36:64 H_2O:MeCN + 0.1% TFA.

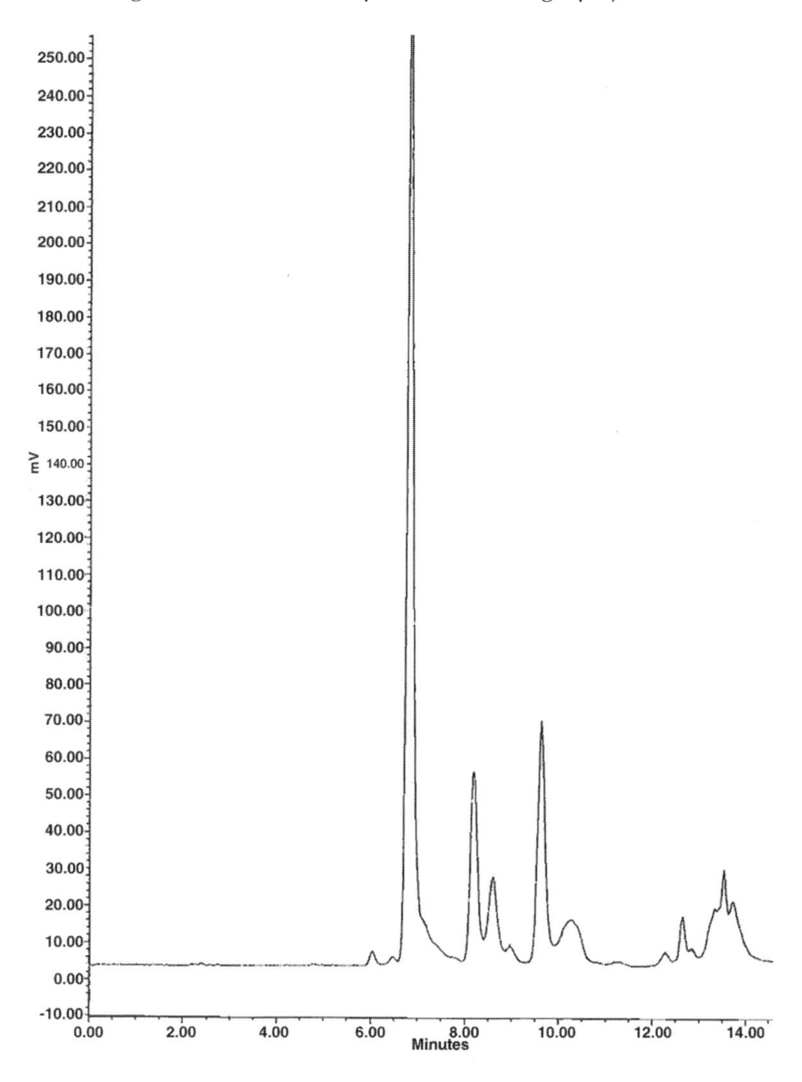

Fig. 3. (*Continued*)

3.9. Sample Work-Up

Once fractions have been collected, the solvent must be removed to yield the purified product. For organic solvents, this may be just a matter of using a rotary evaporator to dry the fraction down. Where aqueous–

organic mixtures are used, the organic phase is evaporated off as mentioned, and the aqueous portion removed by freeze drying. If inorganic buffers have been utilized or the aqueous portion is particularly large, the fraction can be passed through a reversed-phase SPE cartridge to trap the target compounds (also known as desalting). The trapped compounds are then eluted with a small amount of organic solvent, which can be evaporated off much more easily.

3.10. Conclusion

Preparative HPLC is a powerful technique, which has become invaluable in the purification of natural products, regardless of source. The cost of the equipment and solvents can easily be recouped by the power, speed, and robustness of the technique placing it into the frontline for purification of compounds. The key to getting most out of the technique comes from method development at the analytical stage. Once systems have been established, then the transfer from analytical to preparative scale becomes a matter of routine. The sample mixture is the critical component in method development. Poor solubility or too complex a mixture can negate the chromatographer's efforts to find the optimal separation. Hence, the sample must be analyzed thoroughly and pretreated to address these issues to make the chromatographer's job easier and maintain the life-span of the HPLC system.

4. Notes

1. Care must be taken to control the aqueous content of the solvents as water deactivates silica, causing a loss in separation or reproducibility.
2. Natural product mixtures are made up of diverse chemicals and as such one of the biggest problems encountered is trying to inject the mixture onto the column at a concentration high enough to give fractions of meaningful yield. In many cases, the samples will need prefractionation, to make the number of compounds and so the problem simpler. This may be achieved by using flash chromatography or with SPE cartridges to chop up the complex mixture. The use of a reversed-phase SPE cartridge prior to reversed-phase HPLC is also useful to "clean up" the mixture. Many fractions contain compounds that will bind to the stationary phase irreversibly, thereby reducing the lifetime of the column. Therefore, filtering the sample through an SPE cartridge prior to prep HPLC will help to keep the prep column clean. Reversed-phase SPE cartridges can be used to reduce the amount of long chain alkanes, if they are of no interest. An SPE packed with crosslinked polyvinylpyrrolidone can be

used to remove tannins from polar plant extracts. Caution is necessary when cleaning up samples, as it may also remove the compounds of interest.

3. It is important that the detector and fraction collector are synchronized such that the peak is detected and the fraction collector registers it and collects it. This can be achieved by programming a delay of a few seconds into the collector before it starts to collect to allow time for the actual compound to travel from the detector to the collector. The delay can be calculated by injecting a dye into the system (if using UV) while the column is disconnected and pumping solvent at the required flow rate. By measuring the time delay between the compound being detected and actually seeing it being eluted at the collector, the delay can be calculated.
4. Much has been written on this topic and the reader is directed to the study by Snyder et al. *(4)*. Essentially, Snyder categorized a solvent's chromatographic selectivity by various physicochemical characteristics. The three organic "strong" solvents for reversed-phase HPLC to be used as first choice are MeCN, MeOH, and THF. For normal-phase HPLC, the three "strong" solvents are chloroform, dichloromethane, and an ether (usually methyl-*t*-butylether). Isohexane or heptane is usually the "weak" solvent.
5. Gradient solvent systems may have to be adjusted when there are increases in column length and also differences in the system void volume. Reference *8* provides equations that will assist in taking these factors into account.
6. When purchasing an HPLC column, the manufacturers usually include a test chromatogram showing the performance of the column in separating a mixture of standards. It is good practice to make up ones own test mixture to check the performance of the analytical and prep columns (the same mixture for both size columns) and test the columns on a weekly or a monthly basis depending on how heavily they are used. Scaling up chromatography can be problematic, and ensuring that the systems are running properly with a test sample means that any problems can be diagnosed quickly and without the loss of valuable material.

References

1. Eurby, M. R. and Petersson, P. (2003) Chromatographic classification and comparison of commercially available reversed-phase liquid chromatographic columns using principle component analysis. *J. Chromatogr. A* **994,** 13–26.
2. Hostettman, K., Hostettman, M., and Marston, A. (1986) *Preparative Chromatography Techniques: Applications in Natural Product Isolation.* Springer-Verlag, Berlin, Germany, pp. 37–39.
3. Young, C. S. and Dolan, J. W. (2003) Success with evaporative light-scattering detection *LC-GC Eur.* **13,** 132–137.

4. Snyder, L. R., Kirkland, J. J., and Glajch, J. L. (1997) *Practical HPLC Method Development*, 2 ed., John Wiley and Sons, New York.
5. Dolan, J. W. (2000) Starting out right, Part 3—The role of the solvent in controlling selectivity. *LC-GC Eur.* **13,** 148–156.
6. Dolan, J. W. (2000) Starting out right, Part 6—The scouting gradient alternative. *LC-GC Eur.* **13,** 388–394.
7. Dolan, J. W. (2003) How much is enough? *LC-GC Eur.* **16,** 740–745.
8. Neue U. D. (1997) *HPLC Columns*, Wiley VCH, NY, USA.
9. Mazzei, J. L. and d'Avila, L. A. (2003) Chromatographic models as tools for scale-up of isolation of natural products by semi-preparative HPLC. *J. Liquid Chromatogr. Relat. Technol.* **26,** 177–193.
10. Zhang, M., Stout, M. J., and Kubo, I. (1992) Isolation of ecdysteroids from *Vitex strickeri* using RLCC and recycling HPLC. *Phytochemistry* **31,** 247–250.

Suggested Readings

1. Basic Liquid Chromatography (1996–2000) Yuri Kazakevich, Harold McNair on http://hplc.chem.shu.edu/NEW/HPLC_Book/index.html
2. Dolan, J. W., ed (2004) LC troubleshooting, in *LC-GC Europe*. Advanstar, US. Also found at www.lcgceurope.com
3. Hostettman, K., Hostettman, M., and Marston, A. (1986) *Preparative Chromatography Techniques: Applications in Natural Product Isolation*, Springer-Verlag, Berlin, Germany.
4. Katz E. D. ed. (1996) *High Pressure Liquid Chromatography: Principles and Methods and Biotechnology*, Wiley, Chichester, UK.
5. Shelley, P. R. (1996) High performance liquid chromatography, in Downstream Processing of Natural Products (Verall, M. S., ed.), Wiley, Chichester, UK.

9

Hyphenated Techniques

Satyajit D. Sarker and Lutfun Nahar

Summary

The technique developed from the coupling of a separation technique and an on-line spectroscopic detection technology is known as *hyphenated technique*. The remarkable improvements in hyphenated analytical methods over the last two decades have significantly broadened their applications in the analysis of biomaterials, especially natural products. In this chapter, recent advances in the applications of various hyphenated techniques, e.g., GC-MS, LC-PDA, LC-MS, LC-FTIR, LC-NMR, LC-NMR-MS, and CE-MS, in the context of preisolation analyses of crude extracts or fraction from various natural sources, isolation and on-line detection of natural products, chemotaxonomic studies, chemical fingerprinting, quality control of herbal products, dereplication of natural products, and metabolomic studies are discussed with appropriate examples. Particular emphasis is given on the hyphenated techniques that involve liquid chromatography, as the separation tool.

Key Words: Hyphenated technique; LC-MS; LC-NMR; LC-IR; LC-PDA; CE-MS; LC-NMR-MS; natural products.

1. Introduction

A couple of decades ago, Hirschfeld introduced the term *hyphenation* to refer to the on-line combination of a separation technique and one or more spectroscopic detection techniques *(1)*. This technique, developed from a marriage of a separation technique and a spectroscopic detection technique, is nowadays known as *hyphenated technique* (**Fig. 1**). In recent

From: *Methods in Biotechnology, Vol. 20, Natural Products Isolation, 2nd ed.*
Edited by: S. D. Sarker, Z. Latif, and A. I. Gray © Humana Press Inc., Totowa, NJ

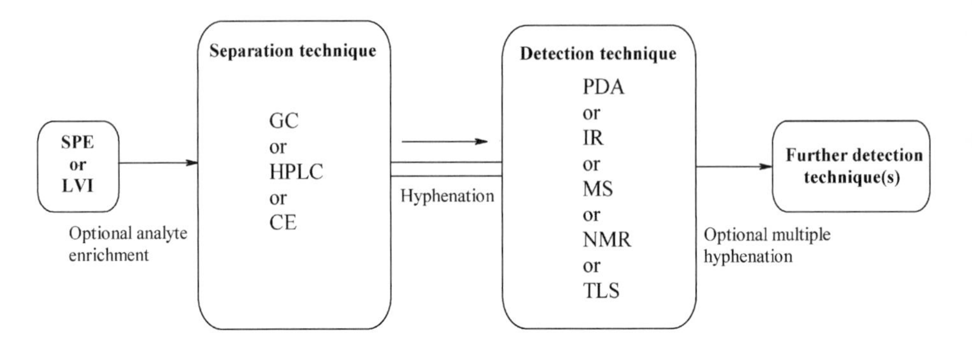

Fig. 1. Hyphenated technique.

years, hyphenated techniques have received ever-increasing attention as the principal means to solve complex analytical problems. The power of combining separation technologies with spectroscopic techniques has been demonstrated over the years for both quantitative and qualitative analysis of unknown compounds in complex natural product extracts or fractions. To obtain structural information leading to the identification of the compounds present in a crude sample, liquid chromatography (LC), usually a high-performance liquid chromatography (HPLC), gas chromatography (GC), or capillary electrophoresis (CE) is linked to spectroscopic detection techniques, e.g., Fourier-transform infrared (FT-IR), photodiode array (PDA) UV–Vis absorbance or fluorescence emission, mass spectroscopy (MS), and nuclear magnetic resonance spectroscopy (NMR), resulting in the introduction of various modern hyphenated techniques, e.g., CE-MS, GC-MS, LC-MS, and LC-NMR. HPLC is the most widely used analytical separation technique for the qualitative and quantitative determination of compounds in natural product extracts. The physical connection of HPLC and MS or NMR has increased the capability of solving structural problems of complex natural products. Because of the greater sensitivity, LC-MS has been more extensively used than LC-NMR. The hyphenation does not always have to be between two techniques; the coupling of separation and detection techniques can involve more than one separation or detection techniques, e.g., LC-PDA-MS, LC-MS-MS, LC-NMR-MS, LC-PDA-NMR-MS, and the like. Where trace analysis is vital, and the analyte enrichment is essential, on-line coupling with solid-phase extraction (SPE), solid-phase microextraction or large volume injection (LVI) can be incorporated to build in a more powerful integrated system, e.g., SPE-LC-MS or

LVI-GC-MS. The two key elements in natural product research are the isolation and purification of compounds present in crude extracts or fractions obtained from various natural sources, and the unambiguous identification of the isolated compounds. Thus, the on-line characterization of secondary metabolites in crude natural product extracts or fractions demands high degree of sophistication, and richness of structural information, sensitivity, and selectivity. The development of various hyphenated techniques has provided the natural product researchers with extremely powerful new tools that can provide excellent separation efficiency as well as acquisition of on-line complementary spectroscopic data on an LC or GC peak of interest within a complex mixture. The main focus of this chapter is to provide an overview of basic operational principles of various modern hyphenated techniques and to present several literature examples of applications of these techniques. Detailed information on the principle, history, instrumentation, and methodology is available in literature *(2–15)*.

2. Available Hyphenated Techniques

2.1. GC-MS

With MS as the preferred detection method, and single- and triple-quadrupole, ion trap and time-of-flight (TOF) mass spectrometers as the instruments most frequently used, both LC-MS and GC-MS are the most popular hyphenated techniques in use today *(1)*. GC-MS, which is a hyphenated technique developed from the coupling of GC and MS, was the first of its kind to become useful for research and development purposes. Mass spectra obtained by this hyphenated technique offer more structural information based on the interpretation of fragmentations. The fragment ions with different relative abundances can be compared with library spectra. Compounds that are adequately volatile, small, and stable in high temperature in GC conditions can be easily analyzed by GC-MS. Sometimes, polar compounds, especially those with a number of hydroxyl groups, need to be derivatized for GC-MS analysis. The most common derivatization technique is the conversion of the analyte to its trimethylsilyl derivative.

In GC-MS, a sample is injected into the injection port of GC device, vaporized, separated in the GC column, analyzed by MS detector, and recorded (**Fig. 2**). The time elapsed between injection and elution is called "retention time" (t_R). The equipment used for GC-MS generally consists of an injection port at one end of a metal column (often packed with a sand-like material to promote maximum separation) and a detector (MS)

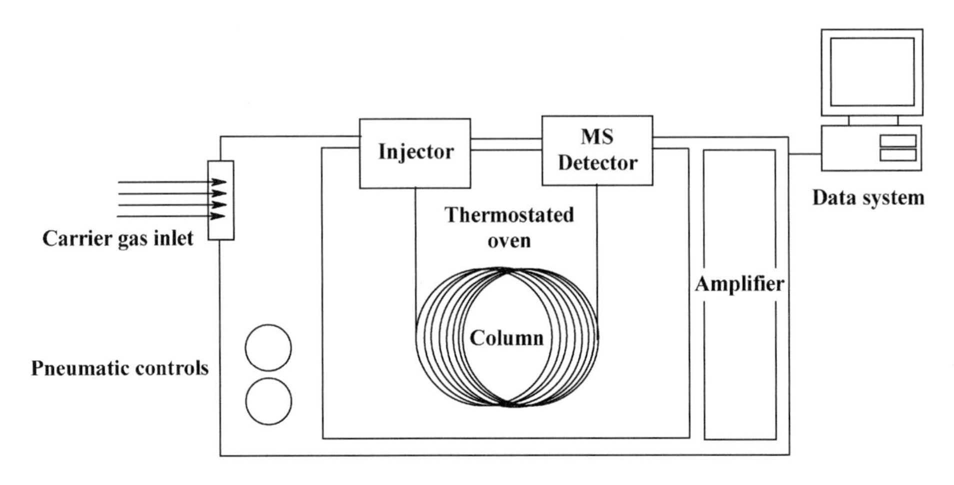

Fig. 2. GC-MS.

at the other end of the column. A carrier gas (argon, helium, nitrogen, hydrogen, to name a few) propels the sample down the column. The GC separates the components of a mixture in time and the MS detector provides information that aids in the structural identification of each component. The GC-MS columns can be of two types: capillary columns, and macrobore and packed columns. The following points need to be considered carefully regarding the GC-MS interface.

1. The interface transports efficiently the effluent from the GC to MS.
2. The analyte must not condense in the interface.
3. The analyte must not decompose before entering the MS ion source.
4. The gas load entering the ion source must be within pumping capacity of the MS.

The most extensively used interfaces for a GC-MS are electron impact ionization (EI) and chemical ionization (CI) modes. However, in modern GC–MS systems, various other types can be used that allow identification of molecular ion. For example, an orthogonal TOF mass spectrometry coupled with GC is used for confirmation of purity and identity of the components by measuring exact mass and calculating elemental composition. Nowadays, a GC-MS is integrated with various on-line MS databases for several reference compounds with search capabilities that could be useful for spectra match for the identification of separated components.

2.2. LC-PDA

Ultraviolet–visible (UV–vis) spectroscopic detector is considered to be a universal detector for any LC system. The PDA detector is an advanced form of UV–vis detector, which can be coupled to a HPLC to provide the hyphenated technique HPLC-PDA, also known as LC-PDA. Over the last couple of decades, on-line PDA detection has been employed for the analysis of crude natural product extracts of various origins. An LC-PDA is extremely useful for the analysis of natural products containing chromophores, such as phenolic compounds including flavonoids, isoflavonoids, coumarins, pterocarpans, and so on. A PDA detector can help to analyze individual LC peaks after a run is finished, and to obtain complete UV–vis spectrum of individual components. The whole chromatogram at multiple wavelengths can be retrieved from the data files after analysis. The HPLC retention time and the UV–vis spectrum for any component (LC peak) can be characteristic of certain compounds. A PDA detector also allows generation of 3D UV data, typically consisting of UV absorption spectra from 190 to 500 nm, for each point along the HPLC chromatogram. The data can be rapidly previewed for unique absorption regions correlating to specific compounds or functional groups. Independent chromatograms can also be constructed for each wavelength to increase the selectivity of the data. The UV data can be complemented by the MS or NMR selective data.

The choice of LC mobile phase is crucial for LC-PDA operation and has to be made according to its inherent UV cutoff point, so that any interference from the mobile phase can be avoided. Modern LC-PDA systems are run by sophisticated software that allow building up of spectral libraries for reference compounds and automated compound search.

2.3. LC-IR

The hyphenated technique developed from the coupling of an LC and the detection method infrared spectrometry (IR) or FT-IR is known as LC-IR or HPLC-IR. While HPLC is one of the most powerful separation techniques available today, the IR or FTIR is a useful spectroscopic technique for the identification of organic compounds, because in the mid-IR region the structures of organic compounds have many absorption bands that are characteristic of particular functionalities, e.g., –OH, –COOH, and so on. However, combination of HPLC and IR is difficult and the progress in this hyphenated technique is extremely slow, bccause the absorption bands of the mobile phase solvent are so huge in the mid-IR region that they often obscure the small signal generated by the sample components. In addition, as a detection technique, IR is much less sensitive compared to various other detection techniques, e.g., UV and MS.

The recent developments in HPLC-IR technology have incorporated two basic approaches based on interfaces applied in HPLC-IR or HPLC-FTIR. One is a flow-cell approach and the other is a solvent-elimination approach. The approach used with the flow cell in LC-IR is similar to that used in UV–vis and other typical HPLC detectors. In this case, absorption of the mobile phase induces the interference of the detection of sample component absorption bands, but some transparent region of the mid-IR range produces detection possibility. For example, if one uses a mobile phase of a deuterated solvent such as heavy water or perdeuterated methanol, IR can monitor many organic compounds that have C–H structures in the molecules. The solvent-elimination approach is the preferred option in most of the LC-IR operations. After the mobile phase solvent is eliminated, IR detection is carried out on some medium that has a transparency for IR light. Generally, KBr or KCl salts are used for the collection of sample components in the eluent, and heating up the medium before IR detection eliminates the volatile mobile phase solvents. There are two types of interfaces for the solvent-elimination approach: diffuse-reflectance infrared Fourier transform (DRIFT) approach and buffer-memory technique ***(16,17)***. A unified interface for GC, HPLC, and SFC hyphenation to FTIR applying IR microscopic technique is also available today ***(18)***.

2.4. LC-MS

LC-MS or HPLC-MS refers to the coupling of an LC with a mass spectrometer (MS) (**Fig. 3**). The separated sample emerging from the column can be identified on the basis of its mass spectral data. A switching valve can help make a working combination of the two techniques. A typical automated LC-MS system consists of double three-way diverter in-line with an autosampler, an LC system, and the mass spectrometer. The diverter generally operates as an automatic switching valve to divert undesired portions of the eluate from the LC system to waste before the sample enters the MS. An LC-MS combines the chemical separating power of LC with the ability of a mass spectrometer to selectively detect and confirm molecular identity. MS is one of the most sensitive and highly selective methods of molecular analysis, and provides information on the molecular weight as well as the fragmentation pattern of the analyte molecule. The information obtained from MS is invaluable

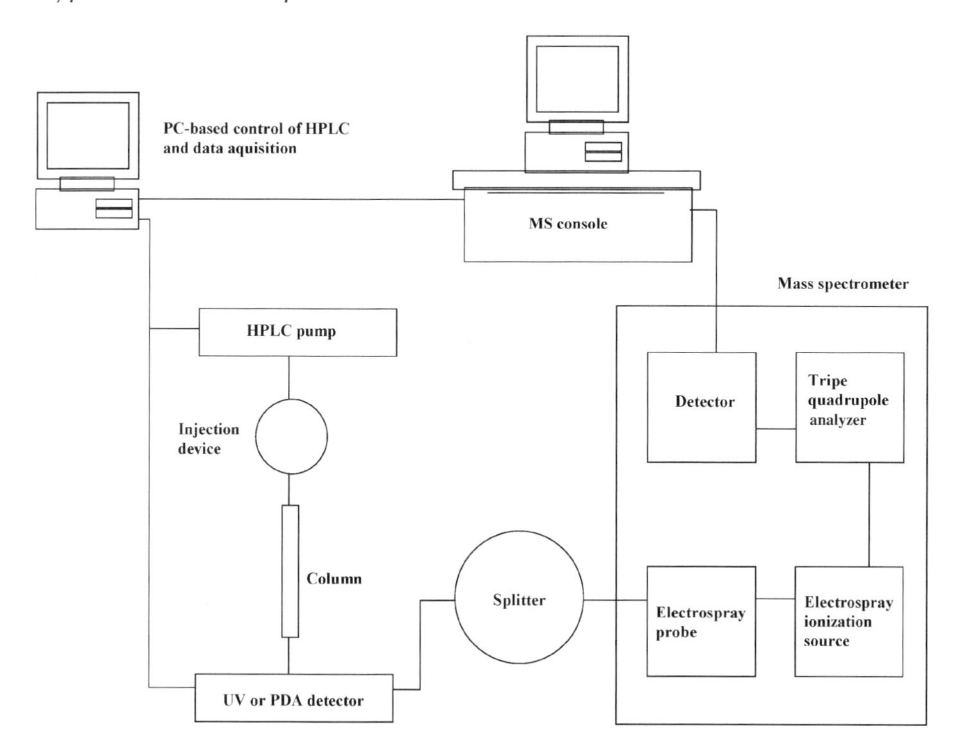

Fig. 3. Schematic of an LC-MS (electrospray ionization interface) system.

for confirming the identities of the analyte molecules. This qualitative analysis makes it possible to reconstruct an unknown compound from MS data. The ionization techniques used in LC-MS are generally soft ionization techniques that mainly display the molecular ion species with only a few fragment ions. Hence, the information obtained from a single LC-MS run, on the structure of the compound, is rather poor. However, this problem has now been tackled by the introduction of tandem mass spectrometry (MS-MS), which provides fragments through collision-induced dissociation of the molecular ions produced ***(19)***. The use of LC-MS-MS is increasing rapidly. Hyphenated techniques such as HPLC coupled to UV and mass spectrometry (LC-UV-MS) have proved to be extremely useful in combination with biological screening for a rapid survey of natural products.

Nowadays, various types of LC-MS systems incorporating different types of interfaces are available commercially. The interfaces are designed in such a way that they offer adequate nebulization and vaporization of the liquid, ionization of the sample, removal of the excess solvent vapor, and extraction of the ions into the mass analyzer. The two most widely used interfaces, especially in relation to natural product analysis, are electrospray (ESI) and atmospheric pressure chemical ionization (APCI). The latter is considered as "the chromatographer's LC-MS interface" because of its high solvent flow rate capability, sensitivity, response linearity, and fields of applicability. With these interfaces, various types of analyzers, e.g., quadrupole, ion trap, or TOF, can be used. Each of these analyzers, however, offers varying degree of mass accuracy and resolution. In the LC-UV-MS mode thermospray (LC-TSP-MS) and continuous-flow FAB (LC-CF-FAB) interfaces can also be applied. For phytochemical analysis, the TSP has been found to be the most suitable interface as it allows introduction of aqueous phase into MS system at a flow rate (1–2 mL/min) compatible with that usually used in phytochemical analysis.

In LC operation for LC-MS, the preferred option is a reversed-phase system using a gradient or isocratic solvent mixture of water, ACN, or MeOH. Small amounts of acetic acid, formic acid, ammonium hydroxide/ammonia solution, or ammonium acetate can also be used in the mobile phase. In conjunction with these interfaces, different types of analyzers, e.g., quadrupole, ion trap, or time of flight, can be used, and they offer various degrees of mass accuracy and MS-MS possibilities.

LC-MS systems do not allow a complete and unambiguous on-line identification of a component, unless it is a well-known natural product, and complementary on-line spectroscopic information is available in databases. One of the main problems associated with LC-MS is that the quality of response strongly depends on various factors, e.g., nature of the compounds to be analyzed, the solvent and buffer used as the mobile phase, the flow rate, and, of course, the type of interface used. For example, a crude natural product extract generally contains a number of various types of compounds that differ considerably in their physicochemical properties, solubilities, molecular size and stability. It is therefore extremely difficult, if not impossible, to optimize the ionization conditions that can be suitable for all those different types of compounds. One way to get around this difficulty is to analyze the extract in different ionization modes *(10)*.

2.5. LC-NMR

Among the spectroscopic techniques available to date, NMR is probably the least sensitive, and yet it provides the most useful structural information toward the structure elucidation of natural products. Technological developments have allowed the direct parallel coupling of HPLC systems to NMR, giving rise to the new practical technique HPLC-NMR or LC-NMR, which has been widely known for more than last 15 years. The first on-line HPLC-NMR experiment using superconducting magnets was reported in the early 1980s. However, the use of this hyphenated technique in the analytical laboratories started in the latter part of the 1990s only. LC-NMR promises to be of great value in the analysis of complex mixtures of all types, particularly the analysis of natural products and drug-related metabolites in biofluids. LC-NMR experiments can be performed in both continuous-flow and stop-flow modes. A wide range of bioanalytical problems can be addressed using 500, 600, and 800 MHz systems with ^{1}H,^{13}C,^{2}H,^{19}F, and ^{31}P probes. The main prerequisites for on-line LC-NMR, in addition to the NMR and HPLC instrumentation, are the continuous-flow probe and a valve installed before the probe for recording either continuous-flow or stopped-flow NMR spectra ***(12)***. A UV–vis detector is also used as a primary detector for LC operation. Using magnetic field strengths higher than 9.4 T is recommended, i.e., ^{1}H resonance frequency of 400 MHz for a standard HPLC-NMR coupling. The analytical flow cell was initially constructed for continuous-flow NMR acquisition. However, the need for full structural assignment of unknown compounds, especially novel natural products, has led to the application in the stopped-flow mode. In fact, the benefits of the closed-loop separation-identification circuit, together with the prospect of using all presently available 2D and 3D NMR techniques in a fully automated way, have prompted the development of stopped-flow modes, e.g., time-slice mode. A typical experimental arrangement of LC-NMR is shown in **Fig. 4**.

Generally, in LC-NMR system, the LC unit comprises autosampler, LC pump, column, and a non-NMR detector (e.g., UV, DAD, EC, refractive index, or radioactivity). From this detector the flow is guided into the LC-NMR interface, which can be equipped with additional loops for the intermediate storage of selected LC peaks. The flow from the LC-NMR interface is then guided either to the flow-cell NMR probe-head or to the waste receptacle. Following passage through the probe-head, the flow is routed to a fraction collector for recovery and further investigation of the various fractions analyzed by NMR. A mass spectrometer can also be attached to the system via a splitter at the output of the LC-NMR interface.

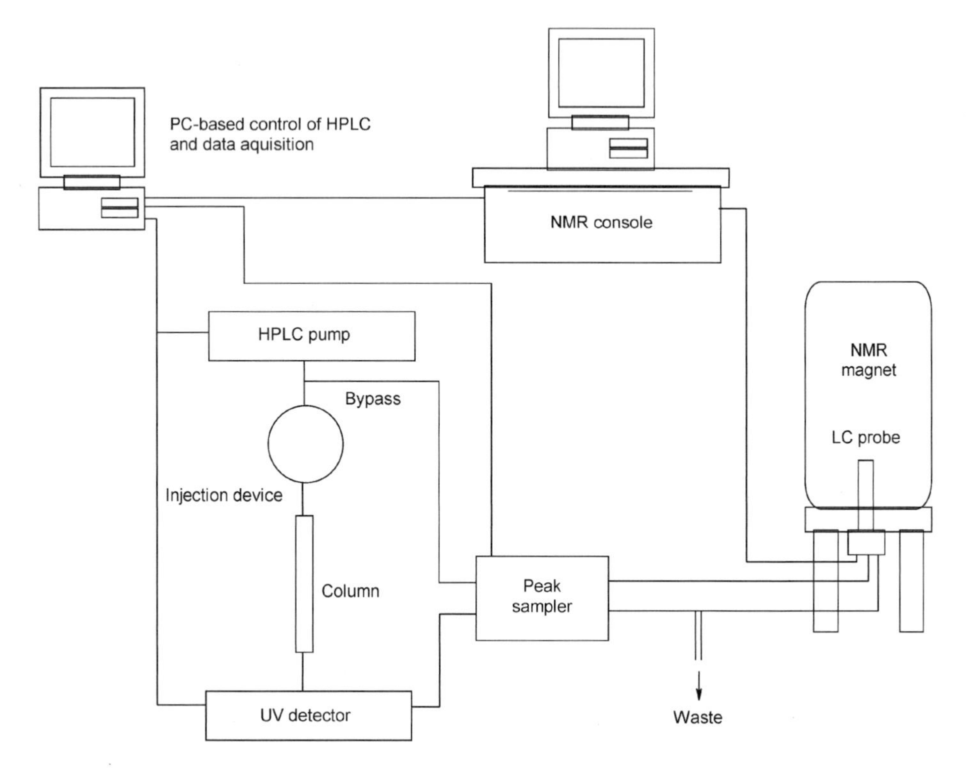

Fig. 4. A typical LC-NMR system.

In most of the LC-NMR operations, reversed-phase columns are used, employing a binary or tertiary solvent mixture with isocratic or gradient elution. The protons of the solvents of the mobile phase cause severe problems for obtaining an adequate NMR spectrum. The receiver of the NMR spectrometer is not quite able to handle the intense solvent signals and the weak substance signals at the same time. To overcome this problem, solvent signal suppression can be achieved by one of the three major methods: presaturation, soft-pulse multiple irradiation, or water suppression enhancement through T_1 effects (WET) presaturation employing a z-gradient ***(12)***. This problem can also be minimized by considering the following guidelines:

1. Using eluents that have as few ^{1}H NMR resonances as possible, e.g., H_2O, ACN, or MeOH.

2. Using at least one deuterated solvent, e.g., D_2O (approx $290/L), ACN-$d_3$ (approx $1600/L), or MeOD (approx $3000/L).
3. Using buffers that have as few ^1H NMR resonances as possible, e.g., TFA or ammonium acetate.
4. Using ionpair reagents that have as few ^1H NMR resonances as possible, e.g., ionpairs with *t*-butyl groups create an additional resonance.

To date, three main types of data acquisition modes have been introduced: continuous-flow acquisition, stopped-flow acquisition, and time-sliced acquisition ***(12)***. Whatever may be the acquisition mode, an optimized HPLC separation is crucial to any LC-NMR analysis. As the sensitivity of LC-NMR is much less than other hyphenated techniques, e.g., LC-MS, or LC-PDA, it is imperative to develop a suitable LC separation where the quantity of the available separated compound is concentrated in the smallest available elution volume.

LC-NMR represents a potentially interesting complementary technique to LC-UV-MS for detailed on-line structural analysis. Indeed, recent progress in NMR technology has given a new impulse to LC-NMR, which is now emerging as a powerful analytical tool. The development of efficient solvent suppression techniques enables the measurement of high-quality LC-^1H-NMR spectra, both on-flow and stop-flow, with reversed-phase HPLC conditions. Nondeuterated solvents such as MeOH or MeCN can be used, while water is replaced by D_2O.

Recent advances in both hardware and software for the direct coupling of LC and NMR have given a new life to this hyphenated technique. These developments include new coil and flow-cell design for high sensitivity, new RF system for multiple solvent suppression and improved dynamic range gradient elution capability, and automatic peak-picking/storing capabilities. As a result, this method is a powerful tool used in many areas such as natural products, organic molecules, biomolecules, drug impurities, by-products, reaction mixtures, and drug degradation products. The potential of HPLC-NMR for the investigation and structural elucidation of novel natural products has been enormously extended by the advent of powerful solvent suppression schemes, and their combination with a series of homo- and heteronuclear 2D NMR experiments such as 2D total correlation spectroscopy (TOCSY) or 2D nuclear Overhauser enhancement spectroscopy (NOESY).

LC-NMR, despite being known for about last two decades, has not quite become a widely accepted technique, mainly because of its lower level of sensitivity and higher cost compared to other available hyphenated techniques. However, the recent advances in technology, especially in relation to the developments in pulse field gradients and solvent suppressions methods, the improvement in probe technology, and the introduction of high-field magnets (800–900 MHz) have offered new impetus to this technique.

2.6. CE-MS

CE is an automated separation technique introduced in the early 1990s. CE analysis is driven by an electric field, performed in narrow tubes, and can result in the rapid separation of many hundreds of different compounds. The versatility and the many ways that CE can be used mean that almost all molecules can be separated using this powerful method. It separates species by applying voltage across buffer-filled capillaries, and is generally used for separating ions that move at different speeds when voltage is applied depending on their size and charge. The solutes are seen as peaks as they pass through the detector and the area of each peak is proportional to their concentration, which allows quantitative determinations. Analysis includes purity determination, assays, and trace level determinations. When an MS detector is linked to a CE system for acquiring on-line MS data of the separated compound, the resulting combination is termed as CE-MS (**Fig. 5**). Separation is achieved through channels etched on the surface of the capillary (connected to an external high-voltage power supply) that delivers sample to ESIMS. This technique runs in full automation and offers high degree of sensitivity and selectivity. A new type of interface, known as coaxial sheath liquid CE-MS interface, has been developed recently, which allows the use of both LC-MS and CE-MS alternatively on the same mass spectrometer ***(20)***. The necessary sheath liquid is delivered by a pump that floats on the ion sprayer of the MS, avoiding any current flow toward ground. LC-MS and CE-MS modes can be switched within minutes. To obtain a stable ion spray and to avoid electrical problems, the CE power supply is used to produce the potential for the CE separation and the ESI sprayer tip simultaneously. ESIMS detection technique is generally used in most of the CE-MS systems because ESI is considered to be one of the most powerful on-line tools for the analysis of biomolecules, including natural products, providing both the molecular weight and structural characterization of analytes ***(21)***. The optimization of the interfacing of CE with MS can be a real challenge because of the low flow rates (10–100 mL/min) required in CE, which is achieved by a make-up liquid.

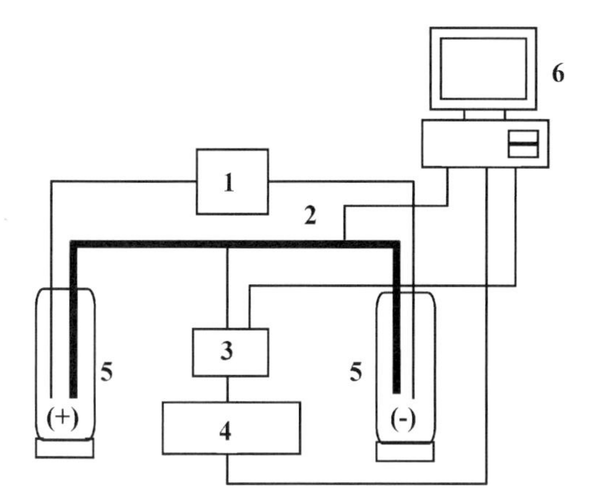

1 = High-Voltage Supply; **2** = Capillary; **3** = UV-vis or PDA detector; **4** = MS detector; **5** = Buffer solution; **6** = PC control

Fig. 5. A typical CE-MS system.

2.7. LC-TLS

HPLC has been recently coupled to thermal lens spectrometry detection technique resulting in the introduction of LC-TLS *(22)*. This method offers sufficient sensitivity and selectivity for studies of photochemical processes in marine phytoplankton involving trace amounts of pigments. A comprehensive review, evaluating the HPLC-TLS methodology and its importance as an analytical tool, has been recently published *(23)*.

2.8. Multiple Hyphenation

Parallel interfacing of the LC-NMR system with an ion trap mass spectrometer (LC-NMR-MS) also gives comprehensive and complementary structural information. The identification of compounds in complex mixtures, like crude natural product extract via multiple hyphenation of chromatography with a range of spectroscopic detectors such as, LC-PDA-NMR-MS, has become an established means for obtaining conclusive information regarding the structures of the compounds present in the mixture *(24–29)* (**Fig. 6**).

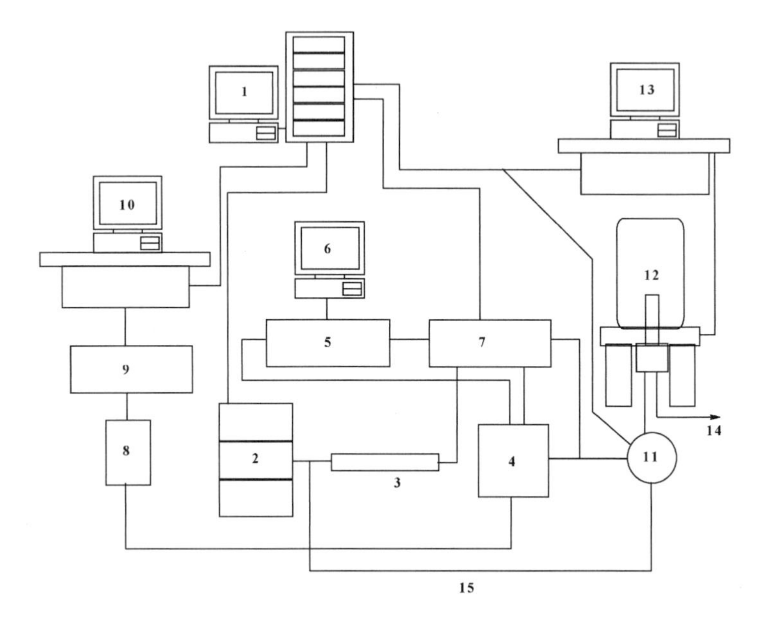

1 = Integrated work station; **2** = HPLC system; **3** = HPLC column; **4** = Splitter; **5** = FT-IR detector; **6** = PC for FT-IR control; **7** = PDA detector; **8** = Pump; **9** = MS detector; **10** = MS Console; **11** = Flow control and peak sampling unit; **12** = NMR detector with LC probe; **13** = NMR console; **14** = Waste or fraction collector; **15** = Bypass

Fig. 6. Multiple hyphenation.

3. Application of Hyphenated Techniques in Natural Product Analysis

Rapid identification and characterization of known and new natural products directly from plant and marine sources without the necessity of isolation and purification can be achieved by various modern hyphenated techniques (**Fig. 7**). Techniques like HPLC coupled to NMR or electrospray ionization tandem mass spectrometry (ESI-MS-MS) have been proven to be extremely powerful tools in natural product analysis, as they permit the fast screening of crude natural product extracts or fractions for detailed information about metabolic profiles, with a minimum amount of material. The combined application of various hyphenated techniques even allows the discovery of new natural product, including complete and conclusive structure elucidation, and relative configurations prior to time-consuming and costly isolation and purification process. Some examples of the application of hyphenated techniques in natural products analysis are discussed here.

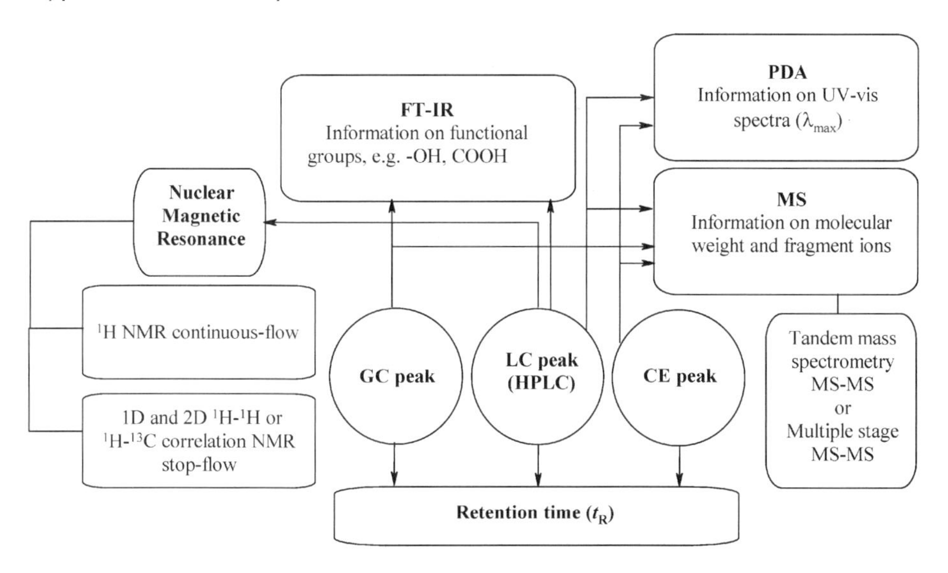

Fig. 7. Summary of on-line information obtained from hyphenated techniques.

3.1. Isolation and Analysis of Natural Products

Crude natural product extracts, which represent extremely complex mixtures of numerous compounds, can be analyzed successfully by using appropriate hyphenated techniques. Among the various hyphenated techniques, LC-PDA and LC-MS are the two most extensively used for natural product analysis. LC-NMR, as well as different multiple hyphenated techniques, LC-PDA-NMR-MS has also become popular most recently. LC-MS, if the ionization technique is chosen appropriately, can be an extremely powerful and informative tool for screening crude plant extracts. The currently available various types of LC-MS systems allow the analysis of small nonpolar compounds to large polar constituents like oligosaccharides, proteins, and tannins present in natural product extracts *(6)*.

3.1.1. Alkaloids

Alkaloids are a large group of nitrogen-containing secondary metabolites of plant, microbial, or animal origin. Various hyphenated techniques have been used in the analysis of several types of alkaloids to date. With the development and wider availability of bench-top systems, GC-MS has become the method of choice for the analysis of various pyrolizidine and quinolizidine types of alkaloids. Quinolizidine alkaloids, the main class of alkaloids found in the family Leguminosae, have been analyzed by GC-MS recently ***(30)***. Most of these alkaloids are sufficiently volatile and thermostable under GC conditions to permit analysis without chemical modification. However, some hydroxylated pyrolizidine alkaloids need to be analyzed as their trimethylsilyl derivatives. Ephedrine-type alkaloids, in dietary supplements containing the Chinese herb *ma huang*, were analyzed by GC-MS and GC-FTIR.

A number of protoberberine metabolites, differing in the number and the placement of various oxygen functions on the aromatic rings, have been identified prior to isolation from the *Corydalis* cell cultures by LC-NMR and LC-MS ***(31)***. This study provided the preliminary evidence for biosynthetic pathways to the formation of these alkaloids, especially the metabolic pathway to 2,3,10,11-oxygenated tetrahydroprotoberberines in cultured cells. An APCI interface was used in the LC-MS system, and the mass spectra were obtained with selected ion monitoring (SIM) and total ion monitoring (TIM) in the positive ion mode. Molecular ion information was obtained on the basis of protonated molecular ion $[M+H]^+$ or a cluster ion $[M+HCF_3]^+$. The LC-NMR analysis was carried out on a Varian UNITY-INOVA-500 NMR spectrometer equipped with a PFG indirect detection LC-NMR probe with a 60 μL flow cell, using stop-flow mode. The LC operation was performed on a Cosmosil 5 C_{18}-AR (4.6 × 150 mm) reversed-phase column, using a mobile phase composed of solvent A =0.1 M NH_4OAc (0.05% TAF) and solvent B = ACN. A gradient elution protocol was adopted as follows: 20–30% B in 10 min, 30% B for 10 min, and 30–155% B in 10 min, flow rate 1 mL/min, detection at 280 nm.

3.1.2. Coumarins

The coumarins are the largest class of 1-benzopyran derivatives that are found mainly in higher plants. HPLC-PDA can be used successfully in the analysis of various phenolic compounds, including coumarins, because of the presence of significant amounts of chromophores in these molecules. The HPLC-PDA determination of coumarins, where absorption spectra are registered with a PDA detector, provide useful information about the identity of the molecule including oxidation pattern. The retention time together with the UV spectrum of individual peaks can be considered characteristic, and can easily be used to detect known coumarins in a crude extract. The coupling of MS to LC-PDA provides further structural information that is helpful for on-line identification of individual coumarins in any crude extract. Various coumarins together with other oxygen heterocyclic compounds, e.g. psoralens and polymethoxylated flavones, present in the nonvolatile residue of the citrus essential oils of mandarin, sweet orange, bitter orange, bergamot, and grapefruit, were analyzed by atmospheric pressure ionization (API) LC-MS system equipped with an APCI probe in positive ion mode ***(32)***. Recording MS spectra at different voltages provided information on molecular weight as well as fragment ions, and this allowed the identification of the main components in the extracts. In this study, cold-pressed citrus oils were analyzed by a Shimadzu LC system coupled with UV and MS detector with an APCI interface. The LC separation was carried out on a C_{18} Pinnacle column (250×4.6 mm, 5 μm), eluted isocratically or using a gradient at a flow rate of 1 mL/min using the solvent mixture: solvent A (THF:ACN:MeOH: water = 15:5:22:58) and solvent B (100% ACN). As coumarins are UV-absorbing compounds, they could be detected at 315 nm. The MS acquisition conditions were as follows: probe high voltage, 4 kV; APCI temperature, 400°C; nebulizing gas (N_2) flow rate, 2.5 L/min; curved desolvation line (CDL) voltage, −25.5 V; CDL temperature, 230°C; deflector voltage, 25 and 60 V; and acquisition mode SCAN, 50–500 m/z.

3.1.3. Carotenoids

This group of natural products includes the hydrocarbons (carotenes) and their oxygenated derivatives (xanthophylls). LC-TLS has been applied successfully to the determination of carotenoids in four marine phytoplankton species, and a good degree of separation of diadinoxanthin, diatoxanthin, and other carotenoids has been achieved by isocratic HPLC elution with a greater sensitivity and selectivity than UV detection. This technique has allowed the monitoring of the interconversion of diadinoxanthin to diatoxanthin, and changes of other carotenoids under different light conditions ***(22)***. LC-TLS has also been found to be an ultrasensitive method for determination of β-carotene in fish oil-based supplementary drugs ***(33)***.

3.1.4. Ecdysteroids

Ecdysteroids are molting hormones of insects and crustaceans, and also occur in a number of plant families. Most have a 2β,3β,14α,20,22-pentahydroxy-5β-cholest-7-en-6-one skeleton with further hydroxylation. Various hyphenated techniques, e.g., LC-PDA, LC-MS, CE-MS, and LC-NMR, have been used successfully for the on-line identification of ecdysteroids in the crude natural product extracts. Most recently, Louden et al. ***(24)*** described the spectroscopic characterization and identification of ecdysteroids in *Lychnis floscoculi* (family: Caryophyllaceae) using HPLC coupled with multiple detectors PDA, FT-IR, NMR, and TOF MS. The TOF MS enabled the determination of molecular formulae of separated compounds via accurate mass measurement. Three ecdysteroids, ecdysone, 20-hydroxyecdysone, and makisterone A, at a concentration of 10–20 mg/mL of D_2O, were used as reference compounds. The experimental protocol applied can be summarized as follows:

1. Air-dried and ground plant material (*L. floscoculi)* was extracted by maceration using EtOH with continuous stirring for 2–3 d. The extract was filtered and the solvent was evaporated.
2. The extract solution was re-constituted in MeOH and centrifuged.
3. MeOH from the clear extract was evaporated prior to HPLC operation, and re-constituted in a small amount of D_2O.
4. The extract (200 μL) was analyzed by a reversed-phase Bruker HPLC system (C_{18} analytical column, eluted isocratically with ACN:D_2O (99.8% isotopic purity) = 20:80, 1 mL/min) coupled with PDA, FTIR, NMR, and MS detectors. The elution order of the reference compounds was: 20-hydroxyecdysone, makisterone A, and ecdysone.
5. The eluent was first allowed to go through the Varian 9065 PDA detector, which enabled the collection of UV spectra over the wavelength range 190–360 nm, which was ideal for most of the ecdysteroids (λ_{max} 238–242 nm).
6. Emerging from the PDA detector, the eluent entered into the Bio-Rad FT-IR model 375C spectrometer fitted with a Spectra Tech Macro Circle Cell ATR (attenuated total reflectance) stainless steel flow cell of 400 μL volume. The IR spectra were acquired with the kinetic software collecting 20 scans per spectrum and at 8 cm^{-1} spectral resolution. The sample was run against a background spectrum of the flowing solvent through the cell prior to injection of the sample solution to subtract out the solvent spectrum from the sample spectra.
7. Following the FT-IR, the eluent passed through a Bischoff Lambda 1000 UV detector set at 254 nm. The solvent stream was then split 95:5, with 5% of the flow directed to the MS and the remainder to the NMR.
8. Mass spectra were obtained on a Micromass LCT TOF mass spectrometer using ESI with a Z spray source. The nebulizer gas flow was set to 85 L/h and the desolvation gas to 973 L/h. Spectra were acquired in +ve ion mode with a capillary voltage of 3.2 kV and a cone voltage of 25 V. The source

temperature was set to 120°C and the desolvation temperature to 350°C. The mass range was 100–900 Daltons.

9. The NMR spectra were obtained using a Bruker DRX 500 NMR spectrometer in the stop-flow mode at 500.13 MHz by a flow through probe of 4-mm i.d. with a cell volume of 120 μL. The spectra were acquired using the NOESYPRESAT pulse (pulses = 90°, relaxation delay = 2 s, and mixing time = 100 ms) sequence to suppress the ACN and residual water signal. The FIDs were collected into 16K data points over a spectral width of 8278 Hz, resulting in an acquisition time of 0.99 s.

The UV spectrum (λ_{max} 238–242 nm) obtained for each separated peak from PDA could identify the separated peaks as possible ecdysteroids, the IR absorption at around 1645 cm^{-1} could be attributed to the carbonyl group of the en-one function in ecdysteroids, the ^1H NMR data could easily pinpoint structural differences among the separated ecdysteroids compared to reference compounds, and finally the MS could provide conclusive information regarding the molecular mass of the separated ecdysteroids. Hence, by using this fully integrated multiple hyphenation, all separated ecdysteroids could be identified rapidly from just a single run.

3.1.5. Essential Oil and Volatile Components

GC-MS has been demonstrated to be a valuable analytical tool for the analysis of mainly nonpolar components and volatile natural products, e.g., mono- and sesquiterpenes. Chen et al. ***(34)*** described a method using direct vaporization GC-MS to determine approx 130 volatile constituents in several Chinese medicinal herbs. They reported an efficient GC-MS method with EI for the separation and structure determination of the constituents in ether-extracted volatile oils of Chinese crude drugs, Jilin *Ginseng*, *Radix aucklandiae,* and *Citrus tangerina* peels. The components, predominantly monoterpenes, of the volatile oil of the oleoresin of *Pestacia atlantica* var *mutica*, have been analyzed recently by GC-MS together with the application of on-line databases ***(35)***.

3.1.6. Flavonoids and Isoflavonoids

The flavonoids (2-phenylbenzpyrone) are a large group of biologically active natural products, distributed widely in higher plants, but also found in some lower plants, including algae. Free flavonoids are moderately polar and their glycosides are obviously polar compounds, and can be separated efficiently by reversed-phase HPLC using ODS C_{18} column by mobile phases of aqueous MeOH or ACN in various proportions with isocratic or gradient elution. As these compounds are UV-active, HPLC-PDA technique has been used extensively for on-line detection and partial identification of flavonoids in plant extracts. Further extensions of this hyphenated technique incorporating MS and NMR detection modes have also been employed for flavonoid analyses. For the LC-PDA-MS analysis of flavonoids, the ESI and APCI interfaces have been found to be the most popular techniques because they allow analysis of flavonoids of a wide range of molecular weights, e.g., simple aglycone, glycosides of various sizes, and malonate or acetate-type conjugates. However, other interfaces, e.g., TSP, FAB, and API, have also been used to a lesser extent. For the precise identification of individual flavonoid in a crude extract, more often an LC-MS-MS or LC-NMR-MS is used. In the ESI mode, in addition to the pseudomolecular ion for the individual flavonoid present in the crude extract or fraction, formation of $[M+H_20]^+$, $[M+Na]^+$, and $[M+MeOH]^+$ has been common in flavonoid analyses. The separated flavonoids can be identified on-line by comparison of UV–vis and MS data with literature data or those available in various on-line databases. During the determination of antioxidant activity of propolis of various geographical origins, a number of flavonoids have been separated and identified quantitatively by LC-PDA-MS ***(36)***. The crude extracts containing polyphenols including flavonoids were dissolved in EtOH (5 mg/mL) and filtered with a 0.45 μm filter prior to 10-μL injection into the LC-PDA-MS system. A Capcell Pak ACR C_{18} column (2.0 × 250 mm, 5 μm) was used and the mobile phase was composed of 0.1% formic acid in water (solvent A) and 0.08% formic acid in ACN (solvent B). The gradient was 20–30% B in 15 min, 30% B from 15 to 35 min, and 30–80% B from 35 to 60 min, at a flow rate of 0.2 mL/min. MS analysis was carried out on an LCQ ion trap mass spectrometer equipped with an ESI interface. The operating parameters were: source voltage, 5 kV; ES capillary voltage, –10 V; and capillary temperature, 260°C. To identify each peak, UV spectra and the SIM of MS spectra of all peaks were compared with those of authentic samples. The quantitative analysis of each component was performed from the calibration curve of the HPLC chromatogram using authentic compounds, as well as from the ion intensity of MS spectrum.

Isoflavonoids belong to a large group of about 700 known plant-derived natural compounds, structures of which are based on 3-phenylbenzpyrone (3-phenylchromone). Like flavonoids, the structures of isoflavones differ in the degree of methylation, hydroxylation, and glycosylation. On-line identification of 14 new isoflavone conjugates (eight isoflavone glycoside malonates and six acetyl glycosides) in *Trifolium pratense* (red clover) was performed by LC-MS (ESI interface) after 2D SPE ***(37)***. The UV spectra, mass spectra of protonated molecular ions and their fragment ions, and subsequent conversion to known glycosides were the basis of the identification process. An HP 1100 LC system was coupled with a PDA and an HP MSD 1100 detectors. The UV spectra were recorded within the range of 190–400 nm because the UV absorption maxima of most of the isoflavones fall within these values. The separation of isoflavones was achieved using a C_{18} (2.0 × 150 mm, 3 μm) column, eluting with a linear gradient from 15% to 25% ACN in aqueous acetic acid (0.2% v/v) in 36 min, and up to 55% ACN in the next 90 min at a flow rate of 0.3 mL/min. The ESI-MS spectra were recorded in positive ion mode (gas temperature, 300°C; drying gas flow, 10.0 L/min; nebulizing gas, 40 psi; capillary voltage, 3.5 kV; scan, 100–800 *m/z;* and fragmentor, 70–100 V). The isoflavones and their conjugates were identified by comparing their retention times, λ_{max}, and the $[M+H]^+$ ions with those of previously isolated and identified reference compounds.

3.1.7. Iridoids and Secoiridoids

Iridoids are cyclopentan-(c)-pyran monoterpenenoids and their glycosides, and constitute a large group of secondary metabolites found in various plant families. Secoiridoids are also plant-derived monoterpene glycosides but originate from their biosynthetic precursor, secologanin. Owing to the lack of important chromophores in iridoids and secoiridoids, except for their acylated aromatic derivatives, HPLC-PDA has limited application in the analysis of these compounds. In most analyses of iridoids and secoiridoids, LC-MS is the preferred option. However, more expensive options, e.g., LC-NMR, LC-MS-NMR, can also be used for these compounds. In the LC-MS analysis, TSP, APCI, and ESI interfaces have been used frequently. A number of secoiridoid glycosides were identified from *Gentiana rhodantha* and *Lisianthius seemannii* by means of LC-MS using TSP interface ***(38,39)***. The retention times and mass spectral information were compared with those of authentic samples. However, to identify three minor secoiridoids including sweroside, it was necessary to perform a second LC-MS analysis using FAB interface.

A combination of CE-MS and HPLC-MS techniques was employed in the determination of iridoid glycosides in *Picrorhiza kurroa* ***(40)***. In the CE-MS mode (applied voltage of 25 kV and thermostating temperature at 30°C), it was possible to achieve baseline separation within 16 min using a fused silica capillary and a borate buffer solution (100 mM, pH 8.6) containing 30 mM SDS and 1% ACN. In the HPLC-MS mode, ESI interface was used, and dominant $[M+Na]^+$ ions could be obtained for HPLC peaks for iridoid glycosides. A good correlation between the iridoid glycoside profiles obtained from CE-MS and LC-MS was observed.

3.1.8. Saponins

Saponins are steroidal or triterpenoidal glycosides that occur widely in plant species of nearly 100 families ***(41)***. As saponins are highly polar compounds and difficult to volatilize, the application of GC-MS is mainly restricted to the analysis of aglycones, known as sapogenins or saponins. Apart from just a few, saponins do not have chromophores that are essential for UV detection. Owing to the lack of chromophores in saponins, it is not helpful to use a UV or PDA as the primary detection technique. An alternative primary detection technique, e.g., refractive index, could be used. Sometimes, precolumn derivatization of saponins can be used to attach a chromophore that facilitates UV detection at higher wavelengths. Among the hyphenated techniques, LC-MS, LC-NMR, and CE-MS could be useful for the rapid initial screening of crude extracts or fraction for the presence of saponins. While in LC-MS analysis TSP interface is used most extensively in phytochemical analysis, saponins having more than three sugars cannot be analyzed using this interface ***(37)***. For larger saponins ($MW > 800$), CF-FAB or ES is the method of choice.

A combination of matrix solid-phase dispersion extraction and LC-NMR-MS was applied in the rapid on-line identification of asterosaponins of the starfish *Asterias rubens* ***(25,26)***. The LC-NMR-MS provided structural information in one single chromatographic run, and was suitable for saponins in the molecular mass range 1200–1400 amu. This technique also allowed semiquantitative LC-NMR measurements through methyl signals (Me-18 and Me-19) of the steroidal skeleton.

3.2. Dereplication

The application of various hyphenated techniques in dereplication and chemical fingerprinting of natural product extracts is discussed in detail in Chapter 12. The discrimination between previously tested or recovered natural product extracts and isolated single components found therein is essential to decrease screening costs by reducing the large collections of isolates that are then subject to further detailed evaluation. Bioassay-guided natural product isolation often leads to already known compounds of limited, or no chemical or pharmacological interest. Hence, appropriate methods that can distinguish at an early stage novel rather than known or already isolated natural compounds are essential for modern cost-effective natural product research. Dereplication strategies employ a combination of separation science, spectroscopic detection technologies, and on-line database searching. Thus, the combination of HPLC with structurally informative spectroscopic detection techniques, e.g., PDA, MS, and NMR, could allow crude extracts or fractions to be screened not just for biological activity but also for structural classes. To perform an efficient screening of extracts, both biological assays and HPLC analysis with various detection methods are used. Techniques such as HPLC coupled with UV photodiode array detection (LC-DAD-UV) and with mass spectrometry (LC-MS or LC-MS-MS) provide a large number of on-line analytical data of extract ingredients prior to isolation. The combination of HPLC coupled to NMR (LC-NMR) represents a powerful complement to LC-UV-MS screening. These hyphenated techniques allow a rapid determination of known substances with only a small amount of source material.

LC-MS-MS spectra are generally reproducible. Therefore, the MS-MS databases of natural products can be used for dereplication purposes. For automated on-line dereplication purposes, Q-DIS/MARLIN^{TM} is one of the comprehensive and powerful analytical knowledge platforms available today, which is ideally suited for natural product dereplications. It permits quick identification of novel chemical classes based on LC-MS data. Q-DIS/MARLIN^{TM} is also perfect to automatically validate proposed chemical structures from combinatorial chemistry experiments of LC-MS data.

Most of the dereplication protocols available to date for natural product analysis predominantly apply LC-PDA-MS. The LC-NMR, despite being able to provide more meaningful structural information, has achieved limited success owing to the lack of sensitivity, the dearth of general access to high-field NMR instruments, and the cost associated with the use of deuterated solvents. However, with the introduction of various solvent suppression techniques, LC-NMR or LC-PDA-MS-NMR has attracted attention of natural product researchers recently.

3.3. Chemical Fingerprinting and Quality Control of Herbal Medicine

The use of hyphenated techniques, e.g., HPLC-PDA, LC-MS, CE-MS, LC-NMR, or LC-NMR-MS, in chemical fingerprinting analysis for quality control and standardization of medicinal herbs has attracted immense interest in recent years. Generally, in the context of drug analysis, fingerprinting method is used to highlight the profiles of the sample matrix, which is often sufficient to provide indications of the source and method of preparation. In herbal medicines, the profile depends not only on the preparation processes but also on the quality of the crude herb source material. The quality of the same herb can vary considerably depending on the geographical origins, sources, harvest times, and so on. The uniformity and stability of the chemical profiles thus represent the quality of the raw herbs. In both good agricultural practice (GAP) and good manufacturing practice (GMP), fingerprinting analysis is used to appraise the quality of the herbal material. In this process, the fundamental objective is to develop links between marker compound-based chromatographic or spectroscopic profiles with the efficacy of herbal products. Thin layer chromatography (TLC) has been the most widely used classical method for fingerprinting analysis in Chinese medicines. In the chemical fingerprinting method, wherever possible, the bioactive compounds or important chemical marker compounds are identified to allow consistent batch-to-batch fingerprinting analysis. For example, in the analysis of valerian (*Valeriana officinalis*) and feverfew (*Tanacetum parthenium*), the two marker compounds are valerenic acid and acetoxyvalerenic acid in the former case and parthenolide and sesquiterpene lactones in the latter ***(42)***. GC-MS or LC-MS can be used to detect and confirm the identity of these trace marker compounds. ESI technique was used in the HPLC-MS-based detailed chemical fingerprinting of danshen, sanqi, and ginkgo. In fingerprinting analysis, it is imperative to optimize all laboratory instrumentations and methodology to avoid any artifacts in the results. The relative intensity of the peaks is important, and chromatographic fingerprints must be specific for the substance being analyzed. Hence, it is necessary to check fingerprints obtained from related botanical products and known adulterants to ensure that the method developed can distinguish true from false identifications. Several analytical protocols based on LC-MS fingerprinting have been developed and integrated into a high-throughput analytical program incorporating standard methods, template structure determination, and structural libraries. For example, LC-MS was used to characterize mixtures of taxanes from *Taxus brevifolia* extracts and to develop a taxane database. The sensitivity of the currently available hyphenated techniques permits minimum sample preparation, thus saving analysis time and reducing unnecessary degradation of the components.

Medicinal properties of herbs used in traditional systems of medicine, e.g., traditional Chinese medicine, or Ayurveda, are attributed to the presence of various types of biologically active molecules. Any variation, either qualitative or quantitative, in the chemical profile of the herb can lead to the total loss of medicinal properties, decreased potency, or even increased toxicity. Therefore, it is essential, for quality control purposes, to ascertain the presence of certain molecules in the herbal preparation or extract, and also to determine the quantity of each of the active principles by applying a suitable method, which allows on-line detection of molecules present in the herbal extract. TLC or paper chromatography used to be the method of choice for the quality control of most of the ancient herbal medicines. Nowadays, with the advent of modern hyphenated techniques, it is possible to obtain comprehensive chemical profiles of herbal medicine preparations or extracts. GC-MS and LC-MS are now being used quite extensively for direct on-line analysis of components present in the herbal preparations and for ensuring the quality of the herb. These techniques have been used in the traditional Chinese medicine ***(43)***. The mass spectra of various components present in the extracts of Chinese medicine have been obtained on-line from the LC-MS run and matched with known standards for structural confirmation. Integrated MS databases have also been useful for identification of these compounds. In this way, GC-MS, LC-MS, and MS-MS fingerprinting profiles of the active ingredients of various Chinese herbal extracts have been obtained, and information has been stored in the form of an electronic database, which can be used for routine comparison of chemical profiles of individual herb extracts for quality control purposes. The GC or LC retention time and mass spectral data are reproducible, provided the chromatographic and spectroscopic conditions are kept constant. LC-NMR and LC-NMR-MS have also been used to this purpose.

A simple protocol for the chemical fingerprinting of *Ephedra* using HPLC-PDA has been recently described ***(44)***. *Ephedra sinica* (family: Ephedraceae), known as "ma huang," is one of the oldest medicinal herbs used in traditional Chinese medicine. In the West, dietary supplements containing *E. sinica* have emerged as one of the top selling weight loss and endurance enhancing products used by over a million consumers.

Six optically active ephedrine alkaloids (0.02–3.40% in aerial parts), (−) -ephedrine, (+)-pseudoephedrine, (−)-methylephedrine, (+)-pseudoephedrine, (−)-norephedrine, and (+)-norpseudoephedrine, are known to be active constituents in this plant. Among these, (−)-ephedrine, which is the major constituent in *E. sinica*, is believed to be the active compound responsible for the claimed pharmacological activities. (−)-Ephedrine and other alkaloids have also been reported to show various adverse effects. Therefore, the presence and level of these alkaloids, especially (−)-ephedrine, in *E. sinica* is crucial for the optimum efficacy as well as the reduced toxicity. The simple HPLC-PDA method incorporating a chemical fingerprinting approach has appeared to be more useful than a number of other available methods, e.g., chiral GC-MS, CE-MS, HPLC-PDA, and LC-MS hyphenated techniques, described for the quality control of *Ephedra*. The overall chemical fingerprinting protocol can be summarized as follows:

1. A total of 25 different species of *Ephedra* were used.
2. Ground plant material (approx 500 mg) was placed in a Falcon Blue Max Jr. 15 mL polypropylene conical tube with 6.0 mL of acetone, and sonicated for 15 min.
3. After sonication, the sample was centrifuged for 10 min, and the supernatant was transferred to a sample vial.
4. The extraction was repeated twice, the respective supernatants were combined, and the solvent (acetone) was removed by rotary evaporation.
5. Absolute ethanol (5.0 mL) was added to the dried extract and the extract allowed to dissolve.
6. Once the extract had dissolved, 2.0 mL was filtered (the first 0.5 mL was discarded) through a 45 mm Nylon filter into an HPLC vial for analysis.
7. HPLC-PDA analysis (sample injection volume = 10 μL) was carried out on a Waters Alliance 2695 Separation module with a Waters 996 PDA detector, using a Waters XTerra RP18 5 μm column (4.6 × 150 mm) and the mobile-phase, isocratic water:ACN = 75:25 for 10 min, gradient water:ACN = 75:25 75:25 to 100% ACN over 45 min, and isocratic 100% ACN for 10 min, flow rate = 1 mL/min. The chromatograms were detected in three different wavelengths, 210, 254, and 320 nm, and were analyzed by Waters Millennium 32 software.
8. The retention time and the UV spectrum obtained for individual peak were matched against known ephedrine alkaloid standards. Quantification was performed on the basis of peak area of individual peaks of known alkaloids, mainly (−)-ephedrine.

9. The fingerprint method was validated by testing a number of populations within a single species of *Ephedra*.

This chemical fingerprinting method was able to verify an *Ephedra* species present in ground plant material and to distinguish between *Ephedra* species grown in different geographical locations.

3.4. Chemotaxonomy

Chemical taxonomy or chemotaxonomy is based on the principle that the presence of certain secondary metabolites is dictated by various enzymes involved in the biosynthesis of these compounds. These enzymes are strictly related to the genetic make up of the organism. Hence, chemical profiling of these secondary metabolites, either by complete isolation and identification, or by separation and on-line identification using modern hyphenated techniques, could provide useful information with regard to the taxonomic or even phylogenetic relationships among various species. Introduction of hyphenated techniques in chemotaxonomic work can reduce the time and cost considerably by allowing on-line detection and identification of secondary metabolites present in extracts. Kite et al. *(30)* described the application of GC-MS in the chemotaxonomic studies based on quinolizidine alkaloid profile in legumes. Using GC-MS, it was possible to obtain data on the quinolizidine alkaloids of less readily available taxa by analyzing crude extracts made from small fragments of herbarium specimens, and thus compile a well-founded knowledge base on the distribution of such compounds in various species of legumes. On the basis of this distribution pattern of quinolizidine alkaloids, various chemotaxonomic inferences could be made. For example, from the GC-MS data analysis, it was observed that Poecilanthe contained quinolizidine alkaloids, and the particular combination of structures present suggested a similarity with members of Brongniartieae *(30)*. As acceptable HPLC separation, using a reversed-phase C_{18} column eluting a gradient of ACN–water or MeOH–water mixture could be achieved for flavonoids and other phenolic compounds. LC-MS was also found to be useful in chemotaxonomic studies based on flavonoid profiles in legumes. Both ES and APCI sources could ionize flavonoids in these mobile phases, and acceptable ionization could be achieved in both positive and negative modes to yield $[M+H]^+$ and $[M-H]^-$ ions, respectively. This technique allowed the analysis of various crude aqueous methanolic extracts of leaves or seeds of several legume species without further purification to obtain metabolomic profiles of the flavonoids produced. A combination of LC-PDA and TSP LC-MS was applied to the chemotaxonomic study on the genus *Epilobium* (family: Onagraceae), where flavonoids were employed as chemotaxonomic markers because chemotaxonomy based on flavonoids allows identification of the different but morphologically similar species ***(44)***. By applying LC-PDA and postcolumn derivatization technique, it was possible to obtain the required structural information of the aglycone moiety of a flavonol glycoside and the corresponding substitution pattern. Further structural information on the molecular weight of separated flavonoids was acquired from a TSP LC-MS. TSP interface was suitable for recording weak pseudomolecular ion, $[M+H]^+$ and intense signals corresponding to the aglycone fragment $[A+H]^+$. In some cases, it was also possible to observe complementary ions for sugar moieties. LC operation was performed on a Shimadzu HPLC, and the conditions applied were as follows: Waters NovaPak RP-18 column (3.9×150 mm, 4 μm) equipped with a NovaPak precolumn, eluted with a gradient of ACN:water = 0 min 10% ACN, 4 min 12% ACN, 12 min 12% ACN, 16 min 18% ACN, and 30 min 25% ACN using a flow rate of 1 mL/min. To avoid the tailing of phenolic compounds, 0.05% TFA was added to the solvents to give a pH of 3.0. LC-NMR was used successfully in the chemotaxonomic studies of several Gentianaceae taxa, based on the distribution of flavones, xanthones, and secoiridoids, using a combination of on-flow and stop-flow modes ***(45)***. The application of hyphenated techniques is not necessarily limited to the chemotaxonomic studies on plant species only. The Frit-FAB LC-PDA-MS technique was used in the chemotaxonomic investigation of micro-organisms based on isoprenoid quinone profiles ***(46)***. In this study, the composition of ubiquinone, menaquinone, rhodoquinone, and their analogs was determined directly using combined information on the HPLC retention time, UV spectrum, and mass spectrum without any standard samples. Frit-FAB technology of LC-MS was found to be highly effective for the analysis of labile and nonvolatile microbial products. A total of 15 microbial strains, including bacteria, actinomycetes, fungi, and yeast, that are known to produce various types of isoprenoid quinines, were analyzed. Appropriate growth media, e.g., nutrient broth, marine broth, zoogloea medium, or ISP no. 2, were used to grow the micro-organisms using standard protocols. After the optimum growth, microbial cell cultures were centrifuged and the resulting pellets were lypophilized and ground to a dry powder. The ground cells were extracted with a mixture of chloroform ($CHCl_3$) and MeOH (2:1 v/v,

100 mL/500 mg of dry cells) and filtered, and the solvent was removed under vacuum to yield dry extract. The residue was dissolved in acetone (100 μL) prior to injection into an HP1090M HPLC-PDA-MS system. The separation of isoprenoid quinones in the extracts was carried out using an Intersil ODS-2 (1.5 × 250 mm) column eluted isocratically with MeOH–isopropyl alcohol (7:3 v/v) at a flow rate of 0.11 mL/min. For postcolumn mixing of matrix and splitting, a solution of 6% *m*-nitrobenzyl alcohol in MeOH was used as a matrix and mixed with HPLC eluent at 1 mL/h. The solution was split at a ratio of 25:1. On-line MS analysis was performed on a JEOL JMS-SX 102 mass spectrometer equipped with a Frit-FAB ion source and a FAB gun. Xe° was employed as a primary beam at 10 mA of emission current.

3.5. Metabolomics

The term "metabolome" refers to the entire complement of low molecular weight metabolites inside a biological cell, and is also used to describe the observable chemical profile or fingerprint of the metabolites in whole tissue ***(47)***. The metabolomes represent the life history of an individual organism, including age and environmental factors such as soil type, moisture content, temperature, and stress factors. The study involving the detailed analysis of these metabolomes is referred to as "metabolomics," which is a newly emerging area in natural product research in the postgenome era. The aim of metabolomics, for example, plant metabolomics, is to provide a better understanding of metabolic or other physiological phenotypes through global genome-related technology. In metabolomics, nontargeted metabolite profiling and linkage of these profiles to genotypes or transcript profiles are regarded as the key issues ***(48)***. The metabolite profiling, without the isolation of individual metabolites, requires sophisticated high-throughput analytical techniques, e.g, various types of hyphenated techniques. The most extensively used hyphenated techniques with regard to metabolomics are GC-MS, LC-MS, LC-PDA-MS, LC-NMR, and LC-NMR-MS.

A typical example of metabolomic study can be demonstrated from the metabolite profiling, which is the first trial of a metabolomic approach of an anthocyanin chemotype, in red and green forms of *Perilla fructescens* using LC-PDA-MS, CE for anion analysis and LC for amino acid analysis ***(47)***. In addition, studies on cell-specific anthocyanin accumulation and localization of anthocyanidin synthase, and gene expression using mRNA differential display of two chemo-varietal forms of *P. fructescensm*, were also carried out. The experimental protocol associated with the hyphenated technique-based metabolite profiling was as follows:

1. The red and green forms of *P. fructescens* were grown on rock wool with nutrient solution of Hyponex (5-10-5) in a plant growth room for 16 weeks with a photoperiod of 18 h light (4500 lx)/6 h dark at 25°C.
2. The leaves and stems of the red and green forms (approx 1 g) were extracted with 6 mL of a solvent mixture (MeOH:AcOH:water = 9:1:10) per 1 g fresh weight of tissues at 4°C overnight.
3. The extracts were filtered through a 0.45-μm nylon filter and applied to HPLC-PDA-ESI-MS system consisting of a Finigan LCQ DECA mass spectrometer and an Agilent HPLC 1100 series.
4. LC separation was carried out on an ODS-A312 column (6.0 × 150 mm) at a flow rate of 0.5 mL/min, using a linear gradient program ACN:water:TFA = 7.5:92.5:0.1 to ACN:water:TFA = 55:45:0.1 over 60 min.
5. A PDA detector was used for detection of UV–visible absorption (200–700 nm), nitrogen was used as sheath gas for the +ve ion ESI-MS (capillary temperature and voltage of 320°C and 5.0 kV, respectively), the tube lens offset was set at 10.0 V, full mass spectra were obtained from 80 to 2000 m/z at 2 scans/s, and Tandem MS analysis was performed with lithium as the collision gas (the normalized collision energy was set to 30%).

Following this protocol, nearly 50 peaks were separated and identified primarily by UV–vis. Absorption spectra and MS fragmentation by Tandem MS analysis were compared with the authentic compounds and previously reported data. A number of anthocyanin pigments, malonylshisonin being the major compound (approx 70%), were detected in the red leaves. However, the green leaves were found not to accumulate these anthocyanins or accumulated only trace levels of anthocyanins. Similar metabolite patterns were observed in the stems of these two varieties of *P. fructescens.* From the results, it was concluded that the accumulation of anthocyanin is strictly specific to red form, and this form specificity is more obvious in leaves than in stems. The levels of flavonoids, luteolin, and apigenin conjugates present in these two forms were similar. This finding indicated that the regulation of production of flavones is different from that of anthocyanin production in two forms of perilla. The amount of rosmarinic acid in the green leaves was slightly higher than that in the red leaves, which is possibly caused by the competition for the common intermediate, 4-coumaroyl-CoA, between biosynthesis of rosmarinic acid and flavonoids. The profile of anion, e.g., nitrate, sulfate, phosphate, was determined by CE and that of amino acids, e.g., aspartic acid, threonine, and serine, by HPLC system with postcolumn derivatization for fluorescence detection. Both profiles were similar in red and green forms.

The hyphenated techniques have become an integral part of studies in the area of metabolomics. Huhman and Summer ***(49)*** used an approach that involved segregation of the metabolome into several subclasses followed by parallel analyses using LC-MS. Triterpene saponins from *Medicago sativa* (alfalfa) and *M. truncatula* roots were separated, profiled, and identified using HPLC-PDA-ESI-MS hyphenated technique.

References

1. Wilson, I. D. and Brinkman, U. A. Th. (2003) Hyphenation and hypernation: the practice and prospects of multiple hyphenation. *J. Chromatogr. A* **1000,** 325–356.
2. Wolfender, J. L., Ndjoko, K., and Hostettmann, K. (1998) LC/NMR in natural products chemistry. *Curr. Org. Chem.* **2,** 575–596.
3. Wolfender, J. L., Rodriguez, S., and Hostettmann, K. (1998) Liquid chromatography coupled to mass spectrometry and nuclear magnetic resonance spectroscopy for the screening of plant constituents. *J. Chromatogr. A* **794,** 299–316.
4. Huber, L. and George, S. A. (1993). *Diode Array Detection in HPLC.* Mercel-Dekker, New York.
5. Niessen, W. M. A. and Tinke, A. P. (1995) Liquid chromatography-mass spectrometry, general principles and instrumentation. *J. Chromatogr. A* **703,** 37–57.
6. Niessen, W. M. A. (1999). *Liquid chromatography-Mass spectrometry.* 2 ed., Dekker, New York.
7. Albert, K. (1995) On-line use of NMR detection in separation chemistry. *J. Chromatogr. A* **703,** 123–147.
8. Lindon, J. C., Nicholson, J. K., Sidelmann, U. G., and Wilson, I. D. (1997) Directly coupled HPLC-NMR and its application to drug metabolism. *Drug. Metab. Rev.* **29,** 707–746.
9. Sudmeier, J. L., Gunther, U. L., Albert, K., and Bachovchin, W. W. (1996) Sensitivity optimisation in continuous-flow FTNMR. *J. Magn. Reson. A* **118,** 145–156.
10. Wolfender, J. L., Abe, F., Nagao, T., Okabe, H., Yamauchi, T., and Hostettmann, K. (1995) Liquid chromatography combined with thermospray and continuous-flow fast atom bombardment mass spectrometry of glycosides in crude plant extract. *J. Chromatogr. A* **712,** 155–168.
11. Bringmann, G., Messer, K., Wohlfarth, M., Kraus, J., Dumbuya, K., and Rückert, M. (1999) HPLC-CD on-line coupling in combination with

HPLC-NMR and HPLC-MS/MS for the determination of the full absolute stereostructure of new metabolites in plant extracts. *Anal. Chem.* **71,** 2678–2686.

12. Albert, K. (2002) *On-line LC-NMR and Related Techniques*. Wiley, London.
13. Jinno, K. (2001) Basics and applications of hyphenated-detection system in HPLC: Part III—hyphenated techniques in HPLC. *Pharm. Stage* **1,** 110–131.
14. Jinno, K. (2001) Basics and applications of hyphenated-detection system in HPLC: Part II—detection systems in HPLC. *Pharm. Stage* **1,** 74–80.
15. Jinno, K. (2001) Basics and applications of hyphenated-detection system in HPLC: Part I—Basics and applications in HPLC. *Pharm. Stage* **1,** 81–94.
16. Jinno, K. (2001) Infrared Detect, in *Encyclopedia of Chromatography* (Cazes, J., ed.), Marcel Dekker, New York, NY, USA.
17. Jinno, K., Fujimoto, C., and Hirata, Y. (1982) An interface for the combination of micro high-performance liquid-chromatography and infrared spectrometry. *Appl. Spectrosc.* **36.** 67–69.
18. Bourne, S., Haefner, A. M., Norton, K. L., and Grifiths, P. R. (1990) Performance-characteristics of a real-time direct deposition gas-chromatography fourier-transform infrared spectrometry system. *Anal. Chem.* **62,** 2448–2452.
19. Herderich, M., Richling, E., Roscher, R., Schneider, C., Schwab, W., Humpf, H. U., and Schreier, P. (1997) Application of atmospheric pressure ionisation HPLC-MS-MS for the analysis of natural products. *Chromatographia* **45,** 127–132.
20. Logar, J. K., Malej, A., and Franko, M. (2003) Hyphenated high performance liquid chromatography-thermal lens spectrometry technique as a tool for investigations of xanthophyll cycle pigments in different taxonomic groups of marine phytoplankton. *Rev. Sci. Instrum.* **74,** 776–778.
21. Dunayevskiy, Y. M., Vouros, P., Winter, E. A., Shipps, G. W., and Carell, T. (1996) Application of capillary electrophoresis-electrospray ionisation spectrometry in the determination of molecular diversity. *Proc. Natl. Acad. Sci.* **93,** 6152–6157.
22. Bernet, P., Blaser, D., Berger, S., and Schar, M. (2004) Development of a robust capillary electrophoresis-mass spectrometer interface with a floating sheath liquid feed. *Chimica* **58,** 196–199.
23. Navas, M. J. and Jimenez, A. M. (2003) Thermal lens spectroscopy as analytical tool. *Crit. Rev. Anal. Chem.* **33,** 77–88.
24. Louden, D., Handley, A., Taylor, S., et al. (2001) Spectroscopic characterisation and identification of ecdysteroids using high-performance liquid chromatography combined with on-line UV-diode array, FT-infrared and 1H-nuclear magnetic resonance spectroscopy and time of flight mass spectrometry. *J. Chromatogr.* **910**, 237–246.

25. Sandvoss, M., Weltring, A., Preiss, A., Levsen, K., and Wuensch, G. (2001) Combination of matrix solid-phase dispersion extraction and direct on-line liquid chromatography-nuclear magnetic resonance spectroscopy-tandem mass spectrometry as a new efficient approach for the rapid determination of natural products: application to the total asterosaponin fraction of the starfish *Asterias rubens*. *J. Chromatogr. A* **917,** 75–86.
26. Sandvoss, M., Pham, L. H., Levsen, K., Preiss, A., Mugge, C., and Wuensch, G. (2000) Isolation and structure elucidation of steroid oligoglycosides from the starfish *Asterias rubens* by means of direct on-line LC-NMR-MS hyphenation and one- and two-dimensional NMR investigations. *Eur. J. Org. Chem.* **7,** 1253–1262.
27. Elipe, M. V. S. (2003) Advantages and disadvantages of nuclear magnetic resonance spectroscopy as a hyphenated technique. *Anal. Chim. Acta.* **497,** 1–25.
28. Bailey, N. J. C., Stanley, P. D., Hadfield, S. T., Lindon, J. C., and Nicholson, J. K. (2000) Mass spectrometrically detected directly coupled high performance liquid chromatography/nuclear magnetic resonance spectroscopy/mass spectrometry for the identification of xenobiotic metabolites in maize plants. *Rapid Commun. Mass Spectrom.* **24,** 679–684.
29. He, X. E. (2000) On-line identification of phytochemical constituents in botanical extracts by combined high performance liquid chromatographic-diode array detection-mass spectrometric techniques. *J. Chromatogr. A* **880,** 203–232.
30. Kite, G. C., Veitch, N. C., Grayer, R. J., and Simmonds, M. S. J. (2003) The use of hyphenated techniques in comparative phytochemical studies of legumes. *Biochem. Syst. Ecol.* **31,** 813–843.
31. Iwasa, K., Kuribayashi, A., Sugiura, M., Moriyasu, M., Lee, D.-U., and Wiegrebe, W. (2003) LC-NMR and LC-MS analysis of 2,3,10,11-oxygenated protoberberine metabolites in *Corydalis* cell cultures. *Phytochemistry* **64,** 1229–1238.
32. Dugo, P., Mondello, L., Dugo, L., Stancanelli, R., and Dugo, G. (2000) LC-MS for the identification of oxygen heterocyclic compounds in citrus essential oils. *J. Pharm. Biomed. Anal.* **24,** 147–154.
33. Luterotti, S., Franko, M., and Bicanic, D. (1999) Ultrasensitive determination of β-carotene in fish oil based supplementary drugs by HPLC-TLS. *J. Pharm. Biomed. Anal.* **21,** 901–909.
34. Chen, Y., Li, Z., Xue, D., and Qi, L. (1987) Determination of volatile constituents of Chinese medicinal herbs by direct vaporization capillary gas-chromatography mass-spectrometry. *Anal. Chem.* **59,** 744–748.
35. Delazar, A., Reid, R. G., and Sarker, S. D. (2004) GC-MS analysis of essential oil of the oleoresin from *Pistacia atlantica var mutica*. *Chem. Nat. Compounds* **40,** 24–27.

36. Oleszek, W. A. (2002) Chromatographic determination of plant saponins. *J. Chromatogr. A* **967,** 147–162.
37. Klejdus, B., Vitamvásová-Štěrbová, D., and Kuban, V. (2001) Identification of isoflavone conjugates in red clover *(Trifolium pratense)* by liquid chromatography-mass spectrometry after two-dimensional solid-phase extraction *Anal. Chim. Acta.* **450,** 81–97.
38. Ma, W. G., Fuzzati, N., Wolfender, J.-L., Hostettmann, K., and Yang, C. (1994) Rhodenthoside A, a new type of acylated secoiridoid glycoside from *Gentiana rhodentha. Helv. Chim. Acta.* **77,** 1660–1671.
39. Hostettmann, K. and Wolfender, J.-L. (2001) Application of liquid chromatography/uv/ms and liquid chromatography/nmr for the on-line identification of plant metabolites, in *Bioactive Compounds from Natural Sources: Isolation, Characterisation and Biological Properties*, (Tringali, C., ed.), Taylor and Francis, London, pp. 33–68.
40. Sturm, S. and Stuppner, H. (2001) Analysis of iridoid glycosides from *Picrorhiza kurroa* by capillary electrophoresis and high performance liquid chromatography-mass spectrometry. *Chromatographia* **53,** 612–618.
41. Kumazawa, S., Hamasaka, T., and Nakayama, T. (2004) Antioxidant activity of propolis of various geographical origins. *Food Chem.* **84,** 329–339.
42. Cai, Z., Lee, F. S. C., Wang, X. R., and Yu, W. J. (2002) A capsule review of recent studies on the application of mass spectrometry in the analysis of Chinese medicinal herbs. *J. Mass Spectrom.* **37,** 1013–1024.
43. Schaneberg, B. T., Crockett, S., Bedir, E., and Khan, I. A. (2003) The rọle of chemical fingerprinting: application to Ephedra. *Phytochemistry* **62,** 911–918.
44. Ducrey, B., Wolfender, J. L., Marston, A., and Hostettmann, K. (1995) Analysis of flavonol glycosides of thirteen *Epilobium* species (Onagraceae) by LC-UV and thermospray LC-MS. *Phytochemistry* **38,** 129–137.
45. Wolfender, J. -L., Rodriguez, S., Hostettmann, K., and Hiller, W. (1997) Liquid chromatography/ultraviolet/mass spectrometric and liquid chromatography/nuclear magnetic resonance spectroscopic analysis of crude extracts of Gentianaceae species. *Phytochem. Anal.* **8,** 97–104.
46. Nishijima, M., Araki-Sakai, M., and Sano, H. (1997) Identification of isoprenoid quinones by frit-FAB liquid chromatography-mass spectrometry for the chemotaxonomy of microorganisms. *J. Microbiol. Methods* **28,** 113–122.
47. Ott, K.-H., Aranibar, N., Singh, B., and Stockton, G. W. (2003) Metabolomics classifies pathways affected by bioactive compounds. Artificial neural network classification of NMR spectra of plant extracts. *Phytochemistry* **62,** 971–985.

48. Yamazaki, M. and Nakajima, J.-C. Yamanashi, M. et al. (2003) Metabolomics and differential gene expression in anthocyanin chemo-varietal forms of *Perilla fructescens*. *Phytochemistry* **62,** 987–995.
49. Huhman, D. V., and Sumner, L. W. (2002) Metabolic profiling of saponins in *Medicago sativa* and *Medicago truncatula* using HPLC coupled to an electrospray ion-trap mass spectrometer. *Phytochemistry* **59,** 347–360.

10

Purification by Solvent Extraction Using Partition Coefficient

Hideaki Otsuka

Summary

Isolation of compounds in a pure state from natural sources is the most important, yet can be a difficult and time-consuming, step in natural product research. It begins with the process of extraction followed by various separation techniques. One such separation technique is the solvent partitioning method, which usually involves the use of two immiscible solvents in a separating funnel. In this method, compounds are distributed in two solvents according to their different partition coefficients. This technique is highly effective as the first step of the fairly large-scale separation of compounds from crude natural product extracts. In this chapter, the generic protocols of this separation technique in the purification of natural compounds, especially phytochemicals of relatively small molecular weights, i.e., in the range of 100–1500 Da, have been outlined.

Key Words: Partition; partition coefficient; solvent extraction; immiscible solvents.

1. Introduction

Isolation and purification of natural products from various sources is the most vital, yet laborious and difficult, step in natural product research. It begins with the process of extraction followed by various separation techniques. Apart from the available chromatographic separation techniques (*see* Chaps. 4–9), solvent partitioning method can be successfully applied

From: *Methods in Biotechnology, Vol. 20, Natural Products Isolation, 2nd ed.*
Edited by: S. D. Sarker, Z. Latif, and A. I. Gray © Humana Press Inc., Totowa, NJ

in the separation of different classes of natural products. In fact, before the introduction of chromatographic techniques, solvent partitioning method and crystallization techniques (*see* Chap. 11) were extensively used for the separation and purification of natural products. A number of biologically active compounds, predominantly alkaloids, were isolated from various plant sources by utilizing these techniques. The separation technique using solvent partitioning involves primarily the use of two immiscible solvents in a separating funnel, and the compounds are distributed in two solvents according to their different partition coefficients. This method is relatively easy to perform and highly effective as the first step of the fairly large-scale separation of compounds from crude natural product extracts. However, for final small-scale separation, more sophisticated and advanced solvent partitioning methods, such as countercurrent distribution (Craig distribution) ***(1)*** and droplet countercurrent chromatography ***(2)*** (*see* Chap. 7), can be applied. This chapter focuses on the general protocols associated with solvent extraction methods for the purification of natural compounds, especially phytochemicals of relatively small molecular weights, i.e., in the range of 100–1500 Da.

2. Partitioning Between Immiscible Solvents

A crude natural product extract is generally an extremely complicated mixture of several compounds possessing varying chemical and physical properties. The fundamental strategy for separating these compounds is based on their physical and chemical properties that can be cleverly exploited to initially separate them into various chemical groups. However, in some cases, from the literature search of the related genera and families, it is possible to predict the types of compounds that might be present in a particular extract. This tentative prediction on the possible identity of the classes of compounds may help choose suitable extraction and partitioning methods, and solvents for extracting specific classes of compounds, for example, phenolics, saponins, alkaloids. Plant natural products are usually extracted with solvents of increasing polarity, for example, first *n*-hexane, diethylether, chloroform ($CHCl_3$), to name a few, followed by more polar solvents, i.e., methanol (MeOH), depending on the chemical and physical nature of the target compounds.

Alcoholic (MeOH or EtOH) extracts of plant materials contain a wide variety of polar and moderately polar compounds. By virtue of the cosolubility, many compounds, which are insoluble individually in pure state in MeOH or EtOH, can be extracted quite easily with these solvents.

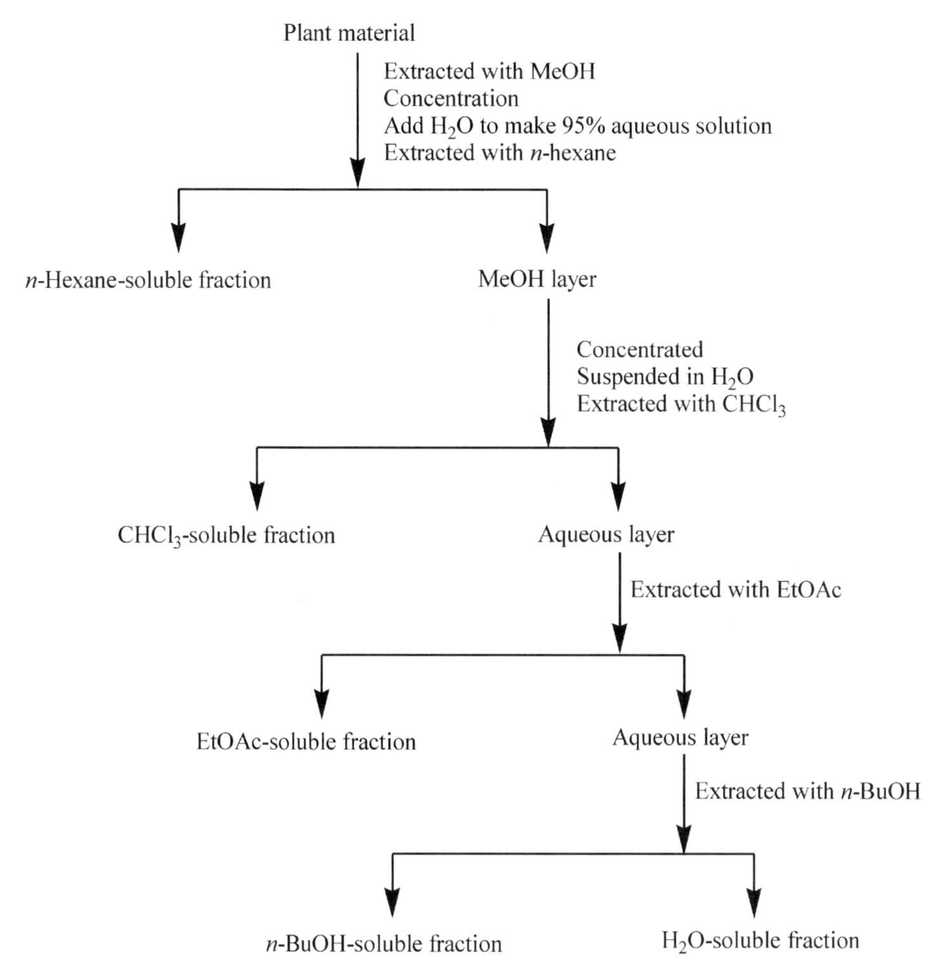

Fig. 1. A typical partitioning scheme using immiscible solvents.

Hence, Nernst's partition law may not be applicable. A dried alcoholic extract can also be extracted directly with a suitable solvent.

A typical partitioning scheme is presented in **Fig. 1** *(3)*. A MeOH extract is concentrated, and the volume is reduced to an appropriate level that can be handled easily with a separating funnel. The concentrated extract is then extracted with an equal volume of *n*-hexane, usually three times, to give a fraction containing nonpolar compounds, such as lipids, chlorophylls, and so on. The process is sometimes referred to as "defatting." Although MeOH and *n*-hexane are not completely miscible, they are miscible to some extent. Sometimes, a small amount of water is added to MeOH to obtain a 95%-aqueous methanolic solution to get two distinct layers with similar volumes. The methanolic layer is evaporated to dryness and then dissolved in water. Occasionally it is not a solution, but a suspension. The solution (suspension) is partitioned between $CHCl_3$, ethylacetate (EtOAc), and *n*-butanol (*n*-BuOH), successively. Partitioning with $CHCl_3$ can be omitted depending on the chemical nature of the target compounds. Less polar compounds are present in the $CHCl_3$-soluble fraction and polar compounds, probably up to monoglycosides, in the EtOAc-soluble one. The *n*-BuOH fraction contains polar compounds, mainly glycosides. Evaporation of the remaining water layer leaves polar glycosides and sugars as a viscous gum. However, separation by solvent partitioning cannot be always performed in a clearcut manner; overlapping of the compounds in successive fractions is usually found. When using EtOAc as an extraction solvent, especially the technical grade solvent, researchers must remember that it contains a trace amount of acetic acid (AcOH), which may cause a *trans*-esterification of acetyl group to the hydroxyl groups, and have a catalytic effect on labile functional groups or delicate structures. When the acetates of some compounds are isolated from the EtOAc-soluble or subsequent *n*-BuOH-soluble fraction, it is suspected that *trans*-esterification may have produced the acetates of the original compounds as artifacts. Chloroform is an ideal solvent for extracting alkaloids owing to its slight acidic nature, because alkaloids tend to be soluble in acidic media. When water layer is to be extracted thoroughly with *n*-BuOH, *n*-BuOH saturated with water is frequently used. Although *n*-BuOH is not miscible with water, 9.1 mL of *n*-BuOH is soluble in 100 mL of water at 25°C. Therefore, when the water layer is extracted with *n*-BuOH unsaturated with water many times, the volume of the water layer drastically decreases. Usage of unbalanced volumes of solvents sometimes causes unexpected partitioning of compounds.

When saponins are the major target, it is advisable that the glycoside fraction (*n*-BuOH layer) is partitioned with a 1%-KOH solution to remove widely distributed phenolic compounds, such as flavonoids and related glycosides. Before concentrating the extract, the *n*-BuOH layer must be washed several times with water. In turn, re-extraction of the acidified alkaline layer gives a fraction rich in phenolic compounds. Some acylated saponins and flavonoids, present in plant extracts, are also hydrolyzed under alkaline conditions. Thus, at least a small-scale pilot experiment, such as tracing the fate of compounds by thin layer chromatography

(TLC), is strongly recommended. However, this method is useful for the isolation of known alkali-resistant saponins on a large scale.

3. Partitioning Between Miscible Solvents

Contrary to what has already been discussed earlier, miscible solvents are sometimes used for partitioning on addition of water ***(4)***. A plant material is extracted with MeOH and evaporated to obtain a residue. The residue is re-dissolved in 90% aqueous MeOH, and the resulting solution is extracted with *n*-hexane. This step seems to be similar to the previous partitioning example. In the next step, an appropriate amount of water is added to the 90%-aqueous MeOH to obtain an 80% aqueous solution, which is then extracted with CCl_4 (MeOH and CCl_4 are miscible). The final step is to make a 65%-aqueous MeOH solution with the addition of water, and the resulting solution is extracted with $CHCl_3$ (MeOH and $CHCl_3$ are miscible). Evaporation of the *n*-hexane, CCl_4, and $CHCl_3$ layers gives three fractions in order of polarity. Concentration of the 65%-aqueous MeOH layer gives the most polar fraction. This fraction is expected to contain glycosides as major constituents as well as a large amount of water-soluble sugars.

References

1. Craig, L. C. (1944) Identification of small amounts of organic compounds by distribution studies. II. Separation by counter-current distribution. *J. Biol. Chem.* **155,** 519–534.
2. Inoue, O., Ogihara, Y., Otsuka, H., Kawai, K., Tanimura, T., and Shibata, S. (1976) Droplet counter-current chromatography for the separation of plant products. *J. Chromatogr.* **128,** 218–223.
3. He, D.-H., Otsuka, H., Hirata, E., Shinzato, T., Bando, M., and Takeda, Y. (2002) Tricalysiosides A–G: rearranged *ent*-kauranoid glycosides from the leaves of *Tricalysia dubia*. *J. Nat. Prod.* **65,** 685–688.
4. Takeda, Y. and Fatope, M. O. (1988) New phenolic glucosides from *Lawsonia inermis*. *J. Nat. Prod.* **51,** 725–729.

11

Crystallization in Final Stages of Purification

Alastair J. Florence, Norman Shankland, and Andrea Johnston

Summary

Methods are described for the laboratory-scale crystallization of "small" organic compounds. The process of crystallization from solution can be used as a purification step in its own right, or to produce crystals for molecular structure determination by single-crystal or powder X-ray diffraction. Both aspects are discussed, with particular emphasis on growing crystals for structure determination in natural product chemistry. The processes detailed for the slow growth of diffraction-quality crystals include solvent selection and solution supersaturation by evaporation, cooling, liquid/vapor diffusion, and thermal gradient methods. Common problems and solutions, including solid-state polymorphism and solvate formation, are highlighted and modern approaches to parallel crystallization and crystal structure determination from X-ray powder diffraction data are also introduced.

Key Words: Crystallization; nucleation; parallel crystallization; single crystal; crystal structure determination; single-crystal diffraction; X-ray powder diffraction; polymorphism; fractional crystallization.

1. Introduction

The hard work has been done and the target compound has been separated from the other organic compounds in the mixture, but is still in solution, and may possibly be "contaminated" by buffer salts or other inorganic compounds. Hence, the next step is to obtain the target compound

From: *Methods in Biotechnology, Vol. 20, Natural Products Isolation, 2nd ed.*
Edited by: S. D. Sarker, Z. Latif, and A. I. Gray © Humana Press Inc., Totowa, NJ

in a useable form. This generally means ending up with a concentrated solution of pure compound, or the pure dry solid, which may or may not be crystalline. The prelude to this final stage is to establish that the purification is complete. Analysis will have been ongoing during the isolation process, and so a suitable analytical system should be in place. However, because no further purification work is anticipated, additional analysis may be useful at this stage to ascertain with greater certainty the level at which contaminants are present.

The definition of "pure" is fairly arbitrary *(1)*. To say a purification process is "complete" and that the target natural product compound is now "pure" does not necessarily imply that it is absolutely free of other chemicals. It simply means that the amount of any impurity present does not exceed some arbitrarily defined "acceptable" level. In this chapter, we focus are on crystallization from solution. This process can be used as a purification step in its own right, or to produce crystals for molecular structure determination by single-crystal or powder X-ray diffraction, and these aspects are discussed here.

1.1. Crystallization

1.1.1. Overview

This section provides guidance on laboratory-scale crystallization of "small" organic compounds. It does not deal with the more specialized area of crystallization of proteins. Crystal structure reports in journals rarely detail crystallization information beyond the solvent used. Textbooks also tend to be limited in value because they vary a lot in scope and focus. What we present here combines information gathered from textbooks with our own experience of growing single crystals for X-ray and neutron diffraction experiments, and also keeps the natural product chemist very much in mind.

1.1.1.1. What is a Crystal?

Crystallinity is synonymous with "order." It is this order that enables us to recognize crystalline material through properties such as a definite melting point, the diffraction of X-rays, and the presence of ordered flat surfaces (faces) with straight edges. The sparkling of crystalline sugar, for example, is caused by light reflecting off the flat faces that bound individual crystals. The term *poly*crystalline describes aggregates of crystals that are really too small for any crystallinity to be instantly obvious to the naked eye. In marked contrast to crystalline solids, amorphous materials, such as liquids, glasses, and rubbers, have no long-range atomic order.

There is an obvious connection between crystal chemistry and natural products. Crystals appear in plant material ***(2,3)***—calcium oxalate is extremely common, potassium acid tartrate in outer perisperm cells of nutmeg, hesperidin and diosmin in cells of many species of Rutaceae, calcium carbonate in cells of Cannabinaceae plants, and silica in the cells of the sclerenchymatous layer of cardamom seeds. The crystallizing temperature of the solid complex between 1,8-cineole and *o*-cresol forms the basis for official assay of oil of eucalyptus ***(3)***. Also of note is the fact that alkaloids, particularly brucine, are widely used to resolve optically active acids by diastereoisomer fractional crystallization ***(4)***.

1.1.1.2. Why Crystallize?

Single-crystal X-ray diffraction allows us to "see" atom positions in three-dimensional space and, as such, is more direct than NMR, which uses radio frequency to "hear" nuclei resonating in a magnetic field. Single-crystal X-ray diffraction can be used to determine, confirm, or complete molecular structures routinely and unambiguously, and thereby establish conformation and relative stereochemistry. For example, the crystal structure of the alkaloid manicoline B ***(5)*** reveals that the molecule crystallizes as diastereoisomers with the same configuration at atom C6 and different ones at atom C1 (**Fig. 1**). Single-crystal X-ray diffraction can also be used to determine absolute configuration. Heavy atom (e.g., bromine) derivatives are preferred if the parent compound contains only the elements C, N, and O because bromine has appreciable anomalous scattering and renders the determination of absolute configuration easier (for an example, *see* **ref. 6**).

Subheadings 1. and **2.** are written from the point of view of the natural product chemist who has compound(s) of acceptable chemical purity, and who wishes to grow crystals, usually for the purposes of molecular structure determination by single-crystal X-ray diffraction. **Subheading 3.**, on the other hand, illustrates how crystallization can be used to achieve separation. One should bear in mind that on a scale smaller than 100 mg, chromatographic techniques are usually more appropriate for separations. However, as the examples in **Subheading 3.** show, crystallization often occurs as part of a concentration step in an extraction.

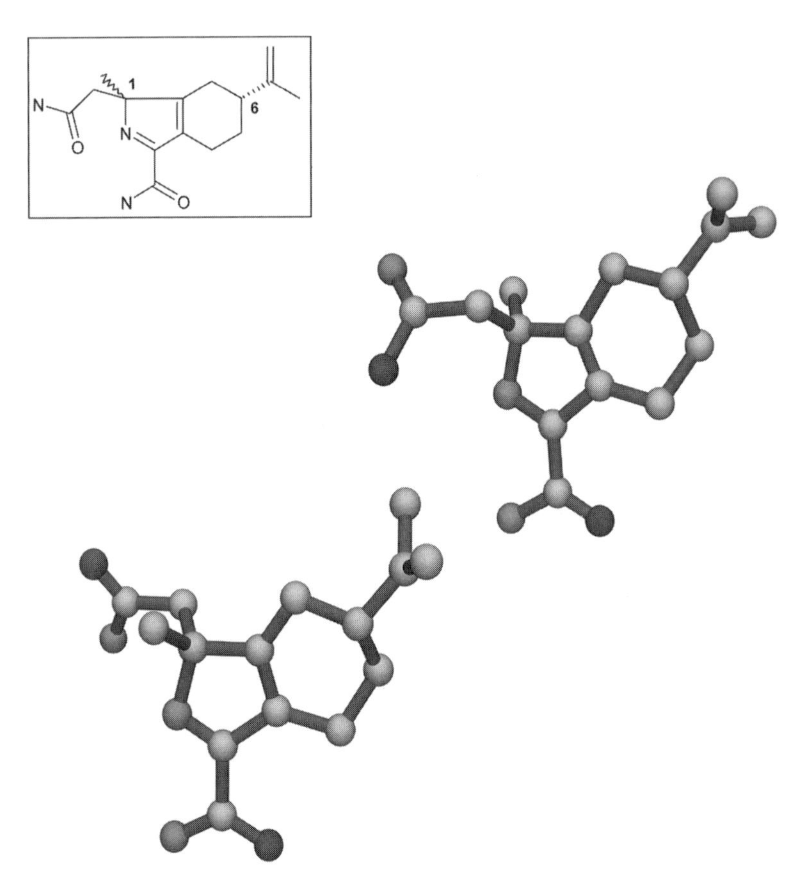

Fig. 1. Diastereoisomers in the crystal structure of Manicoline B, plotted using atomic coordinates retrieved from the *Cambridge Structural Database*, as per "Note 4" in **ref. 5**. The inset identifies chiral centers C1 and C6.

1.1.2. Obtaining Crystals

The most practical method of crystallizing natural products is from solution and the process can be described as one in which:

1. A saturated solution containing one or more compounds of interest becomes supersaturated.
2. Nucleation occurs and crystal growth ensues.

Crystallization is essentially a collision process—molecules collide to form a cluster called the nucleus, which then develops into a crystal with a characteristic internal structure and external shape. It therefore follows that factors such as stirring and degree of supersaturation, which influence the number of molecular collisions in solution, can affect the crystallization process.

1.1.2.1. Solvent Selection

The common recrystallization solvents are listed in **Fig. 2**. A solvent is chosen such that the compound of interest is neither excessively soluble nor insoluble, and it is apparent in **Fig. 2** that there is tremendous flexibility in the choice of solvent polarity and recrystallization temperature. Compound solubility varies significantly with solvent—the solubility of naphthalene, for example, is approximately doubled going from methanol to ethanol, while the addition of water to either drastically reduces solubility *(7)* Solvent mixtures provide a convenient way of tailoring solubility to a required level. Methylated spirits (a ready-made solvent mixture) are commonly used, but it is generally advantageous to cast the net wider and any miscible pair of solvents is worth investigating, especially if compound solubility differs significantly between the two. Ultimately, one should go with the solvent(s) system that yield diffraction-quality crystals.

1.1.2.2. Preparation of Solution and Crystallization

In instances where there is no shortage of sample compound, a saturated solution can be prepared by dissolving some of the sample in the crystallizing solvent, then incrementally adding more solid up to the point where no more will dissolve, i.e., the point of saturation. It must be noted that the maximum concentration of compound that can be dissolved at a particular temperature (the saturation solubility) generally rises with increasing temperature.

When only small quantities of sample compound are available, an alternative approach is to dissolve the compound in solvent 1 and then add a second, miscible solvent dropwise to produce a mixed solvent system in which the compound is less soluble than in solvent 1 alone. When the solution first turns hazy (i.e., when it is just supersaturated), adding a drop of solvent 1 will produce a clear solution that is now close to the point of saturation. Once the solution has been filtered to remove gross particulate contamination—glass wool in a Pasteur pipet is convenient for small volumes—one should proceed to the supersaturation stage. The most common methods of supersaturating a solution (i.e., raising the concentration of a dissolved compound above its saturation solubility) are evaporation and cooling.

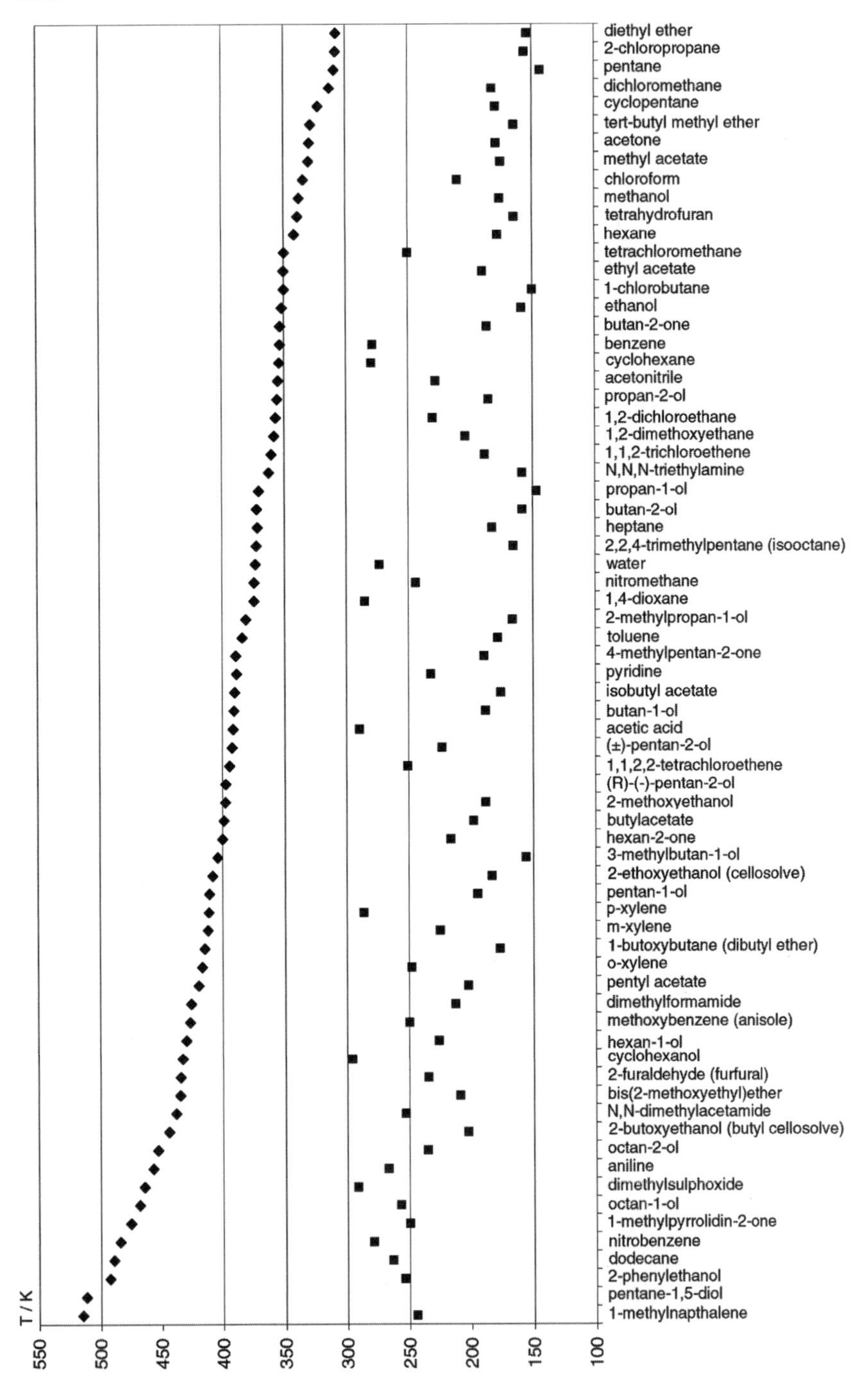

Fig. 2. Common recrystallization solvents ranked in order of decreasing boiling point (◆) and listing the corresponding freezing point (■).

Evaporation: The solution is left open to the atmosphere at constant temperature or:

1. The rate of evaporation is reduced by covering with perforated aluminum foil.
2. Evaporation is enhanced by directing a gentle stream of nitrogen gas over the solution surface.

One obvious question is, "Does temperature matter?" Strictly speaking, the answer is, "Yes." A racemic aqueous solution of sodium ammonium tartrate for example, spontaneously resolves into dextrorotatory and levorotatory crystals when crystallized below 28°C, but yields racemic compound crystals at temperatures above 28°C ***(8)***. Furthermore, the single enantiomer crystals are tetrahydrate (i.e., four molecules of water cocrystallize with each molecule of sodium ammonium tartrate), whereas the racemic compound crystals are monohydrate. This is an example of solid-state polymorphism (**Subheading 1.1.3.2.**) and a historically important one—Louis Pasteur effected the chiral separation of sodium ammonium tartrate by crystallization in 1848 ***(9)***. However, it is by no means unique; quite the opposite in fact. Polymorphism is widespread among organic compounds ***(10)***, and temperature often has an important effect on the polymorphic form of an organic compound obtained by crystallization from solution.

That said, if all that is required from a single-crystal X-ray diffraction experiment is a proof of molecular structure, then the occurrence of polymorphism is not necessarily a problem, other than to note that single crystals of one physical form of a compound may be more suitable for a diffraction experiment than those of a different form.

Cooling: The solubility of "small" organic molecules generally reduces with decreasing temperature. By controlling the rate and extent of cooling, and thereby the degree of supersaturation, it is possible to control the degree of nucleation and the rate of crystal growth. Cooling rate can be conveniently controlled using a water bath, and simply making adjustments based on observations of whether nucleation and growth are proceeding too rapidly or too slowly. Microcrystals are indicative of too rapid a cooling rate—this tends to result in the formation of an excessive number of nuclei and hence a large number of small crystals.

Slowing crystal growth: It follows from what has just been said that it can often be highly advantageous to slow the rate of crystal growth, circumventing excessive nucleation, and hopefully producing a small number of larger single crystals. Important alternatives to slow evaporation and slow cooling include vapor diffusion, liquid diffusion (layering), and thermal gradients. The vapor diffusion method ***(11–13)*** produces a mixed solvent system more slowly than the simple dropwise addition described in **Subheading 1.1.2.2.**:

1. The sample is placed in a small test tube and just adequate quantity of solvent 1 is added to produce a solution.
2. The test tube is placed in a larger sealed beaker containing miscible solvent 2, which should be sufficiently volatile to diffuse into solvent 1, producing a mixed system in which the compound is less soluble than in solvent 1 alone.
3. Evaporation of solvent 1 will help supersaturate the solution, although it can be advantageous when the process actually increases the volume of recrystallizing solution, as this helps avoid "crusting" (**Subheading 2.1.**). Ideally, solvent 2 should be more volatile/dense and a poorer solvent for the compound than solvent 1.

In common with vapor diffusion, liquid diffusion (layering) produces a solvent mixture more slowly than dropwise addition:

1. A small volume of solvent 1 is added to a narrow glass capillary, such as that used for melting point determination, NMR analysis, or X-ray powder diffraction.
2. Using a syringe, a volume of less dense solvent 2 is carefully layered on solvent 1.

The sample can be dissolved in either layer. Provided that some degree of interfacial mixing occurs to induce supersaturation, crystal growth will be observed at the interface. **Fig. 3** illustrates an alternative "thermal gradient" approach. A solution is produced by heating the solid in contact with a relatively poor solvent, and convection/diffusion then transports the solute to a cooler region, where crystallization occurs. The report of the method ***(14)*** cites success in growing crystals as large as 1 mm on edge within one week and also mentions the use of the same apparatus for growing crystals of nitrophenol by sublimation.

1.1.2.3. Summary: Growth of Good-Quality Single Crystals for Diffraction

Growth of diffraction-quality crystals necessitates ***(11)***:

1. A limited number of nuclei.
2. A slow rate of growth—days, rather than hours, of undisturbed growth.

As a general rule, good single crystals are produced by slow growth from unstirred solutions. These conditions limit nucleation, whereas high degrees

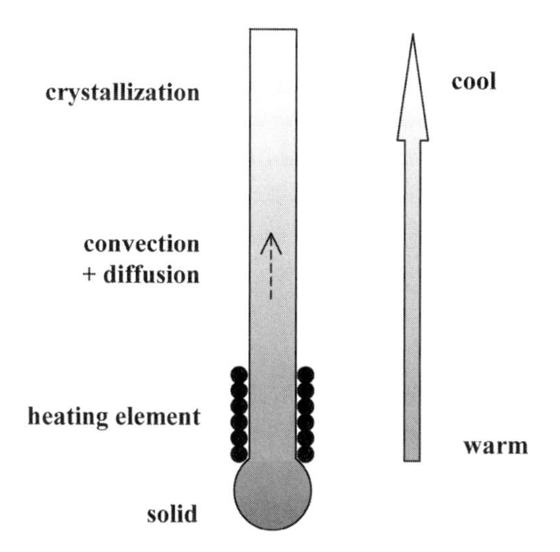

Fig. 3. Small-scale thermal gradient recrystallization apparatus.

of supersaturation result in the rapid formation of a large number of nuclei and tend to yield less suitable crystals. Major multiple nucleation can sometimes be prevented by scratching the inside of the beaker, or seeding the solution with a small crystal of the compound. It must be kept in mind that a seed crystal will still grow in a solution where the degree of supersaturation is sufficiently low so that the rate of nucleation is zero, or close to zero, but the solution is nonetheless supersaturated. Hence, the solution will "fatten up" the seed in preference to depositing "new" crystals.

It is clear from the literature that methodological variations on a theme of slow crystallization are considerable, but it is important to remember that they necessarily revolve around the same elementary set of crystallization principles discussed earlier. The interested reader is referred to **ref. *15***, which tabulates the advantages and limitations of a plethora of different crystallization techniques suitable for growing single crystals.

1.1.3. Selecting a Crystal

1.1.3.1. Crystal Quality

Once the compound of interest has been crystallized, the question then arises, "What makes a single crystal good?" Size, shape, and quality are essential for obtaining good single-crystal X-ray diffraction data ***(12)***. The crystal should be ideally equidimensional, with edges in the range approx 0.1–0.3 mm. Thus, the volume of the crystal should be large enough to ensure adequate diffraction of the X-ray beam, but not too large either, otherwise absorption of X-rays becomes problematic. The crystal must be single in the sense that it is neither twinned (*see* below) nor is it an aggregate of microcrystals. It should not be physically distorted or fractured, but beyond that does not have to have well-developed faces, and hygroscopicity is not necessarily a problem. Crystal quality is conveniently checked using a polarizing microscope ***(12)***, preferably one with a rotating stage:

1. A crystal is placed on a glass microscope slide—a fine paintbrush is good for gently manipulating small crystals.
2. The crystal is rotated about an axis normal to the polarizer, or rotate the polarizer itself.

A single crystal should ideally appear either uniformly dark, irrespective of position, or else undergo a sharp change in appearance from uniformly dark to uniformly bright every 90°. Undulating extinction is indicative of strain in the crystal lattice. A composite crystal, i.e., an intergrowth of two or more crystals, will often show light and dark simultaneously. A twin is a composite of two crystals joined symmetrically about a twin axis or a twin plane, and often appears as a "V," "L," or "+" shape ***(16)***.

Having selected one or more good-quality single crystals for a diffraction experiment, it is essential to store the samples so as to prevent chemical or physical degradation. Common mechanisms of physical degradation include:

1. Moisture gain—hygroscopic crystals are best stored under nitrogen or in a dry atmosphere.
2. Solvent loss—it is common for compounds to crystallize as solvates, i.e., with molecules of solvent incorporated into the crystal structure (*see*, for example, **Subheadings *Evaporation*** and **3.2.1.**). Single-crystal solvates may become opaque and polycrystalline as a result of a desolvation transformation and are best stored in a sealed glass vial.

1.1.3.2. Solid-State Polymorphism

It is not uncommon for microscopic examination to reveal two or more populations of crystals with characteristically different shapes. Assuming the sample is sufficiently pure to rule out the possibility that the populations are in fact different compounds, there is a chance that the compound is polymorphic. Polymorphs of the same chemical compound have different crystal lattice structures, i.e., they have variations in internal molecular packing arrangements that necessarily gives rise to different external crystal shapes. As pointed out earlier, if all that is required is a proof of molecular structure, then polymorphic form is not necessarily a problem, or even important on this scale, other than to note once again that single crystals of one physical form of the compound may be more suitable for a diffraction experiment than those of a different form. However, the existence of polymorphism should always be documented and may well assume significance should the compound reach the stage of industrial production. A comprehensive review is provided in **ref. *10*.**

1.1.4. Parallel Crystallization

Where gram quantities of material are available, parallel crystallization approaches are especially well suited to the rapid screening of crystallization conditions to identify those that yield the optimal crystal size and quality for further structural analysis. There are several commercial systems available for small-scale parallel syntheses that are also very well suited to the task of crystallization within the laboratory without any modification. A typical system (**Fig. 4**) enables the parallel crystallization of solute from a range of solutions to be induced by controlled heating/cooling/evaporation/agitation in up to 24×10 mL vessels. It is worth noting that parallel crystallization methods commonly yield multiple crystalline forms of the same chemical compound (i.e., polymorphs and solvates, **Subheading 1.1.3.2.**) and indeed are used for this express purpose in pharmaceutical development.

2. Common Problems and Solutions

2.1. A Polycrystalline Crust Forms as Solvent Recedes During Evaporation

The rate of solvent evaporation is reduced and preferably one or two seed crystals are introduced into the solution, if available.

2.2. The Product Is Not a Crystalline Solid

"Oiling" is common and can be overcome by varying the crystallizing temperature and/or the composition of the crystallizing solvent. It may be that nucleation is a problem, in which case stirring or seeding may help. Failing that, derivatization is a possibility. Many cyanogenic glycosides are difficult to crystallize, whereas their peracetates form crystals readily

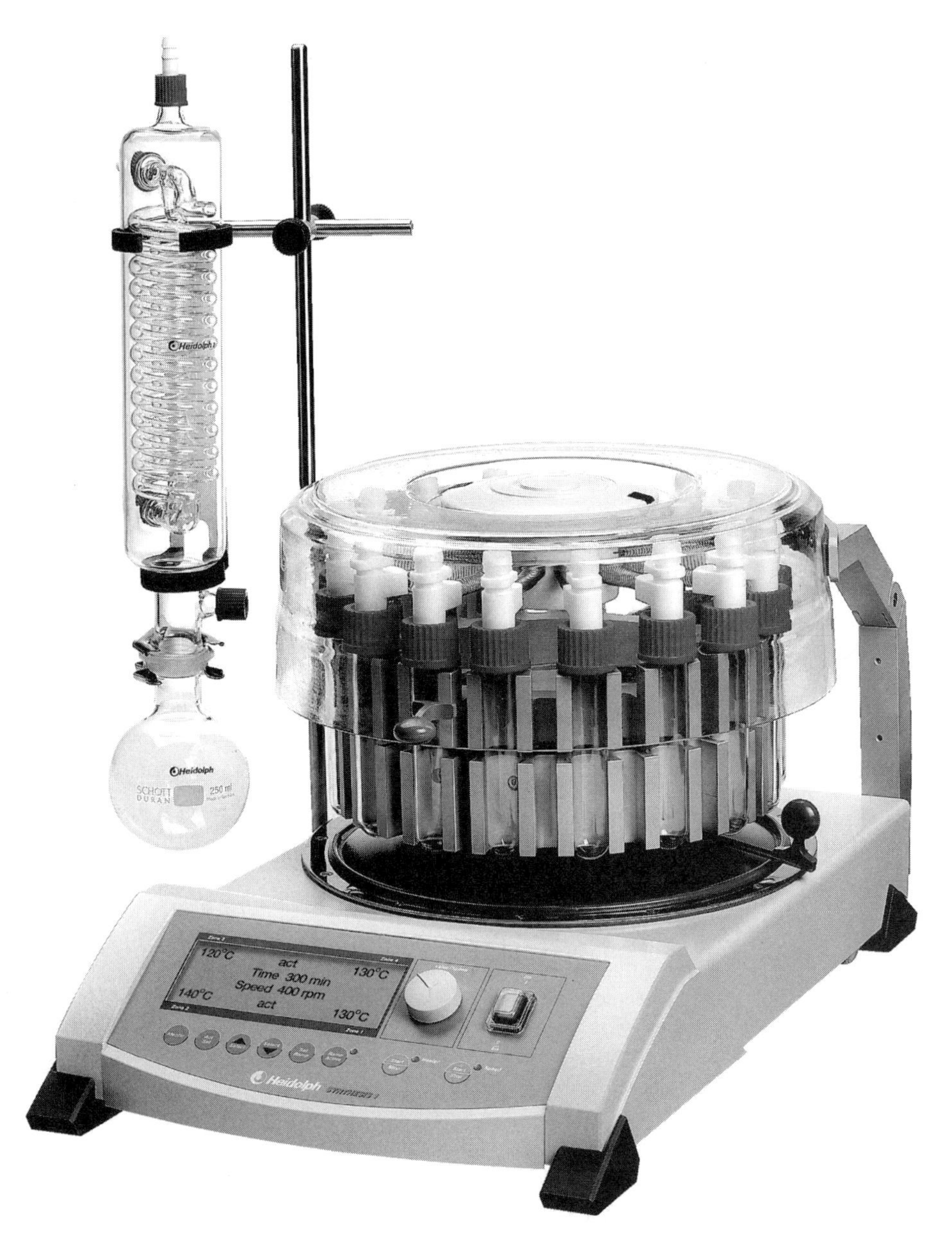

Fig. 4. Parallel crystallization system.

(17). The hydrochloride and hydrobromide salts of organic bases are often easier to crystallize than the parent compound itself. Derivatives are also useful in fractional crystallizations *(18)*, for example, picrates of alkaloids and osazones of sugars.

2.3. Crystallization Proceeds Slowly

Supersaturated solutions can often sit for very long periods of time without producing crystals *(7)*. This seems counterintuitive but stems from the fact that nuclei below a certain critical size find it energetically unfavorable to expand against the surrounding solvent. Chemical impurities also tend to impede crystallization. However, if purity is not a problem, then the following could be tried.

1. Stirring can enhance nucleation by promoting the frequency of collisions in solution, although excessive agitation can have the opposite effect *(19)*.
2. Crystals will grow on foreign surfaces *(20)*, and leaving the flask open to atmospheric dust might be all that is needed to stimulate crystallization; if seed crystals are available, this may be even more favorable.
3. Filtering a hot solution into a cold flask so as to achieve very rapid cooling is usually sufficient to induce rapid crystallization, and roughening the inside of the glass flask with a spatula is also often helpful.
4. Temperature cycling can be advantageous; a supersaturated solution is cooled to fridge or freezer temperatures to reduce solubility and hopefully induce nucleation, and is then warmed to room temperature to encourage the nuclei to develop into crystals. This circumvents the possibility that excessive cooling might actually impede progress by increasing solution viscosity.

2.4. Crystals Grow But Are Unsuitable for Single-Crystal X-Ray Diffraction

It is often the case that single crystals are too small in one or more dimensions to give good-quality diffraction data (**Subheading 1.1.3.1.**). If so, the crystallizing solvent has to be changed. Alternatively, the following can be tried:

1. Reducing the number of nuclei formed by decreasing the degree of supersaturation and by filtering off any gross particulate contamination.
2. If the crystals are large enough to handle, using one or two to seed a separate crystallization.
3. Dispersing a small quantity of the crystals in a drop of saturated mother liquor on a glass microscope slide, and observing the crystals with a microscope, paying particular attention to those dispersed around the periphery

of the droplet. Provided the rate of evaporation is not too high, significant increases in crystal size can often be observed as the droplet evaporates. Water droplets evaporate at a reasonable rate at room temperature, whereas a highly volatile organic solvent such as diethyl ether or acetone tends to evaporate too rapidly for this approach to be useful.

In the event that conditions yielding a crystal for single-crystal diffraction simply cannot be found in reasonable time, a change of tack to X-ray powder diffraction (XRPD) may be necessary. By way of an illustration, **Fig. 5** shows the crystal structure conformation of capsaicin as determined directly from an XRPD data set using the well-established simulated annealing approach to global optimization *(21)*. The data were collected over 9 h from $\approx$ 10 mg of polycrystalline capsaicin mounted in a 0.7 mm borosilicate glass capillary. The principal restriction on the application of global methods in this way is the requirement to know, in advance, the molecular formula and connectivity of the compound of interest. That said, structural ambiguities can often be resolved by optimizing multiple distinct structural models against the measured diffraction data. In summary, XRPD, if properly executed, can be used to determine the crystal structures of polycrystalline organic molecular materials with a high degree of positional accuracy.

2.5. Only a Small Quantity of Compound Is Available for Crystallization

The vapor diffusion technique is useful for milligram amounts of material. In particular, a hanging-drop variation of this method has proved to be a popular and effective technique of crystallizing milligram quantities of proteins *(11–13)*. In principle, the hanging drop is not restricted to macromolecules and a simple modification of the scheme described earlier is worth trying with very small quantities of natural product extract. The compound should be dissolved in a drop of solvent 1 and the drop suspended from a glass coverslip over a reservoir of solvent 2, with the system sealed with grease to prevent solvent escaping. Vapor diffusion results in supersaturation and, hopefully, crystal growth. Crystals growing in the hanging drop can be observed and retrieved easily, and the method has been used to grow protein crystals in drop volumes as low as 5–15 μL, over a reservoir of 500 μL *(11)*. The liquid diffusion and thermal gradient methods are also useful when only small amounts (ca. 5 mg) of the compound are available.

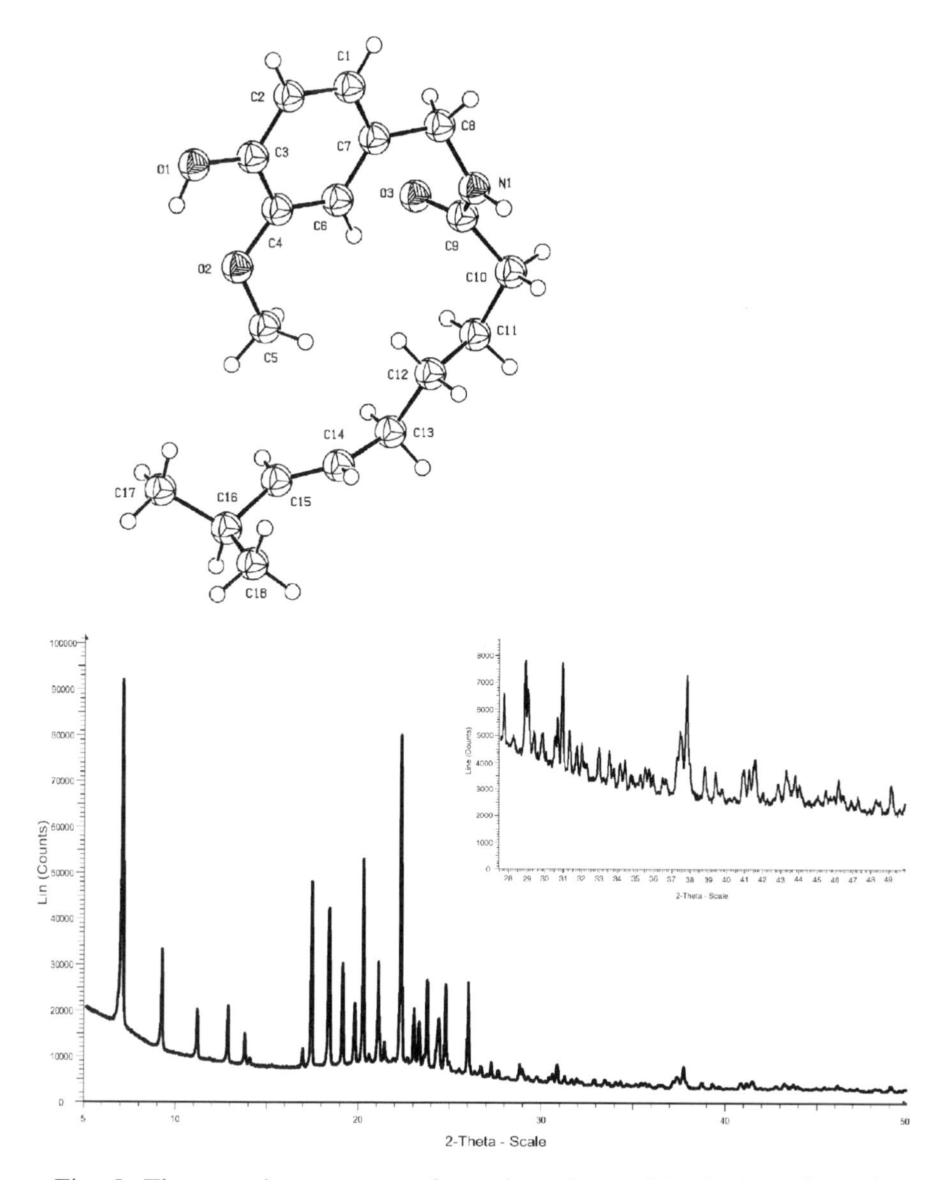

Fig. 5. The crystal structure conformation of capsaicin (top) as determined directly from a lab XRPD data set (bottom) comprising 338 reflections to a spatial resolution of ca. 1.8 Å. Although diffraction intensity is relatively weak beyond 27° 2θ, a utilizable signal-to-noise ratio is maintained (inset).

3. Crystallization as a Separation Method

3.1. General

The principle of crystallization as a separation method is fairly simple and draws on what has already been outlined in **Subheading 1.1.2**. For example, assuming we have a product comprising target component *A* mixed in with impurities *B* and *C*:

1. A sample of the mixture is dissolved in a hot solvent — the solvent is chosen such that *B* and *C* are soluble at any temperature reached in the crystallization, while component *A* is not.
2. Cooling yields a crop of *A*, separated from components *B* and *C*.
3. **Steps 1** and **2** are repeated, using fresh solvent each time, until the required degree of separation is achieved (note that one crystallization step from a mixture of compounds does not guarantee a chemically pure crystal product).

The disadvantage of this approach is that losses can be prohibitive and on a scale smaller than 100 mg, chromatographic techniques are more appropriate for separating the components of a sample. There are variations in the above scheme, designed to recycle the liquid filtrates produced by successive crystallization steps and conserve the target compound. A good general account of these fractional crystallization schemes is available in the literature ***(1)***, and another method is mentioned here for completeness:

1. Crystallizing product and retaining filtrate.
2. Dissolving product in fresh solvent.
3. Recrystallizing product and retaining filtrate.
4. Concentrating the filtrate from **step 1** to yield more product, which is then recrystallized from the filtrate produced in **step 3.**

This simple sequence therefore makes use of both filtrates to obtain twice-crystallized product. It may be, of course, that component *A* is the "impurity," not the target compound—examples of both cases follow.

3.2. Examples of Purification of Natural Products by Crystallization

Concentrated extracts may deposit crystals on standing by virtue of the fact that solvent evaporation, decreasing temperature, or a combination of the two result in supersaturation ***(22)***. For example, good yields of crystals are sometimes obtained when the hot solvent used in a Soxhlet extraction is cooled overnight. The crystals may be either the target compound of interest or impurities.

3.2.1. Target Compound Crystallizes Leaving "Impurities" in Solution

Crude solanine, extracted from the potato plant, is purified by dissolving in boiling methanol, filtering, and concentrating until the alkaloid crystallizes out *(23)*.

Naringin is isolated from grapefruit peel by extracting into hot water, filtering through celite, and concentrating the filtrate to the extent that naringin crystallizes at fridge temperatures as the octahydrate (melting point = 83°C) *(24)*. Recrystallization from isopropanol (100 mL to 8.6 g naringin) yields the dihydrate (melting point = 171°C). The di- and octahydrate compounds are examples of crystalline solvates (**Subheading 1.1.3.2.**).

Piperine is extracted from powdered black pepper with 95% ethanol. The extract is filtered, concentrated, 10% alcoholic KOH added, and the residue formed is discarded. The solution is then left overnight to yield yellow needles of piperine *(25)*.

Capsanthin is isolated from red pepper or paprika. A 20 mL volume of concentrated ether extract diluted with 60 mL petroleum and left to stand for 24 h in a fridge produces crystals of almost pure capsanthin *(26)*.

Salicin is extracted from willow bark into hot water. The solution is filtered and concentrated and the tannin removed by treatment with lead acetate; further concentration and cooling yields salicin crystals *(27)*. It is also worth highlighting the potential use of derivatives in fractional crystallizations *(18)*, for example, picrates of alkaloids and osazones of sugars.

3.2.2. "Impurities" Crystallize Leaving Target Compound in Solution

Concentrated extracts of carotenoids are occasionally contaminated with large amounts of sterols. These can be removed conveniently by leaving a light petroleum solution to stand at −10°C overnight and removing the precipitated sterol by centrifugation *(26)*.

During the purification of plant acids, oxalic acid, which may be present in excessive amounts, can be precipitated out as calcium oxalate by adding calcium hydroxide solution to a concentrated alcoholic plant extract *(28)*.

In the production of medicinal cod-liver oils, saturated acylglycerols such as stearin can be removed from the crude cod-liver oil simply by cooling and filtering them off as precipitate, leaving the unsaturated acylglycerols in the liquid *(29,30)*.

These examples draw on the basic principles of **Subheading 1.1.2.**, particularly evaporation and cooling. It is worth reiterating, however, that using a rotary evaporator to concentrate and supersaturate a solution is

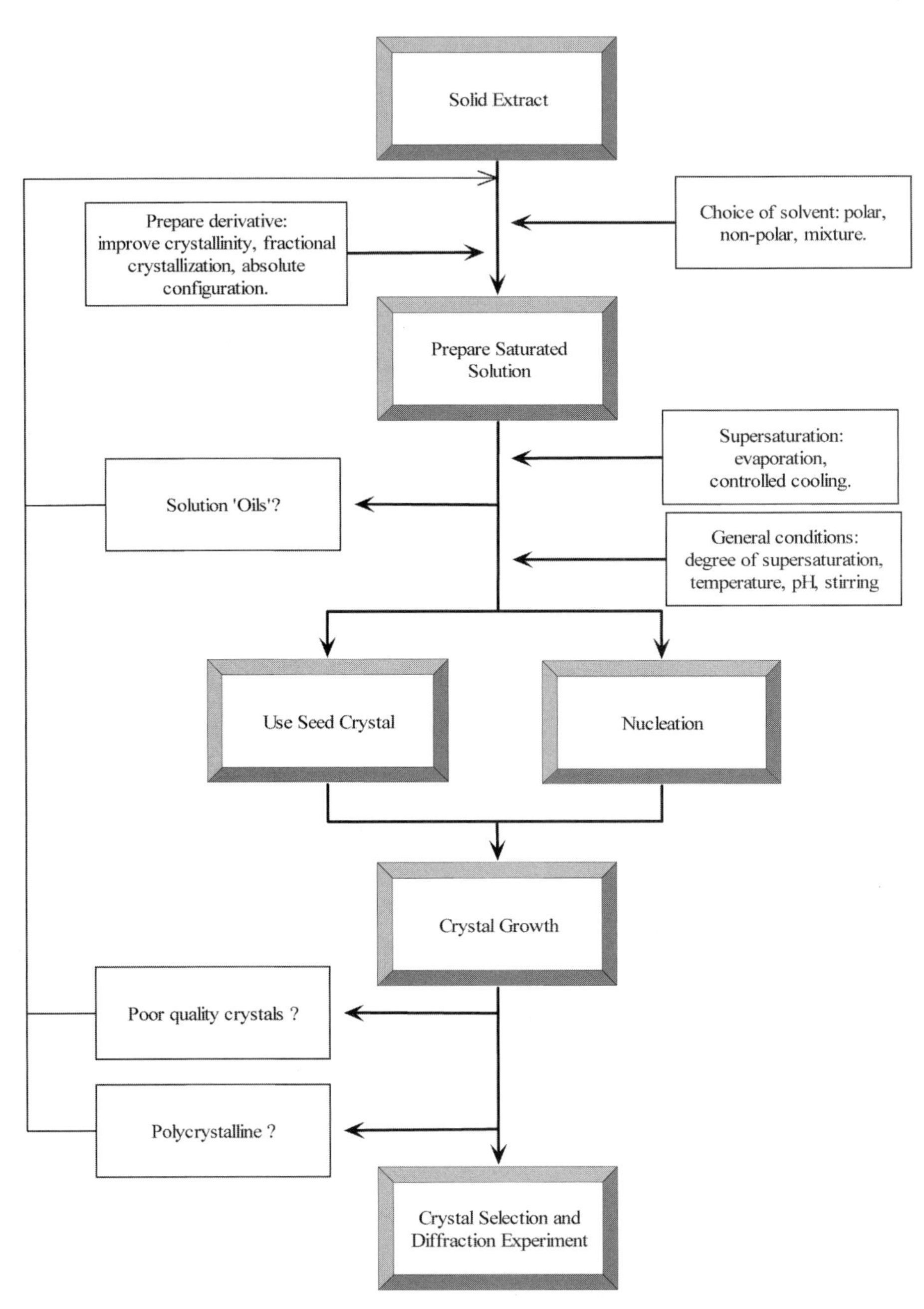

Fig. 6. Crystallization flowchart.

certainly not the best way to obtain single crystals of quality suitable for X-ray diffraction—a more genteel approach is preferable, and the key points are summarized in **Fig. 6**. It must be kept in mind that if one is unsure about whether a single crystal is suitable for an X-ray determination of molecular structure, it should *never* be redissolved without taking advice first. Above all, patience is required!

References

1. Mullin, J. W. (1972) Crystallization techniques, in *Crystallization*, 2 ed., Butterworths, London, pp. 233–257.
2. Trease, G. E. and Evans, W. C. (1996) Cell differentiation and ergastic cell contents, in *Pharmacognosy*, 14 ed., W B Saunders Company Ltd., London, pp. 554–567.
3. Trease, G. E. and Evans, W. C. (1996) Volatile oils and resins, in *Pharmacognosy*, 14 ed., W B Saunders Company Ltd., London, pp. 255–292.
4. Jacques, J., Collet, A., and Wilen, S. H. (1981) Formation and separation of diastereomers, in *Enantiomers, Racemates and Resolutions*. John Wiley & Sons, New York, pp. 251–368.
5. Polonsky, J., Prangé, T., Pascard, C., Jacquemin, H., and Fournet, A. (1984) Structure (X-ray analysis) of manicoline B, a mixture of two diastereoisomers of a new alkaloid from *Dulacia guianensis* (Olacaceae). *Tetrahedron Lett.* **25,** 2359–2362.
6. Ghisalberti, E. L., Jefferies, P. R., Skelton, B. W., White, A. H., and Williams, R. S. F. (1989) A new stereochemical class of bicyclic sesquiterpenes from *Eremophila virgata* W. V. Fitzg. (Myoporaceae). *Tetrahedron* **45,** 6297–6308.
7. VanHook, A. (1961) Crystallization in the laboratory and in the plant, in *Crystallization Theory and Practice*. Reinhold Publishing Corporation, New York, pp. 192–237.
8. Jacques, J., Collet, A., and Wilen, S. H. (1981) Solution properties of enantiomers and their mixtures, in *Enantiomers, Racemates and Resolutions*. John Wiley & Sons, New York, pp. 167–213.
9. Jacques, J., Collet A., and Wilen, S. H. (1981) Resolution by direct crystallization, in *Enantiomers, Racemates and Resolutions*. John Wiley & Sons, New York, pp. 217–250.
10. Threlfall, T. L. (1995) Analysis of organic polymorphs. *Analyst* **120,** 2435–2460.
11. Glusker, J. P., Lewis, M., and Rossi, M. (1994) Crystals, in *Crystal Structure Analysis for Chemists and Biologists*. VCH Publishers Inc., New York, pp. 33–72.

12. Stout, G. H. and Jensen, L. H. (1989) Crystals and their properties, in *X-ray Structure Determination. A Practical Guide.* John Wiley & Sons, New York, pp. 74–92.
13. Glusker, J. P. and Trueblood, K. N. (1985) Crystals, in *Crystal Structure Analysis. A Primer.* Oxford University Press, New York, pp. 8–19.
14. Watkin, D. J. (1972) A simple small-scale recrystallization apparatus. *J. Appl. Crystallogr.* **5,** 250.
15. van der Sluis, P., Hezemans, A. M. F., and Kroon, J. (1989) Crystallization of low-molecular-weight organic compounds for X-ray crystallography. *J. Appl. Crystallogr.* **22,** 340–344.
16. Mullin, J. W. (1972) The crystalline state, in *Crystallization*, 2 ed., Butterworths, London, pp. 1–27.
17. Seigler, D. S. and Brinker, A. M. (1993) Characterisation of cyanogenic glycosides, cyanolipids, nitroglycosides, organic nitro compounds and nitrile glucosides from plants, in *Methods in Plant Biochemistry*, vol. 8, *Alkaloids and Sulphur Compounds* (Waterman, P. G., ed., Dey, P. M. and Harborne, J. B. series eds.), Academic Press, London, pp. 51–131.
18. Trease, G. E. and Evans, W. C. (1996) Introduction and general methods, in *Pharmacognosy*, 14 ed., W B Saunders Company Ltd., London, pp. 119–130.
19. Mullin, J. W. (1972) Crystallization kinetics, in *Crystallization*, 2 ed., Butterworths, London, pp. 174–232.
20. VanHook, A. (1961) Historical review, in *Crystallization Theory and Practice.* Reinhold Publishing Corporation, New York, pp. 1–44.
21. Shankland, K. and David, W. I. F. (2002) Global optimization strategies, in *Structure Determination from Powder Diffraction Data* (David, W. I. F., Shankland, K., McCusker, L., and Baerlocher, Ch. eds.), Oxford University Presss, Oxford, pp. 252–285.
22. Harborne, J. B. (1984) Methods of plant analysis, in *Phytochemical Methods*, 2 ed., Chapman and Hall, London, pp. 1–36.
23. Harborne, J. B. (1984) Nitrogen Compounds, in *Phytochemical Methods*, 2 ed ., Chapman and Hall, London, pp. 176–221.
24. Harborne, J. B. (1984) Sugars and their derivatives, in *Phytochemical Methods*, 2 ed., Chapman and Hall, London, pp. 222–242.
25. Harborne, J. B. (1973) Nitrogen compounds, in *Phytochemical Methods*, 1 ed., Chapman and Hall, London, pp. 166–211.
26. Harborne, J. B. (1984) The terpenoids, in *Phytochemical methods*, 2 ed., Chapman and Hall, London, pp. 100–141.
27. Trease, G. E. and Evans, W. C. (1996) Phenols and phenolic glycosides, in *Pharmacognosy*, 14 ed., W B Saunders Company Ltd., London, pp. 218–254.
28. Harborne, J. B. (1984) Organic acids, lipids and related compounds, in *Phytochemical Methods*, 2 ed., Chapman and Hall, London, pp. 142–175.

29. Trease, G. E. and Evans, W. C. (1996) Vitamins and hormones, in *Pharmacognosy*, 14 ed., W B Saunders Company Ltd., London, pp. 441–450.
30. Trease, G. E. and Evans, W. C. (1996) Hydrocarbons and derivatives, in *Pharmacognosy*, 14 ed., W B Saunders Company Ltd., London, pp. 172–190.

12

Dereplication and Partial Identification of Compounds

Laurence Dinan

Summary

This review summarizes the advances in dereplication technology since 1998, its current status, and the prospects for future development. Developments are being driven by the need to identify novel pharmaceutical and agrochemical lead compounds rapidly and effectively from complex natural matrices, while avoiding spending time or money on the re-isolation and re-identification of known natural products. As many commercial pharmaceuticals and agrochemicals are at least derived from natural products, and only a small proportion of organisms has yet been examined for biologically active compounds, there is still enormous potential for the identification of many novel agents. In conjunction with modern screening methods, dereplication strategies must have high resolution, be sensitive, rapid, reproducible, and robust. Further, they must include effective data processing and information retrieval systems to permit comparison with internal and external sources of information on known compounds.

Key Words: Dereplication; screening; natural products; agrochemicals; pharmaceuticals.

1. Introduction

Dereplication was originally the analysis, without isolation, of a natural product, a fraction, or an extract for spectroscopic, structural, and biological activity information, and comparing the information with internal and/or commercial databases, ascertaining whether novel and/or known

From: *Methods in Biotechnology, Vol. 20, Natural Products Isolation, 2nd ed.*
Edited by: S. D. Sarker, Z. Latif, and A. I. Gray © Humana Press Inc., Totowa, NJ

compounds are present, and determining a strategy for further investigation *(1)*. However, it has now also been extended to the ability to exclude previously tested organisms (mainly micro-organisms) from screening programs *(2)* (**Fig. 1**). Thus, effective dereplication strategies enhance throughput, increase the number of novel compounds identified, and make more efficient use of limited resources *(3)*.

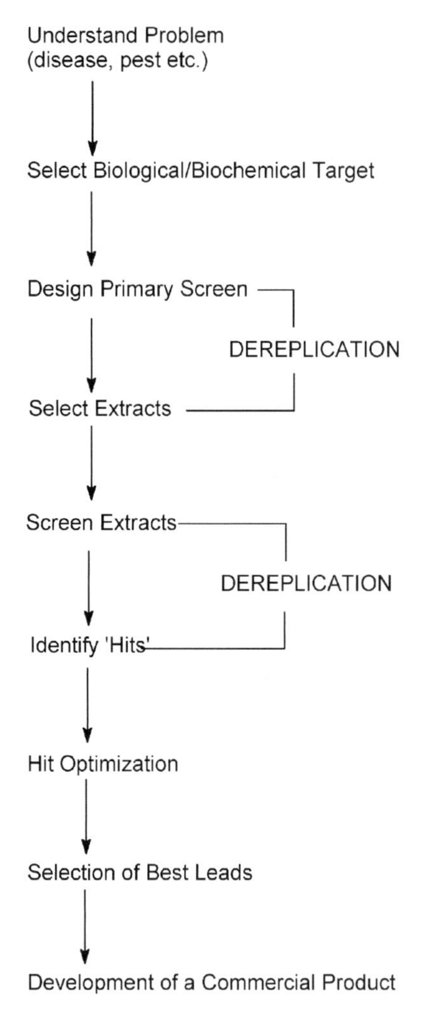

Fig. 1. Summary of a generalized screening strategy, showing the stages where dereplication contributes.

This review focuses on the developments in dereplication strategies and applications since 1998, when the first edition of this book appeared *(4)*. In this time, the importance of dereplication in high-throughput screening (HTS) has increased, and continues to rise, because as more natural products are identified from a wider range of sources, it becomes even more essential to be able to identify rapidly which components present in an extract are already known. Further, with the automation and miniaturization of screening programs, where thousands of samples can be assayed in a day in a variety of assays, it is important to know which positive samples could contain novel compounds and to focus effort on these. Thus, together with improvements in bioassay, chromatographic and spectrometric methods for the detection and partial identification of compounds, improvements in data handling, and information retrieval have been just as important. The structural diversity of natural products, with approx 175,000 known structures, makes them a significant resource for the discovery of potential drug leads. This is not just a theoretical consideration, as it has been estimated that more than 50% of the most prescribed drugs in the United States have a natural product origin *(1)*. For the world as a whole, 80% of the human population relies upon herbal remedies and the annual sale of botanical medications is greater than US $70 milliard; plants still form the basis of the traditional medical treatments in countries such as China and India *(5)*. Appropriate sources for natural products are bacteria, fungi, algae, plants, invertebrates, and marine organisms, of which fungi and plants have the most diverse mixtures of secondary compounds, although marine organisms also possess many unique chemistries. Terrestrial invertebrates, including insects, which are the most abundant animals on the earth, have hardly been investigated with regard to the active compounds they might contain. Although natural sources are still regarded as inexhaustible sources for novel chemicals *(3)*, the effort required to find each new one increases. It has been estimated that it takes US $50,000 and 3 mo work to isolate and identify an active compound from a natural product source *(6)*. Hence, it is exceedingly important to recognize previously known compounds early on, not only for the saving in time and money, but also for the opportunity cost, which means that resources are not applied to more profitable extracts. As screening programs are truly multidisciplinary efforts, a clear strategy and coordination are required to make the most of the contributions from chemists, botanists, microbiologists, pharmacologists, and the like. Despite the clear advantages of screening natural products for lead compounds, there has been a tendency for the pharmaceutical industry to reduce investment in natural product research over recent years, probably because of the longer time needed for natural product isolation and identification relative to, for example, the generation of combinatorial chemistry libraries. This time span is not so compatible with HTS-driven industrial drug discovery processes (>100,000 test substances per week). The tide is now turning back in favor of natural products, as the expected surge in productivity from the combinatorial chemistry approach has not materialized.

It is beyond the scope of this chapter to describe the theory of the techniques used in dereplication. Rather, recent developments are mentioned along with relevant applications from the literature, with the overall aim of providing the reader with a summary of the current status of dereplication and an indication of how this area will develop in the future. Details of techniques are given in other chapters in this book (*see* Chaps. 4–9). Progress is not only being facilitated by developments in screening and analytical techniques, but also by the dramatic advances in bioinformatics.

2. Species and Taxonomic Information

Owing to the significant amount of chemotaxonomic information available, especially for plants, even the choice of organism(s) selected for study can contribute to dereplication, as literature sources can provide much information on the chemical classes present or absent in a particular species. However, among the estimated 300,000–500,000 plant species, very few have even been minimally screened for biologically active components. The assessment of microbes in this respect is no more extensive; little or nothing was published on 17.5% of bacterial species between 1991 and 1997, and the publication rate for a further 56% of species was extremely low ***(7)***. As the search for new organisms for investigation is being widened, it is extended not only to marine organisms and invertebrates, but also to endophytic micro-organisms in plants. According to Strobel ***(8)***, there is a high probability that any novel micro-organism will contain novel secondary products. As the endophytic bacteria and fungi associated with very few plant species have been investigated, one is faced with the problem of which plant species to investigate. Strobel ***(8)*** suggests that plants growing in moist, warm climates in geologically isolated regions should be given preference because microbial competition would be fierce. It may be easier to culture the endophytic organisms for the production of secondary products, rather than collect the possibly rare plant, which is the normal host. This has received a boost from the finding that endophytic micro-organisms from *Taxus brevifolia* also produce taxol ***(9)***.

Screening approaches to natural product extracts may be random or nonrandom. Often screening is initially random, testing as wide a diversity of species as possible, but as hits are identified, the testing becomes more focused, concentrating on related species to the positive hits. Several commercial companies now specialize in providing extract collections (microbial, plant, marine) for screening and can supply bulk samples and information to assist in dereplication.

Having discovered a hit in a natural product-containing extract screen, it is necessary to be able to go back and collect more of the source organism ("recollection"). Satellite navigation technologies assist considerably in the specification of the exact location (latitude and longitude) of a site for the collection of an organism. In the future, this technology will assist in avoiding the duplication of past collections (geographical dereplication). Dereplication of known compounds from the same source is increasingly important in discovery screens. Even partial structural information can be used to search a substructure-based database; it is considerably easier to analyze a positive extract if one has a list of chemical structures isolated previously from that source as possible candidates, rather than having to identify each component from first principles.

It has been argued ***(10)*** that the chances of detecting new fungal metabolites are enhanced by culturing the fungus in solid state fermentation (rather than liquid cultures) because this is the matrix where fungi have evolved. While agar plates can be used to produce enough fungus for structural determination of metabolites, the handling of many plates is time-consuming and may pose safety problems. Nielsen et al. ***(11)*** have proposed the use of the porous inert support (lightweight expanded clay aggregate [LECA]), coated with absorbed medium, as a way of enhancing microfungal sporulation and associated metabolite production; yeasts and bacteria are more suited to growth in liquid culture.

Correct identification of microbial or plant species is essential. A rapid, fully automated, method which is now being developed is pyrolysis mass spectrometry (PyMS), whereby the biological material is thermally degraded in a vacuum, analyzed by mass spectrometry (MS), and can be used to discriminate between closely related microbial strains ***(2)*** and higher plant species ***(12)***. The method is advantageous because it is rapid, reproducible, uses small samples, and can be automated. The sample is heated in a vacuum at a high temperature (530°C) and the volatilized breakdown products analyzed by MS. Consequently, the spectra generated do not represent specific chemical components, but rather the bond strengths of the chemical constituents of the biological sample. Interpretation is usually made by principal component analysis and canonical variate analysis. The validity of relating PyMS spectra to taxonomic status has been evaluated favorably for actinomycetes ***(2,13)***.

Since different subspecies or strains within a species may produce different secondary metabolites, it can be necessary to dereplicate isolates. An automated fingerprinting method ("Riboprinter") has been successfully applied to *Streptomyces* soil isolates, whereby genomic DNA is digested with a restriction endonuclease, the fragments separated by gel electrophoresis, transferred onto a membrane before being hybridized with an rDNA probe by Southern blotting. The pattern of hybridization can be related to the particular strain ***(14)***.

Less than 5% of all naturally occurring micro-organisms have been isolated and cultured. Prokaryotic actinomycetes have proved to be excellent sources of natural products, as are the myxobacteria. Anaerobic organisms and yeasts are far less productive. It is now realized that marine-organisms are also rich sources of novel compounds, not only in themselves, but for the micro-organisms they harbor. However, only a small proportion of micro-organisms living in the sea have been isolated. The isolation of many more micro-organisms will require the development of new collection techniques and novel culture media and conditions. Molecular methods are being developed to rapidly identify micro-organisms, develop databases and for dereplication. Useful reviews in this area are available in the literature ***(15–17)***.

3. Bioassays and Immunoassays

The method of detecting the active compounds is at the core of all screening and dereplication strategies. Bioassays detect compounds with particular biological activities and should be specific, sensitive, simple to perform, robust, economical and, preferably, suitable for automation. When it is known which class of compound is responsible for activity, the bioassay can be replaced or supported by biochemical assays detecting structurally related compounds, such as immunoassays. For example (**Fig. 2**), in a search for insect steroid hormone (ecdysteroid) receptor agonists and antagonists, a microtiter plate-based bioassay ***(18)*** based on the ecdysteroid-responsive *Drosophila melanogaster* B_{II} permanent cell line has

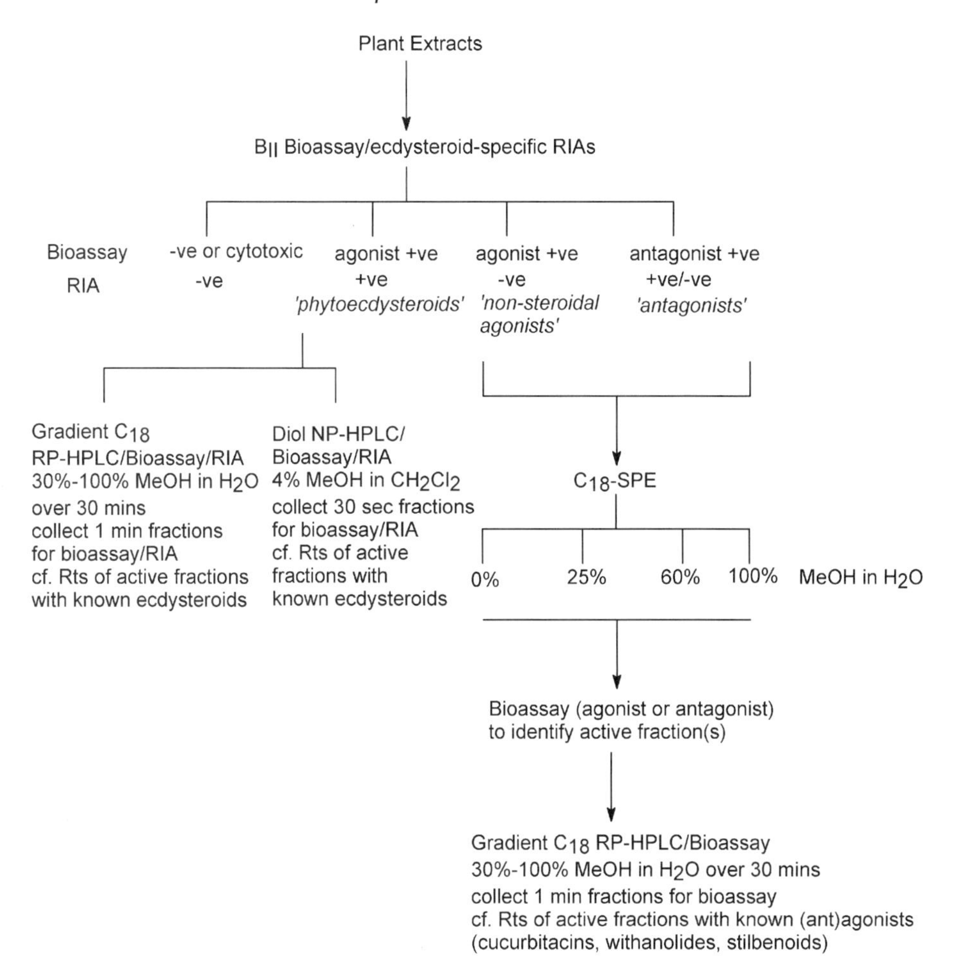

Fig. 2. Dereplication strategy for ecdysteroid agonists and antagonists from plat sources.

been used to detect compounds present in approx 5000 crude plant extracts which either act like ecdysteroids (agonists), or which block the action of 20-hydroxyecdysone (antagonists) ***(19)***. Since it has been known for almost 40 years that some terrestrial plants produce steroidal analogs of ecdysteroids (phytoecdysteroids) ***(20)***, the extracts were also screened with two ecdysteroid-specific radioimmunoassays (RIAs) (using antisera with differing specificities) to identify the ecdysteroid-containing extracts. Thus, extracts could be allocated to four classes:

1. Inactive in both the bioassay and the RIAs (the vast majority of extracts).
2. Positive in the agonist version of the bioassay and the RIAs (phytoecdysteroid-containing).
3. Positive in the agonist version of the bioassay, but not in the RIAs (containing nonsteroidal agonists).
4. Positive in the antagonist version of the bioassay (containing ecdysteroid antagonists corresponding to various chemical classes).

With this approach, it has been possible to identify several novel phytoecdysteroids and to identify certain cucurbitacins, withanolides, stilbenoids, phenylalkanes, and limonoids as ecdysteroid antagonists ***(21,22)***.

The difficulty of all such assays is the potential for the occurrence of false negatives and false positives. Clearly, the specificity of the assay plays an important role in minimizing the number of these, so that proper verification of the assay before using it for screening, although time-consuming, pays dividends in the long run. When screening a large number of extracts to identify those possessing a particular activity, the false positives cause a larger problem than the false negatives. A false negative may represent a missed opportunity, but at least one does not lose time over it. However, following up a false positive, if not identified as such rapidly, can result in the loss of much precious time. Thus, an additional aim of any dereplication strategy should be to identify false positives (and false negatives, if possible).

Ramakrishna et al. ***(3)*** have used a layered screening approach for 38,000 extracts from micro-organisms and plants to identify novel compounds with antibacterial activity against vancomycin-resistant *Enterococcus faecium*. A primary screen tested extracts qualitatively against a panel of four species of bacteria. The secondary screen assessed the positives from the primary screen (1022) against four bacteria, from which 111 extracts were selected. These were partially purified by chromatography and then subjected to dereplication using the hyphenated technique, high-performance liquid chromatography-photodiode array-mass spectrometry (HPLC-PDAD-MS), in association with a literature search based on Chapman & Hall's *Dictionary of Natural Products*, which identified 55 extracts containing known compounds with known activities, 15 with known compounds with previously unreported activity, and four containing new compounds.

Simplicity is a highly desirable property in a bioassay. A good example of a bioassay fulfilling this criterion was described by Antonio and Molinski ***(23)***, and used to screen marine invertebrates for ergosterol-sensitive antifungal compounds. It is an agar disc diffusion assay using the human pathogenic yeast *Candida albicans* as the test organism. It detects the ability of extracts to inhibit the growth of *C. albicans* in the absence or presence of ergosterol at various concentrations (0.001–100 ppm). It was thus possible to identify extracts that produce a zone of inhibition, and to determine if each activity is specifically affected or unaffected by the presence of ergosterol (i.e., ergosterol would overcome the inhibitory effect if the inhibition prevented normal levels of ergosterol in the yeast). Using this method, the authors found that extracts of 10 marine sponges (from 116 tested) showed antifungal activity, of which three were unaffected by ergosterol, while the activities of a further three were significantly reduced in the presence of ergosterol.

With the recent explosive growth in knowledge concerning cellular signaling pathways, their importance in regulation of biological processes within and between cells and their implication in numerous disease states, many new screening assays based on these signal transduction processes are being developed ***(24)*** to identify novel agonists and antagonists. This trend, with its implications for dereplication, will continue for the foreseeable future.

4. Solvent Partition

Even simple solvent partitions provide some information about the polarity and ionizability of the biologically active components within an extract. This is an important preliminary information that influences the choice of subsequent chromatographic separation methods. Although solvent partitions do not result in the identification of compounds, they do permit the exclusion of classes to which the active principles cannot belong.

5. Chromatography

5.1. Countercurrent Chromatography (CC) and Centrifugal Partition Chromatography (CPC)

Countercurrent liquid–liquid partition systems possess several advantages: considerable choice of solvent systems, high capacity, reproducibility, mild chromatographic separation conditions, and high recovery of sample components (*see* Chap. 7). CPC has the additional advantage of being faster than CC.

Ingkaninan et al. ***(25)*** have proposed a CPC method for the one- or two-step analysis and dereplication of ethanolic extracts of plant material, using an initial solvent system of heptane/ethyl acetate/methanol/

water (6:1:6:1, v/v/v/v) and, if necessary, a second separation using ethyl acetate/methanol/water (43:22:35, v/v/v), coupled to four differing bioassays. They verified the procedure with plant extracts known to contain compounds active in each of the bioassays. The procedure also proved useful for the separation, identification, and dereplication of nonselective compounds affecting the assays, such as fatty acids and phenolics, because the reproducible separations could be characterized by determining in which fractions such compound classes eluted.

CPC coupled to a PDA detector has also been used to dereplicate strains of *Streptomyces violaceusniger* producing the known compounds elaiophylin and/or geldamycin ***(26)***. CPC-PDAD was used for initial fractionation of ethyl acetate extracts of microbial cultures. Identification of the components in biologically active fractions was achieved by liquid chromatography–mass spectrometry (LC-MS). This proved to be an efficient strategy for the recognition and elimination of extracts containing the known compounds.

5.2. Solid-Phase Extraction (SPE)

SPE methods are simple, reproducible and commercially available with a wide range of stationary phases but have low resolution (*see* Chap. 5). They have, to a large extent, replaced solvent partition systems in providing initial information about the polarity of biologically active components in natural product extracts. A dereplication strategy based on various SPE phases has been developed for the dereplication of HIV-inhibitory compounds in aqueous extracts of a wide variety of organisms ***(27)***. Surprisingly, 15% of the aqueous extracts were positive in the primary screen, indicating the presence of one or more recurring compounds. In part, the high proportion of positives was a consequence of the presence of anti-HIV-active sulfated polysaccharides, which could be eliminated by precipitation. The remaining active supernatants were then assessed for general chemical nature on Sephadex G-25 cartridges (molecular size and weight), C_4 wide-pore cartridges (polarity + size and shape), C_{18} narrow-pore cartridges (polarity + size and shape), and polyamide cartridges (to identify plant extracts containing tannins, which are known to be HIV inhibitory, and accounting for 90% of the positive plant extracts). The recovery of activity was also monitored to provide information about stability. The G-25, C_4, and C_{18} cartridges were each eluted to generate four fractions and these were bioassayed. Elution profiles on each cartridge type were characterized with known anti-HIV-active compounds, e.g., AZT (3′-azido-3′-deoxythymidine) and cyclosporin, which generated a characteristic 12-unit activity matrix (3 cartridge types × 4 fractions), which could be compared to the patterns obtained with each biological extract.

5.3. Thin-Layer Chromatography (TLC)

In theory, TLC could be used for dereplication of natural product-containing extracts, because there are a number of advantages to the method:

1. It is simple and cost-effective.
2. One has the complete polarity range between the origin and solvent front and a wide choice of mobile-phase mixtures.
3. The strong UV absorbances of certain solvents are not a problem because the solvent is evaporated from the plate before it is viewed.
4. Certain bioassays could be performed directly on the plate, e.g., coating with agar containing a test micro-organism, incubating and looking for zones of inhibition of growth; bioautography ***(28)***.

However, in practice, TLC is rarely used because it is time-consuming, and not a high-resolution method (when compared to HPLC). Recovery of compounds is also less efficient. Consequently, there appear to be no examples where TLC has been used for dereplication in the literature of the last 6 years.

5.4. Gas Chromatography–Mass Spectrometry (GC-MS)

Owing to its limited general applicability to natural products, GC is not commonly used for dereplication. It is appropriate for nonpolar, volatile classes of compounds and this range can be somewhat extended by derivatization (*see* Chap. 9). GC is, however, readily compatible with MS. GC-MS has been used for the dereplication of extracts of sweet-tasting plant species ***(29)***. The strategy was used to analyze and quantify TMS-derivatives of sugars/polyols, to identify species that were sweet, but contained low levels of free sugars or polyols.

5.5. High-Performance Liquid Chromatography (HPLC)

HPLC is frequently a component of current dereplication strategies, owing to its general availability, simplicity, flexibility, and good resolution. C_{18} is the most commonly used stationary phase, with UV (ultraviolet) monitoring or, increasingly, PDA detectors to monitor the separations, allowing the comparison of extract chromatograms with known compounds on the basis of their retention times and UV-spectra.

As part of the study to identify ecdysteroid antagonists ***(19)***, it was observed that a methanolic extract of *Iberis umbellata* (Cruciferae) possessed significant biological activity in the B_{II} cell bioassay ***(30)***. Bioassay-guided HPLC fractionation of the extract resulted in the identification of cucurbitacins B and D as the active principles. Subsequent studies have revealed that several antagonistic extracts from other plant species also contain cucurbitacins ***(19)***. Since the HPLC behavior of cucurbitacins has only been investigated in a limited way ***(31)***, the analysis of each positive extract in the same RP-HPLC system has been beneficial for dereplication in identifying those extracts that contain previously identified antagonistic cucurbitacins (**Fig. 2**).

A more extreme example, arising from the search for ecdysteroid agonists, is the identification of extracts containing novel phytoecdysteroids, where over 250 analogs are already known from the literature ***(32)***. Phytoecdysteroids occur in 5–6% of terrestrial plant species and may reach levels of 2–3% of the dry weight, although 0.01–0.1% is far more common. Phytoecdysteroid-containing species generally possess 1–3 major ecdysteroids, accounting for 90–95% of the total ecdysteroid, and a plethora of minor ecdysteroids. This diversity provides a wonderful natural resource for structure–activity studies ***(21,22)***, and almost every investigated positive extract yields at least one new analog, possessing a novel permutation of structural variations, e.g., 21-hydroxy ecdysteroids ***(33)***. The identification of which HPLC peaks possibly correspond to new analogs has been considerably facilitated by the use of standard reversed-phase (RP)- and normal-phase (NP)-HPLC systems, which had previously been characterized by the separation of reference samples of approx 30 of the most frequently encountered phytoecdysteroids ***(34)*** (**Fig. 2**). In view of the dramatic advances in the sensitivity and resolution of nuclear magnetic resonance spectroscopy, 1–2 mg of a phytoecdysteroids is usually fully sufficient for its unambiguous spectroscopic identification, and this can often be achieved with much smaller amounts ***(35)***. Owing to the strong UV-chromophore present in most ecdysteroids (λ_{max} at 242 nm in methanol or ethanol; $\varepsilon = 12{,}400\ L\ mol^{-1}\ cm^{-1}$), 1 mg samples are sufficient for accurate quantification and also for bioassay in cell-based assays, such as the B_{II}bioassay ***(18)***.

5.6. Capillary Electrophoresis (CE)

CE is a relatively new method (*see* Chap. 9) for the separation of natural products, but owing to its high resolution, sensitivity, and compatibility with UV-detection and MS, it has considerable potential for dereplication. At this stage, researchers are establishing the suitability of CE for different classes of natural products ***(36)***. The amount of compound required for detection is small (approx 1 ng), but the sample concentration must be relatively high (μg/mL) because only nanoliter amounts can be loaded.

6. Spectroscopy and Associated Hyphenated Techniques

Several recent reviews cover this aspect ***(37–40)***. Also Chapter 9 is devoted to various hyphenated techniques available for natural product research. A summary of the applications of available spectroscopic and hyphenated techniques in dereplication of natural products is presented here.

6.1. Ultraviolet/Visible Spectroscopy (UV/Vis)

The UV/Vis spectrum is often the first generally available piece of spectroscopic data that a researcher obtains for an unknown natural product and, with the rise of PDA detectors, it is becoming even easier to obtain this information. The form of the UV/Vis spectrum is often common to the class of natural product, because the multiplicity of the peaks, the wavelengths of maximal absorbance (λ_{max}), and the extinction coefficients (ε) are characteristic of the chromophoric groups present in the molecule. However, UV/Vis spectra are not very specific; they are at best indicative. Despite this fact, many dereplication strategies incorporate UV/Vis spectra as part of the information. Until recently, a problem has been that several of the structural databases could not be searched by UV/Vis spectrum ***(41)***. However, this situation is now improving.

6.2. MS and HPLC-MS

MS is probably the most commonly used dereplication technique, because of its high accuracy and sensitivity ***(42–44)***. It permits characterization of natural products based on molecular weight, elemental composition (with high-resolution MS, HRMS), and/or fragmentation patterns. The method is extremely sensitive (fmol–pmol). Soft ionization techniques, such as electrospray ionization (ESI) and atmospheric pressure chemical ionization (APCI), have considerably improved the applicability of MS to natural product analysis, as they allow the method to be applied to more polar, higher molecular weight compounds than more traditional modes of MS, such as electron impact ionization (EI) or chemical ionization (CI). APCI is generally suitable for moderately polar natural products, such as those found in chloroform or ethyl acetate extracts, while ESI is generally suitable for polar compounds, such as those present in methanol, ethanol, or acetone extracts ***(45)***. These techniques generally generate pseudomolecular ions (protonated or deprotonated in positive- and negative-ion modes, respectively), with little fragmentation. While molecular weight information is important, fragmentation patterns are important for structure determination. Consequently, collision-induced dissociation (CID) MS/MS is often performed to generate fragmentation.

There is considerable interest in the identification of new taxoid analogs from *Taxus* species other than *T. brevifolia*, both for the provision of further analogs for structure-activity-relationships (SAR) studies and as potential precursors for the chemical generation of specific derivatives. Stefanowicz et al. ***(45)*** developed a CID-MS (ESI-MS, MS/MS, and MS/MS/MS) method with which they could identify 57 basic taxoids from the ethyl acetate extract of *T. wallichiana* without prior chromatographic separation, thus allowing dereplication of known and unreported analogs. The method could be applicable to other classes of natural products. Its only major limitation is in distinguishing between isobaric (same molecular mass) and isomeric (same molecular formula) compounds (although fragmentation patterns can help), but coupling of the MS to HPLC would resolve this problem and allow fractions to be collected for bioactivity profiling.

Owing to the relatively high structural diversity (>50 known compounds) among citrus limonoids and the range of biological activities they possess, including anticancer activity, there is interest in the development of dereplication methods to identify rapidly the limonoid analogs present in various species and varieties of citrus. UV detection is hampered by the lack of a uniform chromophore within the class. Tian and Schwartz ***(46)*** have used APCI-MS/MS for four common limonoid aglycones and ESI-MS/MS for four common limonoid glucosides, taking advantage of the applicability of these soft ionization techniques to nonvolatile compounds.

Wolfender et al. ***(47)*** have examined CID MS/MS with hybrid quadrupole time-of-flight (Q-TOF) and ion-trap (IT) analyzers for the dereplication of flavonoids present in crude plant extracts after passage through an RP-HPLC column. Flavonoids are one of the most widespread constituents in the plant kingdom. The authors found that APCI was more robust at generating ions than ESI and optimized the collision energy to provide adequate fragmentation for all flavonoids. Positive- and negative-ion spectra were recorded. It was concluded that it is necessary to generate instrument-specific spectral libraries for effective dereplication of flavonoids.

The dereplication strategy used at Dow Agrochemicals is based on HPLC-PDAD-MS ***(48)***. Extracts are separated by gradient C_8-HPLC, with 95% of the effluent being collected into microtiter plates for subsequent bioassay(s), and 5% for positive-ion and negative-ion ESI-MS (alternating on a scan-to-scan basis). Once the biologically active fractions have been identified, portions of these fractions are further analyzed by MS, HRMS, MS/MS, and MSn techniques. Molecular weight assignment and UV-spectral data are used to search in Chapman & Hall's *Dictionary of Natural Products*. Potential identifications are then tested by means of the further MS data.

HPLC/ESI-MS linked with a bioassay has been used in a dereplication strategy for the rapid identification of two alkenyl catechols that are responsible for the cytotoxic activity present in fruits of *Semecarpus anacardium* (Anacardiaceae). It was possible to deduce the probable structures based on the correlation of the biological activity profile with the HPLC chromatogram (UV absorbance), m/z values of pseudomolecular ions, and reference to the *Dictionary of Natural Products*. The identities were confirmed by isolation and thorough spectroscopic analysis ***(49)***.

MS (APCI or ESI) may be coupled to other chromatographic methods, generating, for example, affinity chromatography-MS or gel-permeation chromatography-MS, or non-chromatographic methods, such as pulsed ultrafiltration-MS, which all have particular application in the identification of natural product binding to receptor species ***(50)***. These are currently niche applications and do not appear to have yet been applied to dereplication.

As part of continuing studies to identify compounds from deep-water, sponge-derived saltwater fungal cultures, Gautschi et al. ***(51)*** used RP-HPLC-MS in conjunction with a disk-diffusion soft agar-based bioassays (based cytotoxicity against various murine and human cell lines) to screen and dereplicate three separate libraries of deep-sea collections and identify cytotoxic fungal cultures producing metabolites with unique molecular formulas. From a mixed culture of *Penicillium coryophilum*, the authors identified seven serinones, of which two were identified without full prior purification.

While most studies of this sort are predominantly qualitative (i.e., which compounds are present and which are novel), there is increasing interest in quantitative aspects. For example, Bily et al. *(52)* have used an HPLC-PDA-APCI/MS method to identify and quantify phenylpropanoids in the grains of various cereals (maize, wheat, and sorghum). Thus, this method is suitable for the screening of cereal genotypes.

6.3. Infra-Red Spectroscopy

Infra-red (IR) spectrometers are available that are compatible with HPLC (in-line), but they are rarely used currently for dereplication purposes. IR spectra provide information about the presence or absence of functional groups, but they are not as useful or information-rich as NMR or mass spectra. Consequently, literature reports on the identification of natural compounds often do not contain IR data, which further limits their use for dereplication.

6.4. NMR Spectroscopy and HPLC-NMR

Since HPLC has become the method of choice for compound purification, and NMR is the main technique of structural elucidation, one can argue that HPLC-NMR should be the most useful of the hyphenated techniques for dereplication, because NMR provides information-dense spectra. Developments in the area of LC-NMR have recently been reviewed by Wolfender et al. *(53)*. They identified recent improvements in pulse-field gradients, solvent suppression, probe technology, and high-field magnets, which are giving impetus to the application to the analysis of natural product mixtures. Modern high-resolution NMR (400–900 MHz) spectrometers have increased sensitivity significantly (requiring 100–500 μg of the component to be identified), and the availability of solvent-suppression systems have meant that it is not necessary to use deuterated solvents for HPLC separations, although D_2O is often used as it is not so expensive. Acetonitrile/D_2O mixtures make solvent suppression easier, because the solvent mixture generates only one signal (acetonitrile–CH_3), which can be avoided by using CD_3CN, if an analyte signal occurs near the CH_3-signal of the undeuterated solvent *(54)*. To obtain adequate NMR spectra, machines operate in stop-flow mode, where the rising UV-absorbance corresponding to a peak activates a cessation of flow until enough scans have been obtained for the sample in the flow-probe, the volume of which has been reduced to 60 μL (2–4 mm i.d.) to give better sensitivity, while maintaining chromatographic performance. The alternative on-flow method, where the column effluent is scanned while maintaining the usual flow rate, is much less sensitive. Lack of sensitivity is the main limitation to the application of LC-NMR to natural product mixtures, especially because these normally contain some components at much lower concentrations than others. This can be partially overcome by high loading of the HPLC column. The problem is, of course, much more extreme for ^{13}C-NMR, and it is often the case that even the major components of the extract will not give adequate ^{13}C-spectra. Suppression of solvent signals will result in the simultaneous suppression of sample signals under the solvent peaks, necessitating the separation of the sample in two different solvent systems to obtain the full NMR spectrum. Further, chemical shifts will generally be somewhat different when determined in HPLC solvent mixtures rather than pure, deuterated NMR solvents. Much of the recent literature on the application of HPLC-NMR to natural products has been summarized ***(55,56)***.

Williamson et al. ***(57)*** have applied diffusion-edited NMR for the identification of components in a partially purified biologically active fraction derived from marine cyanobacteria. The translational diffusion of organic molecules depends upon their molecular size. In the NMR experiment, the spin systems with the same diffusion coefficient are grouped together as part of the same molecule, thus allowing the signals arising from a mixture to be assigned to individual compounds and interpreted. As many secondary products have formula weights of approx 400, conditions can be selected that qualitatively suppress all low MW impurities. If compounds have similar diffusion coefficients, they can be resolved once accurate values for the diffusion coefficients have been determined for each component in the mixture from a series of experiments to determine how the peak integration varies with maximum gradient strength. This approach permitted the deconvolution of the active fraction to identify symplostatin 1, even though it only made up approx 20% of the fraction.

As many sponge polycyclic terpenoids are described in the literature, dereplication is important, but MS and NMR spectra of such compounds are not readily diagnostic. Stessman et al. ***(58)*** have identified the sesquiterpenes present in *Lendenfeldia frondosa* and *Hyrtios erectus* by combining analysis of the low-field ^{13}C signals, using gradient 1D total correlation spectroscopy (TOCSY), and substructure searching of databases

(MARINLIT), which proved suitable for both dereplication and identification of novel compounds.

A word of warning has been sounded by Pauli et al. ***(59)***, who demonstrate, using caffeoyl quinic acid derivatives as an example, that the NMR solvent may significantly affect chemical shifts and coupling patterns, which complicates structure determination and dereplication.

7. Databases

Perhaps the most significant development in the area of dereplication over the last two decades has been the impact of the bioinformatics revolution. A (perhaps the) key aspect of dereplication is comparing one's findings with the vast and disparate body of existing literature to determine if a compound is already known or using the literature to go from a partial structure to a putative complete identification. Several commercial databases exist to facilitate these processes:

1. *Chemical Abstracts* (http://cas.org/), including NAPRALERT.
2. Beilstein (http://www.beistein.com/).
3. *Bioactive Natural Products Database* (Bérdy Antibiotic Database; 23,000 compounds).
4. DEREP (7000 compounds).
5. MARINLIT (www.chem.canterbury.ac.nz/marinlit/marinlit.shtml; 6000 compounds).
6. *Marine Natural Products Database* (4000 compounds).
7. Chapman & Hall's *Dictionary of Natural Products* (182,000 entries, available on CD-ROM and updated every 6 mo).

These exist on-line, on CD-ROM, or in book form, but increasingly it is the on-line versions that are the most useful, in part because of the sheer size of such databases these days. The largest number of natural product structures is held in *Chemical Abstracts* (CA), and entries can be searched according to formula weight, carbon count, full or partial structures, bioactivity, taxonomy, and spectral data ***(41)***. CA Services Registry File is a component of the Scientific and Technical Information Network (STN), which also incorporates NAPRALERT, Beilstein, Specinfo, Medline, Embase, Biosis, and JICST, thus providing access to other databases, some of which are more specific to natural products (e.g., NAPRALERT; >140,000 natural products), while others provide access to abstracts of the relevant chemical and biological literature. Some of the other databases have grown out of researchers' own databases (e.g., MARINLIT), and all researchers in natural products chemistry sooner or later start to create their own databases focusing on the types of compounds, the organisms, and/or the literature. The different databases in STN are crosslinked; hence, it is possible to switch between them. For example, one can switch between NAPRALERT and the CA Registry by using the CAS number of an entry, but only about half of the NAPRALERT entries possess CAS numbers ***(41)***. Unfortunately, some attributes that are important to natural product chemists (e.g., UV-spectra) are not searchable in all databases.

A simplified strategy for dereplication of compounds using commercial databases has been proposed by Bradshaw et al. ***(60)***, in which a search of a textfile linking structures to their FW and counts of the number of methyl, methylene, and methine groups (readily determined by MS and NMR) is performed. This proved to be highly discriminating for the identification of known compounds, because only a few compounds (generally one to four) fit all four criteria.

8. Prospects: Recent Developments in Hyphenated Techniques

Bioassays and biochemical detection methods have traditionally been performed off-line from the chromatography, generally because of the difficulties of automation and the inhibitory effects of most HPLC solvents. This results in the need for considerable manual handling of fractions and delay in identifying active fractions. A method has been developed for the on-line biochemical screening of natural products separated by HPLC for their binding to estrogen receptors (ER) α and β ***(61)***. In the latest application of this method, Van Elswijk et al. ***(62)*** developed a method for the dereplication of phyto-estrogens (flavonoids) from the peel of pomegranate (*Punica granatum*) using RP-HPLC coupled on-line to a biochemical (β-estrogen receptor; β-ER) assay and negative-ion APCI MS/MS, such that the column effluent is split with 1/25 going to biochemical detection and 24/25 for MS. For biochemical detection, the column effluent was diluted to a constant 10% acetonitrile. The receptor assay assessed the ability of compounds eluting from the HPLC to compete with the fluorescent ligand coumestrol for the ER. Coumesterol undergoes fluorescent enhancement on binding to ER. Consequently, the column effluent was mixed in continuous flow with a solution of β-ER (100 nM) for 40 s, then coumestrol (500 nM) was added and mixed for a further 40 s. The effluent was then passed through a flow-through fluorescence detector (excitation, 340 nM; emission, 410 nM). Mass spectra were compared to a database generated for a wide variety of estrogenic compounds. Although effective, the sensitivity of the method is reduced by dilution and band-broadening inherent in the biochemical assay. Nevertheless, it was possible for the authors to identify luteolin, quercetin, and kaempferol as new estrogenic compounds from pomegranate peel.

On-line, HPLC-compatible, biochemical assays have also been developed for natural products acting as phosphodiesterase inhibitors ***(63)*** and inhibitors of phosphate-consuming or -releasing enzymes ***(64)***.

Multiple hyphenation of techniques for the identification of chromatographically separated compounds, including natural products, has been reviewed by Wilson ***(54,65)***, covering HPLC-NMR-MS, HPLC-NMR-IR, and HPLC-NMR-IR-MS. Thus, these systems are capable of delivering a considerable amount of spectroscopic information, approaching all that normally used for identification. NMR and MS components of the system are generally in parallel, with the vast majority (approx 95%) of the column effluent going to the NMR and only a small proportion (approx 5%) going to MS, reflecting the relative sensitivities of the two techniques. The application to plant extracts has been demonstrated through successful analysis of the major phytoecdysteroids present in an extract of *Silene otites* (Caryophyllaceae) and various natural products in *Hypericum perforatum* (St. John's wort).

There will be a transition from HPLC-MS to CE-MS. The slow separations on HPLC are a bottleneck in the processing of natural product extracts. CE is much faster and TOF-based MS can acquire complete spectra at compatible rates, and these will tend to replace the scanning instruments.

There is increasing interest in the genetic manipulation of plants or cultured plant cells to either enhance the production of a particular secondary metabolite or modification of the secondary metabolite profile through combinatorial genomics or metabolic engineering ***(66)***. Similarly, short biosynthetic pathways could be functionally expressed in appropriate microbes to produce secondary metabolites. Dereplication strategies will play a major role in identifying lines that produce the optimal yield of the desired product in the absence of interfering compounds or processes.

References

1. Jia, Q. (2003) Generating and screening a natural product library for cyclooxygenase and lipoxygenase dual inhibitors, in *Studies in Natural Products Chemistry*, vol. 29 (Atta-ur-Rahman, ed.), Elsevier, pp. 643–718.

2. Colquhoun, J. A., Zulu, J., Goodfellow, M., Horikoshi, K., Ward, A. C., and Bull, A. T. (2000) Rapid characterisation of deep-sea actinomycetes for biotechnology screening programmes. *Antonie van Leeuwenhoek* **77,** 359–367.
3. Ramakrishna, N. V. S., Nadkarni, S. R., Bhat, R. G., Naker, S. D., Kumar, E. K. S. V., and Lal, B. (1999) Screening of natural product extracts for antibacterial activity: early identification and elimination of known compounds by dereplication. *Ind. J. Chem.* **38B,** 1384–1387.
4. VanMiddlesworth, F. and Cannell, R. J. P. (1998) Dereplication and partial identification of natural products. *Methods Biotechnol.* **4,** 279–327.
5. Newman, D. J., Cragg, G. M., and Snader, K. M. (2000) The influence of natural products upon drug discovery. *Nat. Prod. Rep.* **17,** 215–234.
6. Cordell, G. A. and Shin, Y. G. (1999) Finding the needle in the haystack. The dereplication of natural product extracts. *Pure Appl. Chem.* **71,** 1089–1094.
7. Bull, A. T., Ward, A. C., and Goodfellow, M. (2000) Search and discovery strategies for biotechnology: the paradigm shift. *Microbiol. Mol. Biol. Rev.* **64,** 573–606.
8. Strobel, G. A. (2002) Useful products from rainforest microorganisms. Part 2. Unique bioactive molecules from endophytes. *Agro-Food-Ind. Hi-tech.* **13,** 12–17.
9. Baloglu, E. and Kingston, D. G. I. (1999) The taxane diterpenoids. *J. Nat. Prod.* **62,** 1448–1472.
10. Gloer, J. B. (1995) The chemistry of fungal antagonism and defense. *Can. J. Bot.* **73,** S1265–S1274.
11. Nielsen, K. F., Larsen, T. O., and Frisvad, J. C. (2004) Lightweight expanded clay aggregates (LECA), a new up-scaleable matrix for production of microfungal metabolites. *J. Antibiot.* **57,** 29–36.
12. Kim, S.W., Ban, S.H., Chung, H.J. et al. (2004) Taxonomic discrimination of higher plants by pyrolyis mass spectrometry. *Plant Cell Reports* **22,** 519–522.
13. Brandão, P. F. B., Torimura, M., Kurane, R., and Bull, A. T. (2002) Dereplication for biotechnology screening: PyMS analysis and PCR-RFLP-SSCP (PRS) profiling of 16S rRNA genes of marine and terrestrial actinomycetes. *Appl. Microbiol. Biotechnol.* **58,** 77–83.
14. Ritacco, F. V., Haltli, B., Janso, J. E., Greenstein, M., and Bernan, V. S. (2003) Dereplication of *Streptomyces* soil isolates and detection of specific biosynthetic genes using an automated ribotyping instrument. *J. Ind. Microbiol. Biotechnol.* **30,** 472–479.
15. Cordell, G. A. (2000) Biodiversity and drug discovery—a symbiotic relationship *Phytochemistry* **55,** 463–480.
16. Donadio, S., Monciardini, P., Alduina, R. et al. (2002) Microbial technologies for the discovery of novel bioactive metabolites. *J. Biotechnol.* **99**, 187–198.

17. Knight, V., Sanglier, J.-J., DiTullio, D. et al. (2003) Diversifying microbial natural products for drug discovery. *Appl. Microbiol. Biotechnol.* **62**, 446–458.
18. Clément, C. Y., Bradbrook, D. A., Lafont, R., and Dinan, L. (1993) Assessment of a microplate-based bioassay for the detection of ecdysteroid-like or antiecdysteroid activities. *Insect. Biochem. Mol. Biol.* **23,** 187–193.
19. Dinan, L., Savchenko, T., Whiting, P., and Sarker, S. D. (1999) Plant natural products as insect steroid receptor agonists and antagonists. *Pestic. Sci.* **55,** 331–335.
20. Dinan, L. (2001) Phytoecdysteroids: biological aspects. *Phytochemistry* **57,** 325–339.
21. Dinan, L. (2003) Ecdysteroid structure–activity relationships, in *Studies in Natural Products Chemistry*, vol. 29 (Atta-ur-Rahman, ed.), Elsevier, Amsterdam, pp. 3–71.
22. Dinan, L. and Hormann, R. E. (2005) Ecdysteroid agonists and antagonists, in *Comprehensive Molecular Insect Science*, vol. 3 (Gilbert, L. I., Iatrou, K., and Gill, S. S., eds.), Elsevier/Pergamon, Oxford, pp. 197–242.
23. Antonio, J. and Molinski, T. F. (1993) Screening of marine invertebrates for the presence of ergosterol-sensitive antifungal compounds. *J. Nat. Prod.* **56,** 54–61.
24. Umezawa, Y. (2002) Assay screening methods for bioactive substances based on cellular signalling pathways. *Rev. Mol. Biotechnol.* **82,** 357–370.
25. Ingkaninan, K., Hazelkamp, A., Hoek, A. C., Balconi, S., and Verpoorte, R. (2000) Application of centrifugal partition chromatography in a general separation and dereplication procedure for plant extracts. *J. Liq. Chromatogr. Rel. Technol.* **23,** 2195–2208.
26. Alvi, K. A., Peterson, J., and Hofmann, B. (1995) Rapid identification of elaiophylin and geldanamycin in Streptomyces fermentation broths using CPC coupled with a photodiode array detector and LC-MS methodologies *J. Ind. Microbiol.* **15,** 80–84.
27. Cardellina, J.H., Munro, M.H.G., Fuller, R.W. et al. (1993) A chemical screening strategy for the dereplication and prioritization of HIV-inhibitory aqueous natural products extracts. *J. Nat. Prod.* **56**, 1123–1129.
28. Hamburger, M. O. and Cordell, G. A. (1987) A direct bioautographic TLC assay for compounds possessing antibacterial activity. *J. Nat. Prod.* **50,** 19–22.
29. Chung, M.-S., Kim, N.-C., Long, L. et al. (1997) Dereplication of saccharide and polyol constituents of candidate sweet-tasting plants: isolation of the sesquiterpene glycoside mukurozioside IIb as a sweet principle of *Sapindus rarak*. *Phytochem. Anal.* **8**, 49–54.
30. Dinan, L., Whiting, P., Girault, J.-P. et al. (1997) Cucurbitacins are insect steroid hormone antagonists acting at the ecdysteroid receptor. *Biochem. J.* **327,** 643–650.

31. Dinan, L., Harmatha, J., and Lafont, R. (1997) Chromatographic procedures for the isolation of plant steroids. *J. Chromatogr.* **935,** 105–123.
32. Lafont, R., Harmatha, J., Marion-Poll, F., Dinan, L., and Wilson, I.D. (2003) Ecdybase [*The Ecdysone Handbook*, 3 ed.] (http://ecdybase.org).
33. Dinan, L., Sarker, S. D., Bourne, P., Whiting, P., Šik, V., and Rees, H. H. (1999) Phytoecdysteroids in seeds and plants of *Rhagodia baccata* (Labill.) Moq. (Chenopodiaceae). *Arch. Insect Biochem. Physiol.* **41,** 18–23.
34. Dinan, L. (1995) A strategy for the identification of ecdysteroid receptor agonists and antagonists from plants. *Eur. J. Entomol.* **92,** 271–283.
35. Girault, J.-P. (1998) Determination of ecdysteroids structure by 1D and 2D NMR *Russian J. Plant Physiol.* **45,** 306–309.
36. Tomás-Barberán, F. A. (1995) Capillary electrophoresis: a new technique in the analysis of plant secondary metabolites. *Phytochem. Anal.* **6,** 177–192.
37. Strege, M. A. (1999) High-performance liquid chromatographic-electrospray ionization mass spectrometric analyses for the integration of natural products with modern high-throughput screening. *J. Chromatogr. B* **725,** 67–78.
38. Wolfender, J.-L., Terreaux, C., and Hostettmann, K. (2000) The importance of LC-MS and LC-NMR in the discovery of new lead compounds from plants *Phamaceut. Biol.* **38 (supp),** 41–54.
39. Hostettmann, K., Wolfender, J.-L., and Terreaux, C. (2001) Modern screening techniques for plant extracts. *Pharmaceut. Biol.* **39 (supp),** 18–32.
40. Wolfender, J.-L., Ndjoko, K., and Hostettmann, K. (2003) Liquid chromatography with ultraviolet absorbance-mass spectrometric detection and with nuclear magnetic resonance spectroscopy: a powerful combination for the on-line structural investigation of plant metabolites. *J. Chromatogr. A.* **1000,** 437–455.
41. Corley, D. G. and Durley, R. C. (1994) Strategies for database dereplication of natural products. *J. Nat. Prod.* **57,** 1484–1490.
42. Potterat, O., Wagner, K., and Haag, H. (2000) Liquid chromatography-electrospray time-of-flight mass spectrometry for on-line accurate mass determination and identification of cyclodepsipeptides in a crude extract of the fungus *Metarrhizium anisopliae*. *J. Chromatogr. A* **872,** 85–90.
43. Waridel, P., Wolfender, J.-L., Ndjoko, K., Hobby, K. R., Major, H. J., and Hostettmann, K. (2001) Evaluation of quadrupole time-of-flight tandem mass spectrometry and ion-trap multiple-stage mass spectrometry for the differentiation of *C*-glycosidic flavonoid isomers. *J. Chromatogr. A* **926,** 29–41.
44. Nielsen, K. F. and Smedsgaard, J. (2003) Fungal metabolite screening: database of 474 mycotoxins and fungal metabolites for dereplication by standardised liquid chromatography-UV-mass spectrometry methodology. *J. Chromatogr. A* **1002,** 111–136.
45. Stefanowicz, P., Prasain, J. K., Yeboah, K. F., and Konishi, Y. (2001) Detection and partial structure elucidation of basic taxoids from *Taxus*

wallichiana by electrospray ionization tandem mass spectrometry *Anal. Chem.* **73,** 3583–3589.
46. Tian, Q. and Schwartz, S. J. (2003) Mass spectrometry and tandem mass spectrometry of citrus limonoids. *Anal. Chem.* **75,** 5451–5460.
47. Wolfender, J.-L., Waridel, P., Ndjoko, K., Hobby, K. R., Major, H. J., and Hostettmann, K. (2000) Evaluation of Q-TOF-MS/MS and multiple stage IT-MS^n for the dereplication of flavonoids and related compounds in crude plant extracts. *Analusis* **28,** 895–906.
48. Gilbert, J. R., Lewer, P., Duebelbeis, D. O., Carr, A. W., Snipes, C. E., and Williamson, R. T. (2003) Identification of biologically active compounds from nature using liquid chromatography/mass spectrometry. *ACS Symp. Ser.* **850,** 52–65.
49. Shin, Y.G., Cordell, G.A., Dong, Y. et al. (1999) Rapid identification of cytotoxic alkenyl catechols in *Semecarpus anacardium* using bioassay-linked high performance liquid chromatography-electrospray/mass spectrometric analysis. *Phytochem. Anal.* **10**, 208–212.
50. Shin, Y. G. and van Breemen, R. B. (2001) Analysis and screening of combinatorial libraries using mass spectrometry. *Biopharm. Drug Dispos.* **22,** 353–372.
51. Gautschi, J. T., Amagata, T., Amagato, A., Valeriote, F. A., Mooberry, S. L., and Crews, P. (2004) Expanding the strategies in natural product studies of marine-derived fungi: a chemical investigation of *Penicillium* obtained from deep water sediment. *J. Nat. Prod.* **67,** 362–367.
52. Bily, A.C., Burt, A.J., Ramputh, A.L., et al. (2004) HPLC-PAD-APCI/MS assay of phenylpropanoids in cereals. *Phytochem. Anal.* **15**, 9–15.
53. Wolfender, J.-L., Ndjoko, K., and Hostettmann, K. (2001) The potential of LC-NMR in phytochemical analysis. *Phytochem. Anal.* **12,** 2–22.
54. Wilson, I. D. (2001) Chromatography with spectroscopy on line, can you have it all? *CAST* 10–13.
55. Bobzin, S. C., Yang, S., and Kasten, T. P. (2000) LC-NMR: a new tool to expedite the dereplication and identification of natural products. *J. Ind. Microbiol. Biotechnol.* **25,** 342–345.
56. Bobzin, S. C., Yang, S., and Kasten, T. P. (2000) Application of liquid chromatography-nuclear magnetic resonance spectroscopy to the identification of natural products. *J. Chromatogr. B* **748,** 259–267.
57. Williamson, R.T., Chapin, E.L., Carr, A.W., et al. (2000) New diffusion-edited NMR experiments to expedite the dereplication of known compounds from natural product mixtures. *Org. Lett.* **2**, 289–292.
58. Stessman, C. C., Ebel, R., Corvino, A. J., and Crews, P. (2002) Employing dereplication and gradient 1D NMR methods to rapidly characterize sponge-derived sesterterpenes. *J. Nat. Prod.* **65,** 1183–1186.

59. Pauli, G. F., Kukzkowiak, U., and Nahrstedt, A. (1999) Solvent effects in the structure dereplication of caffeoyl quinic acids. *Magn. Reson. Chem.* **37,** 827–836.
60. Bradshaw, J., Butina, D., Dunn, A.J. et al. (2001) A rapid and facile method for the dereplication of purified natural products. *J. Nat. Prod.* **64**, 1541–1544.
61. Schobel, U., Frenay, M., Van Elswijk, D.A. et al. (2001) High resolution screening of plant natural product extracts for estrogen receptor α and β binding activity using an online HPLC-MS biochemical detection system. *J. Biomol. Screening* **6**, 291–303.
62. van Elswijk, D. A., Schobel, U. P., Lansky, E. P., Irth, H., and van der Greef, J. (2004) Rapid dereplication of estrogenic compounds in pomegranate (*Punica granatum*) using on-line biochemical detection coupled to mass spectrometry. *Phytochemistry* **65,** 233–241.
63. Schenk, T., Breel, G.J., Koevoets, P. et al. (2003) Screening of natural products extracts for the presence of phosphodiesterase inhibitors using liquid chromatography coupled online to parallel biochemical detection and chemical characterization. *J. Biomol. Screening* **8**, 421–429.
64. Schenk, T., Appels, N. M. G. M., van Elswijk, D. A., Irth, H., Tjaden, U. R., and van der Greef, J. (2003) A generic assay for phosphate-consuming or releasing enzymes coupled on-line to liquid chromatography for lead finding in natural products. *Anal. Biochem.* **316,** 118–126.
65. Wilson, I. D. (2000) Multiple hyphenation of liquid chromatography with nuclear magnetic resonance spectroscopy, mass spectrometry and beyond. *J. Chromatogr.* **892,** 315–327.
66. Guttman, A., Khandurina, J., Budworth, P., Xu, W., Hou, Y.-M., and Wang, X. (2004) Analysis of combinatorial natural products by HPLC and CE. *LC. GC Europe* **17,** 104–111.

13

Extraction of Plant Secondary Metabolites

William P. Jones and A. Douglas Kinghorn

Summary

Plant secondary metabolites are currently the subject of much research interest, but their extraction as part of phytochemical or biological investigations presents specific challenges that must be addressed throughout the solvent extraction process. Successful extraction begins with careful selection and preparation of plant samples, and thorough review of the appropriate literature for indications of which protocols are suitable for a particular class of compounds or plant species. During the extraction of plant material, it is important to minimize interference from compounds that may coextract with the target compounds, and to avoid contamination of the extract, as well as to prevent decomposition of important metabolites or artifact formation as a result of extraction conditions or solvent impurities. This chapter presents an overview of the process of plant extraction, with an emphasis on common problems encountered and methods for reducing or eliminating these problems. In addition to generally applicable extraction protocols, methods are suggested for more or less selectively extracting specific classes of compounds, and phytochemical methods are presented for detection of classes of compounds commonly encountered during plant extraction, including selected groups of secondary metabolites and interfering compounds.

Key Words: Plant extracts; plant secondary metabolites; percolation; maceration; extraction artifacts; interfering compounds; phytochemical detection methods.

From: *Methods in Biotechnology, Vol. 20, Natural Products Isolation, 2nd ed.*
Edited by: S. D. Sarker, Z. Latif, and A. I. Gray © Humana Press Inc., Totowa, NJ

1. Introduction

Researchers from a variety of scientific disciplines are confronted with the challenge of extracting plant material with solvents, often as a first step toward isolating and identifying the specific compounds responsible for biological activities associated with a plant or a plant extract. The impetus for this research arises largely because plants form the foundation of traditional pharmacopeias, and because many of our currently important pharmaceutical drugs are obtained from plants ***(1–3)***. Further interest arises from the growing awareness that many of the secondary metabolites of organisms, including plants, serve important biological and ecological roles, mainly as chemical messengers and defensive compounds ***(4,5)***. Investigators engaged in the isolation of secondary metabolites from plants soon discover the need for considerable laboratory finesse in the apparently routine "sample preparation" steps that convert crude plant material into an extract suitable for chemical analysis, biological testing, or chromatographic separation. This chapter outlines the steps involved in plant extraction, beginning with approaches to selecting, collecting, processing, and documenting plant samples. Procedures for extraction of plant material are described, techniques designed to eliminate some of the most common "nuisance" compounds that are often extracted along with compounds of interest are presented, and recommendations are given for tailoring extraction protocols to suit specific purposes. Common sources of contamination and potential causes of artifact formation during extraction are discussed, and suggestions are given for reducing or eliminating these problems. Techniques for the detection of selected classes of plant secondary metabolites of potential interest (and common contaminants and "nuisance" metabolites that interfere with chemical and biological assays) are presented as well. Emphasis will be placed throughout on practical means to recognize and avoid common pitfalls, and to overcome specific problems that may be encountered during the extraction of plant secondary metabolites.

2. Methods

2.1. Selection, Collection, and Identification of Plant Material

The methods employed in the selection, collection, and identification of plant material directly affect the reproducibility of phytochemical research, and carelessness at this stage of an investigation may greatly reduce the scientific value of the overall study. Plant secondary metabolites often accumulate in specific plant parts. Thus, unless it is already known which part contains the highest levels of the compounds of interest, it is prudent to collect multiple plant parts, or the whole plant, to ensure the extracts prepared are representative of the range of secondary metabolites produced by the plant (*see* **Note 1**). Specific secondary metabolites also vary both quantitatively and qualitatively among closely related species, within a single species, and among members of a population *(6,7)*. Caution should therefore be exercised when making broad inferences about the presence or absence of specific compounds in a species under investigation, and when recollecting samples with the intention of isolating more of a specific metabolite (*see* **Note 2**).

For drug discovery from plants, samples may be selected using a number of approaches, for example, (1) the investigation of plants traditionally used by humans for food, medicine, or poison based on review of the literature or interviews conducted as part of the investigation *(2,8)*(*see* **Note 3**); (2) the random or systematic collection of a biodiverse set of plant samples, typically from an ecological region that is comparatively uncharted as regards secondary metabolite production *(3,9)*; (3) the selection of species based on phylogenetic relationship to a species known to produce a compound or compound class of interest *(10)*; and (4) the study of species based on reports of biological activity in the literature (including chemical ecology, toxicology, and veterinary reports). Databases may be used to aid in selecting species that meet one or a combination of specified criteria *(11,12)*. Biodiversity-based approaches have logistical advantages for large-scale screening programs, and they are expected to yield a higher percentage of novel, biologically active chemical structures *(3,13)*. In addition to phytochemical considerations, ethical and legal issues associated with intellectual property are increasingly important and contentious, particularly when plant material or extracts will cross international borders *(14,15)* (*see* **Note 4**).

Many logistical considerations related to the collection of plant material from the field have been addressed elsewhere *(2,16,17)*, and so only general aspects are mentioned here. Prior to field collection, it is advisable to review the flora of the region; to compile a list of which species, genera, or families are of particular interest; and to determine which taxa are to be avoided *(7)*. For a biodiversity-based collection, taxa endemic to the region concerned are of high priority, pandemic weedy species are probably of little interest, and rare or endangered species are to be strictly avoided (*see* **Note 5**). It is advisable to attempt field identification of the samples collected (at least to the level of genus) (*see* **Note 6**). To aid taxonomic experts in confirming or refining the field identification, and as a permanent scientific record, voucher specimens (including reproductive organs, when feasible) should be prepared and deposited in herbaria, including at least one major institution and, if applicable, in a local herbarium in the source country. The voucher code should be retained for possible inclusion in publications resulting from the collection. A note card affixed to the voucher specimen should include observations such as local uses of the species, its habitat, microenvironment (e.g., shaded vs. sunny location), state of overall plant health (*see* **Note 7**), stage in the reproductive cycle, and other facts that may be useful for future investigations.

2.2. Drying and Grinding

In general, plant material should be dried at temperatures below 30°C to avoid decomposition of thermolabile compounds. Likewise, it should be protected from sunlight because of the potential for chemical transformations resulting from exposure to ultraviolet radiation. To prevent the buildup of heat and moisture, air circulation around the plant material is essential. Hence, it should not be compacted, and it may be necessary to use a fan or other means to provide air flow around or through the drying sample (*see* **Note 8**).

When fresh plant material is required for study, it is advisable to extract it as soon as possible using organic solvents, such as methanol (MeOH) or ethanol (EtOH), that will deactivate enzymes present in the plant. Usually, the resulting extract will contain a significant portion of water, and, if desired, it can be partitioned with an organic solvent that is immiscible with hydroalcoholic mixtures. If the plant material is to be ground before extraction, it may be necessary to grind the material in a buffered solvent, thus preventing hydrolysis reactions because of pH changes that can occur when organic acids present in cellular compartments are released. Alternatively, the plant material could be frozen and transported on dry ice, or preserved in alcohol.

Small quantities of plant material can be milled using an electric coffee or spice mill, or in a mortar and pestle, in which case, addition of a small amount of sand may aid in the process. Milling of large quantities of plant material is usually best carried out using industrial-scale comminution equipment (*see* **Note 9**). Grinding improves the efficiency of extraction by increasing the surface area of the plant material. It also decreases the amount of solvent needed for extraction by allowing the material to pack more densely. Although it might seem that milling plant material to a fine powder would be ideal, if the particles are too fine, solvent cannot flow easily around them. Furthermore, the friction of milling generates heat (the finer the particle produced, the more heat), potentially causing volatile constituents to be lost, and thermolabile components to degrade and oxidize. Plants containing volatile components may be extracted by steam distillation of coarsely chopped plant material.

2.3. Extraction

A range of techniques, varying in cost and level of complexity, may be used for extraction of plant material. For most applications, relatively simple techniques, such as percolation and maceration, are effective and economical. Some specific applications, however, require more sophisticated and costly extraction techniques using specialized equipment (*see* **Note 10**), such as that utilized in large-scale steam distillation and supercritical-fluid extraction (SFE) (*see* Chap. 3). Methods of solvent extraction can be classified as continuous or discontinuous. In continuous methods (e.g., percolation and Soxhlet extraction), solvent flows through the plant material continuously. As constituents diffuse from the plant material into the surrounding solvent, the solvent becomes increasingly saturated, but because the solvent is continually flowing, the saturated solvent is replaced with less-saturated solvent. In the case of discontinuous methods, the solvent is added and removed in batches. Hence, once equilibrium is reached between the concentration of solute inside the plant material and the concentration in the solvent, extraction essentially stops until the solvent is decanted and replaced with new solvent.

Regardless of the extraction technique used, the resulting solution should be filtered to remove any remaining particulate matter. Plant extracts should not be stored in the solvent for long periods at room temperature, or in sunlight, because of the accompanying increased risk of artifact formation and decomposition or isomerization of extract constituents (*see* **Subheading 2.4.**). Extracts can be concentrated at reduced pressure on a rotary evaporator or dried under a stream of nitrogen. If a rotary evaporator is used, it is advisable to keep the water bath temperature below 40°C to prevent decomposition of thermolabile components. Especially when extracting large amounts of a single sample, the solvent collected from the rotary evaporator condenser during the concentration of one extraction batch may be recycled for further extraction of that same sample, but use of recovered solvent in extraction of other samples is not advised, because it may lead to cross-contamination of later extracts.

2.3.1. Percolation and Soxhlet Extraction

Percolation is an efficient method of extraction, suitable for medium to large sample sizes. A variety of different vessels can serve as percolators. The main requirements are that they have a wide opening at the top to accommodate addition and removal of plant material, and a valve at the base to regulate solvent flow (*see* **Note 11**). With the valve in the closed position, plant material is added to the container, leaving room for expansion. Then, enough solvent is added to cover the sample. If the plant material is too loosely packed, it will not percolate efficiently, and it may need to be compressed to reduce the amount of solvent needed to cover the sample (glass or nonporous ceramic labware may be used as weights for compressing plant material). The plant material is allowed to soak for several hours or overnight, with adequate amounts of the solvent being added to keep the plant material covered. Next, the valve is opened slightly to allow solvent to flow slowly into a container. The flow rate is regulated to ensure that the solvent exiting is nearly saturated with solute, and solvent is added at the top of the percolator to replace that lost from the bottom. Although in principle more efficient and economical of solvent than maceration (*see* **Subheading 2.3.2.**), percolation is usually not practical for small amounts of plant material or for large numbers of samples. Soxhlet extraction, using commercially available devices, is a convenient method for extraction of small to moderate volumes of plant material. Because the extraction takes place in a closed system in which the solvent is continually recycled, the amount of solvent needed for Soxhlet extraction is minimal. In the most commonly used extractors, however, the heat needed to drive the extraction will likely cause thermolabile constituents to form artifacts or decomposition products.

2.3.2. Maceration

Maceration is a common method for extraction of small amounts of plant material in the laboratory because it can be carried out conveniently in Erlenmeyer flasks (the flasks can be covered with parafilm or aluminum foil to prevent evaporation of solvent). As a rough guideline, after each addition of fresh solvent, the plant material should be left to macerate overnight. The solvent should be decanted through a screen or filter, and fresh solvent added to the flask. The sample is then mixed with the fresh solvent by stirring or swirling, and left to macerate again. Sonication of the macerating sample or gentle swirling on a fermentation broth table is sometimes used to reduce the time needed for thorough extraction. Experience indicates that after three solvent changes, the plant material is almost completely exhausted (*see* **Note 12**). Large samples may also be macerated, usually in a large container with a tap at the base, as with large-scale percolation, except that the solvent is changed in batches.

2.3.3. Choice of Solvent

Factors that should be considered when choosing a solvent or solvent system for extracting plant material include solubility of the target constituents, safety, ease of working with the solvent, potential for artifact formation, and the grade and purity of the solvent. Following the principle of "like extracts like," it is often possible to tailor the solvent choice to maximize the yields of the compounds of interest, while minimizing the extraction of unwanted compounds. Specific extraction techniques that prevent decomposition of the analytes of interest or that efficiently extract the desired compounds are often reported. A careful review of the literature related to the species and compound classes under investigation often saves time and effort. Even when little information exists about the secondary metabolites from a particular species, phytochemical reports on other species in the genus or family can provide clues regarding which compound classes to expect, and which solvents and procedures may be used to isolate them. In addition, it may be worthwhile to perform several trial extractions using different solvents and techniques. Comparison of total extract yields, yields of metabolites of interest, or intensity of biological activity will indicate which method gives the best results.

Precautions must be taken to minimize the risk of fire and explosion when using and storing highly flammable solvents and solvents that tend to form explosive peroxides (such as diethyl ether and other ether-containing solvents) (*see* **Note 13**). Solvents with low boiling points are generally easier to use from the standpoint that they are more easily concentrated. Acetone, chloroform ($CHCl_3$), dichloromethane (DCM), ethyl acetate (EtOAc), and *n*-hexane/petroleum ether evaporate relatively quickly, whereas water and butanol are more difficult to remove.

2.3.4. Sample Preparation for Large-Scale Biological Screening

When a large number of extractions are to be carried out, it is generally not feasible to extract each plant sample with a tailored solvent system. A general procedure must be developed and validated, giving consideration to the rate of detection of active extracts ("hits") obtained using several extraction methods, and followed by analysis of the rate of false-positive responses. This approach has been used to develop a general extraction protocol to be utilized in extracting plant constituents for in vivo or in vitro biological screening *(18,19)* **(Fig. 1)**. The protocol consists of initial extraction by maceration with MeOH, followed by concentration under vacuum and a series of partitioning steps that serve to first "defat" the extract, removing oils and waxes, long-chain alkanols, carotenoids, and other nonpolar constituents, including some plant sterol aglycones and nonpolar triterpenoids. A second partition step removes the polar constituents, such as sugars and highly glycosylated flavonoids and triterpenoids. The $CHCl_3$ layer is then partitioned with a 1%-NaCl aqueous solution to remove tannins (plant polyphenols). The resulting $CHCl_3$-soluble extract is essentially free of tannins, and may be used in primary screening against a variety of cell lines, in vivo systems, and enzyme-based assays. For these test systems, it has been determined that the $CHCl_3$ extract prepared in this manner retains most of the biological activity of a plant sample, except for activity owing to vegetable tannins or highly polar or nonpolar compounds that tend not to be promising candidates for drug development *(18,19)*.

2.3.5. Preparation of Enriched Extracts

When a particular phytochemical constituent or compound class is to be the target of an investigation, specific extraction procedures may be employed to produce extracts enriched in the constituents of interest. In some instances, the polarity of a solution may be modified to cause particular compound classes to precipitate, leaving unwanted compounds in solution. Compounds containing primary, secondary, or tertiary amines, carboxylic acids, lactones, and phenols may be selectively extracted using pH modifications to manipulate the polarity/solubility of the compounds of interest. It is advisable to test the stability of the target compounds on a small scale prior to submitting a major portion of the plant sample or crude extract to one of these potentially damaging techniques.

Figure 2 shows a procedure for isolating mixtures of crude saponins (i.e., steroidal or triterpene glycosides). The plant material is defatted with *n*-hexane, and extracted with MeOH. The MeOH extract is concentrated under vacuum, and suspended in deionized water (presaturated with *n*-butanol) and partitioned with *n*-butanol. Diethyl ether is added to the butanol partition to precipitate the saponin fraction *(20)*. Selective extraction and fractionation of plant sterols (including sapogenins, bufadienolides, and cardiac glycosides) using chemical manipulations and

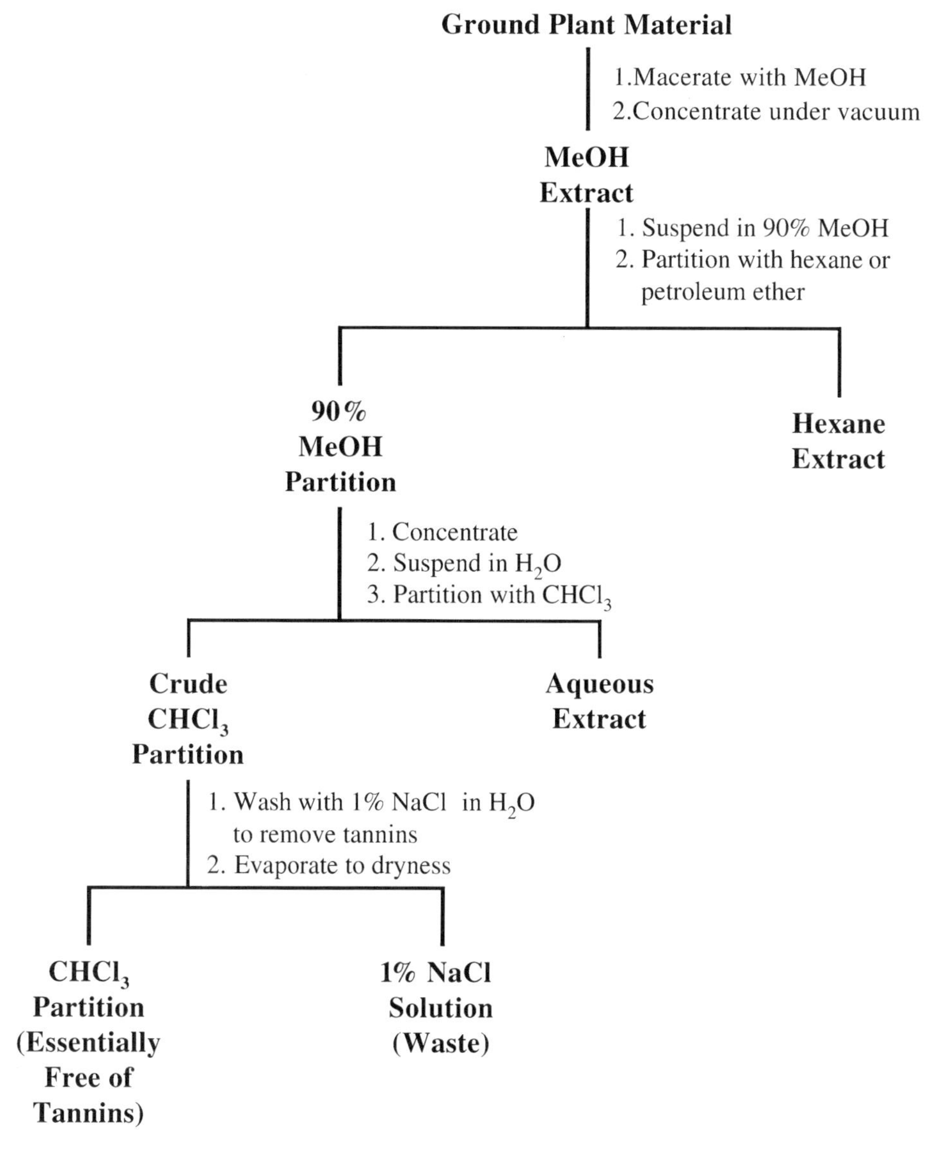

Fig. 1. General procedure for preparing extracts representing a range of polarities, including a virtually tannin-free chloroform extract ***(19)***.

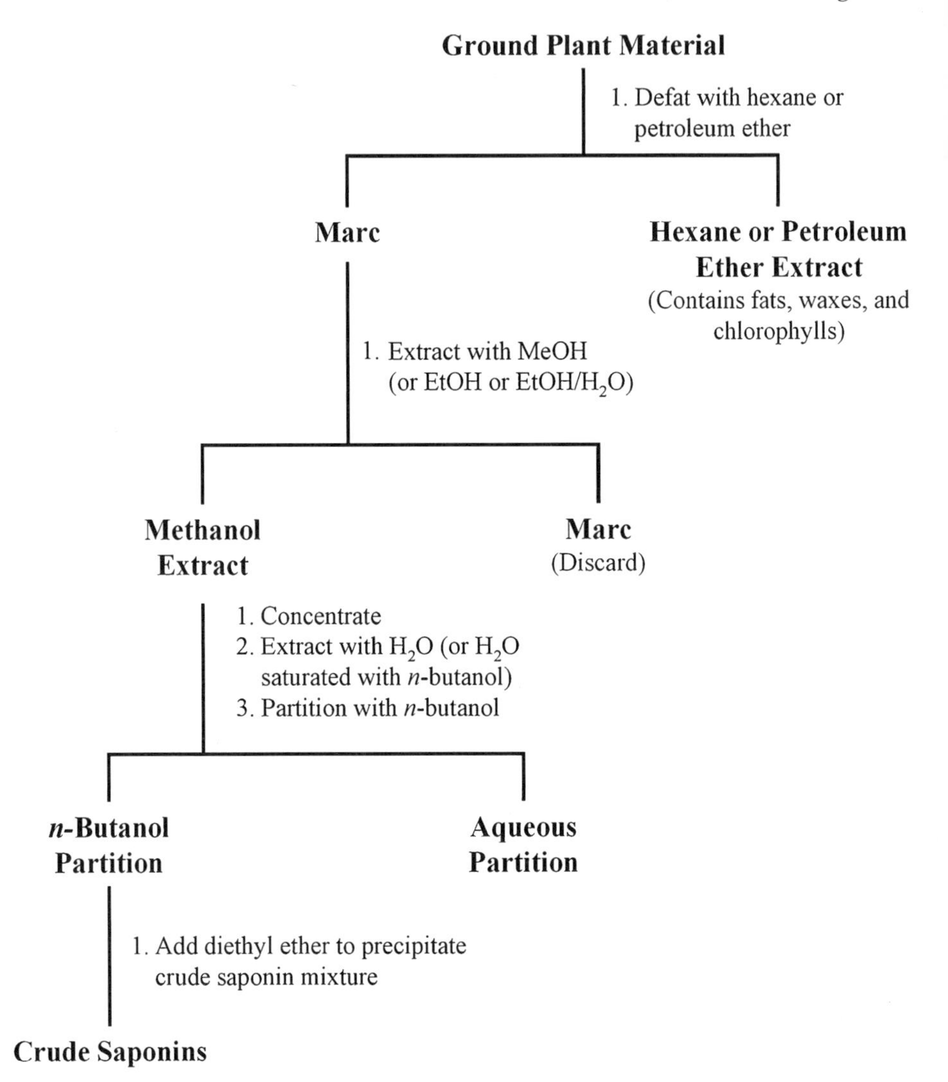

Fig. 2. General fractionation procedure to obtain a precipitate of crude saponin from plants, adapted from the literature *(20)*.

solvent/solvent partitioning have been described *(21)*. Non-alcohols can be separated from alcohols by partitioning between aqueous phthalic anhydride and organic solvent. The alcohols partition into the aqueous layer as half-phthalates and can be regenerated by treatment with sodium methoxide in MeOH. Ketone-containing sterols can be separated from non-ketones by partitioning between organic and aqueous layers with Girard's hydrazide reagents ($H_2N.NH.CO.CH_2.NR_3^+Cl^-$), and the ketones can be regenerated by acid hydrolysis *(21)*.

Alkaloids containing basic amines can be selectively extracted using a modified version of the classic "acid–base shakeout" method (**Fig. 3**).

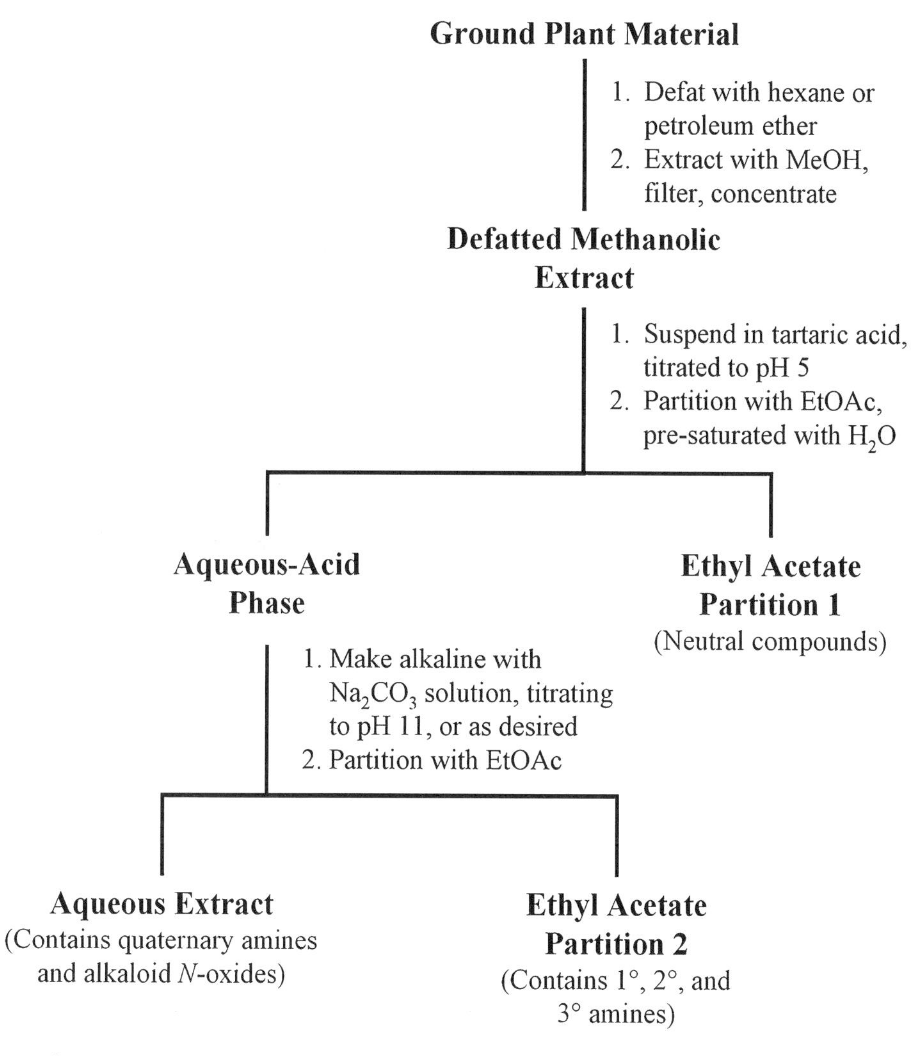

Fig. 3. General procedure to obtain alkaloidal extracts from crude plant material *(22)*.

Mineral acids and strong bases should be avoided in extracting alkaloids (and plant material in general) because of the risk of artifact formation *(22,23)* (*see* **Subheading 2.4.**). The alkaloid extraction scheme outlined employs tartaric acid and Na_2CO_3 to modify the pH during partitioning. First, the pH is adjusted to at least two units below the p*K*a values of the basic amines present, causing them to form salts, and thus to partition into the aqueous layer (along with the polar constituents, including quaternary amines and alkaloid *N*-oxides). The nonpolar constituents partition into the organic layer. After removal of the organic layer, the aqueous layer is made basic, producing the free-base form of the basic alkaloids. Upon partitioning with fresh organic solvent, these alkaloids favor the organic layer, producing an almost pure alkaloid fraction, with the exception of the *N*-oxides and quaternary amines, that would remain in the aqueous layer. This extraction scheme can also be modified for use with a series of partitioning steps, employing increasingly acidic solutions, amounting to a pH gradient for extraction of neutral (or mildly acidic) to increasingly basic alkaloids. Alkaloids can also be extracted with 10%-acetic acid in EtOH, followed by concentration under vacuum to one-quarter the original volume and precipitation of the crude alkaloid fraction by dropwise addition of NH_4OH *(24)*. Alternatively, the plant material may first be wetted with a dilute base ($NaHCO_3$, Na_2CO_3, or $CaCO_3$ solution), followed by percolation or maceration with nonpolar organic solvent.

Most compounds containing carboxylic acid functional groups are lipophilic in the presence of acidic solutions and are hydrophilic in the presence of basic solutions, although the degree of lipophilicity or hydrophilicity is modified by the nature of the rest of the molecule. Similarly, polyphenols can often be removed from an organic extract by partitioning with aqueous base (about pH 12.0). Aqueous alkali also increases the hydrophilic character of flavonoid aglycones that possess free phenolic groups. One method for extracting low polarity flavonoids and anthocyanins from leaf surfaces or flower petals is to crush the fresh plant material in a 1% solution of HCl in MeOH *(25)* (for other flavonoid extraction methods, *see* **Subheading 2.3.6.**).

2.3.6. Water Extraction

Although traditional medicines are often prepared by water extraction as infusions (steeping in hot or cold water) or decoctions (extracting in boiling water), many investigators prefer not to work with aqueous extracts. Water-soluble constituents present challenges for conventional isolation methods such as chromatography over silica gel, and water is not an ideal solvent for extraction of many biologically active plant constituents. Water extracts are difficult to concentrate on a rotary evaporator because of the relatively high boiling point of water and its high surface tension. Water preferentially extracts polar compounds, although some ostensibly hydrophobic compounds are also extracted because of co-solubility, and because the polarity of water decreases somewhat at high temperatures. Hot water extraction of *Silybum marianum* seeds by maceration in water at 85°C extracts proportionally more of the polar constituents, taxifolin and silycristin. Water at 100°C, however, extracts proportionally more of the less-polar compounds, silybinins A and B ***(26)***. Aqueous extracts may be freeze dried and re-extracted with a series of solvents in the order of increasing polarity ***(27)***.

2.4. Extraction Artifacts

Most of the solvents used in extraction of plant material have been implicated in artifact formation, either directly or because of common solvent impurities. Certain extraction procedures may also result in artifact formation. In addition to the specific examples that follow, Middleditch ***(28)*** has compiled a detailed treatment of the subject of analytical artifacts in chromatography. Phillipson and Bisset ***(29)*** reported that brucine and strychnine formed bromochloromethane and DCM adducts during extraction of a species of *Strychnos*, and concluded that traces of HCl and DCM in the $CHCl_3$ used for extraction reacted with the alkaloids to form artifacts. Phosgene ($COCl_2$), a reactive and toxic compound that rapidly forms from the decomposition of $CHCl_3$ in the presence of air and light, is known to react with alkaloids, particularly those with secondary amino groups ***(30)***, and phosgene may combine with alcohols such as MeOH, EtOH, and isopropanol and react with amines to form methyl, ethyl, or isopropyl carbamates ***(31,32)***. Several methods have been described for the removal of phosgene from chloroform (*see* **Note 14**). DCM does not readily form phosgene; however, DCM extraction of tablets of cyproheptadine hydrochloride was found to produce an *N*-chloromethyl adduct, suggesting the potential for an analogous reaction to occur with other compounds in plant extracts ***(33)***. MeOH may directly form methoxy group-containing artifacts by acting as a nucleophile with compounds containing α,β-unsaturated carbonyl groups, as in the case of a number of minor ring-A methoxylated withanolide artifacts from the aerial parts of *Physalis philadelphica* (tomatillo) recently isolated and structurally characterized in our laboratory *(34)*. Use of ammonium hydroxide with acetone may produce artifacts that give false-positive responses in alkaloid screening tests *(35)*.

Compounds containing acidic protons at chiral centers are prone to artifact formation, even under relatively mild basic conditions. A classic example is the racemization of (-)-hyoscyamine under the usual "acid–base shakeout" conditions, but other classes of compounds are also susceptible to base-catalyzed modification. For instance, lignans containing strained lactone rings may epimerize under basic conditions, as has been noted for the conversion of podophyllotoxin to picropodophyllin and for other lignans associated with *Podophyllum* in the presence of organic or inorganic base *(36)*.

In addition to taking steps to reduce the risk of artifact formation, it is important to recognize structural features that indicate that a compound isolated under certain conditions may be an artifact. It is advisable to retain a sample of the original plant material for analysis in case one of the isolates should later be suspected of being an artifact of extraction methods. The reference material can be extracted using nonreactive solvents and mild conditions, and the resulting extract analyzed using LC-MS/MS to determine if the compound is present in the original plant material *(34)*.

3. Interfering Compounds

3.1. Lipids

Fatty acids have been found to give false-positive results in assays for cyclo-oxygenase activity and in some receptor-binding assays *(37,38)*. Iodine vapor in a closed chamber will reveal lipids as brown spots. Other general thin-layer chromatography (TLC) reagents that are appropriate for detecting fatty acids are presented in Chap. 4. Extracts in which fatty acids are thought to be major components can be analyzed by ^{1}H-NMR spectroscopy to determine whether characteristic signals for fatty acids are present. In their ^{1}H-NMR spectra, an intense, broad peak at approx. δ 1.2–1.4 (aliphatic methylenes) is observed, and smaller signals around δ 5.3 (olefinic methines) and δ 0.9 (terminal methyl) are often apparent *(39)*. Liquid chromatography-mass spectrometry (LC-MS) analytical support is also useful in this regard, as many common fatty acids ionize in the negative mode. It is usually possible to infer which major fatty acids are present by examination of the molecular weights of the major components and by comparison with authentic standards *(40)*. Gas chromatography-mass spectrometry (GC-MS) can be used to identify fatty acids in a mixture, without direct comparison with standards ***(39)***.

In addition to the methods mentioned above, fatty acids and other lipid constituents can be separated from the more polar extract components by vacuum-liquid chromatography (VLC, *see* Chap. 5) using petroleum ether or *n*-hexane as eluent over silica gel. In some cases, it is convenient to use reversed-phase chromatographic columns or solid-phase extraction cartridges to remove fats and waxes from a sample. The hydrophobic materials will generally be retained until washed from the column by a strong eluent, such as a mobile phase consisting of 100% organic solvent. A simple technique that is often effective is to add a minimal amount of MeOH or MeOH–water to the lipid "solution," in effect back-extracting the desired compound or compounds. The methanolic partition can then be transferred using a pipet to a separate container, leaving the lipid portion behind.

3.2. Plant Pigments

Plant pigments are present to varying degrees in many plant parts and may interfere with chemical or biological assays. Carotenoids have been reported to interfere with electron capture detection, and a method for removing them by filtering over silver nitrate-impregnated alumina has been described ***(41)***. One or more of the following methods is usually suitable for removal of chlorophylls from a sample. Solvent–solvent partition between *n*-hexane or petroleum ether and 90% MeOH (**Fig. 1**) serves to concentrate chlorophylls in the upper layer, although this is usually not totally effective. VLC over Diaion LH-20 resin has been used effectively in our laboratory for separation of chlorophyll from other extract constituents. Diaion resin previously washed with acetone is equilibrated with 50% MeOH in deionized water. This may be accomplished by washing the resin several times in a beaker before loading the column or can be carried out after loading the resin in a column. The sample is loaded onto the column head dissolved in a minimum amount of MeOH (or other compatible solvent). A step gradient is used to develop the column. Chlorophylls typically begin to elute with 100% MeOH, and can be completely washed from the column using a 1:1 ratio of acetone and MeOH. Chlorophylls can also be removed by passage over (or standing with) activated charcoal ***(42)***, although this carries the risk of loss of important active constituents.

3.3. Vegetable Tannins

Vegetable tannins are plant polyphenols commonly found in plant extracts, often in high concentrations. Two broad groups are recognized, namely polyesters of gallic acid and/or hexahydroxydiphenic acid and their derivatives (the so-called hydrolyzable tannins), and proanthocyanidins (the so-called condensed tannins, containing oligomers of flavan diol units) ***(43)***. Many tannins give false-positive results in various biological assays, usually because of their tendency to form nonselective complexes with proteins (including enzymes, receptors, and structural proteins) through multipoint hydrogen bonding ***(44)***. Aqueous and organic extracts containing tannins inhibit enzymes, such as topoisomerases 1 and 2 (T-1 and T-2), viral reverse transcriptase, and other enzymes, leading to false-positive results ***(19,44–46)***. If tannins are thought to be responsible for the observed bioactivity of a plant extract, the extract can be tested for the presence of tannins (**Subheading 4.4.**) before and after treatment with tannin-removal methods.

To reduce the amount of tannins present in a chloroform extract, the extract may be washed with an equal volume of 1%-aqueous NaCl. The upper phase is discarded, and the chloroform phase is dried with anhydrous Na_2SO_4. Tannins can also be removed from aqueous and nonpolar extracts by passage over polyvinyl pyrrolidone (PVP) or polyamide, collagen, Sephadex LH-20, silica gel, or by precipitation by a gelatin/sodium chloride solution (5% w/v NaCl and 0.5% w/v gelatin), or caffeine ***(47,48)***. PVP/polyamide may be used in batches, or packed in a small column through which the extract is passed, with monitoring of the supernatant or eluate by the ferric-chloride reaction to determine whether the tannins have been removed, although this method is not without a risk of loss of important non-tannin constituents. Phenolic compounds can also be removed by partitioning with a 1%-NaOH solution ***(49)***.

3.4. Plasticizers

Plasticizers (*see* **Subheading 2.3.1.**) may contaminate solvents, filter papers, plastic apparatus, and chromatographic stationary phases stored in plastic containers. When some grades of MeOH are used in bulk for extraction, followed by concentration of the extract, plasticizers can become a significant portion of the extract ***(50)***. Phthalate esters are the plasticizers perhaps most likely to be encountered. Dioctylphthalate sometimes contaminates extracts from plants. Pure dioctylphthalate is a yellow oil that exhibits discernible cytotoxic activity for P-388 murine lymphocytic leukemia cells. Diethylhexylphthalate, reportedly isolated from *Aloe vera*, was found to induce apoptosis in several human cancer cell lines *(51)*. On TLC plates, dioctylphthalate shows a pink-violet spot when sprayed with concentrated sulfuric acid or concentrated sulfuric acid/ acetic acid (4:1), and heated at 110°C for 5 min with $R_f = 0.4$ (petroleum ether/ethyl acetate, 19:1) (*see* **Note 15**).

Plasticizers can be eliminated or greatly reduced by distilling solvents used for extraction and chromatography. Instructions for constructing a distillation apparatus suitable for laboratory-scale purification of solvents can be found in practical chemistry texts *(52)*. A convenient and effective plasticizer removal method that can be used for small amounts of solvent is to "distill" the solvent using a rotary evaporator. The condenser and collecting flask must be clean and free of condensate from other solvents.

4. Techniques for Detection of Phytochemical Groups in Extracts

The phytochemical screening reagents and procedures presented in this section are suitable for use without chromatographic separation. General issues related to phytochemical screening have been discussed in detail elsewhere *(53,54)*. A positive reaction should not be taken as proof of the presence of a certain type of secondary metabolite because other compound types may give false-positive reactions. Despite this caveat, these detection methods are often effective for generating hypotheses about what types of secondary metabolites may be present in a mixture of "unknowns," and of monitoring the presence of compounds of interest. In addition to these colorimetric procedures, the use of HPLC in analysis of plant extracts is becoming increasingly important, both for metabolite profiling studies and for dereplication of active constituents (*see* **Note 16**, and Chap. 12).

4.1. Alkaloids

Some alkaloids that might be present in a plant extract may not give a positive reaction because of structural idiosyncrasies. When the following reagents are used for phytochemical screening, it is often desirable to use at least two different reagents to reduce the risk of false-positive and false-negative results *(53,55,56)*. Secondary screening methods may be used to confirm the positive results (*see* **Note 17**). Several of these reagents may be used to precipitate alkaloids producing a semipurified alkaloid extract upon regeneration of the free alkaloid.

1. Mayer reagent—Solution I: dissolve 1.36 g $HgCl_2$ in 60 mL water. Solution II: dissolve 5 g KI in 10 mL water. Procedure: combine the two solutions and dilute with water to 100 mL. Add a few drops to an acidified extract solution (diluted HCl or H_2SO_4), and if alkaloids are present, a white to yellowish precipitate will appear. Care should be taken not to agitate the test system, because the precipitate may be redissolved.
2. Dragendorff reagent—Solution I: dissolve 8.0 g bismuth subnitrate [$Bi(NO_3)_3 \cdot H_2O$] in 30% w/v HNO_3. Solution II: dissolve 27.2 g KI in 50 mL water. Procedure: combine the solutions and let stand for 24 h, filter, and dilute to 100 mL with deionized water. In acid solutions, an orange-brownish precipitate will appear. The alkaloids may be recovered by treatment with Na_2CO_3 and subsequent extraction with diethyl ether. This reaction may also be performed on a filter paper or on a TLC plate by adding a drop of the reagent onto a spot of the sample.
3. Wagner reagent—Solution: dissolve 1.27 g I_2 (sublimed) and 2 g KI in 20 mL water, and make up with water to 100 mL. Procedure: a brown precipitate in acidic solutions suggests the presence of alkaloids.
4. Ammonium reineckate—Solution: add 0.2 g hydroxylamine to a saturated solution of 4% ammonium reineckate $\{NH_4[Cr(NH_3)_2(SCN)_4] \cdot H_2O\}$, and acidify with dilute HCl. Procedure: when added to extracts, a pink precipitate will appear if alkaloids are present. The precipitate is soluble in 50% acetone, which may also be used for compound recrystallization.

4.2. Sesquiterpene Lactones and Cardiac Glycosides

Many compounds containing α,β-unsaturated lactones may also give a positive reaction with the tests below. Other detection reagents for sesquiterpene lactones and cardiac glycosides may be found in the literature ***(21,53)***.

1. Kedde reagent—Solution I: dissolve 2% of 3,5-dinitrobenzoic acid in MeOH. Solution II: 5.7% aqueous KOH. Procedure: add one drop of each solution to 0.2–0.4 mL of the sample solution, and a bluish to purple color will appear within 5 min. The solution should not contain acetone, which gives a deep bluish color.
2. Baljet reagent—Solution I: dissolve 1 g picric acid in 100 mL EtOH. Solution II: 10 g NaOH in 100 mL water. Procedure: combine solutions I and II (1:1) before use and add two to three drops to 2–3 mg of sample; a positive reaction is indicated by an orange to deep red color.

4.3. Flavonoids

Certain flavonoids may cause false-positive results in assays using 3-(4,5-dimethylthiazol-2-yl)-2,5-diphenyl-tetrazolium bromide (MTT) for the quantification of cell viability. Kaempherol was found to directly reduce MTT to the colored product in a cell-free system, constituting a "positive" result that could be interpreted as indicating the presence of living cells ***(57)***. The reagents below are generally described in published monographs ***(25,53)***.

1. Shinoda test—Procedure: to an alcoholic solution of the sample, add magnesium powder and a few drops of concentrated HCl. Before adding the acid, it is advisable to add *t*-butyl alcohol to avoid accidents from a violent reaction; the colored compounds will dissolve into the upper phase. Flavones, flavonols, the corresponding 2,3-dihydro derivatives, and xanthones produce orange, pink, red to purple colors with this test. By using zinc instead of magnesium, only flavanonols give a deep-red to magenta color; flavanones and flavonols will give weak pink to magenta colors, or no color at all.
2. Sulfuric acid—Procedure: flavones and flavonols dissolve into concentrated H_2SO_4, producing a deep yellow colored solution. Chalcones and aurones produce red or red-bluish solutions. Flavanones give orange to red colors.

4.4. Other Polyphenols

Vegetable tannins are loosely defined by a combination of structural and functional characteristics as polyphenolic compounds that precipitate protein. The following detection procedures are taken from the literature ***(19,58)***.

1. Ferric chloride—Solution: dissolve 5% (w/v) $FeCl_3$ in water or EtOH. Addition of several drops of the solution to an extract produces a blue, blue-black, or blue-green color reaction in the presence of polyphenols. This is not a specific reagent for tannins, as other phenolic compounds will also give a positive result.
2. Gelatin-salt test—Procedure: for the detection of tannins in solution, dissolve 10 mg of an extract in 6 mL of hot deionized, distilled water (filtering if necessary), and the solution is divided between three test tubes. To the first is added a 1% solution of NaCl, to the second is added a 1%-NaCl and 5%-gelatin solution, and to the third is added a $FeCl_3$ solution. Formation of a precipitate in the second treatment suggests the presence of tannins, and a positive response after addition of $FeCl_3$ to the third portion supports this inference ***(19)***.

4.5. Sterols

The following may give positive responses with compounds other than sterols. Additional sterol-detecting reactions have been described in the literature ***(21,53,59)***.

1. Liebermann–Burchard test—Solution: combine 1 mL acetic anhydride and 1 mL $CHCl_3$, and cool to 0°C, and add one drop concentrated H_2SO_4. Procedure: when the sample is added, either in the solid form or in solution in $CHCl_3$, blue, green, red, or orange colors that change with time will indicate a positive reaction; a blue-greenish color in particular is observed for Δ^5-sterols, with maximum intensity in 15–30 min. (This test is also applicable for certain classes of unsaturated triterpenoids.)
2. Salkowski reaction—Procedure: dissolve 1–2 mg of the sample in 1 mL $CHCl_3$ and add 1 mL concentrated H_2SO_4, forming two phases, with a red color indicating the presence of sterols.

4.6. Saponins

When shaken, an aqueous solution of a saponin-containing sample produces foam, which is stable for 15 min or more. An additional test for saponins makes use of their tendency to hemolyze red blood cells ***(20,58)***, although this tendency may be inhibited by the presence of tannins in the extract, presumably because tannins crosslink surface proteins, thereby reducing the cell's susceptibility to lysis ***(60)***.

5. Notes

1. Small herbaceous plants are usually processed as the whole plant. This is not feasible for large plants, especially woody species, and a number of plant parts (leaf, twig, fruit, to name a few) must be collected, taking care not to destroy the entire plant. Preliminary bioassay or analytical-scale chromatographic evaluation or "dereplication" (*see* Chap. 12) of separate parts may suggest high priority parts for further investigation. It is the normal practice for our current research program on plant-derived anticancer agents to collect up to four separate anatomical plant parts, as different phytochemical profiles may occur ***(61)***. For example, in a recent investigation, the leaves and roots of *Picramnia latifolia* (Simaroubaceae) were collected in Peru and examined separately. Two previously known anthraquinones were isolated from both these plant parts. Moreover, anthraquinone glycosides and some new benzanthrones occurred exclusively in the roots, while some previously uncharacterized anthrone- and oxanthrone *C*-glycosides could be isolated from the leaves but not the roots ***(62)***. However, for the same plant

part, the local habitat conditions at the site of collection appear to be important. Thus, when samples of the bark of *Diospyros maritima* were obtained from different regions in Indonesia, naphthoquinones predominated in a sample collected in West Java at sea level, while methoxylated and methylated naphthoquinone monomers predominated in the second specimen collected at approx 350 m altitude on Lombok Island ***(63)***.

2. For maximum assuredness that a recollected sample will contain the same constituents as the original collected plant material, recollections should as far as possible be carried out in the same location, on the same plant part, at the same time of the year. However, care should be taken not to destroy all of the specimens growing at a particular collection location.
3. If herbal medicines are studied, the decision must be made whether they will be collected from the wild, purchased from a local store or wholesaler, grown in a greenhouse or research farm, or obtained through collaboration with a manufacturer. Regardless of the source, ensuring the authenticity of samples is essential. Often, it is possible to purchase minimally processed material, such as whole or coarsely milled root or leaf, which can be authenticated using macroscopic and microscopic characteristics, in addition to chemical characteristics ***(64,65)***. Monographs published by the American Herbal Pharmacopoeia (PO Box 66809, Scotts Valley, CA 95067) and other organizations (including the United States Pharmacopeia and National Formulary) are becoming available for a growing list of herbal remedies. If the purpose of the study is to compare the levels of specific active constituents or marker compounds in commercially available products, it may only be feasible to identify the commercial source of the products. In the absence of the deposit of a voucher specimen in a herbarium, some scientific journals (including the *Journal of Natural Products* and *Planta Medica*) require proof of the identity of a herbal remedy through the presentation of a standard HPLC chromatogram showing the presence of known marker compounds.
4. It is the responsibility of investigators to ensure that the necessary permits and agreements have been obtained, and that the stipulations therein are followed. Examples of the types of documentation that may be required include official permits for collecting in a national park, proof of prior-informed consent when interviews are conducted as part of the selection and collection process, and benefit-sharing agreements between official representatives from a collaborating institution or government agency in the source country and from the investigator's institution (and potentially other legal parties).
5. The International Union for Conservation of Nature and Natural Resources *Red List of Threatened Species* lists many species considered to be endangered or threatened, and may be searched online (www.redlist.org, *2003 IUCN Red List of Threatened Species*, accessed May 2004).

6. In many cases, it will be necessary to collaborate with an experienced botanist, because reproductive organs are often missing at the time of collection, and sterile features, such as leaf morphology, smell, and bark characteristics, must be used for identification. Nonetheless, even skilled botanists may only be able to complete a partial identification for some species, in which case specimens should be sent to taxonomic experts specializing in the family or genus in question.
7. Herbivore damage and microbial infection may lead to increased levels of specific defensive secondary metabolites.
8. An effective technique is to place plant material in bags made of loose-weave muslin or synthetic mesh with drawstring closures ***(16)***. The use of such bags aids in labeling samples and prevents accidental mixing of samples, and the porous fabric allows air circulation, speeding the drying process, and preventing the buildup of heat and moisture and the growth of mold. These bags can be spread on the ground to dry and can be conveniently moved under shelter, suspended from hooks, and transported, as necessary.
9. Various mills suitable for comminution of plant material are commercially available. In a typical arrangement, plant material is introduced into a chamber that contains a set of rotating knives, which chop the plant material and traject it against a screen. The degree of milling is controlled by selection of a screen that will give the desired particle size. Screen size is expressed as the number of holes per linear inch (termed "mesh size"), or as the diameter of the holes in millimeters. Milling to a 3 mm mesh size is a practical compromise between increasing surface area and minimizing adverse effects.
10. Steam distillation is used when the compounds of interest are volatile, mainly in the preparation of fragrance and flavoring agents ***(66)***. SFE using carbon dioxide as the extraction solvent shows great promise as a "green alternative" to conventional extraction methods, because it uses an essentially nontoxic solvent, exhibits minimal potential for artifact formation, and CO_2 can be obtained in high purity suitable for production of food-grade extracts (*see* Chap. 3). The addition of polarity modifiers such as EtOH, and the development of SFE equipment capable of producing pressures in excess of 600 bar, have made possible the extraction of some compounds of intermediate polarity, but polar compounds, including those with phenolic and glycosidic groups, are still poorly extracted ***(67)***. For this reason, SFE is too selective for use in general extractions of plant material for bioassay-guided isolation.
11. Simple conical glass percolators are useful for amounts of plant material of a kilogram or less. Stainless steel percolators are useful for larger sample sizes. Deagen and Deinzer ***(68)*** have described a percolator system using

55-gallon drums, a solvent pump, electronic flow regulation, and parallel ion-exchange columns for large-scale extraction of pyrrolizidine alkaloids.

12. The question of when extraction is complete can be answered empirically. It is advisable to dry an aliquot of the extract. If a large amount of residue remains after drying, or if specific chemical detection methods indicate that a significant amount of a compound (or compound class) of interest is present, additional extraction may be warranted, or it may be necessary to switch to a more suitable solvent.
13. Solvents that produce toxic vapors should be manipulated in a fume hood or other approved area. Although odor is sometimes a reliable indicator of unsafe air concentrations of solvents, the safe exposure limits of some solvents, including chloroform, can be considerably exceeded without being detected by smell ***(69)***.
14. Phosgene can build to dangerous concentrations in unsealed bottles of $CHCl_3$ ***(70)***. Although EtOH or other stabilizers are usually added to $CHCl_3$, they are not entirely effective. Distillation of $CHCl_3$ removes a portion of the stabilizer, but is not effective in removing phosgene. Phosgene can be removed by passing $CHCl_3$ over activated alumina or standing over CaOH for 18 h ***(31)***. If it is desired to remove both phosgene and EtOH from $CHCl_3$, $CHCl_3$ can be partitioned with aqueous H_2SO_4, and dried over anhydrous CaCl for 18 h, followed by distillation ***(71)***. Chloroform thus treated should be protected from extended exposure to light and air, and it is advisable to prepare only as much as is needed for immediate use, as the stabilizing EtOH has been removed.
15. The following phthalate-specific spray reagent combination is described in the literature ***(72)***. Spray solution I: add zinc powder to a 20% ethanolic resorcinol solution. Spray solution II: 2 M H_2SO_4. Spray solution III: 40% aqueous KOH solution. Procedure: spray with I, heat for 10 min at 150°C, spray with II, heat 10 min at 120°C, and spray with III. Phthalate esters will appear as orange spots on a yellow background. Spectroscopic data: UV λ_{max} 275 nm (log ε 3.17), shoulder at 282 nm; ^{1}H NMR (δ, $CDCl_3$) 7.70 (2H, dd, H-3, and H-6 aromatic protons), 7.52 (2H, dd, H-4, and H-5 aromatic protons), 4.20 (4H, dd, H-1′ and H-1″ 2-ethylhexyl residue), 1.2–1.8 (14H, m, CH and CH_2s of 2-ethylhexyl residue), 0.90 (12H, CH_3 groups); electron impact mass spectrometry (m/z) 279, 167, and 149 (100%) ***(50)***. Common ions owing to plasticizers detected in positive-ion electrospray and atmospheric pressure chemical ionization mass spectra include m/z 391, representing $[M + H]^+$ for dioctylphthalate, and m/z 550, 522, 371, and 282 for various other plasticizers from polyethylene and other sources ***(73)***.
16. Analysis of crude or semipure plant extracts by HPLC early in a phytochemical investigation, or indeed throughout the investigation is becoming

commonplace. The following are a few practical tips and considerations that have been employed in our laboratory, but it is advisable to consult with experienced HPLC users during method development to ensure compatibility with the specific system being used. The sample should be dissolved in the mobile phase, or another suitable solvent (i.e., one that will not dissolve the filter, or the column stationary phase), and filtered through a 0.45-μm filter. A precolumn is essential for protecting the more expensive HPLC column, and the precolumn should be changed regularly. The replacement schedule will depend on the level of "sticky" compounds in the extracts being analyzed. It is often desirable to employ sample preparation techniques prior to HPLC analysis of plant extracts to "clean up" the sample, and to reduce the sample complexity.

17. As a secondary screening procedure, duplicate TLC plates can be prepared and sprayed with two different alkaloid visualization reagents, such as Dragendorff and iodoplatinate. Electrospray-ionization mass spectrometry in the positive mode can be used for confirmation of the presence of alkaloids, because most alkaloids (those with an odd number of nitrogen atoms) will produce a strong even-integer ion, corresponding to the $[M + H]^+$ ion, distinguishing them from other natural product classes, which generally give an odd-integer ion. Thus, if an extract contains one or more compounds that conform to this pattern, it can be concluded that alkaloids are present. It should be noted that this method cannot be used to rule out the presence of alkaloids, as alkaloids with an even number of nitrogen atoms will produce an odd-integer ion.

References

1. Farnsworth, N. R. (1994) Ethnopharmacology and drug development, in *Ethnobotany and the Search for New Drugs*, (Prance, G. T., Chadwick, D. J., and Marsh, J., eds.), Ciba Foundation Symposium Series, vol. 185, Wiley, New York, 42–51.
2. Balick, M. J. and Cox, P. A. (1996) *Plants, People, and Culture: The Science of Ethnobotany.* Scientific American Library, New York.
3. Kinghorn, A. D. (2001) Pharmacognosy in the 21st century. *J. Pharm. Pharmacol.* **53,** 135–148.
4. Wink, M. (1999) Function of secondary metabolites, in *Functions of Plant Secondary Metabolites and Their Exploitation in Biotechnology* (Wink, M., ed.), Annual Plant Reviews, vol. 3, Sheffield Academic Press, Sheffield, UK, 1–16.
5. Caporale, L. H. (1995) Chemical ecology—a view from the pharmaceutical industry. *Proc. Natl. Acad. Sci. USA* **92,** 75–82.

6. Ayres, D. C. and Loike, J. D. (1990) *Lignans: Chemical, Biological and Clinical Properties. Chemistry and Pharmacology of Natural Products* (Phillipson, J. D., Ayres, D. C., and Baxter, H., eds), Cambridge University Press, Cambridge, UK.
7. Barclay, A. S. and Perdue, R. E., Jr. (1976) Distribution of anticancer activity in higher plants. *Cancer Treat. Rep.* **60,** 1081–1113.
8. Soejarto, D. D. (1996) Biodiversity prospecting and benefit-sharing: perspectives from the field. *J. Ethnopharmacol.* **51,** 1–15.
9. Calderon, A. I., Angerhofer, C. K., Pezzuto, J. M., et al. (2000) Forest plot as a tool to demonstrate the pharmaceutical potential of plants in a tropical forest of Panama. *Econ Bot.* **54,** 278–294.
10. McKee, T. C., Covington, C. D., Fuller, R. W. et al. (1998) Pyranocoumarins from tropical species of the genus *Calophyllum*: a chemotaxonomic study of extracts in the National Cancer Institute collection. *J. Nat. Prod.* **61,** 1252–1256.
11. Cordell, G. A., Beecher, C. W. W., and Pezzuto, J. M. (1991) Can ethnopharmacology contribute to the development of new anticancer drugs? *J. Ethnopharmacol.* **32,** 117–133.
12. Loub, W. D., Farnsworth, N. R., Soejarto, D. D., and Quinn, M. L. (1985) NAPRALERT: computer handling of natural product research data. *J. Chem. Inf. Comput. Sci.* **25,** 99–103.
13. Cragg, G. M. and Newman, D. J. (2002) Chemical diversity: a function of biodiversity. *Trends Pharmacol. Sci.* **23,** 404–405.
14. Baker, J. T., Borris, R. P., Carté B. et al. (1995) Natural product drug discovery and development: new perspectives on international collaboration. *J. Nat. Prod.* **58**, 1325–1357.
15. Cantley, M. F. (1997) International instruments, intellectual property and collaborative exploitation of genetic resources, in *Phytochemical Diversity: A Source of New Industrial Products* (Wrigley, S., ed.), Royal Society of Chemistry, Cambridge, UK, 141–157.
16. Soejarto, D. D. (1993) Logistics and politics in plant drug discovery: the other end of the spectrum, in *Human Medicinal Agents from Plants* (Kinghorn A. D. Balandrin M. F., eds.), American Chemical Society, Washington DC, 96–111.
17. Perdue, R. E., Jr. (1976) Procurement of plant materials for antitumor screening. *Cancer Treat. Rep.* **60,** 987–998.
18. Statz, D. and Coon, F. B. (1976) Preparation of plant extracts for antitumor screening. *Cancer Treat. Rep.* **60,** 999–1005.
19. Wall, M. E., Wani, M. C., Brown, D. M. et al. (1996) Effect of tannins on screening of plant extracts for enzyme inhibitory activity and techniques for their removal. *Phytomedicine* **3,** 281–285.

20. Hostettmann, K., Hostettmann, M., Marston, A. (1991) Saponins, in *Terpenoids* (Charlwood B. V., Banthorpe D. V., eds.), *Methods in Plant Biochemistry* (Dey, P. M. and Harborne, J. B., eds.), vol. **7,** Academic Press, San Diego, CA, pp. 435–471.
21. Klyne, W. (1957) *The Chemistry of the Steroids.* Wiley, New York.
22. Cordell, G. A. (1981) *Introduction to the Alkaloids: A Biogenetic Approach.* Wiley-Interscience, New York.
23. Hesse, M. (2002) *Alkaloids: Nature's Curse or Blessing*? Wiley-VCH, Weinheim, Germany.
24. Harborne, J. B. (1998) *Phytochemical Methods: A Guide to Modern Techniques of Plant Analysis.* 3 ed., Chapman & Hall, New York.
25. Markham, K. R. (1982) *Techniques of Flavonoid Identification.* Academic Press, New York.
26. Barreto, J. F. A., Wallace, S. N., Carrier, D. J., and Clausen, E. C. (2003) Extraction of nutraceuticals from milk thistle: I. Hot water extraction. *Appl. Biochem. Biotechnol.* **105–108**, 881–889.
27. Vvedenskaya, I. O., Rosen, R. T., Guido, J. E., Russell, D. J., Mills, K. A., and Vorsa, N (2004) Characterization of flavonols in cranberry (*Vaccinium macrocarpon*) powder. *J. Agric. Food. Chem.* **52,** 188–195.
28. Middleditch, B. S. (1989) *Analytical Artifacts: GC, MS, HPLC, TLC, and PC.* Journal of Chromatography Library, vol. 44, Elsevier, New York.
29. Phillipson, J. D. and Bisset, N. G. (1972) Quaternization and oxidation of strychnine and brucine during plant extraction. *Phytochemistry* **11,** 2547–2553.
30. Babad, H. and Zeiler, A. G. (1973) The chemistry of phosgene. *Chem. Rev.* **73,** 75–91.
31. Cone, E. J., Buchwald, W. F., and Darwin, W. D. (1982) Analytical controls in drug metabolic studies. II. Artifact formation during chloroform extraction of drugs and metabolites with amine substituents. *Drug Metab. Dispos.* **10,** 561–567.
32. Moody, J. D., Heinze, T. M., and Cerniglia, C. E. (2001) Fungal transformation of the tricyclic antidepressant amoxapine: identification of *N*-carbomethoxy compounds formed as artifacts by phosgene in chloroform used for the extraction of metabolites. *Biocatal. Biotransform.* **19,** 155–161.
33. Li, M., Ahuja, E. S., and Watkins, D. M. (2003) LC-MS and NMR determination of a dichloromethane artifact adduct, cyproheptadine chloromethochloride. *J. Pharm. Biomed. Anal.* **31,** 29–38.
34. Gu, J. Q., Li, W. K., Kang, Y.-H. et al. (2003) Minor withanolides from *Physalis philadelphica*: structures, quinone reductase induction activities, and liquid chromatography (LC)-MS-MS investigation as artifacts. *Chem. Pharm. Bull.* **51,** 530–539.

35. Housholder, D. E. and Camp, B. J. (1965) Formation of alkaloid artifacts in plant extracts by the use of ammonium hydroxide and acetone. *J. Pharm. Sci.* **54,** 1676–1677.
36. Hartwell, J. L. and Schrecker, A. W. (1958) The chemistry of *Podophyllum*. *Progr. Chem. Org. Nat. Prod.* **15,** 83–166.
37. Ingkaninan, K., Ijzerman, A. P., Taesotikul, T., and Verpoorte, R. (1999) Isolation of opioid-active compounds from *Tabernaemontana pachysiphon* leaves. *J. Pharm. Pharmacol.* **51,** 1441–1446.
38. Ringbom, T., Huss, U., Stenholm, A. et al. (2001). COX-2 inhibitory effects of naturally occurring and modified fatty acids. *J. Nat. Prod.* **64**, 745–749.
39. Su, B.-N., Cuendet, M., Farnsworth, N. R., Fong, H. H. S., Pezzuto, J. M., and Kinghorn, A. D. (2002) Activity-guided fractionation of the seeds of *Ziziphus jujuba* using a cyclooxygenase-2 inhibitory assay. *Planta Med.* **68,** 1125–1128.
40. Su, B. N., Jones, W. P., Cuendet, M. et al. (2004). Constituents of the stems of *Macrococculus pomiferus* and their inhibitory activities against cyclooxygenases-1 and -2. *Phytochemistry*, **65,** 2861–2866.
41. Holmes, D. C. and Wood, N. F. (1972) Removal of interfering substances from vegetable extracts prior to the determination of organochlorine pesticide residues. *J. Chromatogr.* **67,** 173–174.
42. Lee, I. S., Ma, X. J., Chai, H. B. et al. (1995) Novel cytotoxic labdane diterpenoids from *Neouvaria acuminatissima. Tetrahedron* **51**, 21–28.
43. Haslam, E. (1989) *Plant Polyphenols: Vegetable Tannins Revisited. Chemistry and Pharmacology of Natural Products* (Phillipson, J. D., Ayres, D. C., Baxter, H., eds.), Cambridge University Press, Cambridge, UK.
44. Spencer, C. M., Cai, Y., Martin, R. et al. (1988). Polyphenol complexation—some thoughts and observations. *Phytochemistry* **27,** 2397–2409.
45. Tan, G. T., Pezzuto, J. M., Kinghorn, A. D., and Hughes, S. H. (1991) Evaluation of natural products as inhibitors of human immunodeficiency virus type 1 (HIV-1) reverse transcriptase. *J. Nat. Prod.* **54,** 143–154.
46. Tan, G. T., Pezzuto, J. M., Kinghorn, A. D. (1992) Screening of natural products as HIV-1 and HIV-2 reverse transcriptase (RT) inhibitors, in *Natural Product Antiviral Agents* (Chu, C. K. and Cutler, H. G., eds.), Plenum, New York, pp. 195–222.
47. Wall, M. E., Taylor, H., Ambrosio, L., and Davis, K. (1969) Plant antitumor agents. III. A convenient separation of tannins from other plant constituents. *J. Pharm. Sci.* **58,** 839–841.
48. Hagerman, A. E. and Butler, L. G. (1980) Condensed tannin purification and characterization of tannin-associated proteins. *J. Agric. Food Chem.* **28,** 947–952.
49. Gosmann, G., Guilliaume, D., Taketa, A. T. C., and Schenkel, E. P. (1995) Triterpenoid saponins from *Ilex paraguariensis*. *J. Nat. Prod.* **58,** 438–441.

50. Banthorpe D. V. (1991) Classification of terpenoids and general procedures for their characterization, in *Terpenoids* (Charlwood, B. V. and Banthorpe, D. V., eds), *Methods in Plant Biochemistry* (Dey, P. M. and Harborne, J. B., eds.), vol. 7, Academic Press, San Diego, pp. 1–41.
51. Lee, K. H., Hong, H. S., Lee, C. H., and Kim, C. H. (2000) Induction of apoptosis in human leukaemic cell lines K562, HL60 and U937 by diethylhexylphatalate isolated from *Aloe vera* Linne. *J. Pharm. Pharmacol.* **52,** 1037–1041.
52. Vogel, A. I., Furniss, B. S., Hannaford, A. J., Rogers, V., Smith, P. W. G., and Tatchell, A. R. (1978) *Vogel's Textbook of Practical Organic Chemistry.* 4 ed., Longman, New York.
53. Farnsworth, N. R. (1966) Biological and phytochemical screening of plants. *J. Pharm. Sci.* **55,** 225–276.
54. Marini-Bettolo, G. B., Nicoletti, M., Patamia, M., Galeffi, C., and Messana, I. (1981) Plant screening by chemical and chromatographic procedures under field conditions. *J. Chromatogr.* **213,** 113–127.
55. Furgiuele, A. R., Farnsworth, N. R., and Buckley, J. P. (1962) False-positive alkaloid reactions obtained with extracts of *Piper methysticum*. *J. Pharm. Sci.* **51,** 1156–1162.
56. Farnsworth, N. R., Pilewski, N. A., and Draus, F. J. (1962) False-positive alkaloid reactions with the Dragendorff reagent. *Lloydia* **25,** 312–319.
57. Bruggisser, R., von Daeniken, K., Jundt, G., Schaffner, W., and Tullberg-Reinert, H. (2002) Interference of plant extracts, phytoestrogens and antioxidants with the MTT tetrazolium assay. *Planta Med.* **68,** 445–448.
58. Wall, M. E., Krider, M. M., Krewson, C. F., et al. (1954) Steroidal sapogenins. VII. Survey of plants for steroidal sapogenins and other constituents. *J. Am. Pharm. Assoc., Sci. Ed.* **43,** 1–7.
59. Dinan, L., Harmatha, J., and Lafont, R. (2001) Chromatographic procedures for the isolation of plant steroids. *J. Chromatogr. A* **935,** 105–123.
60. Segelman, A. B., Farnsworth, N. R., and Quimby, M. W. (1969) Biological and phytochemical evaluation of plants. III. False-negative saponin test results induced by the presence of tannins. *Lloydia* **32,** 52–58.
61. Kinghorn, A. D., Farnsworth, N. R., Soejarto, D. D. et al. (2004). Novel strategies for the discovery of plant-derived anticancer agents. *Pharm. Biol.*, **415,** 53–67.
62. Diaz, F., Chai, H. B., Mi, Q. et al. (2004) Anthrone and oxanthrone C-glycosides from *Picramnia latifolia* collected in Peru. *J. Nat. Prod.* **67,** 352–356.
63. Gu, J. Q., Graf, T. N., Lee, D. et al. (2004). Cytotoxic and antimicrobial constituents of the bark of *Diospyros maritima* collected in two geographical locations in Indonesia. *J. Nat. Prod.* **67,** 1156–1161.

64. Stahl, E. (1973) *Drug Analysis by Chromatography and Microscopy. A Practical Supplement to Pharmacopoeias.* Ann Arbor Science Publishers, Ann Arbor, MI.
65. Wagner, H. and Bladt, S. (1996) *Plant Drug Analysis: A Thin Layer Chromatography Atlas,* 2 ed., Springer-Verlag, New York.
66. Starmans, D. A. J. and Nijhuis, H. H. (1996) Extraction of secondary metabolites from plant material: a review. *Trends Food Sci. Technol.* **7,** 191–197.
67. Hamburger, M., Baumann, D., and Adler, S. (2004) Supercritical carbon dioxide extraction of selected medicinal plants—effects of high pressure and added ethanol on yield of extracted substances. *Phytochem. Anal.* **15,** 46–54.
68. Deagen, J. T. and Deinzer, M. L. (1977) Improvements in the extraction of pyrrolizidine alkaloids. *Lloydia* **4,** 323–328.
69. Amoore, J. E. and Hautala, E. (1983) Odor as an aid to chemical safety: odor thresholds compared with threshold limit values and volatilities for 214 industrial chemicals in air and water dilution. *J. Appl. Toxicol.* **3,** 272–290.
70. Turk, E. (1998) Phosgene from chloroform. *Chem. Eng. News*, Mar. 2, 6.
71. Anonymous (1974) *Laboratory Techniques Manual.* vol. 1, Department of Chemistry, Massachusetts Institute of Technology, Cambridge, MA, pp. 15. 40–15. 41.
72. Peereboom, J. W. C. (1960) The analysis of plasticizers by micro-adsorption chromatography. *J. Chromatogr.* **4,** 323–328.
73. Tong, H., Bell, D., Tabei, K., and Siegel, M. M. (1999) Automated data massaging, interpretation, and e-mailing modules for high throughput open access mass spectrometry. *J. Am. Soc. Mass Spectrom.* **10,** 1174–1187.

14

Isolation of Marine Natural Products

Wael E. Houssen and Marcel Jaspars

Summary

Marine organisms offer a wealth of chemically diverse compounds that have been evolutionarily preselected to modulate biochemical pathways. Many industrial and academic groups are accessing this source using advanced technology platforms. The purpose of this chapter is to offer some practical guidance in the process of extraction and isolation of marine natural products. Special attention is given to the isolation procedures adapted to the physical and chemical characteristics of the compounds isolated. Recent advances in isolation technology, including automation and direct integration into high-throughput screening systems, are also discussed.

Key Words: Marine natural products; marine invertebrate collection; extraction; fractionation; purification; isolation; chromatography; automated fractionation; HPLC-SPE coupling.

1. Introduction

Marine organisms have evolved to colonize a large variety of ecological niches. To cope with such a wide range of habitats, they have developed diverse secondary metabolic pathways that produce a vast number of unusual chemical moieties to accommodate their lifestyles. These compounds encompass a wide variety of chemical classes, including terpenes, shikimates, polyketides, acetogenins, peptides, alkaloids, and many unidentified and uncharacterized structures. In the past decade alone, the structures of over 5000 marine natural products have been published *(1–10)*. Many of these compounds have proved their potential in several

From: *Methods in Biotechnology, Vol. 20, Natural Products Isolation, 2nd ed.*
Edited by: S. D. Sarker, Z. Latif, and A. I. Gray © Humana Press Inc., Totowa, NJ

fields, particularly as potential therapeutic agents for a variety of diseases. A few decades ago, the purification of almost any natural product was an enormous undertaking. As a result of advances in technology, this process has become almost routine. Moreover, the trend toward faster and automated separations has generated widespread interest, especially in the pharmaceutical industry.

Several reviews ***(11–14)*** have discussed the techniques involved in the isolation of marine natural products. This chapter is organized to highlight the hurdles encountered in the isolation process, and the most recent technologies and strategies applied both on laboratory and industrial scales. A brief discussion on the various applicable chromatographic techniques is also included. However, more details on the basic chromatography theory and applications can be found in Chapters 4–9. The last section in this chapter is devoted to a discussion of automated fractionation and the direct integration into high-throughput screening (HTS) systems. Notes with more specific information on certain topics are also provided where deemed necessary, to be of further assistance to the interested readers.

1.1. Hurdles in the Isolation of Marine Natural Products

Some factors that can complicate the isolation of marine natural products have been discussed in detail in the review on isolation of marine natural products by Amy Wright ***(14)***. We re-emphasize some of these factors beside others that may add extra challenges to the separation process.

1.1.1. Taxonomic Uncertainty

Taxonomic information can facilitate literature searches on compounds reportedly produced by the species under investigation and of course their methods of purification. This has some impact on the selection of the best purification scheme for new metabolites. Taxonomic identification of marine organisms is challenging, and incorrect or incomplete taxonomic assignments can lead to difficulties if assumptions are made about the chemistry that an organism may contain. The use of chemotaxonomy to predict the types of compounds that an organism may contain is not always successful.

1.1.2. Small Quantities of Metabolites

The presence of a highly potent metabolite in trace amounts can complicate the extraction and isolation process. A large amount of the organism is needed for the isolation of the active metabolite at the level that can facilitate the subsequent structure elucidation (*see* Chap. 17).

One example is the isolation of only 10.7 mg of the highly potent antitumor macrolide spongistatin 4 (**Fig. 1**) from about 2.5 ton of the South African marine sponge *Spirastrella spinispirulifera*. Trials to reduce the sponge biomass were extremely laborious. At one point in this endeavor, it was necessary to use high-performance liquid chromatography (HPLC) columns that were nearly 3 m in length and 15 cm in diameter ***(15)***.

Another example is the isolation of nearly 1 mg of the exceptionally important antitumor peptide dolastatin 10 (**Fig. 2**). Almost 2 ton of the sea hare *Dolabella auricularia* was collected off the island of Mauritius in the Indian Ocean for this purpose. The collection step needed over 10 years. The separation process was extremely challenging. The simplest way of isolation involved about 20,000 fractions and some 23 separate chromatographic steps using various techniques. Subsequently, it became apparent that dolastatin 10 might be produced by a cyanobacterium growing on *D. auricularia* ***(16,17)***.

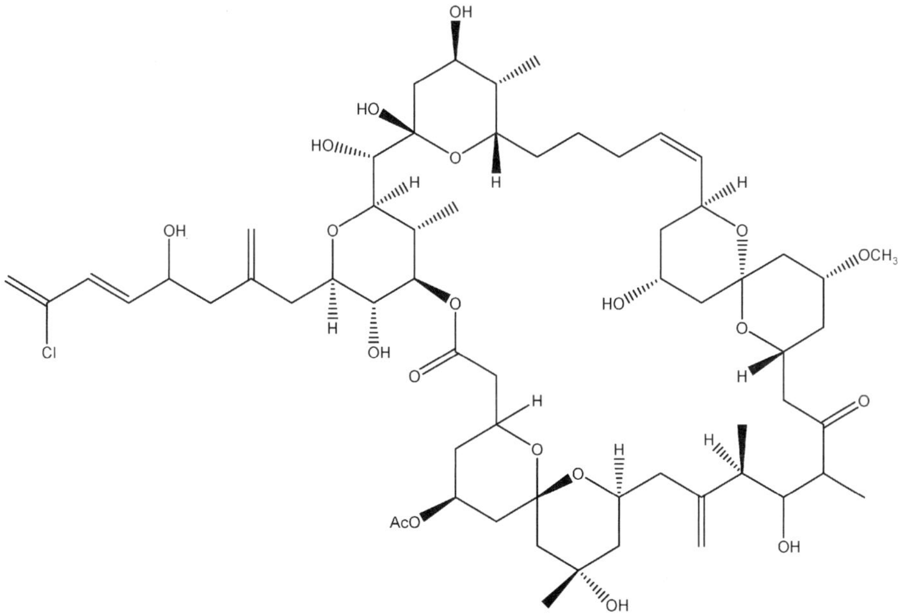

Fig. 1. Structure of spongistatin 4.

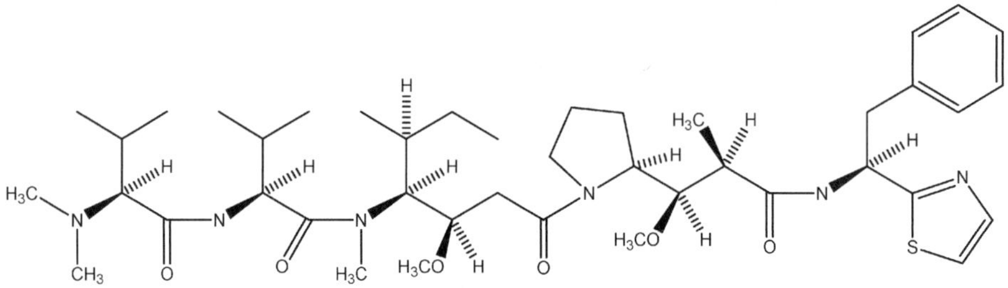

Fig. 2. Structure of dolastatin 10.

1.1.3. Instability of Metabolites

Marine extracts may contain extremely labile compounds. Decomposition of these compounds may occur at any step during the purification process. Heat, light, air, and pH are among other factors that may lead to the degradation of compounds. Materials used for separation may also activate some reactions. Alumina can catalyze the aldol condensation, rearrangement, hydration and dehydration reactions, while silica can enhance oxidation, rearrangement, and *N*- and *O*-demethylation. Some solvents such as acetone, methanol (MeOH), ethylene glycol, and dimethylformamide (DMF) may give rise to adducts ***(18)***. The slightly acidic nature of some NMR solvents (e.g., $CDCl_3$) may cause degradation of highly pH-sensitive compounds.

1.1.4. Purification of Water-Soluble Compounds: Effects of High Water and Salt Content

There are many difficulties associated with the isolation and purification of water-soluble compounds (*see* Chap. 16). Because target compounds are extremely polar, aqueous media or strongly polar solvents such as MeOH must be used for extraction. In the case of aqueous solutions, an inevitable problem is bacterial and fungal growth, which often degrades the active components or gives false results in bioassays because of endotoxins produced by the micro-organisms. The concentration of aqueous extracts also creates problems because of the high temperature of evaporation of water. Moreover, aqueous extracts often contain surface-active agents. These surfactants can cause foaming and bumping during the concentration process. The abundance of salts carried over from seawater into aqueous extracts makes the isolation process more difficult. In most cases, the extraction and fractionation of water-soluble compounds necessitate the use of buffer solutions. However, the separation of buffer salts from the compound is not an easy task ***(11,14,19)***.

1.1.5. Compounds Lacking UV Chromophores

Ultraviolet (UV) detection is the preferred detection technique for HPLC analysis of natural products because of its ease of use and high sensitivity. However, one shortcoming of UV detectors is the inability to detect compounds that lack UV chromophores. Moreover, some of the solvents used in normal-phase chromatography are themselves strong absorbers of UV light, which means that low detector wavelength settings are not possible (*see* **Table 5** for the UV cutoff values of different solvents).

The refractive index (RI) detector was, until recently, the only available alternative to UV detection. However, RI detectors are much less sensitive and are limited to isocratic elution. Gradient elution involves the mixing of solvents of differing RI, thus giving rise to large baseline drifts ***(20)***. These drawbacks make the detection of low concentrations of nonchromophoric metabolite notoriously difficult.

1.1.6. Cost and Time Effectiveness

Until recently, purification of marine natural products was a time-consuming, tedious, and an expensive process. Recent developments in HTS have created a new bottleneck for drug discovery from nature. A steady supply with a great number of structurally interesting samples in sufficient concentration is a necessity to operate HTS systems optimally. The integration of a marine natural product library into the HTS systems was, not so long ago, completely unfeasible.

2. Chromatography

This section gives a brief description of the different chromatographic techniques commonly applied in the isolation of marine natural products. An outline of the properties of the most commonly used solvents and stationary phases is also included. The ultimate goal of this section is to facilitate the process of choosing the proper technique, solvent, and stationary phase.

All chromatographic techniques involve the distribution of extract components between two phases—a moving mobile phase that is passed over an immobile stationary phase. Separation depends on the difference in affinity of the components toward stationary phase and mobile phase.

According to the nature of the mobile phase, chromatography can be classified into gas chromatography (GC), liquid chromatography (LC), and supercritical fluid chromatography (SFC). GC and SFC are widely used analytical techniques but cannot be used for preparative isolation of natural products, and are not discussed in this chapter. LC, however, is widely used and comes in many different forms.

2.1. Classification of LC

LC can be classified in a number of ways ***(21)***:

1. Classification according to the mode of operation.
   Chromatography can be carried out in four main forms: countercurrent chromatography (CCC), thin-layer chromatography (TLC), solid-phase extraction (SPE), and column chromatography (*see* **Subheading 2.3.**, for details).
2. Classification according to the mode of separation (**Fig. 3**).
   There are five mechanisms by which separation can occur and more than one may be responsible during a given separation. These modes of separation are discussed here.

2.1.1. Adsorption

Adsorption chromatography is based on the ability of solute molecules to physically interact with the stationary phase (also *see* Chap. 5). The nature of the interactions can be hydrogen bonding, van der Waals, or dipole–dipole. The stronger the interaction is, the longer the solute will

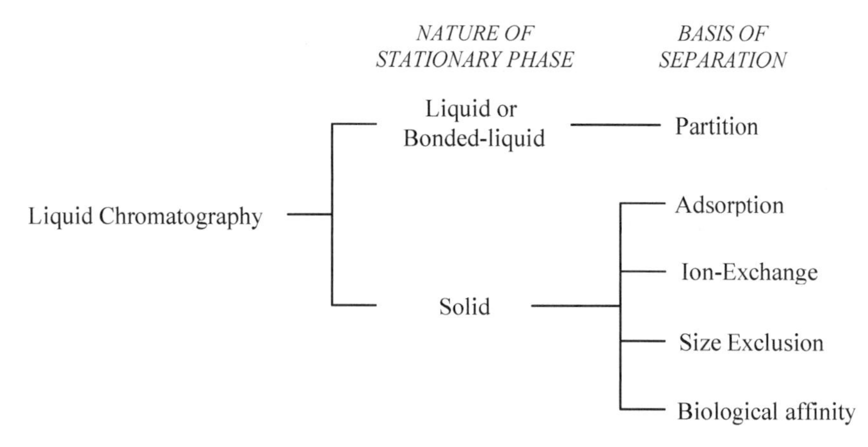

Fig. 3. Classification of LC systems.

be retained on the stationary phase. Inappropriate choices of stationary phases can lead to irreversible adsorption of solutes onto the packing material, or at least very poor recovery. Moreover, mobile-phase molecules compete with solute molecules for adsorption sites. The more strongly the mobile phase interacts with the adsorbent, the faster the solute will be eluted. Because solutes must interact with the stationary phase to be adsorbed, the greater the surface area offered by the stationary phase, the greater the number of possible interactions. Thus, adsorbents that offer a high ratio of surface area to packing volume give rise to better separation. Surface area is a function of the particle size, pore size, and the degree of porosity. The smaller the particle size of the adsorbent, the higher the surface area that will be available. Very small particle sizes are generally only reserved for HPLC, as they give rise to substantial backpressure when a mobile phase is passed through. The porosity of a particle reflects the ratio of the volume of surface pores to the total volume of the particle. Compounds larger than the pore size will not bind efficiently because they will be unable to penetrate the pore. Fortunately, pore sizes can vary tremendously from approx 50 nm to macroscopic size. **Table 1** contains detailed information on the most commonly used adsorbents.

2.1.2. Partition

Partition chromatography is based on the ability of solutes to distribute between two liquid phases. Most commonly, the liquid stationary phase is chemically bound to an inert support, e.g., porous silica gel to give a "bonded phase." This involves the formation of a hydrolytically stable bond, between the surface silanol group of a silica support and a chlorosilane. The silane usually carries an organic functional group and this is in effect, the liquid stationary phase ***(21)***. Bulky organo-silane ligands cannot react with all available silanol groups because of the steric hindrance. The unreacted silanol groups can interfere with the separation process and produce undesirable effects such as reaction, adsorption, tailing, and so on. To minimize these secondary interactions, the bonded phase is typically subjected to an "endcapping" reaction in which the residual silanols are methylated, typically with trimethylsilyl (TMS) groups. Therefore, pore size of the silica gel and the amount of liquid phase or functional groups bonded to the silica and the extent of endcapping are among the factors to be considered in the selection of the proper bonded phase. Some

Table 1
Stationary Phases Commonly Used in Adsorption Chromatography

Adsorbent	Notes
Silica 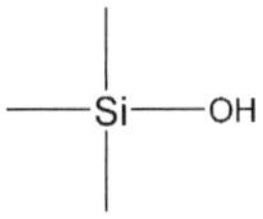	Normal phase (stationary phase is more polar than mobile phase). Polar compounds are retained while nonpolar ones are not. Polar solvents have better elution power. Cannot be used with aqueous solvents as water deactivates the surface. Soluble at pH values higher than 7.0. Reactive—some natural products are unstable on it.
Alumina Al_2O_3	Normal phase. Acidic, basic, and neutral forms are available to be used in the separation of acidic, basic, and relatively nonpolar compounds, respectively. Quite reactive.
Styrene–divinyl benzene polymer 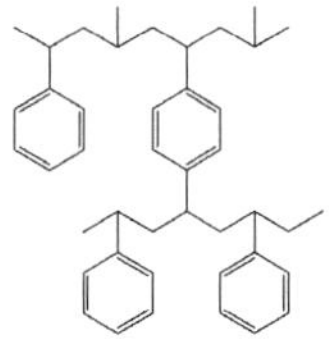	Reversed phase (stationary phase is less polar than the mobile phase). Stable at low and high pH values. Less reactive (problems owing to exposed silanol groups are avoided). Has higher capacity for polar compounds than C_{18} because of its higher carbon percentage. Provide lower resolution than C_{18} because of its bigger bead size. Ideally used for desalting and adsorption–elution of natural products from aqueous extracts. More economic than C_{18}. Bromination makes the polymer more retentive.

of the properties of commonly used bonded phases are summarized in **Table 2**.

However, a liquid can also act as a stationary phase even when it is not bound to a support, as in the case of CCC (*see* Chap. 7).

2.1.2.1. Ion-Pairing Reagents and pH Effects

Highly water-soluble ionic compounds can also be separated using nonpolar stationary phases, e.g., octadecyl silica. This can be achieved by altering the pH of the mobile phase to suppress the ionization of the compounds so that they can be retained as neutral species on the

Table 2
Bonded Phases Commonly Used in Marine Natural Product Isolation

Stationary phase	Notes
C_{18} (Octadecyl)[a] —Si—$C_{18}H_{37}$	Reversed phase (common solvents are water, MeOH, acetonitrile, and THF). The most retentive for nonpolar (hydrophobic) compounds. Compounds are eluted in order of increasing hydrophobicity (water is the weakest eluent). Some manufacturers offer ODS that can tolerate up to pH 10.0
C_8 (Octyl)[a] —Si—C_8H_{17}	Reversed phase. Similar selectivity to C_{18} but less retentive.
Ph (Phenyl-hexyl) —Si—$(CH_2)_6$—Ph	Reversed phase. Aromatic compounds are retained more.
CN Cyanopropyl —Si—$(CH_2)_3$·CN	Can be used in normal- and reversed-phase modes. Slightly polar. Unique selectivity for polar compounds.

[a]Other alkyl-bonded phases of 1, 2, 3, 4, 5, 6, and 12 carbon atoms are also available. The degree of hydrophobicity and thus retentiveness varies proportionally with the length of the alkyl chain (*see* **Note 1**).

hydrophobic surface of the stationary phase. An alternative way is to add a suitable ion-pairing reagent to the mobile phase. Ion-pairing reagents are usually ionic compounds that contain a hydrocarbon chain imparting hydrophobicity. They can act either by interacting with the sample, forming a reversible neutral ion pair that can be retained on the stationary phase, or by interacting with the stationary phase via its hydrophobic part, forming a stationary phase with charged groups to which ionic compounds can bind reversibly *(21)*. Trifluoroacetic acid (TFA), pentafluoropropionic acid (PFPA), and heptafluorobutyric acid (HBA) are examples of anionic ion-pairing reagents while triethylamine (TEA) is an example of a cationic one. Additional advantages of TFA and TEA include their high volatility (allowing for rapid removal) and compatibility with mass spectrometric analysis *(22)* (*see* **Note 2**).

2.1.3. Ion-Exchange

Ion-exchange chromatography (IEC) is applicable to the separation of ionic species or components that are ionizable at a given pH (*see* Chap. 6). It is commonly used in peptide purification because virtually all of these macromolecules carry surface charges allowing for their adsorption onto a solid support. An advantage of the technique is the fact that biological activity is almost always preserved because the mobile-phase composition is typically aqueous ***(22)***. On the other hand, the use of the technique often introduces a large amount of inorganic buffer salts and thus it requires subsequent desalting steps.

The stationary phase consists of water-insoluble particles bearing covalently bound positively or negatively charged functional groups. Free counterions of the opposite charge are associated with these functional groups. Ionic species in the sample can exchange with these counterions and bind to the stationary phase. Separation is achieved because of the difference in the affinity of ionic components toward the stationary phase. For more details on the mechanism of the ion-exchange process, the reader should consult the highly authoritative review written by Claude Dufresne ***(23)***.

2.1.3.1. Support Matrices

The support matrices making up the stationary-phase particles can be polystyrene resin, carbohydrate polymer, or silica gel. The nature of the support matrix has an impact on the column's flow characteristics, the porosity of the stationary-phase particles, and their resistance to mechanical pressure. However, the exclusion limit (pore size) provided by the manufacturer should be noted, because molecules larger than these limits will not bind efficiently to the stationary phase owing to their limited pore accessibility. **Table 3** summarizes the physicochemical properties of the most common IEC support matrices.

2.1.3.2. Functional Groups

Functional groups attached to the support matrix can be positively or negatively charged to give anion- or cation-exchange chromatography, respectively. The number of these functional groups per unit volume of the matrix determines the column capacity. The p*K*a value of the functional group determines the strength of the exchanger. In anion-exchange chromatography, full ionization occurs at pH values 2.0 units below the p*K*a while full neutralization occurs at pH values 2.0 units above the p*K*a.

Table 3
Support Matrices in IEC

Matrix	Notes
Styrene–divinyl benzene polymer	Hydrophobic surface. Rigid, offers very good mechanical strength. Can tolerate pH range 1.0–14.0. Can interact with (adsorb) solutes leading to poor recovery. These are called "backbone interactions".
Carbohydrate polymer	Examples are crosslinked dextran and crosslinked agarose. Swells readily in water resulting in gel that has very good flow properties. Does not show backbone interactions. Hydrophilic, does not lead to protein denaturation.
Silica gel	Does not swell in water. Offers very small particle sizes—suitable for HPLC. Unstable at pH values above 7.5.

The reverse is true for cation-exchange chromatography. The p*K*a values for various functional groups used in IEC are shown in **Table 4**. It can be seen that strong exchangers are charged over the full pH range, while weak exchangers are only charged over a limited pH range. The p*K*a value of the analyte should be considered as well. Organic acids are negatively charged when the pH of the solution is above the acid's p*K*a, and amines are positively charged at pH values below the base's p*K*a value. However, the pH of the mobile phase should facilitate the ionization of both sorbent functional groups and analyte.

It should be noted that only weak ion exchangers should be utilized to capture strong ionic species (e.g., sulfonic acids and quaternary amines).

Table 4
Functional Groups in IEC

Function group	Notes
Sulfonic acid ($-SO_3H$)	Strong cationic—p*K*a < 1
Carboxylic acid ($-CO_2H$)	Weak cationic—p*K*a $\cong$ 4–5
Quaternary amine ($-NR_3^+$)	Strong anionic—p*K*a > 13
Tertiary amine ($-NR_2$)	Weak anionic—p*K*a $\cong$ 8

This is because the recovery of these analytes is often low on strong ion exchangers owing to the inability to neutralize the quaternary amines or the sulfonic acid groups on either the sorbent or the analyte. On the other hand, analytes containing weak amines or carboxylic acid moieties can be extracted with either weak or strong ion exchangers, as pH adjustment can be used to neutralize the charge on the analyte to bring about elution ***(24)***. However, weak ion exchangers are generally preferred in natural product work, as elution may be achieved through mild, nondestructive conditions.

2.1.3.3. Counterions and Ionic Strength

Counterions in the matrix and the mobile phase compete with sample ions for charged binding sites on the sorbent surface. Therefore, the presence of counterions that have a greater affinity for the sorbent than the analytes or the presence of a high concentration (ionic strength) of virtually any counterion can impede the binding of the analytes to the stationary phase ***(24)***.

Anions can be ranked according to their affinity for anion-exchange matrix as follows:

$$hydroxide < fluoride < acetate < bicarbonate, formate < chloride < phosphate, citrate$$

In a similar manner, cations can be ranked in terms of their affinity to cation exchangers as follows:

$$lithium < hydrogen < sodium < ammonium < potassium < magnesium < calcium$$

As a general rule of thumb, the retention of analytes is facilitated by loading the sample in a low ionic strength buffer comprising low affinity or weak displacer counterions at the proper pH. Elution is promoted by high ionic strength salts and buffers containing strong displacer counterions and/or by altering pH to fully neutralize the charged groups on either the sorbent or the analyte.

2.1.3.4. Tentacle Ion Exchangers

In most ion exchangers, the functional groups are fixed directly on the surface of the matrix. Only a certain number of ionic groups can be attached because of the limited inner surface of the matrix. With tentacle ion exchangers, however, the exchanger groups are covalently anchored to the matrix via linear polymer chains. Therefore, exchange capacity can be increased significantly. Another advantage is that tentacle ion exchangers do not show any backbone interactions as the matrix of the stationary phase is hidden. Hence, tentacle ion exchangers are much more suitable for the separation of large protein molecules where the risk of denaturation and loss of biological activity is significant.

2.1.4. Size Exclusion

Size-exclusion chromatography (SEC), also known as gel filtration or gel permeation chromatography, separates molecules according to their molecular size (*see* Chap. 5). The stationary phase comprises porous particles in which the pore size is strictly controlled. As the mobile phase flows over and through these particles, it carries along with it solutes that, depending on their size and shape, may flow into and out of the pores. Molecules larger than the pore size cannot enter the pores and elute together as the first peak in the chromatogram, a phenomenon called "total exclusion." Smaller molecules that can enter the pores will have an average residence time in the particles that depends on the molecules' size and shape. Different molecules, therefore, have different total transit times through the column. This portion of the chromatogram is called the selective permeation region. The smallest molecules can penetrate the smallest pores and thus have the longest residence time on the column and elute together as the last peak in the chromatogram. This last peak determines the total permeation limit. SEC is a widely used technique for the separation of peptides and proteins because harsh elution conditions are not required. Moreover, it can be used successfully for desalting.

The most commonly used size-exclusion sorbents are polydextran gels, e.g., Sephadex® produced by Pharmacia. These gels are available commercially in a bead form, which has to be immersed in the mobile phase for several hours before use to allow swelling to take place. Sephadex is manufactured by crosslinking of dextran with epichlorohydrin and is available in a range of grades with different pore sizes (*see* Chap. 5).

Sephadex G gels are hydrophilic and their use is limited to aqueous solutions. Thus, they are ideal for fractionation of water-soluble mixtures. Although their main mode of separation is gel filtration, additional adsorption mechanisms may exist, giving rise to good resolution.

Sephadex LH-20 is a hydroxypropylated G-25. The introduction of hydroxypropyl groups docs not alter the number of hydroxyl groups but increases the ratio of carbon to hydroxyl. The resultant gel has, therefore, both hydrophilic and lipophilic properties. The added lipophilicity allows this gel to be used with nonaqueous solvents. Sephadex LH-20 is commonly used for fractionation of organic-soluble natural products. When a single solvent is used, the separation mainly occurs by gel filtration mode. When a solvent mixture is used, the gel will take up predominantly the polar component of the solvent mixture. This results in a two-phase system with stationary and mobile phases of different compositions. Separation then takes place by virtue of partition and size exclusion. Best fractionation is usually obtained when a mixture of a polar and nonpolar solvent is used.

Moreover, it should be considered that phenolic, heteroatomic, and cyclic compounds are preferentially retained on Sephadex gels especially when lower alcohols are used as eluting solvents. These compounds usually stay longer on the gel than would be expected based on their size.

Sephadex gels are fairly inert, so sample recovery is usually very good. They are stable in all solvents, which are not strongly acidic (i.e., below pH 2.0), and do not contain strong oxidizing agents. On the other hand, these gels are susceptible to fungal and bacterial attack. Other disadvantages include the long elution time and the relatively low resolution. The extended operational times are mainly because of the requirement for long narrow columns and slow flow rates to effect adequate resolution.

Other sorbents for size exclusion have been developed using different types of matrices, e.g., styrene–divinyl benzene polymer and silica gel. Both matrices offer a good mechanical strength and can be used in HPLC mode. Silica gel used in SEC is completely endcapped to minimize nonspecific interactions. It is stable in the pH range of 2.0–7.5. Styrene–divinyl benzene polymer, however, offers a wider pH stability range but shows lower efficiency than silica. SEC has a low loading capacity in terms of sample mass and sample volume. Samples should be concentrated up to but not beyond the point of precipitation for optimum performance.

2.1.5. Biological Affinity

Affinity chromatography is far more specific than other purification techniques. It relies on the preparation of a matrix to which the compound of interest, and preferably only this compound, will bind reversibly. The matrix is usually beaded agarose (Sepharose or BioGel A), polyacrylamide (e.g., BioGel P), crosslinked dextran (e.g., Sephacryl), or silica gel to which a biospecific ligand (antibody, enzyme, or receptor protein) has been covalently attached. The immobilized ligand interacts only with molecules that can selectively bind to it. Other compounds in the sample, incapable of biospecific binding, are washed away. The retained compound can be recovered by altering the pH and/or the buffer composition in order to disrupt the solute–ligand interaction. An alternative elution technique involves the addition of a competitive agent that can compete with the solute of interest for the specific binding sites on the column. Some ligands may have an affinity to a group of related substances rather than a single one and can be used to purify several substances.

In spite of its high selectivity, affinity chromatography is not frequently applied to the isolation of marine natural products. This is mainly because the technique is extremely expensive. Many difficulties are encountered in the choice of suitable ligand and the preparation of stationary phase. The ligand should exhibit specific and reversible binding affinity for the substance to be purified. It should also have chemically modifiable groups that allow it to be attached to the matrix without destroying its binding activity. Moreover, the covalent linkages used to immobilize the ligand must be stable in all conditions employed during chromatography.

2.2. Mobile Phases in LC

In adsorption and partition chromatographic systems, the old rule of "like has an affinity for like" holds a special significance in terms of which solvent system can be successfully used for sample elution. A polar solvent is needed to elute a polar analyte from a polar normal-phase column, while a hydrophobic organic solvent is required for the elution of hydrophobic analytes from a hydrophobic reversed-phase column. Other solvent properties like volatility, viscosity, flammability, toxicity, reactivity, compatibility with detection method, and cost should be considered. **Table 5** contains detailed information on the properties of the most common organic solvents.

The addition of various mobile-phase modifiers such as acids (or less commonly, bases), ion-pairing reagents, or inorganic buffer salts could be valuable and gives rise to better resolution. Buffer salts are usually added to the mobile phase in reversed-phase high-performance liquid chromatography (RP-HPLC). The choice of buffer salt, however, is controlled by many factors. It should be transparent at the detection wavelength and free from any organic contaminants. Moreover, it should exhibit complete solubility when organic and aqueous components of the mobile phase are mixed.

Table 5
Properties of Common Solvents Used in Purification of Natural Products[a]

Solvent	Polarity index	RI at 20°C	UV (nm) cutoff at 1 AU	Boiling point (°C)	Viscosity (cPoise)	Solubility in water (%w/w)
Acetic acid	6.2	1.372	230	118	1.26	100
Acetone	5.1	1.359	330	56	0.32	100
Acetonitrile	5.8	1.344	190	82	0.37	100
Benzene	2.7	1.501	280	80	0.65	0.18
n-Butanol	4.0	1.394	254	125	0.73	0.43
Butyl acetate	3.9	1.399	215	118	2.98	7.81
Carbon tetrachloride	1.6	1.466	263	77	0.97	0.08
Chloroform	4.1	1.446	245	61	0.57	0.815
Cyclohexane	0.2	1.426	200	81	1.00	0.01
1,2-Dichloroethane	3.5	1.444	225	84	0.79	0.81
Dichloromethane	3.1	1.424	235	41	0.44	1.6
DMF	6.4	1.431	268	155	0.92	100
DMSO	7.2	1.478	268	189	2.00	100
Dioxane	4.8	1.422	215	101	1.54	100
Ethanol	5.2	1.360	210	78	1.20	100
Ethytl acetate	4.4	1.372	260	77	0.45	8.7
Diethyl ether	2.8	1.353	220	35	0.32	6.89
Heptane	0.0	1.387	200	98	0.39	0.0003
Hexane	0.0	1.375	200	69	0.33	0.001
MeOH	5.1	1.329	205	65	0.60	100
Methyl-*t*-butyl ether	2.5	1.369	210	55	0.27	4.8
Methyl ethyl ketone	4.7	1.379	329	80	0.45	24
Pentane	0.0	1.358	200	36	0.23	0.004
n-Propanol	4.0	1.384	210	97	2.27	100
Iso-propanol	3.9	1.377	210	82	2.30	100
Di-iso-propyl ether	2.2	1.368	220	68	0.37	
Tetrahydrofuran	4.0	1.407	215	65	0.55	100
Toluene	2.4	1.496	285	111	0.59	0.051
Trichloroethylene	1.0	1.477	273	87	0.57	0.11
Water	9.0	1.333	200	100	1.00	100
Xylene	2.5	1.500	290	139	0.61	0.018

[a]Data are from Phenomenex® SPE Manual.

Generally, buffer salts are used at a concentration of 10–100 mM ***(20)***. However, the use of these salts necessitates a subsequent desalting step.

In IEC, the mobile phase is mainly a buffer solution. The critical factors are the pH and the ionic strength of the mobile phase. Traditional inorganic buffers like phosphate, acetate, and formate are usually used in ion-exchange separations. Some authors prefer to use volatile buffers to eliminate the subsequent desalting step. Examples of volatile buffers are ammonium bicarbonate and ammonium acetate. Other volatile buffers mainly contain pyridine, which is highly toxic. However, the evaporation of these buffers is not an easy task, and during evaporation of buffer solution, drastic changes in pH may occur that may affect the desired components ***(19)***.

2.3. Forms of LC

2.3.1. Countercurrent Chromatography (CCC)

Both stationary and mobile phases are liquid. Separation of solutes is achieved by partitioning. Two forms of CCC are available (*see* Chap. 7). The old form is droplet countercurrent and is based on the passage of droplets of the mobile phase through the stationary phase over a long distance. The stationary phase is contained in 200 or more glass tubes connected in series by inert Teflon tubing. The mobile phase is pumped through as a continuous series of droplets that are small enough to rise (or to fall) through the stationary phase without touching the sides of the tube. Centrifugal CCC is an advance on this and gives much faster results. The stationary phase is held in the tube as a thin layer by centrifugal force. Although CCC offers an excellent sample recovery, the limited range of solvent mixtures that can be used and the high cost involved restrict its application ***(25)***.

2.3.2. Thin-Layer Chromatography

Stationary phase is solid and spread on flat sheets. Both adsorption and partition stationary phases are available. The most commonly used stationary phases are silica gel and octadecyl silica. This technique is useful in fractionation, not only as a final process for purification of comparatively small amounts of almost pure compounds, but also as a method for designing some types of column separation and also for monitoring the composition of fractions obtained by other fractionation processes (*see* Chap. 4) ***(25)***.

2.3.3. Column Chromatography

Column chromatography is the most common form of chromatography. All modes of separation are represented. The stationary phase is solid and packed in a column made of glass, stainless steel, or other inert material. The sample mixture is applied to the top of the column and the mobile phase passes through the column either by gravity, vacuum, or pressure to give open-gravity LC, vacuum LC, or medium- or HPLC, respectively (*see* Chap. 5).

2.3.4. Solid-Phase Extraction (SPE)

SPE provides a rapid and simple way for analyte purification and concentration. An advantage of the technique is that it avoids the emulsion problems often encountered with liquid–liquid extraction procedures. The technique can be used with any type of stationary phase, except those used for exclusion chromatography. The most commonly used format consists of a syringe barrel (cartridge) that contains 50 mg to 20 g sorbent material through which the sample solution can be gently forced with a plunger or by vacuum (*see* Chap. 5). When used for sample purification, the stationary phase either retains the desired solutes while passing undesired components (retentive mode) or passes the desired solutes while retaining the interfering components (nonretentive mode). Dilute sample solutions can also be concentrated by extracting the solutes onto the SPE sorbent followed by their elution in a small volume of strong eluent ***(24,26)***.

3. General Approaches for Purification of Marine Natural Products

3.1. Collection and Storage of Marine Organisms

Handling of marine organisms during collection is highly critical. Measures should be taken to avoid decomposition of compounds. Information on the organism and the place of collection should be carefully recorded to facilitate the re-collection and the subsequent taxonomic identification. **Figure 4** presents an example of a collection record sheet that should be completed for each sample collected.

In the first place, each sample should have a special collection number. This number may be chosen to indicate the collection year, expedition number, and specimen number, e.g., collection number 97212 means year 1997, expedition number 2, and specimen number 12. Location should be recorded on a map or chart of the area at a scale suitable to enable re-collection. If possible, global position apparatus (GPS) should be used to obtain coordinates accurate to 10 m. Information on the habitat

COLLECTION RECORD SHEET					
COLLECTION NUMBER: ❑ New sample ❑ Recollect Old collection #: **DATE:** **COMMON NAME:**			**COLLECTOR:**		
PHOTO:		❑ Under water		❑ Above water	
Site: **Depth (m):** **Habitat:**					
Abundance	❑ Abundant	❑ Common	❑ Occasional	❑ Rare	
Weight (Kg):					
APPEARANCE					
Color:	External:	Internal:		In Alcohol:	
Shape:	❑ Encrusting	❑ Tree Like	❑ Mushroom shaped	❑ Furled	❑ Shrub Like
	❑ Brain Like	❑ Cavernous	❑ Ball/globular	❑ Fan	❑ Fingered
	❑ Other (specify):				
Texture:	❑ Slimy	❑ Fleshy	❑ Waxy	❑ Shiny	
	❑ Compressible	❑ Hard	❑ Resilient	❑ Tough	
	❑ Rubbery	❑ Spongy	❑ Leathery		
Surface:	❑ Smooth	❑ Spined	❑ Lumpy	❑ Lobed	
	❑ Fuzzy	❑ Grooved			
Exudates	❑ Mucus	❑ Slime	❑ Sticky	❑ Coloured	
Easy to Tear		❑ Yes		❑ No	
Further Notes:					

[a] This sheet is adapted from one obtained from Phil Crews' research group, University of California, Santa Cruz, USA. It must be adapted for each different phylum under investigation.

Fig. 4. An example of a collection record sheet[a].

(e.g., sample grows on a rock or on the surface of another organism) as well as any ecological observations (e.g., being able to prevent the growth of neighboring organisms) should be recorded. A detailed description of the organism's morphological features, including color, shape, texture, should be written. Closeup photographs of the organism, taken under and above water, are extremely important for later taxonomic identification and should be attached to the collection sheet (*see* **Note 3**).

Voucher specimens for taxonomic purposes should be prepared by taking a small (e.g., 2–5 cm) section of tissue and preserving it in a solution of 10% formalin in seawater. Algae specimens are usually preserved in a solution of 5% formalin in seawater. Specimens should be representative of the entire organism and include as much tissue relevant to taxonomy as possible, e.g., for sponges, both the exo- and endosome are essential for accurate identification. For tunicates and soft corals, often part of the organism or whole organism (if not too large) must be collected, including the "root." After the specimen reaches the lab, formalin solution should be decanted and replaced by 70% ethanol for long-term storage.

The amount of the organism to be collected is usually determined in view of its abundance. An ideal sample size is 1–2 kg wet weight (100–200 g dry weight). Complete harvesting of the organism should be avoided. If only a single large organism is available, a part of it may be collected.

Ideally, the sample should be lyophilized immediately after collection to prevent any chemical degradation. If this is not possible, the sample should be kept at –20°C to 0°C until freeze drying. An alternative approach is to fix the samples by immersing them in a mixture of ethanol–water (50:50 v/v) for approx 24 h after which the liquid is discarded. Damp organisms are then placed in high-density polyethylene bottles (Nalgene® 2 L wide-mouth containers are best) and shipped back to the home lab at ambient temperature *(27)* (*see* **Note 4**). Samples preserved in this way usually remain in good condition for up to 2 wk in tropical conditions with no significant loss of secondary metabolic content. The addition of MeOH should occur immediately after samples reach the lab.

3.2. Extraction

Three extraction strategies are widely used in the field of marine natural products. The choice of a method depends on the aim of the isolation process, the facilities available, as well as the intrinsic advantages and disadvantages of the procedures (*see* **Note 5**).

The first method involves maceration of the sample with solvent, followed by filtration or centrifugation. The tissue residue is returned to the extraction container and extracted again. The process continues until no extractive yield is obtained (*see* **Note 6**). The samples are usually cut into small pieces or ground into fine particles to facilitate solvent penetration.

Stirring or sonication can be applied to increase the diffusion rate. In most cases, MeOH or EtOH is the solvent of choice. However, the use of a series of solvents of increasing polarity is also common to achieve a certain degree of fractionation. Filter aids and vacuum are commonly used to speed up the filtration process. After extraction, the solvent is removed by rotary evaporation at no more than 35°C to avoid degradation of compounds (*see* **Note 7**). This method is simple and does not need any sophisticated equipment. On the other hand, the large amounts of solvents involved and the energy required for their evaporation, as well as the long procedure may restrict its industrial application.

The second extraction scheme was developed by the scientists at the US National Cancer Institute (NCI) as a part of an extensive screening program of natural products to detect compounds with antitumor or anti-HIV activities. Frozen samples are ground with dry ice (CO_2) and extracted with water at 4°C. The aqueous extract is removed by centrifugation and lyophilized. The dry marc is then successively extracted with MeOH–CH_2Cl_2 (1:1 v/v), followed by MeOH (100%). The organic extracts are combined and concentrated under vacuum ***(28,29)***. This method is highly efficient. Moreover, lyophilization of aqueous extracts eliminates the risks of bumping and heat degradation.

The third extraction protocol involves the use of SCFs (*see* Chap. 3). The critical point is defined as the highest temperature and pressure above which there is no difference in density between the liquid and gaseous forms of the substance. At temperatures and pressures above the critical point, a single homogenous fluid is formed and is said to be supercritical. The critical temperature and pressure vary with the substance and with its purity. For water, the values are 374°C and 220 atm, respectively, whereas for carbon dioxide the corresponding values are 31°C and 74 atm, respectively.

SCFs have the advantages of low viscosity, superior mass transfer properties, and good solvation power. They also have the ability to penetrate microporous materials. Thus, their use in natural product extraction is widely appreciated. Supercritical carbon dioxide is the most preferred solvent as low temperature can be employed. It has other advantages, such as nontoxicity, nonflammability, noncorrosiveness, chemical inertness, and cost effectiveness. Moreover, it will easily evaporate into the atmosphere after extraction ***(30)***. Supercritical CO_2 resembles the nonpolar solvents hexane and benzene in their solubilizing power. Its affinity for compounds of higher polarity can be improved by increasing its density (by small changes in temperature and pressure) or adding organic solvents (e.g., MeOH, EtOH, or DCM). However, the addition of these organic modifiers will alter the critical temperature and pressure and will necessitate modifications to the procedure for removing extraction fluid at the end of the process *(25)*. References *(31,32)* include examples for the application of this technique in the extraction of marine algae. Although this method offers a fast and effective way for extraction and subsequent solvent removal, it needs sophisticated equipment and some experimentation to choose the best organic modifier.

3.3. Fractionation of Marine Extracts

Marine extracts are extremely complex, and comprise mixtures of neutral, acidic, basic, lipophilic, and amphiphilic compounds. The nature of the compound(s) of interest may differ according to the aim of the project, and as a consequence there is no general fractionation procedure or recipe that can serve for all eventualities. It should be noted that despite the recent advances in separation technology, experience still plays an indispensable role in the isolation of marine natural products.

Generally, fractionation procedure passes through four stages (**Fig. 5**). The first stage includes collection of information about the chemical content profile and the biological activity of the extract, the nature of compounds of interest, as well as the type of impurities. This information is highly valuable for planning an isolation strategy. In the second stage, dereplication (*see* Chap. 12) usually takes place. The objectives are to identify extracts that contain only known compounds as early as possible before elaborate fractionation steps are undertaken, and to prioritize extracts in terms of their content of interesting new and/or active compounds. It may be useful to point out that all procedures involved in stages 1 and 2 should be carried out for all fractions obtained after any separation step. In this way, interesting components can be tracked until final purification. The aim of the third stage is often to remove the bulk of unwanted materials, e.g., fats and salts using fairly low-resolution separation steps, e.g., liquid–liquid partition, SPE, and SEC. The fourth stage usually involves high-resolution separation steps, e.g., HPLC with the aim of purifying interesting compounds to a degree that enables the subsequent structure elucidation. The procedures involved in the four mentioned stages are discussed below with some details.

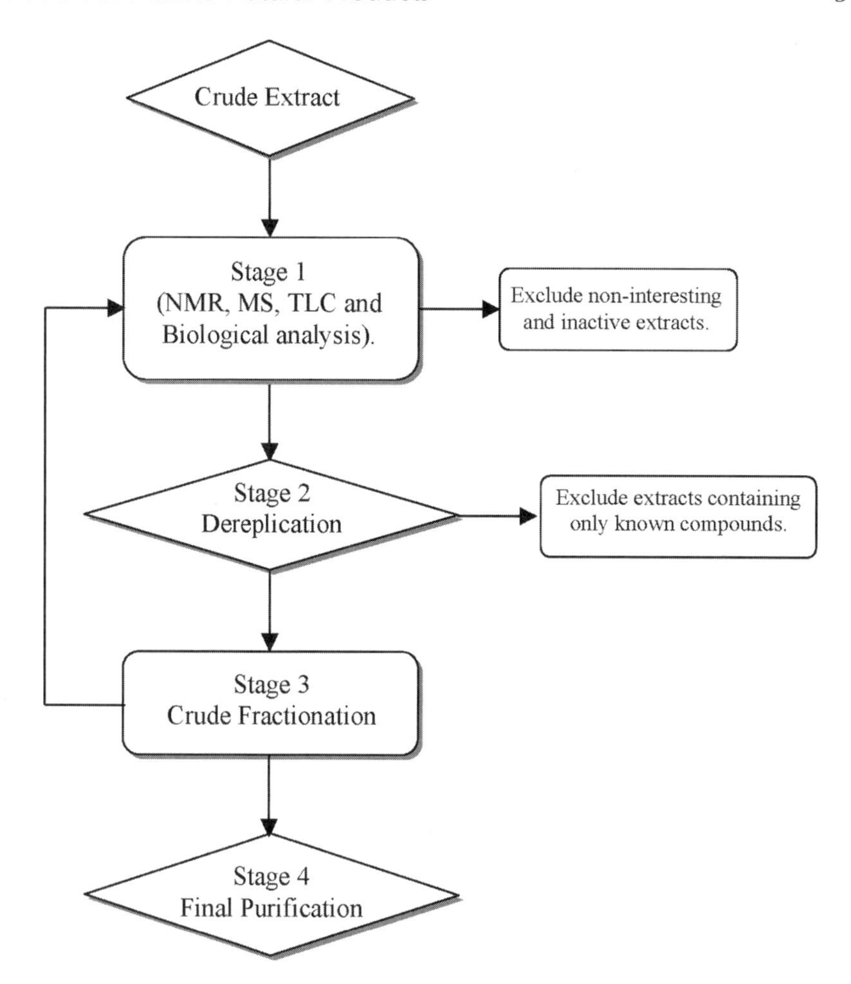

Fig. 5. General approach for fractionation of marine extracts.

3.3.1. Stage 1: Investigation of the Nature of Extract Components

This is probably the most important stage in the isolation process. The judicious planning of a fractionation scheme depends mainly on the information available on the nature of the substances present in the extract. Moreover, any knowledge about the chemical nature and/or the biological activity of the target compounds can efficiently guide the separation process. In effect, some information can be revealed by exact taxonomic identification of the organism under investigation. Literature searches can provide information about compounds previously isolated from the species. If the chemistry of a species has not been studied, useful information may be gained by searching the closely related species within the genus. However, it is worth remembering that the chemical content of a marine organism may completely differ if it is collected from a different locality and/or in another season. The type of solvents used in extraction can also provide some useful information. Aqueous extracts usually contain highly polar compounds and a large amount of inorganic salts. On the other hand, organic extracts often contain less-polar compounds and a lot of fats. Further information can be gleaned by carrying out biological testing, TLC analysis, mass spectrometry (MS), and NMR experiments.

3.3.1.1. Biological Screening

Bioassay-guided fractionation is very appealing for research on drug discovery from natural sources. It is essential to maintain a reference sample of the fractions obtained after each separation step so that it can be biologically tested and serve as a record of the materials recovered at each stage of the process ***(21)***. However, one of the most difficult problems in bioassay-directed fractionation is the possibility of getting false positives and false negatives. False positives usually occur if any of the inactive components in the extract has the ability to interact nonspecifically with the molecular target of the assay (e.g., being able to precipitate proteins and hence show inhibitory activity in many enzyme-based assays). Similarly, some inactive components may give rise to positive hits by interacting with some components of the assay system other than the target. Others may interfere with the assay detection method, e.g., UV quenchers ***(18)***. On the other hand, false negatives usually occur if the active compounds acts on a molecular target other than that of the assay. It should also be noted that many interesting chemical entities might be activated in vivo by metabolic enzymes, a factor that is not considered in many in vitro screening systems. Thus, one can conclude that the fractionation process should not rely solely on biological screening.

3.3.1.2. TLC Analysis

Analytical TLC plates can be used to get an idea about the degree of polarity and/or hydrophilicity of different extract components. They are also widely applied in the detection of compounds through several separation steps. Moreover, they can be used to predict the separation pattern on column chromatography, and thus help in selecting the best column chromatographic systems. They may assist as well in assessing the degree of purity of isolated compounds. In all the above applications, more than one solvent system should be tried as different compounds may have the same R_f values in one system and thus appear as a single spot.

TLC plates may be sprayed with reagents that react specifically with certain classes of compounds. The use of different spraying reagents can give plenty of information about the chemical classes present in the extract. There are numerous spraying reagents listed in several standard texts on the subject ***(33,34)*** and also in Chapter 4. **Table 6** lists the most widely used spraying reagents for marine natural products.

Moreover, direct combination of TLC with bioassay (bioautography) can provide more information about the active component within the extract mixture. This is exemplified in the discovery of antimicrobial agents

Table 6
TLC Spraying Reagents Commonly Used in Most Marine Natural Products Labs

Reagent	Recipe	Treatment	Notes
Vanillin/ sulfuric acid	Dissolve vanillin (4 g) in concentrated H_2SO_4 (100 mL)	Heat at 100°C until coloration appears	Universal spraying reagent
Anisaldehyde/ sulfuric acid	Add 0.5 mL anisaldehyde to 10 mL glacial acetic acid and then add to the mixture 85 mL MeOH and 5 mL concentrated H_2SO_4, in this order	Heat at 100°C until coloration appears	Detect many compounds, especially terpenes, sugars, phenols, and steroids
Dragendorff's reagent	Add 10 mL of 40% aqueous solution of KI to 10 mL of solution of 0.85 g of basic bismuth subnitrate in acetic acid (10 mL) and distilled water (40 mL). Dilute the resulting solution with acetic acid and water in the ratio: 1:2:10	No heat is required	Detect alkaloids and heterocyclic nitrogen-containing compounds
Ninhydrin reagent	Dissolve ninhydrin (30 mg) in 10 mL *n*-butanol. Add 0.3 mL acetic acid to the solution	Heat at 100°C until coloration appears	Detection of amino acids, amines, and peptides

from marine extracts. Following development, TLC plates of extract are dried and then overlaid with a thin layer of agar containing the test organism against which the extract is active. Following an appropriate incubation period, zones of growth inhibition in the agar can be seen in regions of plates containing the active compound *(35)*. The same method can also be applied to detect compounds with antitumor activity *(36)*.

3.3.1.3. NMR Analysis

NMR spectroscopic analysis plays an indispensable role in the structure elucidation of pure compounds. It can also provide a lot of information on the chemical nature of compounds in a mixture. In effect, it is recommended to obtain ^{1}H- and ^{13}C-NMR spectra for marine extracts. The objectives are to detect the presence/absence of common artifacts, e.g., plasticizers (*see* **Note 8**) and to assign the components in the mixture to certain chemical classes (*see* **Note 9**). Combination of fractions after any separation step can be decided on the basis of their similar NMR spectra.

3.3.1.4. MS Analysis

MS is a technique used to identify molecular weights of unknown compounds by ionizing them and detecting the mass-to-charge ratio (m/z) of the resulting molecular ions. Molecules that cannot be ionized will not be detected. One advantage of the technique is its high sensitivity. It can even detect microgram amount of compounds. The problem in generalizing MS to the process of identifying extract components is the lack of a universal ionization mode under which any unknown compound could be ionized. Fortunately, many MS ionization techniques have been introduced under which most marine natural products can be ionized. Generally, electrospray (ESI) is the recommended ionization technique for polar extracts, while atmospheric pressure chemical ionization (APCI) is favored for moderately polar ones *(37)*. MS analysis is difficult to apply to crude marine extracts, but it can be of great value for identifying compounds from semipurified mixtures.

3.3.2. Stage 2: Dereplication

Isolation of pure compounds from marine extracts is a tedious and expensive process. Measures should be taken to avoid isolation of known compounds. Dereplication may be defined as the attempt to remove duplicate leads or compounds (*see* Chap. 12) *(38)*. This process relies mainly on the availability of comprehensive databases for known compounds. Many databases are currently available, including those containing information on the source of the organism, taxonomic identification, and extraction methods, as well as the different chromatographic and spectroscopic characteristics of the isolated compounds. Most of these databases can be accessed via the Internet. Others are available on CDs. **Table 7** lists some of these useful databases as well as their URLs, especially in relation to marine natural products.

3.3.3. Stage 3: Crude Fractionation

The objectives at this stage are to simplify the extract composition by dividing it into groups of compounds sharing similar physicochemical characteristics and/or to remove the bulk of unwanted materials and thus enrich the extract with respect to the target compounds. Procedures commonly employed involve solvent partitioning, defatting, and desalting.

3.3.3.1. Solvent Partitioning

The procedure described in **Fig. 6** is a modification of the method developed by Kupchan *(39)*. It can be used for defatting and desalting as well.

Table 7
Some Useful Databases for Marine Natural Product Dereplication

Database	URL
MarinLit	http://www.chem.canterbury.ac.nz/marinlit/marinlit.shtml
Dictionary of Natural Products and others	http://www.chemnetbase.com/
Chemical Abstracts (CAS)	http://info.cas.org/
NAPRALERT (Natural Product Alert) database	http://www.cas.org/ONLINE/DBSS/napralertss.html
Beilstein CrossFire	http://www.mimas.ac.uk/crossfire/
Silverplatter	http://web5.silverplatter.com/webspirs/start.ws
NCI data search	http://dtp.nci.nih.gov/docs/dtp_search.html
United States Patents	http://www.uspto.gov/patft/
Chemical Database Services	http://cds.dl.ac.uk/cds/
National Institute of Standards	http://webbook.nist.gov/chemistry/
Cambridge Structural Database	http://www.ccdc.cam.ac.uk/products/csd/
Marine Biological Laboratory, Woods Hole, MA, USA	http://www.mbl.edu/
Chromatography application database	http://www.chromatography.co.uk/apps/hplc/dbases/form.htm

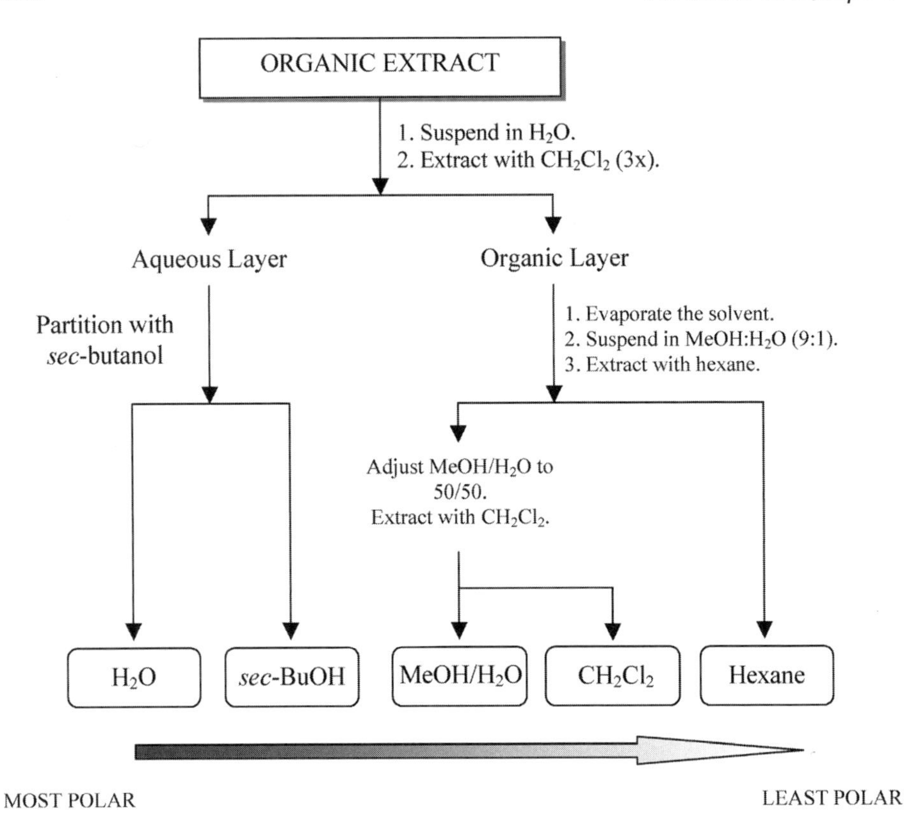

Fig. 6. Modified Kupchan's partition scheme.

Most of fats will go with the *n*-hexane fraction, while inorganic salts will go with the aqueous one. The advantage of the method is total recovery of target compounds. Drawbacks are the problems of emulsion formation, time ineffectiveness, and use of large volumes of solvents.

3.3.3.2. Defatting

A number of procedures have been described for defatting of marine extracts. The use of Sephadex LH-20 and MeOH:CH_2Cl_2 (1:1 v/v) as eluent is one of the commonly used procedures. Fats and large nonpolar organic compounds are usually eluted first. Other common procedures involve the use of an SPE cartridge containing C_{18} silica and MeOH/H_2O as a washing liquid. Owing to their strong hydrophobic nature, fats and lipids are retained on the stationary phase, while other more hydrophilic extract components are eluted. The latter procedure is not suitable if the target compounds show poor recovery from C_{18} silica.

3.3.3.3. Desalting

The most efficient method for desalting of marine extracts has been described by West and Northcote ***(40)*** and West et al. ***(41)***. In their procedure, a methanolic extract is passed through a column of Diaion HP20® resin (styrene–divinyl benzene polymer) pre-equilibrated with MeOH. The eluents are concentrated and passed again through the same column. The resulting eluents are diluted with water and passed through the column. The last step is repeated to ensure that all compounds containing hydrophobic domains are adsorbed on the resin. Desalting can be easily achieved by washing the resin with plenty of water. Different proportions of MeOH or acetone in water can be used to elute the adsorbed compounds and to achieve a certain degree of fractionation. Better results can be obtained by using beads of smaller particle size (e.g., Diaion HP20ss); but, in this case, application of pressure is needed to achieve good flow properties.

3.3.4. Stage 4: Final Purification

Preparative HPLC has been, by far, the most useful tool for separation of complex mixtures. When interfaced with a diode array detector (DAD), HPLC allows an analyst to identify known compounds by comparison of their HPLC retention time and UV spectra. The introduction of the evaporative light scattering detector (ELSD) allows the detection of compounds that lack UV chromophores. In the past few years, tandem or hyphenated analytical techniques such as LC-MS, LC-MS-MS, LC-NMR, and LC-NMR-MS have also been developed (*see* Chap. 9). These techniques provide powerful tools for rapid identification of known compounds and determination of structure classes of novel ones.

3.3.4.1. UV Diode Array Detector (DAD)

UV photodiode array detectors allow the collection of UV-absorbance data across many wavelengths simultaneously and thus enable peak purity assessment. Background impurities can be easily detected by comparing the UV spectra at different time points across the peak of interest. Most modern DADs are supported with libraries containing UV spectra of previously reported compounds. The operating software of these detectors has the capability for spectral library generation and searching and thus enables rapid identification of known compounds ***(20)***.

3.3.4.2. Evaporative Light-Scattering Detector (ELSD)

ELSD has been developed to complement UV detection for weakly UV-absorbing compounds. In ELSD, the HPLC effluent is nebulized and then vaporized in a heated drift tube, which results in a cloud of analyte particles that pass through a beam of light. The analyte particles scatter the light and generate a signal ***(42)***. In contrast to UV detectors, the extinction coefficients of the analytes have no effect on the response of ELSD. Thus, ELSD is now the preferred concentration detection method for LC. When ELSD is connected to preparative HPLC, the effluent from the column is split and only small proportion is directed to the detector.

3.3.4.3. LC-MS and LC-MS-MS

LC-MS is the most widely applied tool for dereplication of natural products (*see* Chaps. 9 and 12) ***(43,44)***. This is mainly because nominal molecular weight can be used as a search query in nearly all databases. Using LC-MS-MS, certain molecular ions are separated and subjected to a second round of fragmentation. The fragmentation pattern produced can give a lot of information about the parent structure. The technique is well suited for identification of fragments from molecules formed of several individual units such as peptides, depsipeptides, oligosaccharides, and saponins ***(13)***.

3.3.4.4. LC-NMR

Recent advances in NMR spectroscopy have allowed its direct coupling with HPLC systems (*see* Chap. 9). The use of high-field magnets (500 MHz or greater), capillary microliter-volume flow cells, and digital signal processing system has dramatically increased the sensitivity to trace quantities of analytes. In addition, the new probe designs have facilitated the efficient and specific suppression of NMR signals of HPLC solvents. However, the technique is still slow and highly expensive. LC-NMR is especially useful in instances where the data from LC-MS do not allow confident identification of a compound (e.g., isomers that have the same molecular weights). The technique has been successfully applied in the identification of the alkaloid aaptamine in the extract of marine sponge *Aaptos* species ***(45)***. The use of HPLC-NMR-MS, in which the separation system is coupled with both NMR and MS, has also been reported ***(46)***.

4. New Vistas in Separation Technology

In spite of their high diversity, natural products have been dropped from many pharmaceutical companies' research programs. This is mainly because of the traditional time-consuming and cost-intensive isolation and identification procedures, and the lack of availability of natural products in a format suitable for modern HTS technologies. In effect, the integration of natural products' crude extracts in HTS systems is difficult. This is mainly because of the high possibility of false positives and recurring known bioactive compounds and the tedious work needed for hit identification. Recently, methods have been described for preparation of large and diverse natural products libraries optimized for HTS *(47)*. One method relies on generating a huge library of semipurified fractions. Advantages of this method include the increase in reliability of biological testing results and the sharp decrease in the subsequent workload for hit identification and dereplication. Another strategy depends on the preparation of a library of pure natural compounds. Although this kind of library offers the highest reliability in biological testing, the huge amount of work needed for sample preparation is still the major drawback. Both libraries rely on the availability of rapid and automated fractionation techniques. Recently, Sepiatec GmbH (Berlin, Germany) in close cooperation with Aventis Pharma Deutschland GmbH (Frankfurt, Germany) has designed systems to boost productivity of sample preparation for HTS. A brief description of two representative examples of these systems is given here.

4.1. 8X Parallel HPLC

This system (**Fig.** 7), which can run unattended for 24 h a day, is able to simultaneously fractionate eight complex extract mixtures using a single high-pressure gradient pump system, a multi-channel (UV, DAD, or ELSD) detector, and expert software. Samples are separated by an array of HPLC columns, providing one column for each sample. It should be noted that liquid injections of samples with a broad polarity range are a challenging task. Dimethyl sulfoxide (DMSO) completely dissolves most samples for injection, but may also disturb the subsequent chromatographic separation. An on-line SPE injection module *(26)* solves this difficulty. Using an autosampler, samples dissolved in DMSO are injected in the module, and water or buffer is simultaneously pumped before the inlet of the SPE column *(48)*. Owing to this instant increase in the gradient polarity, the extract components remain adsorbed onto the SPE column

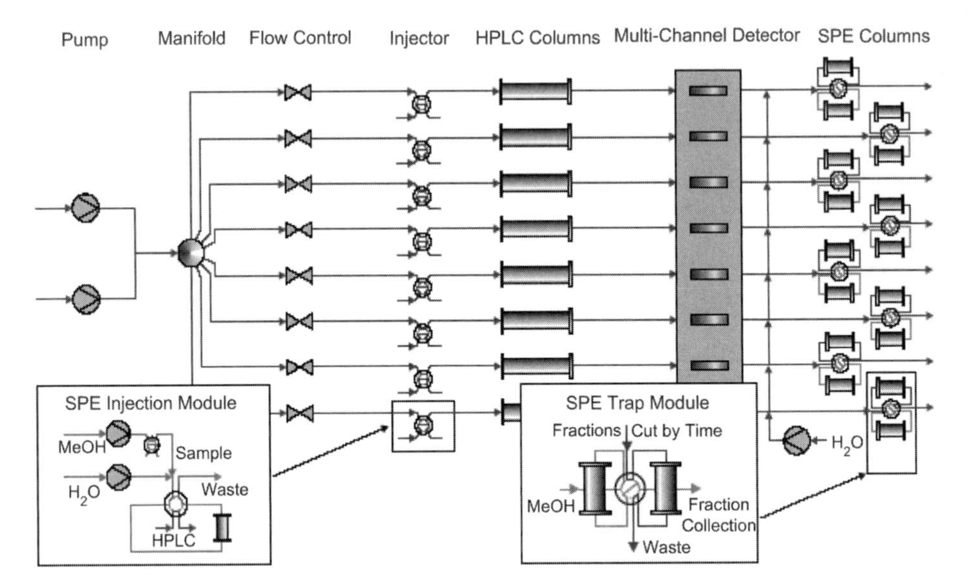

Fig. 7. Schematic outline of plumbing of 8X Parallel HPLC device. Figure reproduced with permission from Sepiatec Company Brochure.

and DMSO/water is flushed into waste. Subsequently, a suitable organic solvent injects the extract onto the separation column.

Eluting fractions undergo an on-line and automated workup before collection. They are again adsorbed onto SPE columns using the same principle as the previously described sample injection module. SPE adsorbed fractions are flushed with water and then eluted with pure organic solvent into 96-well plate. The system is able to separate more than 100 natural product extracts a day into several thousands of water- and buffer-free fractions. The quality and the purity of the fractions obtained fit the demands of HTS.

4.2. sepbox®: An HPLC-SPE-HPLC-SPE Arrangement

This system (**Fig. 8**) is based on the combination of HPLC and SPE. In this HPLC-SPE-HPLC-SPE arrangement, the polarity of the eluent is increased by the addition of water to such an extent that the fractions eluted from separation column I are adsorbed onto the trap columns I. These trapped fractions are then passed through the separation column II where final separation is completed. The individual components eluted are adsorbed onto the trap columns II, separated from buffer, and flushed into

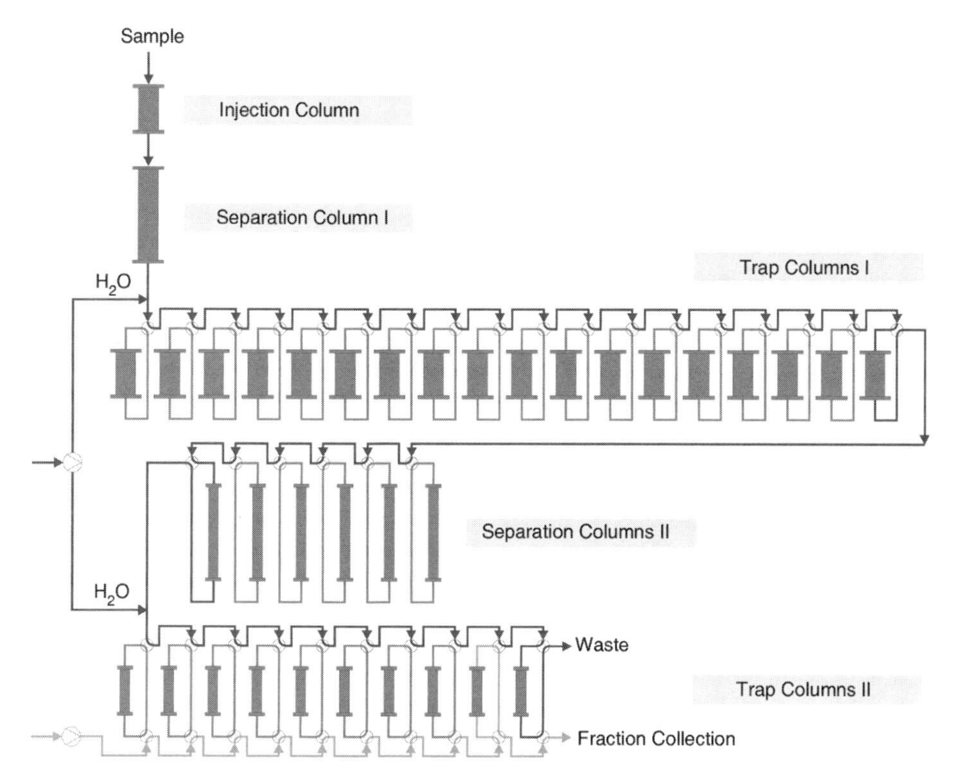

Fig. 8. Schematic outline of plumbing of sepbox®. Figure reproduced with permission from Sepiatec Company Brochure.

the fraction collector. This system is much more suitable for generating a library of almost pure compounds.

5. Notes

1. It is highly recommended to use C_{18} or C_8 bonded phases for separation of hydrophilic peptides and C_4, C_5, or cyanopropyl bonded phases for hydrophobic ones.
2. Concentration of ion-pairing reagent in the mobile phase should be considered so that it does not affect the stability of either the stationary phase or the sample components. TFA (0.02–0.1%) and TEA (0.1–1%) are commonly used concentration ranges.
3. An identifying tag should be included in the photographs (plastic tags with collection numbers on them are useful, a clothes-peg may be used to secure the tag to the closeup frame). If possible, a digital camera should be used

to enable photographs to be processed quickly. If this is not available, Polaroid® slide film and processors provide a good alternative to allow processing of photographs on site.

4. If Nalgene® bottles are not available, glass containers with metal lids can be used instead. Plastic bottles or lids should be avoided to prevent the possible contamination with plasticizers.
5. If the target of the isolation process is to get a sufficient amount of compound known to be produced by the organism, then the extraction process should take advantage of any reported extraction schemes as well as any unusual chemical features of the target compound that may make the extraction process more selective. Most commonly, the isolation aims at getting as many new compounds as possible to introduce them in a battery of biological assays.
6. Maceration for long time with organic solvents at room temperature may have deleterious effects on the substances extracted. It is advisable to remove solvents as quickly as possible. Extraction is better done in batches. In each batch, the sample is macerated for maximally 24 h, after which the solvent is removed and replaced by another fresh solvent. All extraction procedures should be carried out at the coldest temperature possible and away from direct sunlight. Concentrated extracts should be stored in the freezer at −20°C.
7. If water or aqueous alcohol is used for extraction, bumping of the extract during concentration under vacuum is likely. It is highly recommended to use a large pear-shaped flask (at least 1000 mL capacity) for rotary evaporation. Small amounts of extracts are added to the flask and the spinning of the flask is adjusted to the fastest rate to minimize foaming. It may be useful to add small amounts (10–20 mL) of isopropanol or *sec*-butanol, to facilitate the removal of water through the formation of azeotropes ***(14)***. After rotary evaporation, concentrated extracts are transferred to wide-mouth vials of suitable capacity. Removal of solvents can be achieved using a high-vacuum pump or a high-purity N_2 blowdown apparatus. Turbovaps using air should be avoided because of the risk of oxidation.
8. Dialkylphthalate esters represent the most common plasticizers found. They can be easily detected in the ^{1}H-NMR spectra, which show the following chemical shift values: (δ, $CDCl_3$) 7.70 (2H, *dd*), 7.52 (2H, *dd*), 4.2 (4H, *dd*), and 1.2–1.8 (*m*) ***(38)***. Other common artifacts and impurities have been reviewed by Middleditch ***(49)***.
9. Many classes of compounds can be detected by their specific NMR pattern. Lipids appear as a high broad peak at δ 1.2–1.4. Peptides can be detected by the characteristic chemical shift values of their α-protons (δ 4–5), α-carbons (δ 40–70), β-protons (δ 1.5–4), and β-carbons (δ 20–40). Sugars can be detected by the presence of anomeric protons ($\approx\delta$ 5) and anomeric carbons

(≈δ 100). Compounds belonging to other chemical classes can be detected and the reader should consult the literature for their specific NMR key features.

References

1. Faulkner, D. J. (1995) Marine natural products. *Nat. Prod. Rep.* **12,** 223–269.
2. Faulkner, D. J. (1996) Marine natural products. *Nat. Prod. Rep.* **13,** 75–125.
3. Faulkner, D. J. (1997) Marine natural products. *Nat. Prod. Rep.* **14,** 259–302.
4. Faulkner, D. J. (1998) Marine natural products. *Nat. Prod. Rep.* **15,** 113–158.
5. Faulkner, D. J. (1999) Marine natural products. *Nat. Prod. Rep.* **16,** 155–198.
6. Faulkner, D. J. (2000) Marine natural products. *Nat. Prod. Rep.* **17,** 7–55.
7. Faulkner, D. J. (2001) Marine natural products. *Nat. Prod. Rep.* **18,** 1–49.
8. Faulkner, D. J. (2002) Marine natural products. *Nat. Prod. Rep.* **19,** 1–48.
9. Blunt, J. W., Copp, B. R., Munro, M. H. G., Northcote, P. T., and Prinsep, M. R. (2003) Marine natural products. *Nat. Prod. Rep.* **20,** 1–48.
10. Blunt, J. W., Copp, B. R., Munro, M. H. G., Northcote, P. T., and Prinsep, M. R. (2004) Marine natural products. *Nat. Prod. Rep.* **21,** 1–49.
11. Shimizu, Y. (1985) Bioactive marine natural products, with emphasis on handling of water-soluble compounds. *J. Nat. Prod.* **48,** 223–235.
12. Quinn, R. J. (1988) Chemistry of aqueous marine extracts: isolation techniques, in *Bioorganic Marine Chemistry*, vol. 2 (Scheuer, p., Coll J. C., Elyalov, G. B., and Quinn, R. J., eds.), Springer-Verlag, Berlin, Heidelberg, pp. 1–41.
13. Riguera, R. (1997) Isolating bioactive compounds from marine organisms. *J. Marine Biotech.* **5,** 187–193.
14. Wright, A. E. (1998) Isolation of marine natural products, in *Methods in Biotechnology, vol. 4: Natural Products Isolation* (Cannell, R. J. P., ed.), Humana, Totowa, NJ, pp. 365–408.
15. Pettit, G. R., Cichacz, Z., Gao, F., et al. (1994) Antineoplastic agents 300. Isolation and structure of the rare human cancer inhibitory macrocyclic lactones spongistatins 8 and 9. *J. Chem. Soc. Chem. Commun.* **58**:1605–1606.
16. Pettit, G. R. (1996) Progress in the discovery of biosynthetic anticancer drugs. *J. Nat. Prod.* **59,** 812–821.
17. Pettit, G. R., Kamano, Y., Herald, C. L., et al. (1987) The isolation and structure of a remarkable marine animal antineoplastic constituent: dolastatin 10. *J. Am. Chem. Soc.* **109**, 6883–6885.
18. VanMiddlesworth, F., Cannell, R. J. P. (1998) Dereplication and partial identification of natural products, in *Methods in Biotechnology, vol. 4:*

Natural Products Isolation (Cannell, R. J. P., ed.), Humana, Totowa, NJ, pp. 279–327.

19. Shimizu, Y. (1998) Purification of water-soluble natural products, in *Methods in Biotechnology, vol. 4: Natural Products Isolation* (Cannell, R. J. P., ed.), Humana, Totowa, NJ, pp. 329–341.
20. Stead, P. (1998) Isolation by preparative HPLC, in *Methods in Biotechnology, vol. 4: Natural Products Isolation* (Cannell, R. J. P., ed.), Humana, Totowa, NJ, pp. 165–208.
21. Cannell, R. J. P. (1998) How to approach the isolation of a natural product, in *Methods in Biotechnology, vol. 4: Natural Products Isolation* (Cannell, R. J. P., ed.), Humana, Totowa, NJ, pp. 1–51.
22. Bradshaw, T. P. (2000) *A User's Guide: Introduction to Peptide and Protein HPLC.* Phenomenex, USA.
23. Dufresne, C. (1998) Isolation by ion-exchange chromatography, in *Methods in Biotechnology, vol. 4: Natural Products Isolation* (Cannell, R. J. P., ed.), Humana, Totowa, NJ, pp. 141–164.
24. *Solid Phase Extraction Reference Manual and User Guide.* Phenomenex, USA.
25. Houghton, P. J. and Raman, A. (1998) *Laboratory Handbook for the Fractionation of Natural Extracts*, Chapman & Hall, London, UK.
26. Thurman, E. M. and Mills, M. S. (1998) *Solid Phase Extraction: Principles and Practice*, John Wiley & Sons, New York.
27. Rodriguez, J., Nieto, R. M., and Crews, P. (1993) New structures and bioactivity patterns of bengazole alkaloids from a Choristid marine sponge. *J. Nat. Prod.* **56,** 2034–2040.
28. Rashid, M. A., Gustafson, K. R., Cartner, L. K., Pannell, L. K., and Boyd, M. R. (2001) New nitrogenous constituents from the South African marine ascidian *Pseudodistoma* sp. *Tetrahedron* **57,** 5751–5755.
29. Chang, L. C., Otero-Quintero, S., Nicholas, G. M., and Bewley, C. A. (2001) Phyllolactones A–E: new bishomoscalarane sesterterpenes from the marine sponge *Phyllospongia lamellose. Tetrahedron* **57,** 5731–5738.
30. Venkat, E. and Kothadarama, S. (1998) Supercritical fluid methods, in *Methods in Biotechnology, vol. 4: Natural Products Isolation* (Cannell, R. J. P., ed.), Humana, Totowa, NJ, pp. 91–109.
31. Gao, D., Okuda, R., and Lopez-Avila, V. (2001) Supercritical fluid extraction of halogenated monoterpenes from the red alga *Plocamium cartilagineum. J. AOAC Int.* **84,** 1313–1331.
32. Subra, P. and Boissinot, P. (1991) Supercritical fluid extraction from a brown alga by stage wise pressure increase. *J. Chromatogr.* **543,** 413–424.
33. Stahl, E. (1965) *Thin Layer Chromatography—A Laboratory Handbook*, Springer-Verlag, Berlin.

34. Wagner, H. and Bladt, S. (1995). *Plant Drug Analysis—A Thin Layer Chromatography Atlas*, Springer-Verlag, Berlin.
35. Hamburger, M. O. and Cordell, G. A. (1987) A direct bioautographic TLC assay for compounds possessing antibacterial activity. *J. Nat. Prod.* **50,** 19–22.
36. Burres, N. S., Hunter, J. E., and Wright, A. E. (1989) A mammalian cell agar-diffusion assay for the detection of toxic compounds. *J. Nat. Prod.* **52,** 522–527.
37. Zhou, S. and Hamburger, M. (1996) Application of liquid chromatography-atmospheric pressure ionisation mass spectrometry in natural products analysis: evaluation and optimisation of electrospray and heated nebulizer interfaces *J. Chromatogr. A* **755,** 189–204.
38. Silva, G. L., Lee, I., and Kinghorn, A. D. (1998) Special problems with the extraction of plants, in *Methods in Biotechnology, vol. 4: Natural Products Isolation* (Cannell, R. J. P., ed.), Humana, Totowa, NJ, pp. 343–364.
39. Kupchan, S. M., Britton, R. W., Ziegler, M. F., and Siegel, C. W. (1973) Bruceantin, a new potent antileukemic simaroubolide from *Brucea antidysenterica. J. Org. Chem.* **38,** 178–179.
40. West, L. M. and Northcote, P. T. (2000) Peloruside A: a potent cytotoxic macrolide isolated from the New Zealand marine sponge *Mycale* sp. *J. Org. Chem.* **65,** 445–449.
41. West, L. M., Northcote, P. T., Hood, K. A., Miller, J. H., and Page, M. J. (2000) Mycalamide D, a new cytotoxic amide from the New Zealand marine sponge *Mycale* species. *J. Nat. Prod.* **63,** 707–709.
42. Allgeier, M. C., Nussbaum, M. A., and Risley, D. S. (2003) Comparison of an evaporative light-scattering detector and a chemiluminescent nitrogen detector for analysing compounds lacking a sufficient UV chromophore. *LCGC N. Am.* **21,** 376–381.
43. Shigematsu, N. (1997) Dereplication of natural products using LC/MS. *J. Mass Spectrom Soc. Jpn.* **45,** 295–300.
44. Pannell, L. K. and Shigematsu, N. (1998) Increased speed and accuracy of structural determination of biologically active natural products using LC-MS. *Am. Lab.* **30,** 28–30.
45. Bobzin, S. C., Yang, S., and Kasten, T. P. (2000) LC-NMR: a new tool to expedite the dereplication and identification of natural products. *J. Indust. Microbiol. Biotech.* **25,** 342–345.
46. Pullen, F. S., Swanson, A. G., Newman, M. J., and Richards, D. S. (1995) Online liquid chromatography/nuclear magnetic resonance/mass spectrometry—a powerful spectrometric tool for the analysis of mixtures of pharmaceutical interest. *Rapid Commun. Mass Spectrom.* **9,** 1003–1006.
47. Adel, U., Kock, C., Speitling, M., and Hansske, F. G. (2002) Modern methods to produce natural-product libraries. *Curr. Opin. Chem. Biol.* **6,** 453–458.

48. God, R., Gumm, H., Heuer, C., and Juschka, M. (1999) Online coupling of HPLC and solid-phase extraction for preparative chromatography. *GIT Lab J.* **3,** 188–191.
49. Middleditch, B. S. (1989) Analytical artifacts: GC, MS, HPLC, TLC and PC. *J. Chromatogr. Library* **44**.

15

Isolation of Microbial Natural Products

Russell A. Barrow

Summary

Microbial fermentations conducted with the express purpose to generate organic molecules are invariably carried out in a liquid medium, and the organic components produced can be distributed between both the solid and liquid phases. A major challenge facing the practitioner attempting to isolate a microbial natural product, where the organic component is extracellular and has been exuded into the medium, is the separation of the desired organic components from the aqueous broth that constitutes the majority of the mass. This chapter places natural products in a medicinal context from a historical perspective, demonstrating that biological activity in natural compounds has a long history. It subsequently goes on to summarize, using appropriate examples, two methods invaluable in the isolation of microbial natural products. The techniques considered are liquid–liquid extraction concentrating on countercurrent methods of separation and liquid–solid extraction focusing on the use of polymeric adsorbents in separation.

Key Words: Natural products; fermentation; isolation; microbial; countercurrent chromatography; countercurrent distribution; liquid ion-exchange extraction; polymeric adsorbents; diaion.

1. Natural Products in Context: From Asclepius to Ehrlich

The isolation of natural products as marketable commodities is not a new phenomenon and has a history dating back many millennia. The fermentation of fruit and grain to produce the catabolic product we know

From: *Methods in Biotechnology, Vol. 20, Natural Products Isolation, 2nd ed.*
Edited by: S. D. Sarker, Z. Latif, and A. I. Gray © Humana Press Inc., Totowa, NJ

as ethanol has its origins in prehistory, while more recently it is known that many cultural groups, including the Celts, the Greeks, and the Egyptians, kept records of production and consumption of alcoholic beverages. Until 1856, when William Perkins established what was to become the synthetic dye industry, the production of dyes was predominantly based on natural products. In addition to the multitude of plant-based colorings, a dyestuff called Tyrian purple, although not natural itself but rather a degradation product based on the naturally occurring tyrindoxyl sulfate, formed a major industry over 3000 yrs ago. Perhaps the most remarkable natural product traded in its pure form is sugar. In 327 BCE, Alexander the Great introduced sugar-cane and de facto sucrose into the Mediterranean. Sucrose is one of the most common and well-documented natural products known today with over 130 billion kgs produced in over 100 countries around the world each year.

Today the driving force behind natural product chemistry as a discipline is still largely predicated on the attainment of an economically viable product, often a therapeutic. As a consequence, much of the recent scientific research directed toward the isolation and structural elucidation of natural products has revolved around the quest for a cure. Until modern times, every civilization throughout history had relied upon the biological properties of natural products to stave of disease and prevent illness. This approach changed with the beginning of the 20th century, when Ehrlich and Hata ushered in a new paradigm for drug discovery, that of chemotherapy, involving an ordered search of a library of pure chemicals for a predetermined activity. Through this structured technique, they developed the synthetic chemical arsphenamine, an arsenic-based compound traded as Salvarsan, which was capable of destroying the syphilis pathogen, *Treponema pallidum*. Today such is the perceived success of this modern discovery paradigm that natural medicines have been largely relegated, at least in most of Western society, to that of complementary status, and to a large extent there is a perception that medicines available presently are synthetic chemicals.

A vast proportion of the population consider drug discovery to be a recent concept that evolved from modern science in the 20th century, whereas in reality it dates back many centuries and has its origins in nature. From a Westcentric perspective, the foundations of science and ergo medicine are considered to have been laid during the "Greek Golden Age," which reached its peak around the 5th century BCE. Indeed, the very word *medicine* along with *hygiene* and *panacea* are derived from the

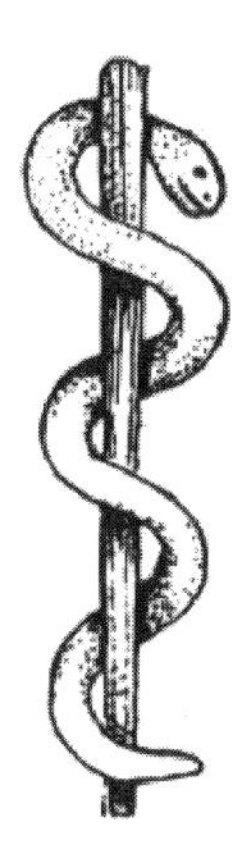

Fig. 1. A representation of the "staff of Asclepius."

names of three of the daughters of Asclepius, Meditrina (Iaso), Hygieia, and Panacea, respectively. Asclepius himself was a physician and was considered to be the son of Apollo. He was worshipped as the Greek god of healing. The medical symbol frequently used today depicting a single snake coiled around a pole is based on the staff of Asclepius (**Fig. 1**).*

During the age of Greek domination of the sciences, a new mode of thought concerning nature was developed, and this became the foundation of modern observational science resulting in a more rational and logical approach to medicine and consequently drug discovery. Empedocles (504–443 BCE) extended the philosophy of Thales (639–544 BCE) and introduced the concept that four fundamental elements—air, earth, fire, and water—were the basis of all things *(1)*. It was into this society where reasoning was centered on the four "humors" as postulated by Empedocles that Hippocrates (460–370 BCE) was born. Considered the Father of Medicine, his methods of observation, scientific assessment, and therapeutic practice were the first of their kind, and he produced a range of medical texts, including those on treatments using many naturally derived drugs,

* There are two frequently used symbols for medicine. One is the staff of Asclepius (**Fig. 1**), which depicts a single snake coiled around a staff, while the second is the caduceus of Hermes and involves two serpents coiled around a staff capped by a pair of wings. Given that we associate the symbol with medicine, it would seem more appropriate that the staff of Asclepius be used as the symbol. On the other hand, the caduceus of Hermes (Hermes is the messenger of the gods and conductor of the dead) became associated with healing around the 7th century CE and was linked to alchemy, from which chemistry and chemotherapy evolved. Alchemists at the time were frequently referred to as Hermetists.

that were unsurpassed in their field for many centuries *(1)*. Theophrastus (372–287 BCE), who upon the death of Aristotle (322 BCE) inherited both his library and most importantly the botanical garden, is credited with creating the first-written herbal, an extensive and precise description of 455 plants and their medicinal properties. Another significant work, which built upon those of his forebears, was *De Materia Medica* written by Pedanius Dioscorides, who lived in the first century of the current era. The work was a study on the preparation, properties, and testing of drugs, and included over 1000 drugs of natural origin. Dioscorides' *De Materia Medica* was republished many times up until the 16th century and was the basis for the transmission of the recorded knowledge of medicine and drugs throughout the Dark Ages of Europe *(1)*. Arguably, the final great European figure in the drug discovery field before the Renaissance was Galenus (131–201 CE), better known as Galen. The first leading figure in experimental physiology, Galen produced more medical texts than any other ancient medical author. He created "galenicals," mixtures of herbs that were used as remedies until the advent of drugs that consisted of a single agent. With his death in 201 CE, scientific medicine declined in Europe for more than a thousand years. It was during this time that the Arab culture of northern Africa, Moorish Spain, and the Middle East established comprehensive libraries and founded hospitals and schools of learning, thereby advancing the scientific knowledge that was to be adopted in Europe with the Renaissance. One of the great Moorish scholars, al-Baitar (1188–1248 CE), built on the knowledge recorded by Dioscorides and Galen in the field of medicinal plants and assembled an extensive textbook on pharmaceuticals documenting over 1400 drugs *(2)*.

The science of natural products as distinct chemical entities with known molecular architectures is a relatively new field having commenced with the birth of organic chemistry as a discipline. The biggest breakthrough in isolation and identification of natural products in the last three decades has been a combination of the widespread availability of high-performance liquid chromatography (HPLC) instrumentation (*see* Chaps. 8 and 9), a topic recently reviewed in the first edition of this book by Stead *(3)*, and the advances in analytical instrumentation. Recent progress in nuclear magnetic resonance (NMR) spectroscopy and mass spectrometry (MS) has revolutionized natural product chemistry, improving dereplication methods (*see* Chap. 12) and allowing novel compounds to be targeted and identified in a time frame of days on as little as a single milligram of compound. The isolation of microbial natural products involves a multitude of techniques, and as such a single chapter can never be hoped to accurately reflect the breadth of techniques involved. However, in many ways, the isolation of microbial natural products does not differ significantly from the isolation of secondary metabolites from plants or animals. This chapter looks at two aspects, countercurrent methods and polymeric adsorbents, used extensively in microbial natural product chemistry.

2. The Isolation of Penicillin: The Beginning of Counter Current Chromatography

In any review concerning the isolation of microbial natural products, mentioning penicillin is not unexpected. While penicillin was not the first compound to be isolated from a micro-organism, indeed in the context of drug discovery that honor would probably belong to the isolation of mycophenolic acid obtained from *Penicillium brevicompactum* ***(4)***, its discovery did promote intense interest into the isolation of microbial natural products, which led to the discovery of many therapeutically useful drugs and the development of many valuable isolation protocols. The original isolation of penicillin by the team headed by Florey and Chain utilized a countercurrent method (*see* Chap. 7), which was crucial to their success where others had failed ***(5)***.

2.1. The Isolation of Therapeutic Penicillin

Penicillin, rather than representing a distinct compound, is a collective term used to describe the class of compounds containing the fused bicyclic β-lactam and thiazolidine moiety (**Fig. 2**). The major penicillin produced by the Oxford team was 2-pentenyl penicillin (penicillin I or penicillin F), while alteration of the media on which the fungus was cultured in the large-scale fermentations conducted in the United States resulted in the major penicillin produced being identified as benzyl penicillin (penicillin II or penicillin G, **Fig. 2**).[†]

The instability of penicillin presented challenges to the Oxford team, which had already defeated at least two previous groups. In 1932, a group led by Raistrick grew *Penicillium notatum* and isolated from it the non-antibiotic pigment, chrysogenin ***(6)***. While the presence of penicillin was recognized, it was not isolated. Three years later in 1935, Reid published

[†] The reader is directed to the Mitsubishi Chemical corporation website, http://www.m_kagaku.co.jp/index_en.htm.

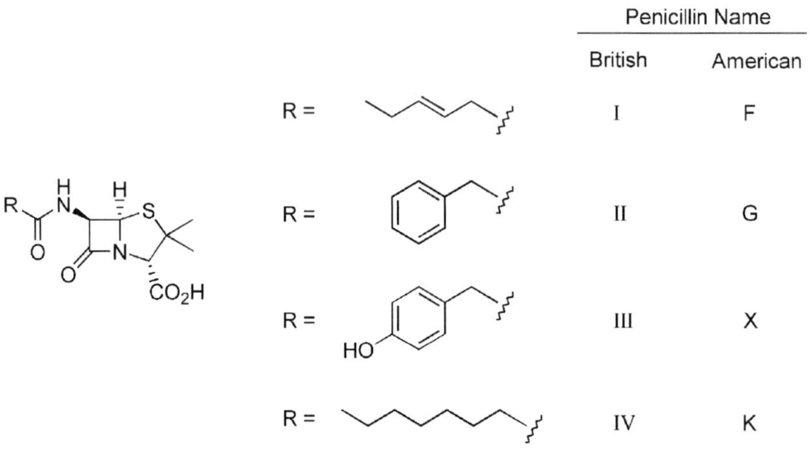

Penicillin Name	
British	American
I	F
II	G
III	X
IV	K

Fig. 2. The structure of some natural penicillin antibiotics isolated from *P. notatum*.

work detailing the inhibitory effects of *P. notatum*; but again the substance responsible for the activity, already named penicillin by Fleming, remained elusive ***(7,8)***. The reason for the chemical instability was the β-lactam moiety, which is susceptible to cleavage in either acidic or basic conditions. The Oxford team satisfactorily overcame this problem, producing penicillin in an active stable form through a combination of liquid–liquid and liquid–solid chromatography. It was the inventiveness of Norman Heatley, one of the team members, who developed a countercurrent extractive separation technique, that resulted in the capacity to produce large amounts of penicillin required for the in vivo experiments that ultimately resulted in the saving of countless lives and won the leaders, Chain and Florey, the 1945 Nobel prize in Physiology or Medicine.[§] The isolation protocol, represented in the text here, is taken from the seminal paper on penicillin production and summarizes the successful strategy employed ***(5)***.

1. Penicillin can be extracted by ether, amyl acetate, and certain other organic solvents from an aqueous solution whose pH has been adjusted to 2.0. From the organic solvent, the penicillin may be re-extracted by shaking with phosphate buffer or with water, the pH of which is maintained at 6.0–7.0.

[§] The 1945 Nobel prize was shared with Alexander Fleming, who while definitely not the first person to recognize the antibiotic effect of the *Penicillium* sp., was the first to demonstrate that a solution injected into an animal was not toxic, thereby laying the foundation of the efforts led by Florey and Chain in the discovery of penicillin.

2. A continuous countercurrent extraction apparatus is used. The crude penicillin having been filtered and acidified, is passed through special jets that break it up into droplets of uniform size. These are allowed to fall through a column of amyl acetate, to which the penicillin is given up (*see* **Note 1**).
3. Fresh solvent is continuously fed into the bottom of the column, from the top of which an equal amount of penicillin-rich solvent is collected for further working up.
4. Batches of 3 L each of the penicillin-containing solvent as delivered from the extraction apparatus are extracted with five successive amounts of 300 mL each of water, using baryta to adjust the pH to 6.5–7.0.
5. The strongest aqueous extract is partially decolorized by shaking with about 8% of animal charcoal and filtering.
6. The partially decolorized solution is cooled, acidified, and extracted into successive amounts of ether; the strongest of the ether extracts is then passed through an adsorption column of Brockmann alumina. The column is eluted with a phosphate buffer (pH 7.2) and the fractions containing the most penicillin are extracted back into ether.
7. Finally, the penicillin is extracted back into water using sodium hydroxide to adjust the pH (*see* **Note 2**).

The penicillin thus obtained, given that it was isolated under British culture conditions would have been mainly penicillin I (2-pentenyl penicillin, **Fig. 2**), was described as a deep reddish-orange fluid with a faint smell and a bitter taste, indicating that while used therapeutically it was not 100% pure.

2.2. The Principle of Countercurrent Chromatography

Countercurrent chromatography has its origins in liquid–liquid extraction (*see* Chap. 10), whereby two immiscible liquids are used to separate organic compounds based on their differential solubility in each solvent. The distribution of the solute molecules between the two phases is governed by the partition constant (K), which is a constant at any given temperature, and theoretically solutes possessing differing partition constants can be separated. Separation occurs in an analogous manner to solid–liquid chromatography or gas–liquid chromatography, where solute molecules are separated on the basis of an equilibrium established between the two phases involved

$$K = C_A/C_B$$

where C_A and C_B represent the concentration of the solute (S) in the two solvents.

While details on the general methodology, especially with respect to natural product isolation, can be found in Chapter 7, an outline of various aspects of countercurrent methods in relation to the isolation of microbial natural products is presented here.

2.2.1. Countercurrent Distribution (CCD)

Craig ***(9)*** recognized the fact that liquid–liquid extraction, for example in a separatory funnel, allowed only limited resolving power, and that the method could be extended to multiple equilibrium events thereby increasing the resolving potential. Multiple solutes could be resolved based on their respective partition constants, and he showed that this method was applicable for a wide variety of binary mixtures. Craig also demonstrated that when the ratio of partition constants between two solutes was greater than 4, separation was easily achieved but also showed that more challenging separations were feasible utilizing his methods. For example, the partial purification of α- and β-naphthoic acids, where the ratio of partition constants was only 1.08, and purification of *p*-toluic and benzoic acids, where the ratio of partition constants is 2.48, was readily achieved. Further to these experiments, Craig was able to show that simply by increasing the number of equilibrium events (transfers), he was able to dramatically raise the resolving power in a system. The separation of a mixture of C12, C14, C16, and C18 fatty acids was readily achieved by increasing the number of transfers ***(10)***. Theoretically, an infinite number of transfers could be utilized to achieve a resolution given that the method does not suffer from adsorption losses, as is frequently the case in other forms of chromatography. The isolation of natural products, many of which were microbial in origin, was achieved using this countercurrent method as illustrated through the isolation of the antibiotic active component, conocandin (**Table 1**) from cultures of the fungus *Hormococcus conorum* ***(11,12)***.

2.2.2. Droplet Countercurrent Chromatography (DCCC)

DCCC best resembles the countercurrent technique utilized by Heatley in the original work on the isolation of penicillin ***(5)***. Whereas Heatley's equipment consisted of a single column, commercial DCCC apparatus may contain as many as 600 columns connected by capillaries through which a mobile phase is pumped. The mobile phase flow can be discretionally reversed to account for the relative density of the stationary phase solvent and the solute molecules can be added in either solvent. The method has been

Table 1
Representative Microbial Natural Products Isolated Using Countercurrent Methods

Structure/name/organism	Countercurrent method	Ref.
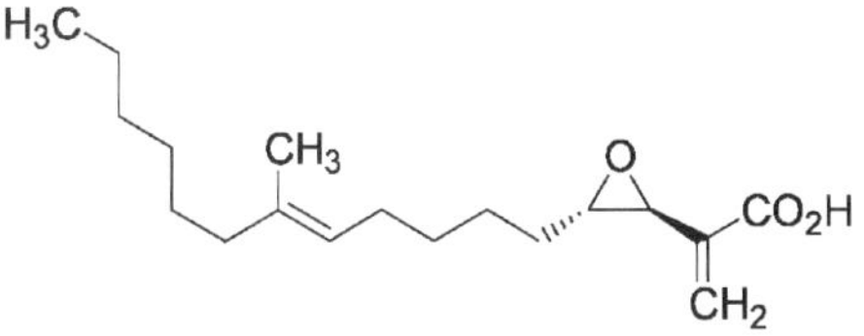 Conocandin *H. conorum*	CCD	*(12)*
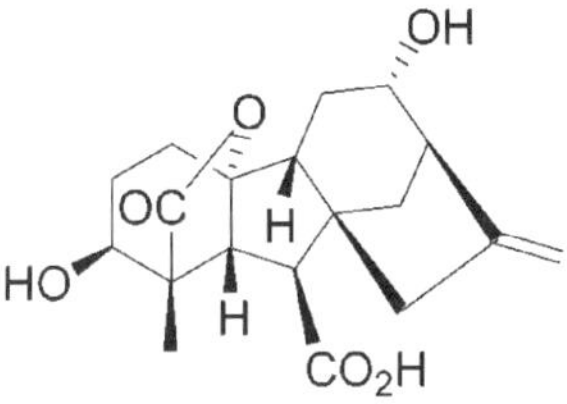 Gibberellin A58 *G. fujikuroi*	DCCC	*(14)*
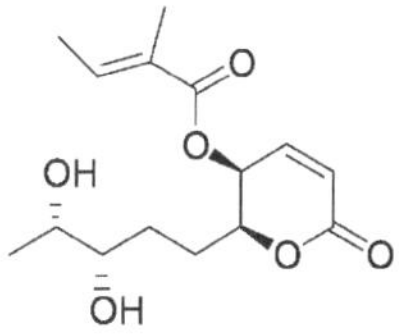 6,7-Dihydrophomopsolide B *Penicillium* sp.	HSCCC	*(17)*

widely used in the isolation of natural products and has been the subject of an excellent review by Hostettmann and Marston ***(13)***. A multitude of natural products have been isolated, and solvent systems suitable for a wide variety of structural classes have been established, thereby lessening the ordeal of finding a suitable solvent system. Both aqueous and nonaqueous systems have been established, allowing virtually any class of compound to be separated. Among the many microbial natural products isolated by this method are members of the gibberellin class of plant growth regulators isolated from the fungus *Gibberella fujikuroi* (**Table 1**) *(14)*.

2.2.3. High-Speed Countercurrent Chromatography (HSCCC)

High-speed countercurrent chromatography represents the most advanced model in the evolution of countercurrent chromatography. It has overcome many of the pitfalls associated with earlier methods and allows the rapid separation of solute molecules from often complex mixtures. For this reason, it has been routinely used as a separation tool *(15)*, and examples of its application in the isolation of natural products can be found in Chapter 7. Among the many examples where HSCCC has been used are those where the technique has been utilized in microbial natural product isolation for the separation of zaragozic acids (squalestatins) *(16)*, phomopsolides *(17)*, and gibberellins, to name a few (**Table 1**).

2.3. Liquid Ion Exchange Extraction

The separation of molecules has been achieved chromatographically by the partition of solute molecules between immiscible phases in a variety of countercurrent techniques. When the solute molecule possesses ionizable functionality, e.g., carboxylic acids or amines, then the possibility of separation based on ion-exchange (*see* Chap. 6) becomes a reality *(18)*. Examples of the application of this technique in the isolation of microbial natural products, e.g., cephamycin A *(19)*, gualamycin *(20)*, and zaragozic acid A *(21)*, are summarized in **Table 2**.

Another ion-exchange process useful in the isolation of microbial natural products is the method of liquid ion-exchange. The *IUPAC Compendium of Chemical Terminology* defines this process as a liquid–liquid extraction process that involves a transfer of ionic species from the extractant to the aqueous phase in exchange for ions from the aqueous phase. The method has found widespread use in the mining industry where it is used for the removal of metal ions from aqueous solution. For example, zinc has been recovered by treating aqueous solutions with organic solutions containing 2-hydroxybenzophenoneoxime and substituted 8-hydroxyquinolines. The zinc ions are transferred into the organic phase, forming an organic soluble ion pair. The organic phase is then separated and

Table 2
Representative Microbial Natural Products Isolated Using Classical Ion-Exchange Methods

Structure/name/organism	Ion-exchanger	Ref.
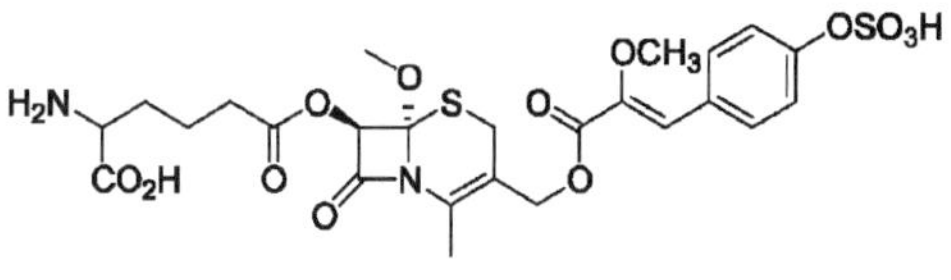 Cephamycin A *S. griseus*	Amberlite IRZ-68(Cl^-) and DEAE Sephadex A-25	*(19)*
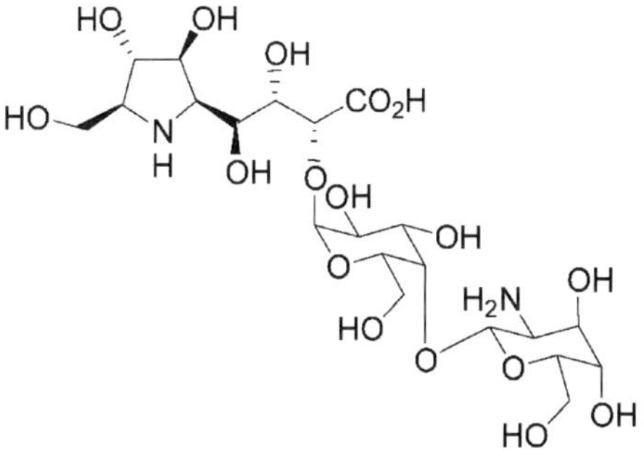 Gualamycin *Streptomyces* sp., NK11687	Dowex 50W (H^+) and CM-Sephadex C-25 (Na^+)	*(20)*
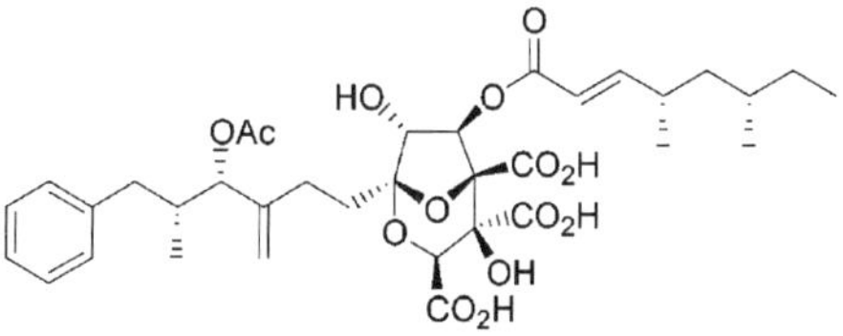 Zaragozic acid A unidentified fungal culture, ATCC 20986	Dowex 1-X2 (Cl^-)	*(21)*

subsequently acidified releasing the zinc ions *(22)*. The technique, while not so widely used in drug discovery operations, has found application in the extraction of microbial natural products, where the isolation of β-lactam antibiotics from *Streptomyces olivaceus* and that of the polyene, amphotericin B from *S. nodosus*, have been achieved and serve as examples (**Table 3**) *(23,24)*.

The isolation of amphotericin B is notable, given that it is an antifungal drug used to treat systemic mycoses which, despite the side effects of its use, remains the drug of choice for life-threatening fungal infections. The isolation of amphotericin B is difficult because it is only sparingly soluble in organic solvents. However, the use of the liquid ion-exchange method greatly enhances its solubility presenting an elegant and efficient method, as outlined later (**Fig. 3**), for the large-scale isolation of this valuable compound.

3. Liquid–Solid Chromatography

A plethora of liquid–solid chromatographic processes have been developed and used in the isolation of natural products, be they produced microbially or otherwise, including gel permeation, adsorption, ion-exchange, and affinity chromatographic methods (*see* Chaps. 4–10). It is not desirable here to review these methods but merely to briefly explore a method used widely in microbial natural product isolation.

3.1. Polymeric Adsorbents

Fermentation broths are predominantly composed of water, and therefore the isolation of exocellular microbial natural products that have been exuded into the growth medium presents problems unique to the field. While intracellular metabolites are readily concentrated by mechanical removal of the cellular biomass and subsequent extraction, the enrichment of secondary metabolites present in fermentation broths requires special attention. Liquid–liquid extraction methods have proven useful in some circumstances. However, they require specialized equipment, and the use and recovery of large volumes of organic solvents, imposing adverse economic and environmental impacts on an operation. As such, alternative strategies have been sought to address these issues, and liquid–solid chromatographic processes have been developed resulting in the creation of a variety of synthetic polymeric adsorbents that have a high affinity for organic molecules.

Polymeric adsorbents represent a large group of available products, including resins supplied by the Mitsubishi Chemical Corporation and the XAD range of Amberlite resins produced by the Rohm and Haas

Table 3
Representative Microbial Natural Products Isolated Using Liquid Ion-Exchange Methods

Structure/name/organism	Organic phase/ion-exchanger	Ref.
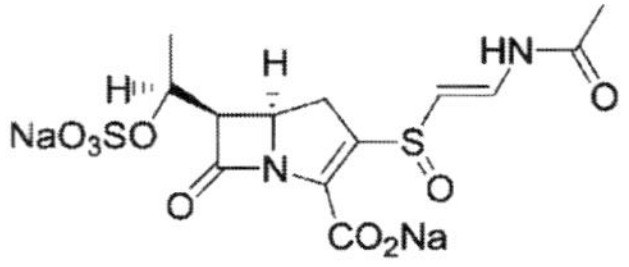 MM 4550, *S. olivaceus*	CH_2Cl_2 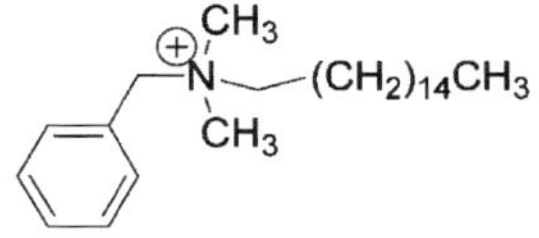	*(23)*
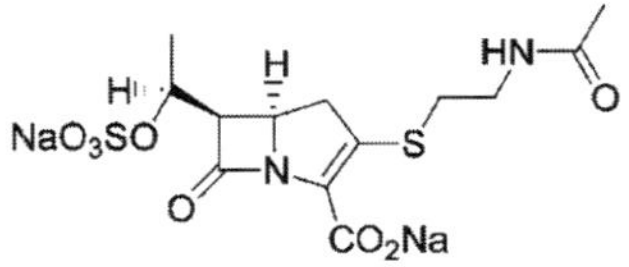 MM 17880, *S. olivaceus*		
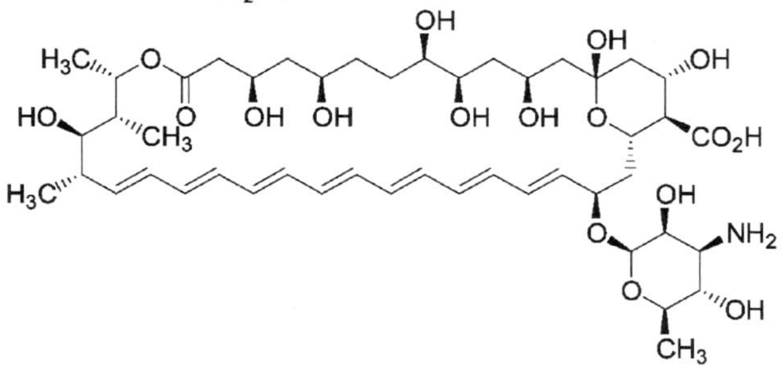 Amphotericin B *S. nodosus*	BuOH 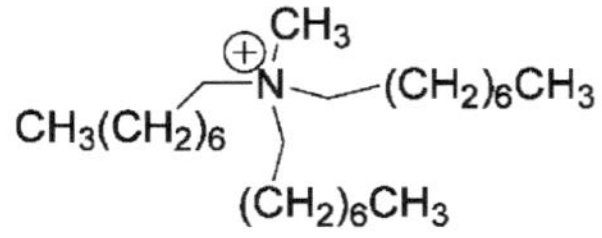 Aliquat 336	*(24)*

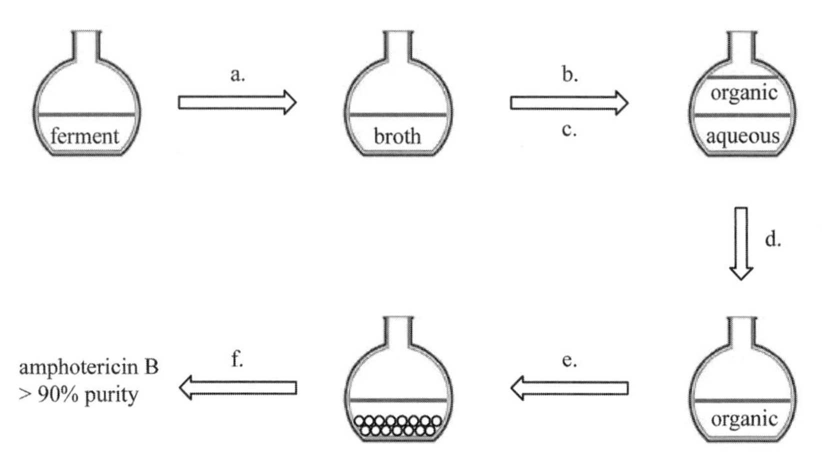

Key steps involved:

a) the fermentation is filtered and the broth retained.

b) pH is adjusted to 10.5 using 7% NaOH solution.

c) broth is mixed with 1-butanol containing 7% w/v aliquat 336.

d) phases are separated and the organic phase is retained. At this stage amphotericin content can be estimated by monitoring at 405 nm.

e) 1% v/v EtOAc is added resulting is gradual hydrolysis of ethyl acetate by aqueous alkaline solution dissolved in the organic phase. The subsequent lowering of the pH (as neutrality is approached hydrolysis stops) results in precipitation of amphotericin B as spherulites (10-40 μm in diameter).

f) filtration recovers over 70% of available amphotericin B obtained in over 90% purity as determined by NMR and HPLC.

Fig. 3. Isolation procedure for amphotericin B using the liquid ion-exchange protocol.

company.¶ While each of the resins has its own special characters and suggested applications, the production of resins is based on similar technology. A variety of materials are used to manufacture the resins, with the most common polymeric supports for use in fermentation applications being based on crosslinked polystyrene matrices that are referred to as aromatic resins. Other chemical structures for synthetic adsorbents are

¶The reader is directed to the Mitsubishi Chemical Corporation website, http://www.m-kagaku.co.jp/index_en.htm.

modified aromatic resins, e.g., SP207 and methacrylic resins such as HP2MG (**Table 4**). The adsorbents are presented as spherical particles characterized by large surface areas. Surface areas of up to $1000\,m^2/g$ are common and possess a highly porous structure. This combination of high surface area and porous nature of the material allows a high uptake of organic solute molecules from aqueous solutions, such as those encountered in fermentations. Owing to the author's experience with diaion products, they are reviewed here. The Amberlite series of XAD resins have similar applications, and a visit to the Rohm and Haas website provides information to the reader with regard to the range and breadth of application of these materials.

Pore size is an important parameter that determines the adsorptive characteristics, and as such resins are provided with varying pore structures. A molecular sieving effect can be achieved by using resins with small pores, thereby favoring adsorption of small molecules while larger ones are excluded. For example, Diaion HP20 is recommended for the adsorption of solute molecules with molecular weights less than 20–30 kDa, while Sepabeads SP825 possessing a smaller pore radius is recommended for the adsorption of solute molecules with a molecular weight less than 1 kDa. If the nature of the solute molecule is known, the most appropriate type of adsorbent can be selected according to the pore size of adsorbent. While in the initial isolation of a natural product this is seldom the case, repeat isolations when the target molecule has been identified can be greatly accelerated and simplified by judicious selection of the most appropriate resin.

The concentration of organic solutes is achieved using these resins by either chromatographic filtration of the fermentation broth, or the polymeric adsorbent can be added to the broth and the resulting suspension stirred gently to allow adsorption. In either case, the fermentations should have been treated to remove cellular biomass prior to adsorbent application to avoid physical impediments to flow. In the author's laboratories, broth is delivered to a column of resin by peristaltic pumping, ensuring a constant flow rate thereby allowing adequate time for diffusion of the solute molecules into the porous support. Typically, a flow rate of 5 bed volumes per hour is employed for the loading of resins, although the raffinate should be monitored to ensure retention of the solute molecules. Once adsorbed, solute molecules are desorbed by applying a suitable concentration gradient to the resin. The synthetic adsorbents are stable in acidic and alkaline solutions and most organic solvents, and they can be easily

Table 4
Properties of Selected Resins From Mitsubishi Chemical Corporation

Resin	Diaion HP20	Diaion HP21	Sepabeads SP825	Sepabeads SP207	Diaion HP2MG
Partial structure	CH, CH_2, CH, CH, CH_2, CH_2			CH, Br, CH_2, CH, CH, CH_2, CH_2	CH_3, CH_3, $-CH_2$, CH_2, C=O, CO_2CH_3, O, $(CH_2)_2$, O, C=O, $-CH_2$, CH_3
Category		Aromatic		Modified aromatic	Methacrylic
apparent density (g/L-R)	680	625	690	780	720
Suggested use	Natural product isolations, extraction of antibiotics from fermentation broths, separation of peptides			Suitable for adsorption of organic solutes at low concentration or of highly hydrophilic substances. Hydrophobic resin	This type of adsorbent is suitable for adsorption of polyphenols and surfactants. Relatively hydrophilic resin
Particle size (>250 μm)		>90%			> 95%
Surface area (m^2/g)	600	570	1000	600	500
Pore volume (mL/g)	1.3	1.1	1.4	1.3	1.2
Pore radius (μm)	>20	8	5.7	11	20

Table has been produced using data from the Mitsubishi Chemical Corporation website (see **Footnote ¶**).

Table 5
Microbial Natural Products Where Synthetic Adsorbents Have Been Utilized in the Isolation Process

Structure/name/organism	Resin	Eluting solvent	Reference
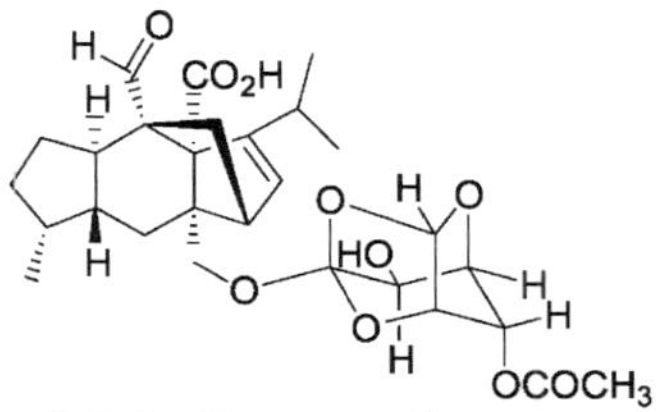 BE-31405, *Penicillium minioluteum*	Diaion HP-20	Aqueous MeOH (50-100% in MeOH)	*(25)*
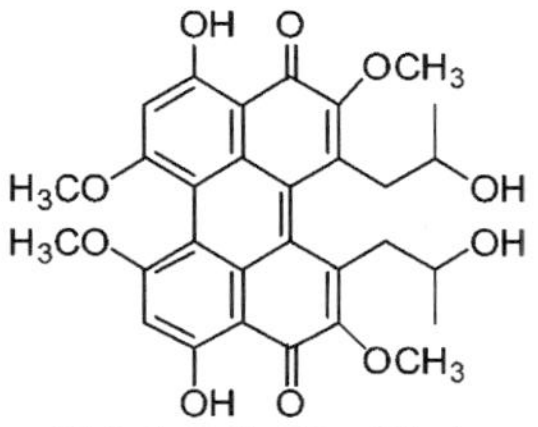 Calphostin D, *Cladosporium cladosporioides*	Diaion HP-20SS	Aqueous MeOH (90-95% in MeOH)	*(26)*

(*Continued*)

Table 5
Microbial Natural Products Where Synthetic Adsorbents Have Been Utilized in the Isolation Process (*Continued*)

Structure/name/organism	Resin	Eluting solvent	Reference
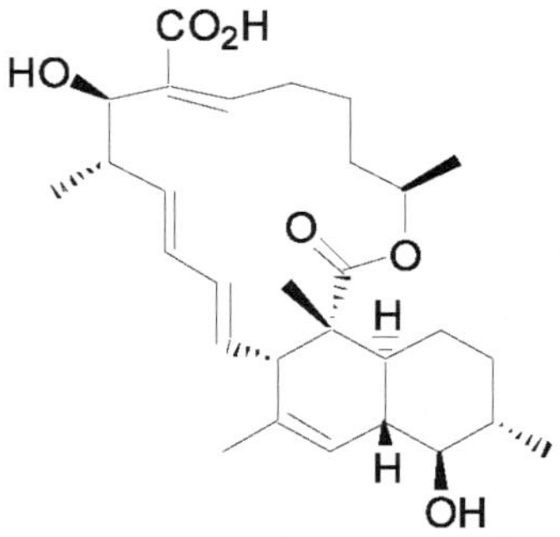 Tubelactomicin A, *Nocardia sp.*	Diaion HP-20	50% aqueous MeOH 50% aqueous acetone	*(27)*
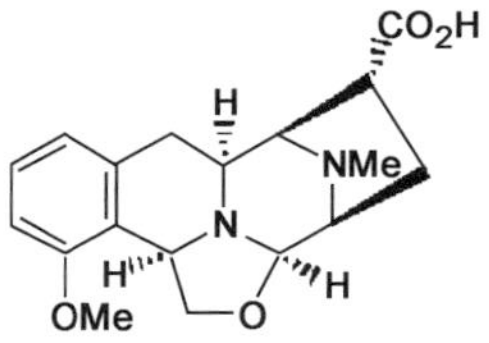 DC-52; 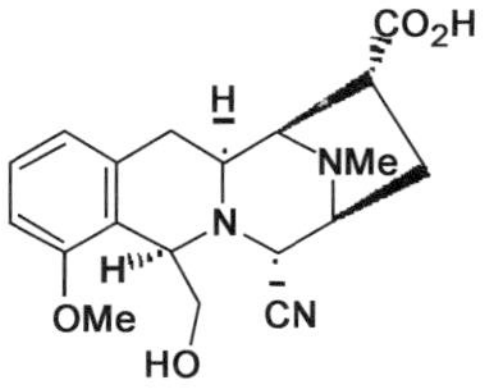 DX-52-1 (artifact) *S. melanovinaceus*	Diaion, SP-207 HP-10	6% aqueous acetone	*(28)* *(29)*

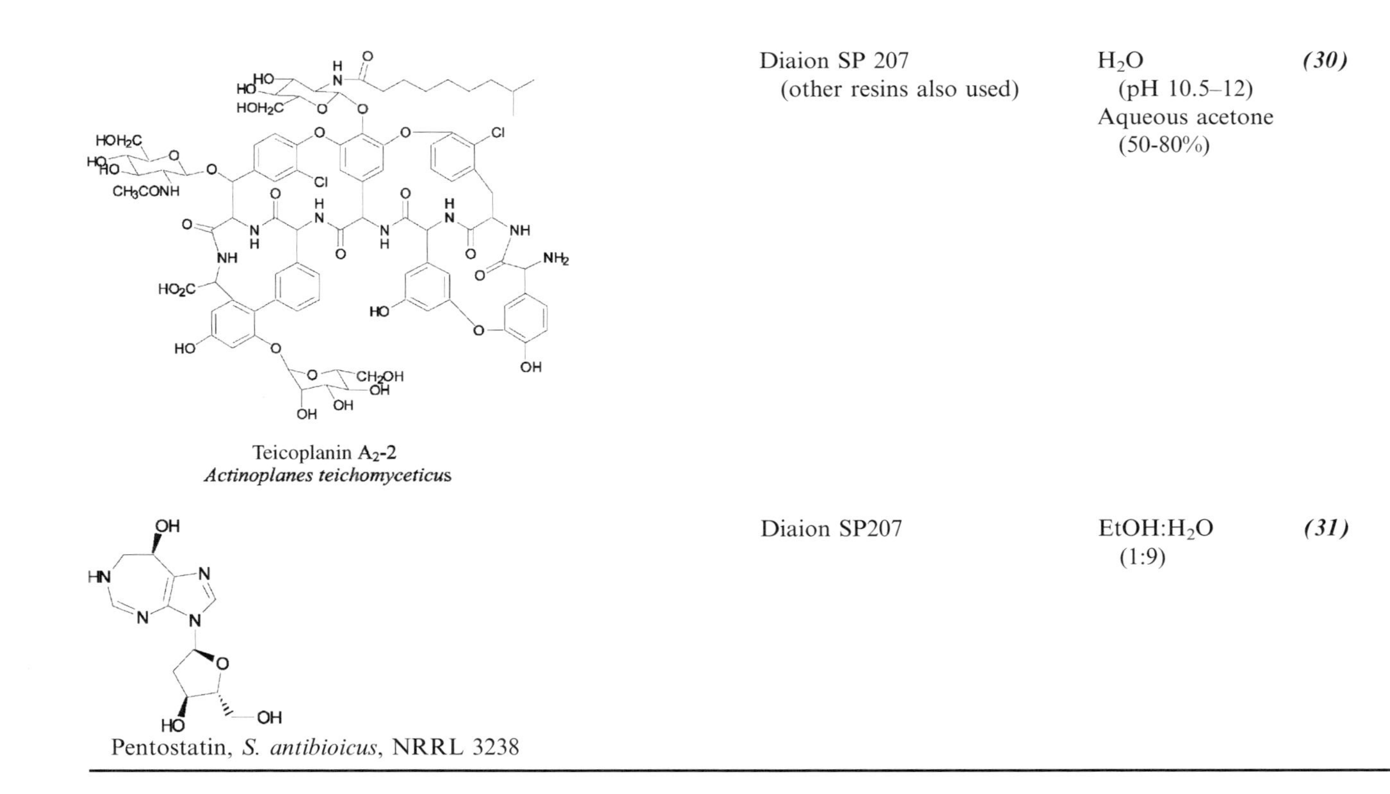 Teicoplanin A_2-2 *Actinoplanes teichomyceticus*	Diaion SP 207 (other resins also used)	H_2O (pH 10.5–12) Aqueous acetone (50-80%)	***(30)***
Pentostatin, *S. antibioicus*, NRRL 3238	Diaion SP207	$EtOH:H_2O$ (1:9)	***(31)***

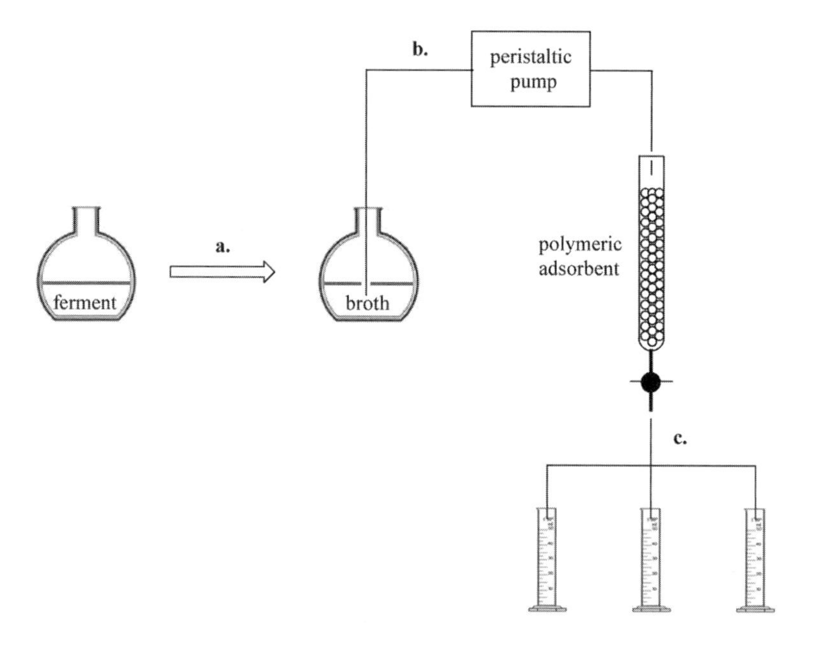

Key steps involved:

a) the fermentation is filtered and the broth retained while the mycelial matt which also contains the target compounds is extracted in EtOH:CH_2Cl_2 (9:1)
b) peristaltic pumping delivers broth at a flow rate of ~5 bed volumes/hr.
c) column was eluted at 1.8 bed volumes per hour using a step gradient from H_2O (fraction discarded), 50% MeOH/H_2O, 100% MeOH and a final elution with acetone (this fraction contained target compounds). Each fraction represented ~1.5 bed volumes.

Fig. 4. Partial purification of ustilaginoidin antibiotics from a cultured entomogenous fungus.

regenerated under mild conditions for repeated use. The aromatic adsorbents are most frequently desorbed by applying an increasing methanol gradient at a flow rate of 1–2 bed volumes per hour and, if necessary, a final elution with acetone will generally ensure complete removal of solute molecules. In the case of ionizable functional groups existing in the solute molecules, solutions of weak acids or bases can be utilized to ensure

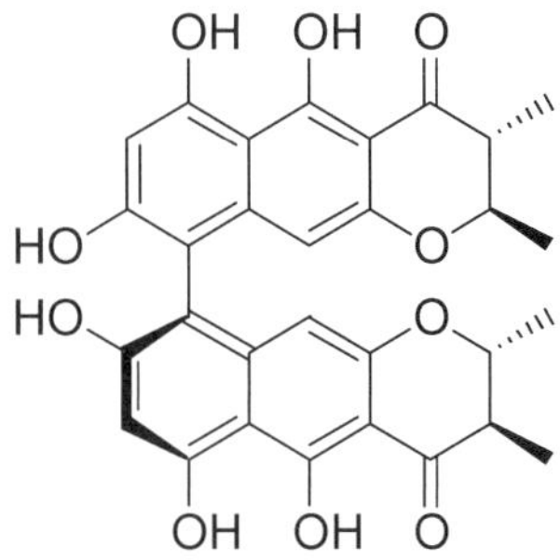

Fig. 5. Structure of ustilaginoidin D isolated from an unidentified entomogenous fungus.

conversion to the ionized species, which then allows rapid elution owing to the repulsive interactions with the support. Examples of isolations where synthetic resins have been utilized in the isolation process and the solvents used to effect elution from the resins are shown for BE-31405 ***(25)***, calphostin D ***(26)***, tubelactomicin A ***(27)***, DC-52 and the artifact derived from it DX-52-1 ***(28,29)***, teicoplanin A_2-2 ***(30)***, and pentostatin ***(31)*** in **Table 5**.

By way of example, the isolation of a series of ustilaginoidins obtained from an (Metarhizium anisopliae) entomogenous fungi has been optimized recently in the author's laboratory ***(32)***. The filtered broth (10 L) was applied to a column of Diaion SP207 (5 cm i.d × 22 cm) at a rate of approximately 2 L per hour. Elution as indicated in Fig. 4 was carried out and the fractions assayed by biological activity and NMR, which subsequently showed that the acetone eluate was composed entirely of a mixture of three ustilaginoidins (>95% of a mixture of three compounds by NMR), including the known compound ustilaginoidin D (**Fig. 5**) and two previously unreported compounds. Final resolution of these three components was achieved by reversed-phase HPLC.

4. Notes

1. Acidifying the cooled broth containing the penicillin seconds before it was added to the countercurrent setup elegantly solved the problem of instability encountered in the acidification step. Extraction into amyl acetate then meant exposure to acid was of the order of seconds.
2. As the solution is not buffered, maximum care must be taken when adding the alkali, as penicillin is rapidly destroyed in alkaline solution.

References

1. Riddle, J. M. (1985) *Dioscorides on Pharmacy and Medicine.* University of Texas Press, Austin, TX.
2. Mez-Mangold, L. (1971) *A History of Drugs.* F. Hoffmann-La Roche, Basle, Switzerland.
3. Stead, P. (1998) Isolation by preparative HPLC, *in Natural Products Isolation*, . (Cannell, R.J.P., ed.), Humana, Totowa, NJ, pp. 165–208.
4. Reiner, R. (1982) *Antibiotics: An Introduction.* Georg Thieme Verlag, Stuttgart.
5. Abraham, E. P., Chain, E., Fletcher, C. M. et al. (1941) Further observations on penicillin. *The Lancet* **16,** 177–189.
6. Clutterbuck, P.W., Lovell, R., and Raistrick, H. (1932) Studies in the biochemistry of micro-organisms. XXVI. The formation from glucose by members of the *Penicillium chrysogenum* series of a pigment, an alkali-soluble protein and "penicillin"—the antibacterial substance of Fleming. *Biochem. J.* **26,** 1907–1918.
7. Reid, R.D. (1935) Some properties of a bacterial-inhibitory substance produced by a mold. *J. Bacteriol.* **29,** 215–221.
8. Fleming, A (1929) On the antibacterial action of cultures of a Penicillium, with special reference to their use in the isolation of *B. influenzae. Br. J. Exp. Pathol.* **10,** 226–236.
9. Craig, L. C. (1944) Identification of small amounts of organic compounds by distribution studies. II. Separation by counter-current distribution. *J. Biol. Chem.* **155,** 519–534.
10. Craig, L. C. (1950) Partition chromatography and countercurrent distribution. *Anal. Chem.* **22,** 1346–1352.
11. Mandava, N. B. and Ruth, J. M. (1988) The origins of countercurrent chromatography, in *Countercurrent Chromatography—Theory and Practice*, Chromatographic Science Series 44 (Mandava, N. B. and Ito, Y. ed.), Marcel Dekker Inc., New York, pp. 27–78.
12. Mueller, J. M., Fuhrer, H., Gruner, J., and Voser, W. (1976) Metabolites from microorganisms. 160th Communication. Conocandin, a new fungistatic antibiotic from *Hormococcus conorum* (Sacc. Et Roum.). *Roback. Helv. Chim. Acta* **59,** 2506–2514.
13. Hostettmann, K. and Marston, A. (1988) Natural products isolation of droplet countercurrent chromatography, in *Countercurrent Chromatography—Theory and Practice*, Chromatographic Science Series 44 (Mandava N. B. and Ito Y. ed.), Marcel Dekker Inc., New York, pp. 465–492.
14. Bearder, J. R. and MacMillan, J. (1980) Separation of gibberellins and related compounds by droplet counter-current chromatography. *Monograph—Br. Plant Growth Regulator Group* **5,** 25–30.

15. McAlpine, J. (1998) Separation by high-speed countercurrent chromatography, in *Natural Products Isolation*, (*Cannell, R.J.P.*, ed.), Humana, Totowa, NJ, pp. 247–260.
16. Dawson, M. J., Farthing, J. E., Marshall, P. S. et al. (1992) The squalestatins, novel inhibitors of squalene synthase produced by a species of *Phoma*. I. Taxonomy, fermentation, isolation, physico-chemical properties and biological activity. *J. Antibiot.* **45,** 639–647.
17. Stierle, D. B., Stierle, A. A., and Ganser, B. (1997) New phomopsolides from a *Penicillium* sp. *J. Nat. Prod.* **60,** 1207–1209.
18. Dufresne, C. (1998). Isolation by ion-exchange methods, in *Natural Products Isolation* (Cannell, R. J. P., ed.,) Humana, Totowa, NJ, pp. 141–164.
19. Miller, T. W., Goegelman, R. T., Weston, R. G., Putter, I., and Wolf, F. J. (1972) Cephamycins, a new family of β-lactam antibiotics. II. Isolation and chemical characterisation. *Antimicrob. Agents Chemother.* **2,** 132–135.
20. Tsuchiya, K., Kobayashi, S., Harada, T. et al. (1995) Gualamycin, a novel acaricide produced by *Streptomyces* sp. NK11687 I. Taxonomy, production, isolation, and preliminary characterization. *J. Antibiot.* **48,** 626–629.
21. Bergstrom, J. D., Kurtz, M.M., Rew, D. J. et al. (1993) Zaragozic acids: a family of fungal metabolites that are picomolar competitive inhibitors of squalene synthase. *Proc. Natl. Acad. Sci. USA* **90,** 80–84.
22. Sudderth, R. B. and Jensen, W. H. (1975) Liquid ion-exchange extraction of zinc. S. African Patent, 19751125.
23. Hood, J. D., Box, S. J., and Verrall, M. S. (1979) Olivanic acids, a family of β-lactam antibiotics with β-lactamase inhibitory properties produced by *Streptomyces* species. II. Isolation and characterisation of the olivanic acids MM 4550, MM 13902 and MM 17880 from *Streptomyces olivaceus*. *J. Antibiot.* **32,** 295–304.
24. Rees, M. J., Cutmore, E. A., and Verrall, M. S. (1994) Isolation of amphotericin B by liquid ion-exchange extraction. Separations in Biotechnology 3, Special Publication—*Roy. Soc. Chem.* **158,** 399–405.
25. Okada, H., Kamiya, S., Shiina, Y. et al. (1998) BE-31405, a new antifungal produced by *Penicillium minioluteum*. I. Description of producing organism, fermentation, isolation, physico-chemical and biological properties. *J. Antibiot.* **51,** 1081–1086.
26. Kobayashi, E., Ando, K., Nakano, H. et al. (1989) Calphostins (UCN-1028), novel and specific inhibitors of protein kinase C. I. Fermentation, isolation, physico-chemical properties and biological activites. *J. Antibiot.* **42,** 1470–1474.
27. Igarishi, M., Hayashi, C., Homma, Y. et al. (2000) Tubelactomicin A, a novel 16-membered lactone antibiotic, from *Nocardia* sp. I. Taxonomy, production, isolation and biological properties. *J. Antibiot.* **53,** 1096–1101.

28. Ishii, S., Katsumata, S., Arai, Y., Fujimoto, K., and Morimoto, M. (1987) Salt of DC-52 and a pharmaceutical composition containing the same. US Patent 4,649,199.
29. Nagamura, A., Fujii, N., and Tajima, K. (2000) Manufacture of antitumor antibiotic substance, DX-52–1, with *Streptomyces melanovinaceus. Jpn. Kokai Tokkyo Koho*; Japanese Patent 20000125896.
30. Kang, T.-W., Choi, B.-T., Choi, G.-S., Choi, Y.-R., and Hwang, S.-H. (2004) Method for purifying teicoplanin A2. US Patent Appl. 20040024177.
31. French, J. C., Edmunds, C. R., McDonnell, P., and Showalter, H. D. H. (1995) Process for purifying pentostatin. US Patent 5,463,035.
32. Barrow, R. A., McCulloch, M. W. B., and Thompson, C. D. unpublished results.

16

Purification of Water-Soluble Natural Products

Yuzuru Shimizu and Bo Li

Summary

Despite the remarkable advancement in purification methods, the isolation of small water-soluble molecules still remains a mystery for natural-product researchers. A general approach to extract and fractionate water-soluble compounds from biological materials and fluids including seawater is presented in this chapter. Some important techniques and special caution needed to deal with water-soluble compounds are also included. The chapter also presents information and discussions about various matrices, such as C_{18} silica gel, resins, and size-exclusion materials with respect to extraction and desalting of water-soluble compounds. In particular, applications and limitations of the increasingly popular nonionic resins are added in this edition. Four typical examples of isolation of water-soluble compounds, which were carried out in the authors' laboratory, are presented in detail. They are: (1) the isolation of amesic shellfish poison, domoic acid, from cultured diatom cells, (2) the isolation of tetrodotoxin derivatives from a newt, (3) the isolation of an aminotetrasaccharide from cultured cyanobacterium cells, and (4) isolation of glycosidic polyether antitumor compounds from the culture broth of a marine dinoflagellate.

Key Words: Water-soluble natural products; extraction methods; purification methods; desalting; reversed phase; ionic resins; nonionic resin; size exclusion; gel filtration.

From: *Methods in Biotechnology, Vol. 20, Natural Products Isolation, 2nd ed.*
Edited by: S. D. Sarker, Z. Latif, and A. I. Gray © Humana Press Inc., Totowa, NJ

1. Introduction

Recent advancement in separation technology and structure elucidation methods has been remarkable. It is not unusual for submicrograms of substances to be purified from vast amounts of biological material, and their structures elucidated. The major contributing factors are progress in high-performance chromatography technique and computer-aided modern spectroscopy, such as high-resolution nuclear magnetic resonance and advanced mass spectroscopy. With the help of these techniques, a number of minute biologically important substances have been brought to light, which would have been inconceivable a few decades ago. The micronization in natural-product research also contributes to the conservation of rare species and protection of global ecology as a whole. Despite all this, the purification of small water-soluble molecules is still considered to be difficult and shunned by most researchers. For example, a glance at the natural-products section of Chemical Abstracts shows that majority of new substances reported are in the lipid-soluble category. There may be more lipid-soluble compounds than water-soluble compounds in nature, but this ratio seems to be greatly disproportionate. Clearly, this reflects the difficulty in purifying water-soluble small molecules. About 20 yrs ago, one of us had written about the purification of the water-soluble compounds with an emphasis on marine natural products *(1)*. It became a quite popular guide to those who dared to work on water-soluble components. In the first edition of this book, we had written a similar chapter to present a general procedure to work with water-soluble compounds *(2)*. In this edition, it is updated in response to some criticism presented by the readers.

2. General Methods

2.1. General Extraction Procedure

In line with the principle "like dissolves like," one might think that the most suitable solvent for the extraction of water-soluble compounds is water. However, more often than not, simple maceration of fresh biological material with water or aqueous buffers fails to extract water-soluble compounds because they are mostly stored in protected states. The mechanism of such protection is varied: binding to the membranes, compartmentalization, and protection by lipophilic material, to name a few. This problem is usually not encountered in the extraction of lipophilic compounds, because organic solvents break up the compartmental structures. For that reason, polar organic solvents, such as methanol (MeOH) or ethanol (EtOH), are often used in the extraction of water-soluble compounds, even if they may not be the best solvents to dissolve the target molecules. Also, these organic solvents are much easier to evaporate than water. Other methods to break up the compartmentalization are sonication, freezing thawing, freeze drying, heating, and enzyme digestion. These techniques can be used singularly or in combination.

Use of organic solvents also prevents microbial growth, which is one of the most serious problems associated with the isolation of water-soluble compounds. Generally, it is imperative that all isolation work be done in a cold room. Use of antimicrobial agents, such as sodium azide, has sometimes been recommended, but the authors discourage the use of azide in isolation, because there is evidence that azide can react with certain molecules to form triaza derivatives. If azide is used for the preservation of gel columns or other supports, care should be taken to wash it out thoroughly before use. Another problem associated with aqueous systems is the activation of enzymes, such as peptidases, glycosidases, sulfatases, and oxidases by solubilization. If the desirable compounds are susceptible to these enzymes, the native natural products will not be isolated. A well-known example is cardiac glycosides from *Digitalis*. The genuine glycosides are purpurea glycoside A and purpurea glycoside B. In the usual extraction process, it is difficult to isolate the native glycosides because the terminal glucose unit is lost by the action of indigenous β-glucosidase, and only digitoxin and gitoxin can be isolated. Deactivation or denaturation of the enzymes can be carried out by quick heating or treatment with alcohols, but alcohol treatment does not necessarily deactivate the enzymes. Therefore, it is strongly recommended that the large molecular weight fraction containing liberated enzymes be removed as quickly as possible by methods, such as ultrafiltration or gel filtration.

2.2. Desalting and Choice of Buffer Solutions

Desalting is probably the most important step in the handling of water-soluble compounds. It is not necessarily difficult to fractionate water-soluble compounds to a single chromatographic peak. In fact, strongly hydrophilic compounds are routinely analyzed by high-performance liquid chromatography (HPLC) without any trouble. However, what makes the preparative separation of water-soluble compounds so difficult is the separation of inorganic salts, often from minute amounts of the target molecule. This is the step where natural-product chemists may simply give up their efforts, because most of them are used to removing inorganic salts by simple partition between water-immiscible organic solvents and aqueous solutions.

If the compounds have some lipophilicity, the standard method for desalting is the use of reversed-phase columns, such as C_{18} silica gel, or various other organic polymers, such as XAD-2, –4, and –7 (vide infra). Usually, the aqueous solution is passed through the column and salts are washed out with water, and subsequently, the organic compounds are eluted with a solvent system containing organic solvents. This method cannot be applied to strongly polar or ionized compounds, e.g., polar amino acids or sugars. Another common desalting method involves the use of size-exclusion columns, such as Sephadex G-10 and Bio-Gel P-2. Inorganic ions are presumed to be the smallest in the mixture, and hence their elution should be expected to be retarded the most. This, however, is not always the case because there are unpredictable interactions between the compounds and column support, so that the predicted elution order can often be reversed. These interactions can hold advantages for the extractor and can be used to isolate specific types of compounds. The purification of paralytic shellfish toxins, e.g., saxitoxin and gonyautoxins, was carried out by taking advantage of their specific adsorption on Bio-Gel P-2 and Sephadex G-10 ***(1)***.

In most cases, the extraction and fractionation of water-soluble compounds necessitate the use of buffer solutions. However, the key to the isolation of water-soluble compounds is finding a suitable pH and ionic concentration to keep the molecules as individual molecular entities separable on a given matrix. However, the separation of the buffer salts from the compound is not an easy task. Several volatile buffers are suggested to avoid this problem (**Table 1**). They are mostly combinations of weak bases and acids, which are bound only by weak ionic interactions, and are volatile under reduced pressure or lyophilization. However, in reality, these buffers are not so easy to handle. For example, the organic buffers containing pyridine and other aromatic bases have intolerable, noxious odors, and are difficult to use in a closed cold room. They may also not be acceptable in the modern laboratory for reasons of safety. They retain lipohilic nature and interact with the solutes just like alcohols. Ammonium acetate and ammonium carbonate buffers are removable by evaporation or lyophilization, but in practice, it is not an easy task. The first lyophilization often results in a sticky mixture of ammonium acetate or carbonate and solutes. Sometimes, this process has to be repeated several times. It is usually necessary to redissolve the residue in a large amount of water, freeze,

Table 1
Examples of Volatile Buffers

Buffer	pH
Ammonium bicarbonate	5.0–7.0
Ammonium acetate	7.0–8.0
Pyridine–acetic acid, 16.1:278.5	3.1
Pyridine–acetic acid, 161.2:143.2	5.0
Pyridine–acetic acid–α-picoline, 11.8:0.1:28.2	8.0
Pyridine–acetic acid-2,4,6-collidine, 10:0.4:10	8.3
Pyridine–acetic acid–*N*–ethylmorpholine, 7.5:0.1–0.5:12.5	9.3

and lyophilize again. In either case, one has to expect drastic pH changes during evaporation, which may affect the desired compounds.

2.3. Selection of Chromatographic Supports

Without the help of some kind of chromatographic method, the purification of water-soluble compounds would be difficult to accomplish. There are a variety of column packing materials with many choices of interacting functional groups (*see* Chaps. 5 and 6). Their interaction mechanisms include weak, moderate, and strong, anion and cation exchange, ligand exchange, reversed-phase and ion- and size exclusion, among others. Here, special attention must be paid to the possible irreversible adsorption of certain substances onto the support material.

We all know from experience that the recovery of substances from chromatography is rarely 100%. Some compounds tend to "disappear" on the column. In case of water-soluble compounds, the tendency is greatly enhanced, especially with silica gel-based supports. Some compounds, including amino acids, are virtually impossible to separate on a preparative scale on silica gel-based supports. The mechanism for such irreversible retention is not fully understood, but trapping by hydrogen bonding in suitably shaped pores or matrices may be the greatest contributing factor. For some compounds, covalent bonding by Michael addition or hemiketal formation is also suspected. With respect to irreversible adsorption, there seems to be a strong relationship between the shape of the molecule and pore size. For example, linear polycyclic ether brevetoxins are easily trapped in silica gel-60, and never eluted even after washing with very polar solvents. The utilization of used silica gel of large pore silica gel can lessen this irreversible adsorption, but it also results in poor separation. Similar observations were made with some cyclic peptides. Use of material with larger pore sizes is sometimes helpful. Materials based on a range of silica gels having a different pore size are available. For example, EM Corporation sells silica gel-based columns of pore size up to 1000 Å. Needless to say, however, the larger the pore size, the smaller the surface area and, consequently, the lower the number of theoretical plates. Capping of free hydroxyl groups on the silica helps to prevent undesirable adsorption, and many such "end-capped" silica-based HPLC columns are commercially available. Fast chromatography or elevated temperature can also help to improve recovery, but often the problem still remains. Fortunately, C_{18} and other reversed-phase columns, which could be used for the separation of water-soluble compounds with greater success, are now widely available. They include matrices, such as polystyrene/divinylbenzene and hydroxyethyl methacrylate/dimethylacrylate copolymers with various functional groups. Recovery rates are generally much better on these polymeric materials and are accompanied by much longer column life. These alternative supports are also more tolerant to extreme pH and high concentrations of buffers, such as ammonium acetate.

2.4. General Fractionation Scheme

Before 1950s, water-soluble natural products isolated were mostly limited to molecules that are stable in strongly acidic and basic conditions. In case of amphoteric compounds, such as amino acids, a combination of strongly acidic and basic ion-exchange resins can easily separate them from other components. The procedure, however, is not applicable for esters and other acid- or alkali-labile compounds that do not survive in the extreme pH conditions of the resins or buffers. Epimerization of chiral centers can also be a problem. Thus, for an unknown compound, a fractionation scheme involving any drastic conditions should be avoided. On the other hand, weakly acidic or basic ion-exchange resins can be used under less drastic conditions. The combination of ionic interaction and reversed phase and/or gel filtration (size-exclusion) properties especially exerts tremendous separation capability. Recently, a large number of this type of column materials have become available. **Table 2** shows the

Table 2
Chromatographic Supports Often Used in the Author's Laboratory for Separation of Water-Soluble Components

Type of Compounds	Column Packing
Mono-and oligosaccharides	Sephadex G-10, G-15, polymer-based gels, e.g., Bio-Gel P-2, Ca-loaded gels (e.g., Supelcogel Ca), strong cation exchange (-SO_3H) resins, weakly basic anion-exchange resins
Polysaccharides	Sephadex G-50, G-100, G-200, Bio-Gel P-6 ~ P-100, styrene divinyl benzene-based size-exclusion resins (e.g., Bio-Gel SEC) and their DEAE-bonded material
Oligopeptides	Sephadex G-10, G-15, Bio-Gel P-2 ~ P-6 and other polymer-based size-exclusion gels and reversed phase gels
Proteins and glycoproteins	Sephadex G-15 ~ G-200, Bio-Gel P-6 ~ P-100, other polymer-based size exclusion material and their DEAE-bonded material, C18 on large-pore (300 Å) support and C18 on polymer-based material (e.g., HEMA and Hamilton PRP)
Amine, Guanidine and amino acid derivatives	Strongly acidic action exchange resins (-SO_3-), weakly acidic(-COOH) cation exchange resins, especially, polymer supports (e.g., HEMA CM and Bio-Rex 70), RP (C18, C8 and Hamilton PRP), Bio-Gel P-2
Polar carboxylic acids	Anion-exchange resins, RP (C18 and Hamilton PRP)
Glycosides	RP (C18, C8, and polymer-based), Sephadex G-10, LH-20
Desalting	Bio-Gel P-2, Sephadex G-10, RP (C18, PRP)

Sephadex, Pharmacia Fine Chemicals; Bio-Gel, Bio Rad Laboratories; HEMA, Altech Assoc., Inc., and Supelcogel, Supelco, Inc.

chromatographic supports often used in the authors' laboratory and their applications. **Figure 1** illustrates the general isolation procedure for the first trial. Once the chemical nature of the target molecule is known, the procedure can be altered or shortened.

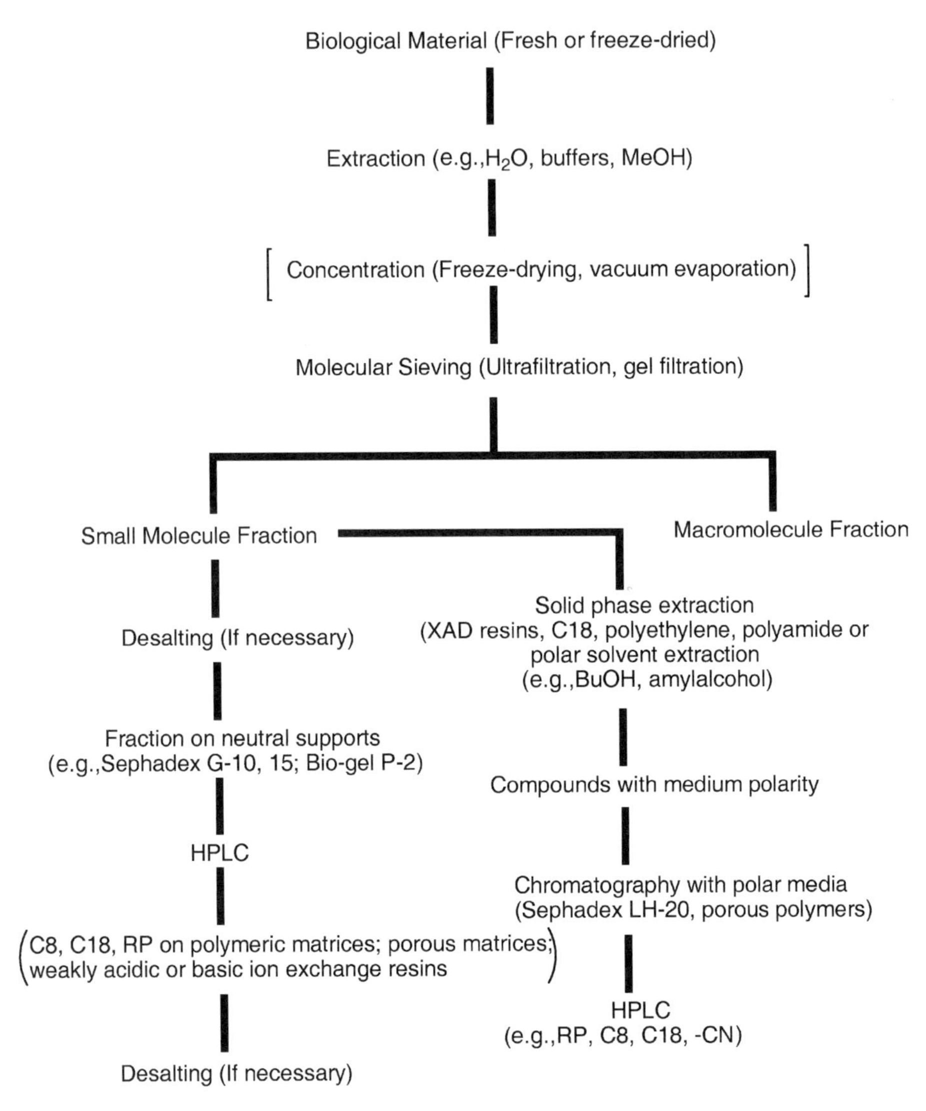

Fig. 1. General isolation scheme for unknown small water-soluble molecules.

2.5. Extraction of Organic Compounds From Aqueous Phase With Nonionic Resins

After publication of our chapter in the first edition of this book ***(2)***, a number of readers questioned the lack of in-depth treatise of this powerful method. There was a good reason for the omission. We felt that the method is not for "true" water-soluble compounds. In this second edition, however, we would like to discuss the merits and limitations of the nonionic resins and clarify some serious misunderstandings especially among the researchers in the nonchemist communities.

The use of nonionic resins to isolate organic compounds in aqueous phases has become quite popular in recent years. There are a couple of reasons behind this enormous popularity. First, we can avoid the use of environmentally hazardous organic solvents, such as chloroform ($CHCl_3$) and dichloromethane (DCM). Second, it is easy to handle a large amount of aqueous phase by just passing through a small cartridge, instead of extracting with a solvent in a separating funnel. It is also suitable for processing a large number of samples in routine extraction and analysis. However, there are also misconceptions of this popular method. Most importantly, the extraction with the nonionic resins does not recover all organic compounds dissolved in water. For example, the extraction of organic material in seawater by this method does not constitute the quantification of the entire carbon compounds in the seawater, as sometimes referred to, but is rather limited to compounds with certain hydrophobicity, including large molecules, such as humic acids and peptides ***(3–7)***. The strongly polar or ionized molecules, such as free amino acids and carbohydrates, are not recovered by the nonionic resins, at least in entirety ***(8)***.

The nonionic resins have macroreticular structures made of aromatic and alkyl matrices. The nonionic resins adsorb compounds by a combination of hydrophobic interactions with the aromatic rings and alkyl chains (van der Waals and $\pi - \pi$ interaction), and sieving effect by the resin matrices. Therefore, the adsorbing strength is greatly influenced by the matrix structure and pore sizes. Those extractable molecules have to possess certain hydrophobic structural features to be extracted even by the most powerful resins, such as Diaion HP-20. Thus, although most of the popular nonionic resins are copolymers of styrene and divinylbenzene, the adsorption capability varies by the products, and it is important to select a proper resin that fits one's target compounds. The representative products are shown in **Table 3**.

2.6. The Application of Nonionic Resins

The nonionic resin extraction has been used for decades for the extraction of pesticides and other synthetic organic molecules dissolved in minute quantities in water ***(9,10)***. In this case, most of the target

Table 3
Polymeric Adsorbent Resins Often Used in the Isolation of Natural Products From Aqueous Phase

Type[a]	Matrix[b]	Particle size	Ave. pore diameter (Å)	Surface area (m^2/g)	Application
Amberlite adsorbents					
XAD-2	pStyDVB	20–60 mesh	90	300	Hydrophobic compounds up to 20,000 MW, removal of antibiotics, organic nitrogen, grease, and various aromatic compounds from aqueous stream
XAD-4	pStyDVB	20–60 mesh	40	725	Small hydrophobic compounds, ~1,000 MW
XAD-7HP	Acrylic ester	20–60 mesh	90	450	Compounds up to 60,000 MW, plant extract, enzyme purification, peptides
XAD-16	pStyDVB	20–60 mesh	100	800	Hydrophobic compounds up to 40,000, more efficient than XAD-2
XAD-1180	pStyDVB	20–60 mesh	300	600	Plant extracts, large MW product recovery
XAD-1600	pStyDVB	400 µm	100	800	Antibiotic recovery, chromatography

Diaion/MCI Gel adsorbents					
Diaion HP-20	pStyDVB	20–30 mesh	260	500	Extraction of antibiotic intermediates from fermentation broth, separation of peptides, or food additivies, debittering of citrus juice, etc
Diaion HP-20SS	pStyDVB	75–150 μm	260	500	Small and medium proteins, hydrophobic compounds, chromatographic separation of peptides, amino acids, reversed phase
MCI Gel CHP-20P	pStyDVB	75–150 m	400–600	500	Biopharmaceuticals, aromatic compounds, peptides, steriods, desalting, reversed-phase application. Good for nonaqueous use
Dowex adsorbents					
Dowex L-323	pStyDVB	16–50 mesh	100	650	Fine chemical, wastewater, product isolation, product decolorization and organic removal. This is an industrial grade resin, not suited for drinking water or food applications
Dowex L-493	pStyDVB	20–50 mesh	46	1100	Removal of certain organics from water. Because of its unique pore size distribution, it has high capacity for organics and good desorption characteristics

[a]XAD, both Diaion and MCI, and Dowex are trade names of Rohm and Haas Co, Mitsubishi Chemical Inc. and Dow Chemical Co, respectively. All data are taken from the manufacturers product information.
[b]pStyDVB: polystyrene/divinylbenzene.

compounds are not water-soluble compounds in true sense, but they are actually quite nonpolar compounds with high hydrophobicity. However, water-soluble compounds having some hydrophobic moieties in the molecules can be also extracted by the nonionic resins. For example, glycosides including saponins and iridoids, short peptides, and some hydrophobic proteins can be extracted. The nonionic resins also provide excellent matrices for chromatographic fraction ***(11,12)***. They are not generally applicable for the ordinary sugars, including aminoglycosides with exceptoins of some carbohydrate derivatives. It should be noted that neutral carbohydrates can be adsorbed on both strongly acidic and basic resins, and they are used to analyze mono- and oligosaccharide. However, the method may not be useful for preparative extraction of sugars from aqueous phase and separation from salts. There is also a stability problem at a very high or low pH. An alternative way to isolate sugars from an aqueous phase is to bring the entire solution to dryness and acetylate or benzoylate, extract with organic solvents, and regenerate it by alkaline hydrolysis. Again, the stability of the sugars at alkaline pH and final removal of salts may present problems. The nonionic resin can be used by batch, and even put directly into on-going culture, or placed in an external medium recycling line. They may enhance the metabolite productivity by reducing the feedback inhibition or self-inflicting toxicity ***(13–15)***.

2.7. Choice of Resins and Solvents

Table 3 displays a list of commercially available popular nonionic resins. Selection of resins with proper pore sizes to match the molecular size of the compounds to be separated is important. If the molecular size of the target compounds is not known or is intended to extract all sizes of molecules, the use of a larger pore size resin is recommended. For the ordinary extraction and fractionation, XAD-4, XAD-16, and Diaion HP-20 are often used. They have pore sizes of 40, 100, and 260 Å, respectively, and are suitable for molecular sizes of around 60,000 or less. Although the XAD resins and Diaion HP-20 closely resemble in their adsorbability, it is generally conceived that Diaion-20 is stronger than XAD-4 or XAD-16. An increasing number of people are using Diaion HP-20. However, the stronger adsorbability can lead to the poorer recovery rates of compounds from the resins. Normally, MeOH or aqueous MeOH is recommended to elute the compounds, but for tightly bound compounds, it may be necessary to use stronger solvents, such as acetone, 1-propanol, or 2-propanols. One common problem with the resins is the leaching of the resin materials. Even well-washed resins leach small amounts of organic matters: oligomers or degraded polymers, and a stronger solvent mean an increasing amount of leaching material. Use of a stronger solvent (e.g., DCM) also results in softening and swelling of the resins. Here, extreme care has to be taken to prevent the breakage or clogging of the column. When a strong organic solvent is applied, it is advisable to use an extremely slow elution speed (space velocity: 1–3) to prevent the rapid expansion of the volume.

2.8. Preparation of Resins

It is strongly recommended to clean the resin before use. With exceptions like prewashed XAD-2 for pesticide analysis, all commercial resins need precleaning. First, the resins should be soaked in water in a beaker, and small particles removed by repeated decantation. This process is especially important, because the small particles drastically slow down the solvent speed. The commercial resins are supposedly prepared in certain ranges of mesh size, but frictions during transportation and spontaneous cracking of the beads produce finer particles. Regarding the breakage of the particles, it is important not to dry the resins. The resins should always be kept moist with either water or MeOH. The resins should be washed thoroughly with water and MeOH, and the solvent that one wishes to use. The use of a Soxhlet extractor is also a choice to wash the resins efficiently. Before use, the resins should be washed well, and equilibrated with water to make it completely free from organic solvents. When washing or equilibration is done in a column, the water elution speed of about 10 times of the resin volume (i.e., space velocity 10) is recommended. When the resin is to be put directly into the on-going fermentation medium, it has to be autoclaved. Most nonionic resins can withstand a high temperature (approx 180°C), well above the sterilizing temperature (approx 120°C). It should be noted that it is not the case with the functionalized resins like strongly basic or acidic ones.

2.9. Heavy Metal Contamination

In modern natural-product research, most purified samples are examined by high-resolution nuclear magnetic resonance (NMR). In this respect, the authors wish to draw special attention to a problem that is unique to water-soluble compounds and rarely encountered with lipophilic compounds: contamination of the sample with paramagnetic heavy metals that can be derived from the original extract, equipment, lab ware, or reagents used in the isolation.

As mentioned earlier, most purification procedures involve chromatographic separation and, thus, use of solvent delivery systems is recommended. Most delivery systems use stainless steel (SUS 13–32) in pumps, lines, and injectors that release Ni and Fe ions when buffer solutions are used. The amount of the leached metal ions may be small but can be enough to cause serious problems in NMR spectroscopic analysis. The contamination can also come from the surface of glassware. The problem is especially troublesome if the final purification step is the concentration of a large amount of an aqueous acidic solution to obtain just a microgram quantity of a substance. In a ^{15}N NMR experiment of biosynthetically enriched small amounts of paralytic shellfish poisoning toxins, it was found to be absolutely necessary to pass the sample solution through Chelex 100, a metal chelating resin. Otherwise, no signals would be observed in ^{15}N NMR *(16)*. **Figure 2** shows an alarming example from our laboratory. The apparently clean ^{13}C NMR spectrum of "chromatographically pure" domoic acid, **1**, obtained by elution with 0.1% acetic acid, has several carbon signals missing or collapsed because of metal chelation. Treatment of the sample with Chelex 100 restored the missing signals. Without treatment, there could have been a serious misinterpretation of the structure or the identity of the compound. Therefore, the authors strongly recommend the use of Chelex 100 for the preparation of all NMR samples of water-soluble compounds. Generally, a remarkable improvement is seen in the spectrum quality.

3. Typical Examples of Isolation of Water-Soluble Compounds

The following are four typical examples of isolation of water-soluble compounds carried out in the authors' laboratory.

3.1. Isolation of 1-Hydroxy-5, 11-Dideoxytetrodotoxin, 2 and Other Tetrodotoxin Derivatives From the Newt, Taricha granulosa (17)

3.1.1. Materials

1. The newts, *T. granulosa* collected in Oregon, were obtained through Carolina Biological Supply Co. Ultrafiltration apparatus: Amicon High Performance

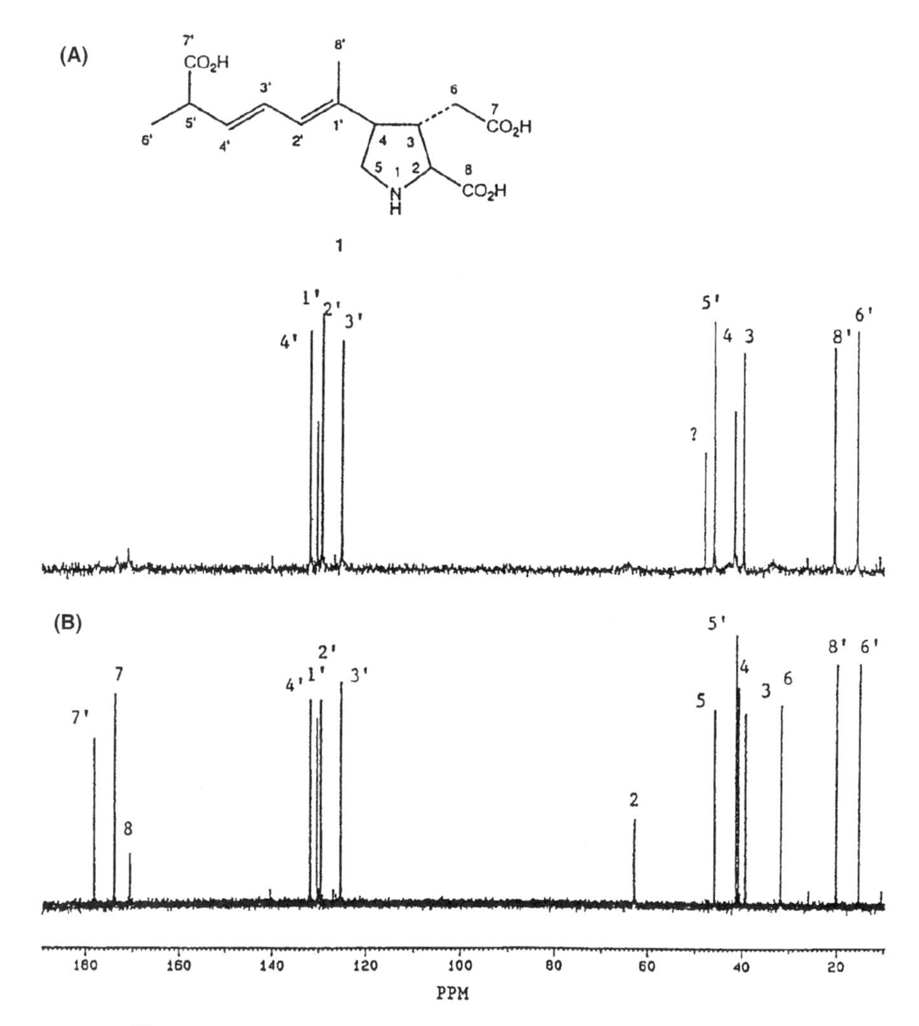

Fig. 2. ^{13}C NMR spectra of domoic acid, (**1**), isolated from culture of the diatom, *Pseudonitzchia*, The spectrum **A** is of a sample without treatment with Chelex 100 (Bio-Rad, Hercules, CA, USA) and **B** after treatment with Chelex 100. In (A) several signals are missing and also chemical shift are seen (spectra of an identical isolate in D_2O).

Ultrafiltration Cell Model 2000B with membrane filters (15 cm diameter), XM100A (100,000 cutoff), and PM10 (10,000 cutoff) was used.

2. Chromatography columns: Bio-Gel P-2, 5.0×70 cm, Bio-Rad Laboratory; Bio-Rex 70 H^+ form, –400 mesh, 2.8×55 cm and 1.8×56 cm, Bio-Rad

Laboratory; Hitachi gel 3013-C H^+ form, 0.4×15 cm, Hitachi Co; and HEMA-IEC BIO CM column, 0.75×25 cm, Alltech Assoc.; Chelex 100, H^+acetate form, -400 mesh, Bio-Rad Laboratory.

3. Solvents: HPLC-grade MeOH and analytical-grade acetic acid were distilled before use. Water used for HPLC was glass-distilled deionized water passed through a high-purity organic remover.

3.1.2. Methods

The live newts (1400 g, 112 bodies) were frozen in liquid nitrogen and stored before extraction. The animal bodies were minced, extracted in a homogenizer with a threefold volume of 0.1% acetic acid in MeOH for 4 min, and centrifuged. The residue was re-extracted twice with a twofold volume of the same solvent. The combined extract was centrifuged, and the supernatant was evaporated to about 500 mL, at which no MeOH was left in the solution. After removal of the precipitates by centrifugation, the supernatant was diluted with 1 N acetic acid to 1000 mL and subjected to ultrafiltration.

The ultrafiltration was performed in two steps: first through XM100A and then through PM 10 membranes at ≈ 50 psi in a cold room (5°C). The dialyzate (<10,000 MW) was evaporated *in vacuo* to 300 mL and loaded onto a Bio-Gel P-2 column (5.0×70 cm) after adjustment of the pH to 5.5 with dilute NaOH solution. The column was first washed with 700 mL of water, and the tetrodotoxin derivatives were eluted with a 0.03–0.06 N acetic acid gradient (total 1000 mL). The combined toxin fractions were re-chromatographed in the same system. Upon completion of the second chromatography, water-soluble components other than tetrodotoxin derivatives were mostly removed.

The toxin fractions were combined and concentrated *in vacuo*. The pH was adjusted to 6.5 with dilute NaOH solution prior to loading onto a Bio-Rex 70 column (2.8×55 cm). The column was washed with a small amount of water, and fractions of 15 mL were collected. The toxin fractions eluted with 0.05 N acetic acid were combined and further purified by chromatography on a Bio-Rex 70 column (1.8×56 cm, linear gradient elution with 0.03–0.06 N acetic acid, 2×250 mL, followed by 0.06 N and 0.5 N acetic acid). Fractions were monitored by TLC (Whatman HP-KF with pyridine/ethyl acetate/acetic acid/water = 10:5:2:3; visualization under UV at 305 nm after spraying with 10% KOH in MeOH and heating at 130°C for 10 min). Fractions of 5 mL were collected, and the fractions containing toxins were combined and lyophilized. In this chromatography, tetrodotoxin, **3**, and three known tetrodotoxin derivatives: 6-epi-tetrodotoxin,

4, 11-deoxytetrodotoxin, **5**, and 4,9 anhydrotetrodotoxin, **6**, were obtained in pure form. The desired compound, **2**, which showed the highest R_f value on TLC, was further purified by HPLC. The new compound fraction was chromatographed on a Hitachi gel 3013-C column with 0.05 N acetic acid as eluting solvent. Fractions of 2 mL were collected and monitored by a UV detector at 225 nm. The toxin was further purified on a HEMAIEC BIO CM column with 0.05 N ammonium acetate (pH 5.75) (**Figs. 3 and 4**). The new tetrodotoxin derivative fractions were collected and lyophilized three times by adding water to eliminate ammonium acetate completely. The combined fractions were then rechromatographed on the same column using 0.1 N acetic acid as eluent. After passing through small Chelex 100 and Bio-Gel P-2 columns (0.5×1.5 cm, elution with 0.1 N acetic acid), the new compound was obtained in pure state with a yield of 1.2 mg.

3.1.3. Comments

The direct application of a crude extract on a small-pore size membrane (e.g., <10,000) usually causes clogging and results in a slow filtration rate.

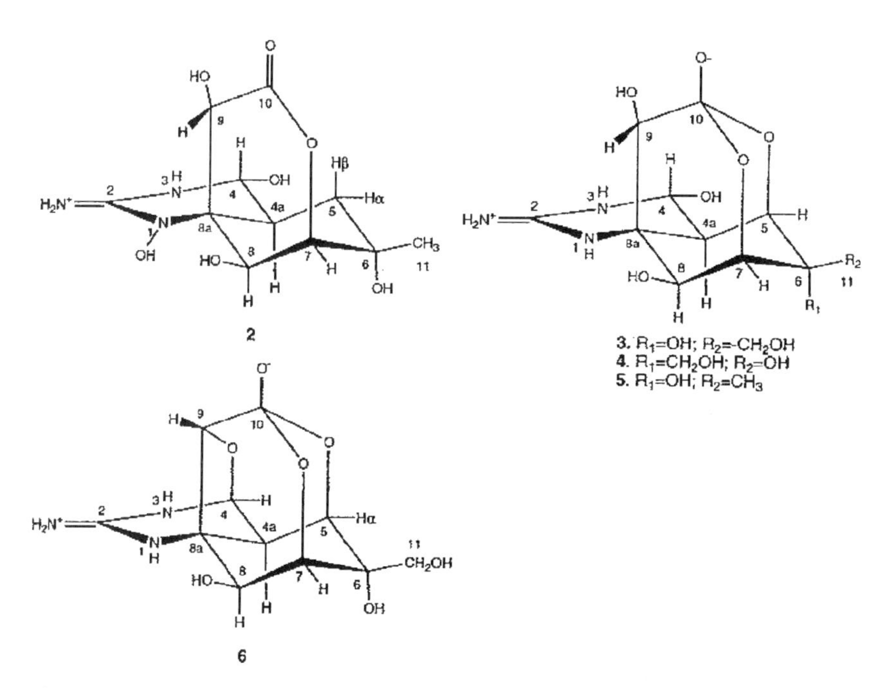

Fig. 3. Structures of tetrodotoxin derivatives isolated from the newt, *T. granulosa* (*17*).

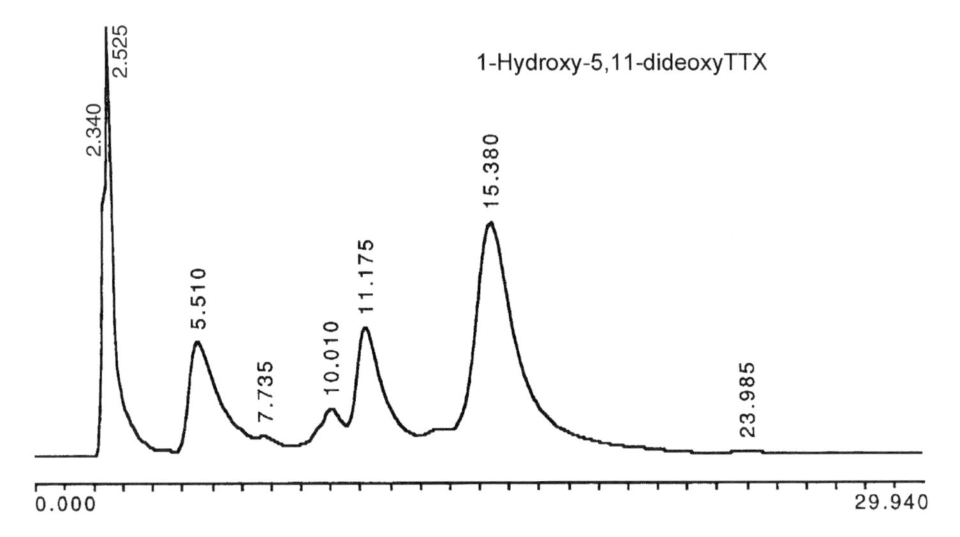

Fig. 4. Chromatographic elution pattern of tetrodotoxin (TTX) derivatives (3 mg total) from a HEMA-IEC Bio 1000 CM column (7.5×250 mm); 3 mL/min, 0.05 M ammonium acetate, pH 5.75; detection at 225 nm.

It is always better to do the filtration in two or more steps, even if it is not necessary to subfractionate the macromolecules. In this experiment, the target molecule has an *N*-hydroxyl guanidine group, which is known to be reduced easily by endogenous enzymes or bacterial action. Low-temperature operation and quick removal of macromolecular fractions raise the yield. Tetrodotoxin derivatives have been found in many different types of organisms. The structural variation among them gives very important clues about the biosynthetic origin of the important toxin. However, separation of the closely related toxins is not so trivial. Silica-based columns used by most researchers give poor yields. We found the carboxylic acid form of HEMA gives an excellent separation and a relatively high yield (**Fig. 4**).

3.2. Isolation of Domoic Acid, a Glutamate Agonist From the Diatom, Pseudonitzschia pungens f. multiseries *Culture Medium (Unpublished Experiment)*

3.2.1. Materials

1. *P. pungens* f. *multiseries,* strain B991-K from Narragansett Bay, RI (Unpublished experiment), was cultured in 8 L of Guillard f/2 medium for 50 d.

2. C_{18} silica gel: Bondesil preparative grade, 40 μm (Analytichem International).
3. Sephadex LH-20 (Pharmacia Fine Chemicals).
4. Solvents: HPLC-grade MeOH and acetic acid were distilled before use. Water used for HPLC was glass-distilled deionized water passed through a high-purity organic remover.

3.2.2. Methods

The 50-d-old *Nitzschia* culture was centrifuged, and the algal cells were separated. After adjustment of the pH to 3.0 with dilute HCI, the supernatant was passed under weak suction through C_{18} silica gel (300 g) packed in a Büchner funnel. The silica gel was washed with 100 mL of distilled water and the adsorbed material eluted with 300 mL of MeOH. The MeOH eluate was evaporated to dryness *in vacuo* and loaded on an LH-20 column (1.5×40 cm). The column was developed with MeOH (approx 200 mL) and monitored by a UV detector (254 nm) and TLC. Domoic acid fractions were combined evaporated and chromatographed on a C_{18} column (1.5×12 cm, medium pressure). The column was developed consecutively with 0.1% acetic acid in 10% MeOH (50 mL), 0.1% acetic acid in 30% MeOH (100 mL), and 0.1% acetic acid in 50% MeOH (50 mL). Domoic acid was mostly eluted with 30% MeOH. The domoic acid fractions were rechromatographed on a C_{18} silica gel column (1×7 cm) using 0.1% acetic acid in 28% MeOH. The domoic acid fractions were combined and evaporated to a crystalline mass with faintly brown color with a yield of 4.76 mg. For NMR measurements, this sample was passed through 2 g of Chelex 100 (prewashed with HCl and distilled water) in a Pasteur pipet using 0.1 N HCI as eluent.

3.2.3. Comments

Domoic acid is known to be the culprit of amnesic shellfish poisoning (ASP) ***(18,19)***. It is a strong glutamate agonist and known to bind to kinate receptors in brain. It is produced by a few species of diatom. It was found that the strain used in these studies exudes most of the domoic acid in the medium. The isolation of the water-soluble amino acid from a large amount of the seawater-based medium was performed by ionization-suppression of 3 carboxylic acid functions by acidification and adsorption on C-18. As mentioned in the introduction, the compound easily forms a chelate with heavy metal ions. Even recrystallized samples contained enough metals to create a disturbance in NMR (**Fig. 2**).

3.3. Isolation of an Aminotetrasaccharide, a Potential Glycosidase Inhibitor, From a Blue-green Alga, Anabaena sp. (20)

3.3.1. Materials

1. The cyanobacterium *Anabaena* sp. URI Strain No.S2-217–2 was cultured at 25–27°C in Fitzgerald medium under fluorescent lighting with a 16 h/8 h light and dark cycle, harvested by centrifugation and freeze dried.
2. Diaion HP-20 column (bed volume: 11×3 cm). The resin was washed in a beaker with deionized water first and small particles were removed by repeated decantation. The resin was then washed with MeOH thoroughly and again with deionized water.
3. C_{18} silica gel column (15×2 cm), Baker brand, 40 μm.
4. Solvent: HPLC-grade MeOH.

3.3.2. Methods

The freeze-dried algal cells (10 g) were extracted with 1% acetic acid (150 mL×3) by heating in a microwave oven for 1 min. The cell suspension was centrifuged at 4000 *g*. The supernatant was passed through the Diaion HP-20 column, which had been equilibrated with deionized water. The column was washed with deionized water (200 mL) and MeOH (300 mL). The MeOH eluate (0.2 g) was flush chromatographed on the C_{18} column with water and mixtures of water and MeOH by increasing the MeOH concentration stepwise. The fractions were monitored by TLC. The tetrasaccharide (37 mg) came out in earlier fractions (approx 20% MeOH) almost in a pure state. It was crystallized from water–EtOH to needles, $C_{28}H_{45}NO_{20}$, 716.2616 $(M+H)^+$ (calcd.: 716.2613). The NMR spectrum suggested a sugar structure with three anomeric protons, two acetyl groups, and one methoxyl group. Subsequent structure studies resulted in a structure, **7**, with glucose, arabinose, glucuronic acid moieties, and a cyclic aminosugar alcohol (**Fig. 5**).

3.3.3. Comments

Usually, mono- and oligosaccharides are not extractable with a nonionic resin. In this case, however, the presence of acetyl group and methoxyl group added some hydrophobicity to the molecule and made it adsorbable

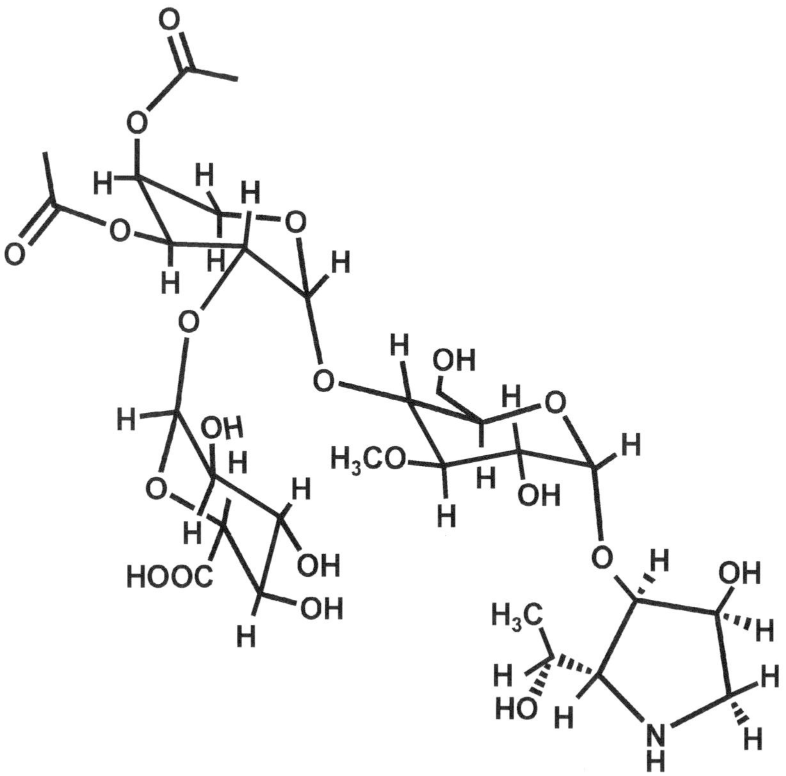

Fig. 5. Structure of the aminotetrasaccharide isolated from the blue-green alga, *Anabaena* sp.

to the resin. But still, the compound, which is freely soluble in water and sparingly soluble in alcohol, was at the borderline.

3.4. Isolation of Protoceratins, Antitumor Metabolites From the Culture Medium of Protoceratium *cf.* reticulatum *(21)*

3.4.1. Materials

1. The photosynthetic dinoflagellate, *P.* cf. *reticulatum,* URI strain S2-191–1, isolated at Sanriku coast, Japan, was cultured in the f/2 enriched seawater medium.
2. Diaion HP-20 column (bed volume: 1.2 L), prepared as described earlier.
3. Silica gel column (Baker, 60 μm, 1.7×38 cm).
4. Sepralyte C_8 HPLC column (1.5×20 cm).
5. Solvents: HPLC grades.

3.4.2. Methods

The 80-d-old culture of *P.* cf. *reticulatum*, URI strain S1-83–6 (23 carboys, total 280 L) was centrifuged continuously, and the supernatant was passed through the column of Diaion HP-20 (1.2 L volume). The resin was washed with 6 L of deionized water and then with MeOH (1.2 L). The MeOH extract was chromatographed on a silica gel column by stepwise increase of MeOH from 99:1 to 50:50. Elution with DCM–MeOH 90:10–85:15 gave semipure protoceratin I, **8** (281.7 mg) (Fig. 6). The subsequent fractions gave a mixture of more polar protoceratins (148.4 mg), which was further separated by HPLC (Seorakyte C_8). The column was eluted with 50%, 60%, and 70% MeOH sequentially. Protoceratin IV, II, and III fractions were eluted with 60% of MeOH in that order, and some protoceratin I with 70% MeOH. The semipure fractions of protoceratin IV (4.1 mg), II (14.0 mg), and III (4.3 mg) were further purified by HPLC (Adsobophere C_{18}) using an eluting solvent: 62% MeOH—38% 0.01 mM

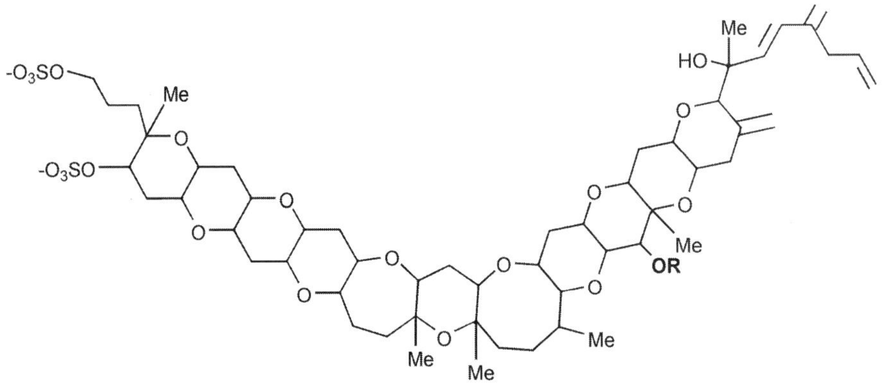

Protoceratin I (homoyessotoxin): R=H

Protoceratin II: R= (Arabinose)$_2$

Protoceratin III: R=Arabinose

Protoceratin IV: R=(Arabinose)$_3$

Fig. 6. Structures of protoceratins, antitumor polyether glycosides isolated from the culture medium of the dinoflagellate, *Protoceratium* cf. *reticulatum*.

pH 7.05 phosphate buffer, UV detection at 220 nm. Those fractions that contained only pure compounds were combined and desalted by C_{18} cartridges (Waters Sep Pak) with 15% water–85% MeOH to give pure protoceratin II (4.7 mg) and two lesser components: protoceratin III, (1.3 mg), and protoceratin IV (1.0 mg). Those compounds were extremely cytotoxic to human tumor cell lines (average $IC_{50} < 0.5$ nMol) and were given the structures, **9**, **10**, and **11**, respectively, as shown in **Fig. 6** on the basis of physico-chemical data and degradation studies.

3.4.3. Comments

The unique cyclic polyether metabolites produced by dinoflagellates possess both hydrophilic and hydrophobic characters. However, their glycosides are considerably more polar and had been overlooked in the past. By the use of the nonionic resin, however, they were recovered from the culture medium for the first time.

References

1. Shimizu, Y. (1985) Bioactive marine natural products, with emphasis on handling of water-soluble compounds. *J. Nat. Prod.* **48,** 223–235.
2. Shimizu, Y. (1998) Purification of water-soluble natural products, in *Methods in Biotechnology, vol. 4: Natural Products Isolation* (Cannell, R. J. P., ed.), Humana Press, Totowa, NJ, pp. 329–341.
3. Lepane, V. (1999) Comparison of XAD resins for the isolation of humic substances from seawater. *J. Chromatogr. A* **845,** 329–335.
4. Weigel, S., Bester, K., and Hühnerfuss, H. (2001) New method for rapid solid-phase extraction of large-volume water samples and its application to non-target screening of North Sea water for organic contaminants by gas chromatography–mass spectrometry. *J. Chromatogr. A* **912,** 151–161.
5. Hubberten, U., Lara, R. J., and Kattner, C. (1994) Amino acid composition of seawater and dissolved humic substances in the Greenland Sea. *Marine Chem.* **45,** 121–128.
6. Leonard, J. D. and Crewe, R. N. (1983) Study on the extraction of organic compounds from seawater with XAD-2 resin. *Marine Chem.* **12,** 222.
7. Slauenwhite, D. E. and Wangersky, P. J. (1996) Extraction of marine organic matter on XAD-2: effect of sample acidification and development of an in situ pre-acidification technique. *Marine Chem.* **54,** 107–117.
8. Fu, T. and Pocklington, R. (1983) Quantitative adsorption of organic matter from seawater on solid matrices. *Marine Chem.* **13,** 255–264.
9. Font, G., Mañes, J., Moltó, J. C., and Picó, Y. (1993) Solid-phase extraction in multi-residue pesticide analysis of water. *J. Chromatogr. A* **642,** 135–161.

10. Tolosa, I., Readman, J. W., and Mee, L. D. (1996) Comparison of the performance of solid-phase extraction techniques in recovering organophosphorus and organochlorine compounds from water. *J. Chromatogr. A* **725,** 93–106.
11. Davankov, V., Tsyurupa, M., Ilyin, M. D., and Pavalova, L. (2002) Hypercross-linked polystyrene and its potentials for liquid chromatography: a mini-review. *J. Chromatogr. A* **965,** 65–73.
12. Adachi, T., Ando, S., and Watanabe, J. (2002) Characterization of synthetic adsorbents with fine particle sizes for preparative-scale chromatographic separation *J. Chromatogr. A* **944,** 41–59.
13. Lee, J C., Park, H. R., Park, D. J., Lee, H. B., Kim, Y. B., and Kim, C. J. (2003) Improved production of teicoplanin using adsorbent resin in fermentations. *Lett. Appl. Microbiol.* **37,** 196–200.
14. Marshall, V. P., McWethy, S. J., Sirotti, J. M., and Cialdella, J. I. (1990) The effect of neutral resins on the fermentation production of rubradirin. *J. Ind. Microbiol.* **5,** 283–288.
15. Gastaldo, L., Marinelli, D., Restelli, E., and Quarta, C. (1996) Improvement of the kirromycin fermentation by resin addition. *J. Ind. Microbiol.* **16,** 305–308.
16. Hori, A. and Shimizu, Y. (1983) Biosynthetic *N*-enrichment and *N*-NMR spectra of neosaxitoxin and gonyautoxin-ll: application to the structure determination. *J. Chem. Soc. Chem. Commun.* 790–792.
17. Kotaki, Y. and Shimizu, Y. (1993) 1-Hydroxy-5, 11-dideoxytetroddotoxin, the first *N*-hydroxyl and ring-deoxy derivative of tetrodotoxin found in the newt *Tricha granulosa. J. Am. Chem. Soc.* **115,** 827–830.
18. Maranda, L., Wang, R., Masuda, K., and Shimizu, Y. (1990) Investigation of the source of domoic acid in mussels; in *Toxic Marine Phytoplankton* (Granèli, E., Sundström, B., Edler, L., and Anderson, D. M., eds.), Elsevier, Amsterdam, pp. 300–304.
19. Wright, J. L. C., Boyd, R. K., de Freitas, A. S. W., et al. (1989) Identification of domoic acid, a neuroexcitatory amino acid in toxic mussels from eastern Prince Edward Island. *Can. J. Chem*. **67,** 481–490.
20. Thammana, S., Suzuki, J., Shimizu, Y., Lobkovsky, E., and Clardy, J. (2002) An aminotetrasaccharide derivative from a filamentous cyanobacterium *Anabena* sp. *43rd Annual Meeting of the American Society of Pharmacognosy Abstracts*, p. 107.
21. Konishi, M., Yang, X., Li, B., and Shimizu, Y. (2004) Highly cytotoxic metabolites from the culture supernatant of the temperate dinoflagellate *Protoceratium* cf. *reticulatum. J. Nat. Prod.*, **67,** 1309–1313.

17

Scale-Up of Natural Product Isolation

Steven M. Martin, David A. Kau, and Stephen K. Wrigley

Summary

Scale-up of natural product isolation involves not only an increase in the scale of the purification of the target compound but also an improved level or scale in its production. As the scale increases, efficiency of operation becomes more important, necessitating process development. The nature of the scale-up challenge inevitably changes throughout the development cycle of a particular compound. This, together with the resources available, ultimately influences the strategy adopted. Factors important in successful scale-up are discussed with particular reference to microbial products and illustrated in case studies of the scale-up isolations of xenovulene A from *Acremonium strictum*, and (6*S*)-4,6-dimethyldodeca-2*E*,4*E*-dienoyl phomalactone from a *Phomopsis* sp.

Key Words: Scale-up; natural products; isolation; purification; fermentation; downstream processing; xenovulene A; (6*S*)-4,6-dimethyldodeca-2*E*,4*E*-dienoyl phomalactone.

1. Introduction

Scale-up is a necessity for any natural product with promise in any field, but a precise definition of the term can be somewhat elusive. A glance through the classical texts and assorted literature would indicate that effective scale-up can be achieved through the logical application of well-characterized biological, chemical, and engineering principles. To a certain extent this is true, but the reality is that scale-up remains part science–part art, where scientific principles provide a useful tool, but where experience

From: *Methods in Biotechnology, Vol. 20, Natural Products Isolation, 2nd ed.*
Edited by: S. D. Sarker, Z. Latif, and A. I. Gray © Humana Press Inc., Totowa, NJ

and, occasionally, a scientist's intuition still have an important role to play.

The breadth of the scale-up field means that even the most definitive texts can do little more than provide a cursory overview of the area. The corresponding chapter in the first edition of this book *(1)* provided the reader with a solid general outline of scale-up principles, and it is not our intention therefore to repeat this exercise, but rather to exemplify how the changing nature of scale-up throughout the development cycle of a compound, together with the resources available, ultimately influence which direction to take. Our field is the discovery and development of microbial metabolites produced by fermentation in the pharmaceutical industry, and we mostly draw on this experience in writing this chapter. Many of the principles discussed are also applicable to the scale-up and isolation of natural products from other sources. Some particular issues relating to scale-up of compounds from plants and marine organisms are addressed in separate sections.

The very diversity of natural products that makes them so valuable in many fields can be seen as both blessing and curse to those involved in their initial and subsequent scale-up isolations. This diversity is reflected in a huge breadth of physicochemical properties that make it difficult to devise generic extraction and purification procedures. When the structure of the target compound is known, reported methods for related compounds can be consulted as starting points for developing scale-up processes. For completely novel compounds, an empirical approach has to be adopted. We concentrate on describing scale-up strategies that have been effective in our experience and illustrate these with reference to two case studies, their associated scale-up issues, and their eventual resolution.

For the purposes of this chapter, we therefore interpret "scale-up" as the generation of increased quantities of a particular compound for further evaluation following an initial demonstration of potential value. Our starting point is thus a natural product that has already been produced and purified in sufficient quantity and purity for its structure to have been elucidated in most instances (although in some particularly challenging cases, scale-up may need to be initiated to enable completion of structure elucidation). With modern spectroscopic techniques, this initial characterization can be achieved with 5 mg or less of most secondary metabolites. Scale-up requires not only efficient purification, often referred to as downstream processing (DSP), but also increased production of the target compound to deliver the quantities and purity appropriate to the intended use or stage of evaluation.

Scale-up programs for microbial products in the pharmaceutical industry are the responsibility of interdisciplinary teams consisting of fermentation scientists and purification chemists. The challenges involved in the scale-up and purification of natural products often change throughout the development lifecycle of individual compounds: from initial discovery and characterization, through structure–activity relationship programs and support for early preclinical studies, to the eventual development and definition of commercially viable production processes capable of rapid transfer into the manufacturing environment. The scale-up and purification scientists must work in parallel at each stage to understand and manage these changing priorities as a compound moves from phase to phase. In the early stages of these programs, there is usually tension between the needs to supply material quickly to expedite evaluation and the desire to develop the fermentation and purification processes to improve the efficiency of isolation. Success in the former activity can justify the latter but also drives the demand for increased material needs. Although priorities change frequently during development projects, there is always a need to adopt pragmatic approaches that balance resources and timelines (so there will occasionally be a need to prioritize a particular fermentation using a non- or a suboptimal process).

The first stage of scale-up will usually encompass the production of 100 mg to multigram quantities of the target compound for various uses ranging from early biological evaluation to initiating a formalized program of semisynthetic derivatization to establish structure–activity relationships and to produce analogs with improved therapeutic properties. This work can be started in laboratory-scale fermenters (1–20 L volumes) but may be completed more efficiently using pilot-scale fermenters (50–500 L volumes). Because microbial metabolites are often produced as families of structurally related metabolites, this stage of scale-up also frequently provides an opportunity to identify minor components related to the original metabolite of interest and, once purified and evaluated, these may also have useful properties.

Further stages of scale-up involve material quantities from 100 g to multiple kilograms and beyond for preclinical and clinical development, which, if successful, will ultimately culminate in full-scale manufacturing. These stages require process-scale fermenters ($>1\ m^3$) and suitably sized DSP equipment and facilities. Production on this scale and the associated considerations for documentation and quality controls encompassed by current Good Manufacturing Practices (cGMP) are largely beyond the scope of this chapter with the proviso that methods adopted in the early stages of development should ideally be robust and adaptable to manufacturing scales. Our chapter mainly focuses on pilot-scale fermentation and downstream process development.

In its simplest form, scale-up can be achieved by merely increasing the scale of compound production and subsequent purification. There are times when this approach is the most practical route to getting a compound quickly. There is an inevitable risk, however, that any increase in scale, without sufficient process development, will not be successful. An alternative and generally preferable approach is to increase the concentration of the target compound relative to unwanted compounds (impurities) in the material to be processed. This improves not only the efficiency of production but also reduces the complexity of the product stream entering purification. This in turn can significantly improve the ease of downstream purification and ultimately result in enhanced yields. The latter course is usually undertaken with microbial products in the form of an integrated program of process development to improve both the fermentation and DSP processes. Costs and process efficiency become more important as the scale increases, and there will always be some scope for improving the purification method itself. The initial compound isolation will, in most cases, have been bioassay guided, with speed of identification the major objective, and is therefore highly unlikely to have been efficient in terms of either compound production levels or purification step yields. As the scale of operation increases, for fermentation products at least, so does the proportion of the overall scale-up process taken up by DSP activities. The importance of the relationship between the production and the DSP of the target compound, and the potential for facilitating the latter activity by improving the former, cannot be stressed too highly.

2. Methods

2.1. Assays and Product Quantitation

The importance of reliable, quantitative assays as tools for directing the development of scale-up processes, by measuring target compound concentrations in fermentations and in concentrated extracts or eluates during various stages of purification, should not be understated. They allow the accurate determination of step yields, thus providing a means of measuring and improving efficiencies. Appropriate assays are usually chromatographic and should be capable of high throughputs (typically 10–100 of samples per day). Reversed-phase high-performance liquid chromatogaphy (HPLC) methods with UV/visible or occasionally mass spectrometric detection are used most frequently. Evaporative light scattering detectors (ELSDs) allow the detection of all compounds in a sample, including those with no UV absorbance, and add an extra dimension when assessing purity. These chromatographic methods are preferred to assays based on biological activity, through which many natural products are first detected and discovered, as such assays measure a total response to all of the active compounds present in a sample. Given that the majority of secondary metabolites are produced as mixtures of closely related compounds with varying degrees of activity, biological activity assays will deliver a composite response that could be misleading in respect of a particular compound. There is also scope for synergistic or antagonistic effects between components of a mixture in such biological assays.

Compound-specific chemical assays are thus useful for separately quantifying different compounds within a series. There is a danger, however, in relying too much on an assay that is highly product specific. For fermentation development, at least, there is merit in running in parallel a product-specific HPLC assay with a steep gradient elution method that allows gross changes in the overall metabolite profile to be monitored. It is relatively common in process development to implement a desirable improvement in process characteristics (e.g., higher titer, better medium composition, to name a few) that also produces some unwanted impact on the process (e.g., a change in the profile of minor components, morphological change in the producing organism, and so on). It is essential, therefore, that the analytical techniques applied be broad enough to pick up both desirable and undesirable changes during the development phase.

2.2. Fermentation Development

The general principles for process development were outlined in the first edition of this volume *(1)*. Hence, we do not reiterate these principles here. Instead, we update and expand upon these initial comments and illustrate with recent references where appropriate. For instance, we cite a case study that exemplifies the more standard approaches to process development, and in addition, illustrates the value of having a taxonomically well-characterized culture collection, which can be explored not only for new secondary metabolites but also for other interesting biological properties such as organisms that may produce related compounds or possess useful enzymic activities that can be utilized to modify certain molecules. We also provide a brief basic introduction to the increasing application of recombinant techniques to process development.

2.2.1. Reproducibility and Inoculum

The first step in process optimization typically involves some investigation into medium composition. Yields of wild-type organisms are often around 1 mg/L or less, and medium optimization can be an effective means to rapidly improve yields to greater than 100 mg/L. These early stage optimization studies are typically carried out in shake flasks and tend to be based around statistical experimental methods such as Placket Burman, surface response, multivariate, and principal component analysis. Process reproducibility is a necessary prerequisite before process development and optimization can begin, or more properly it might be considered the first step, as it provides a reliable baseline against which to gage future improvements. This is particularly important when using statistical experimental design techniques, where process variability may be misinterpreted as a statistically significant improvement.

It is not uncommon for organisms producing natural products to display a certain degree of physiological and metabolic instability which, in practice, may manifest itself in a lack of reproducibility—for instance, in activity in a downstream bioassay. Precious resources and much time can be tied up both in fermentation and chemical isolation by trying to track down these "lost" or "variable" activities from wild-type isolates. It is therefore desirable in the first instance that the initial activity be reproduced in the same way as the original fermentation. A confirmed activity from such a refermentation will provide the scale-up scientist with the confidence to take the hit further.

The first scale-up fermentation should preferably be performed in a system as close as possible to the fermentation system in which the hit was first detected—using multiple units if necessary—to avoid unnecessary complications with scale-up issues. For example, if the initial hit was detected in an extract obtained from a 50 mL shake flask fermentation, then multiple shake flasks could be employed to produce volumes up to 5–10 L. If the referment is successful, it adds to the growing confidence in the activity and increases the likelihood that the activity will scale-up into stirred fermenters. If the referment is unsuccessful, then this likely indicates some issue with poor reproducibility rather than a scale-up issue.

Alternatively, if the equipment is available, bench top fermenters may be used—if this strategy is adopted it is desirable that a "confirmatory referment" be run alongside the fermenter. The confirmatory referment would typically comprise one (or multiple) 250 mL baffled shake flask(s), specifically designed to more closely mimic the physical situation in a fermenter ***(2)***. These would contain 50 mL of culture taken directly from the fermenter post-inoculation and would be incubated in parallel to the fermenter and sampled and harvested similarly. If there is no activity in the fermenter, but activity in the flask, this is a clear indication that there may be a genuine scale-up issue. If activity is lost in both fermenter and flask, it points toward a reproducibility issue or a technical problem. This strategy is not only applicable during the transition from flask to fermenter, but it is also good practice to include a confirmatory referment alongside each increase in fermentation scale to provide an internal quality check on inoculum quality and medium preparation. When moving from 5 L to 50 L fermenters, for instance, the confirmatory referment would be a 5 L fermenter, and so on with each step increase in scale.

Poor reproducibility can frequently be linked to reproducibility issues in inoculum generation ***(3–5)***. For example, the inoculum may not be fully grown or suitably developed morphologically (e.g., sporulation, conidiation, diffuse or pelleted mycelium, hyphal differentiation, and the like), the start material (cryovials, agar slants, and so on) used to inoculate the seed stage might not be consistent, or the seed medium composition might be unsuitable. Understanding the problems with the inoculum can often be guided by simple microscopic examination of the organism growing in the seed stage. For instance, many actinomycetes and fungi will display a tightly pelleted morphology when the medium composition is unsuitable. In other cases, understanding the precise nature of the variability may be difficult—but it is important to be clear about whether the variability in a process arises from the specifics of the production stage itself or from the seed generation stage. In the case study of XR543 (*see* **Subheading 3.2.**), both the problems of reproducibility and an extended lag time were simply eliminated by allowing a longer time for the seed stage to develop (3–6 d), resulting in a well-grown and consistent inoculum for transfer into the production vessel.

2.2.2. Recombinant Approaches

There can be no doubting the value of traditional methods of increasing product titers, such as successive rounds of random mutagenesis and screening that have yielded many notable successes for various natural products, including penicillin, erythromycin, tylosin, and daptomycin

(6). Such empirical techniques are now being complemented by more rational recombinant approaches, and the use of this technology in the process development of natural product processes has increased dramatically over the last 10 yr *(7–9)*. Recombinant techniques have been used not only to increase production *(10,11)*, and reduce impurities *(12)*, but also to create numerous novel compounds structurally related to the initial microbial product characterized *(8,13)*.

One of the drawbacks of this approach is that along with the deliberate modifications introduced into a host during the cloning process, a number of unidentifiable pleiotropic mutations are also occasionally introduced. The consequence is that these recombinant strains may exhibit physiological properties that differ markedly from the parent strain. Product titers are generally significantly lower, and fermentation characteristics such as morphology, oxygen uptake, and growth rate may also be affected *(10)*.

In addition, these techniques lend themselves to the generation of large numbers of strains that need to be assessed for possible process improvements. This means that an effective generic approach to optimizing recombinant strains, while highly desirable, may be difficult to achieve given that each strain may possess varying fermentation characteristics. Several approaches have been attempted with different degrees of success; but recently, successful examples have been reported by researchers using proven industrial production strains as hosts for heterologous expression *(8,10)*. This approach attempts to offset the drop in productivity and any drift in strain properties by using a highly productive and well-characterized host.

2.3. Downstream Processing

As already stated, there is no generic purification process that can be applied to natural products because of their broad chemical and biological diversity; and therefore the purification process must be tailored for the particular compound groups of interest. Scale-up isolation processes have been successfully developed for a wide variety of natural products. As details of extraction and purification methods applicable to large-scale natural products isolation are well covered in a relatively recently published book *(14)*, we focus on DSP strategies and principles that we have found to be useful and illustrate their application with some case studies in **Subheading 3.**

When starting to develop a scale-up purification process, it is important to establish certain parameters and objectives such as the nature and properties of the compound, where the compound is located, how many processing steps will be required from extraction to final purified product, future economics and safety considerations for each process step, and the purity and quantity required for the end product. The last two items will be dependent upon the intended use and the requirements will become more stringent as a compound progresses from preclinical evaluation through clinical trials to eventual active pharmaceutical ingredient (API) status. Scale-up of natural product purification often involves multiple steps encompassing a variety of different techniques. Every technique employed must strike a balance between resolution, speed, capacity, and recovery (yield), and should be selected to match the objectives for each process step. The overall efficiency of the process will be dependent upon the number of unit operations as well as the yield at each step. It is worth emphasizing that the yield (in terms of grams or kilograms of product) from the first step has the biggest impact on the quantity of product obtained at the end of the process. Loss of product during initial recovery is hard to make up for in the subsequent steps, and therefore effort for improvement is generally better prioritized on the early process steps. Elaborate purification methods can be developed and implemented for use at the laboratory scale, but such an approach runs the risk that the subsequent scale-up will be unfeasible. It is generally a sensible strategy to keep the process as simple as possible, minimizing sample handling and the number of purification steps.

Microbial secondary metabolites are biosynthesized within the cell and frequently secreted into the surrounding medium as extracellular products—this is typically the case with actinomycetes. Some products, particularly of fungi, however, either remain within the cell or are secreted but remain tightly associated with the cell wall. These biomass-associated compounds require some form of extraction before they can be processed further. Occasionally, a product will appear in both biomass and supernatant, in which case it is preferable to process just one product stream. The approach here may be to conduct a whole-broth extraction or to try to manipulate the broth so that the product can be recovered in either biomass fraction or the supernatant (*see* **Subheading 3.1.**). Different approaches are taken when developing a purification process for biomass associated or extracellular products, but in general there are four key stages that can be applied to both categories and a basic description is given below.

2.3.1. Clarification

This step provides separation of the biomass from the aqueous fermentation medium prior to the next processing step. The two most common techniques that are used for this stage are filtration (e.g., dead-end filtration, microfiltration, tangential flow filtration) and centrifugation. The net result is to either provide a particulate-free supernatant, or filtrate for product capture, or biomass for subsequent extraction. It is essential that the clarified broth is particulate-free to prevent fouling of resins that may be used during the capture step. Most biomass-associated natural products are extractable with water-miscible solvents such as methanol (MeOH) and acetone, but it is also important for the biomass to be as dry as possible to enhance the extraction efficiency and keep the volume of organic solvent needed to a minimum.

One of the most common centrifuges used for broth clarification is a disk stack centrifuge such as that used for the purification of xenovulene (*see* **Subheading 3.1.**). This type of centrifuge tends to be less efficient in separating pelleted cultures such as those typically seen with fungal fermentations—due in part to the narrow spaces between the stacked discs and in part to the method of slurry discharge. They tend to provide a supernatant that, in our experience, requires further filtration before proceeding to the capture step, and a biomass-rich retentate, which is discharged as a wet slurry and could present difficulties in achieving efficient solvent extraction without further drying. For filamentous cultures, we have found scalable separation systems such as the Carr Powerfuge™ provide a more practical separation solution. This type of semicontinuous centrifuge overcomes many of the problems experienced with the disk stack not only by providing high centrifugal forces (20,000*g*), but also by avoiding the use of disks and by possessing an elegant system of contained discharge. This system more often gives a particulate-free broth and a dry, frequently powdery or friable biomass fraction that greatly facilitates subsequent solvent extraction.

When using filtration as the initial separation step, there are a number of strategies that can be employed to improve the process efficiency of the step such as addition of filter aids such as celite, or adjustment of the fermentation conditions (e.g., pH adjustment or temperature) to induce precipitation or flocculation. Care should be taken, however, when additives are used as they may interfere with the compound of interest and will most probably have to be removed prior to further purification steps.

2.3.2. Product Capture and Concentration

The aim of this step is simply to capture, concentrate, and stabilize the target compound from the clarified product stream. It is important that this is performed as quickly and efficiently as possible so that it will minimize any potential effects from extracellular degradative enzymes and acids that may be present in the fermentation. For extracellular products, the capture step depends on the chemical nature of the target compound. For hydrophobic compounds, a water-immiscible solvent such as ethyl acetate (EtOAc) or butanol might be used. But handling large volumes of these solvents can be problematic and impractical on scale-up. In our experience, it is generally preferable to use a hydrophobic adsorption resin, whereby the compound is retained and concentrated on the resin and polar contaminants flow through unbound thereby achieving a crude purification. The resin is then washed with water to remove weakly binding impurities before eluting the product from the resin with a water-miscible solvent such as MeOH or acetone. A number of adsorption resins that we have found to be useful are listed in **Table 1**. These differ in particle size, porosity, and selectivity. It is possible to achieve a higher level of purification if a more selective resin is used at this stage, or by employing a more specific solvent elution regime where the elution strength is only gradually increased. But it is often more pragmatic to use a relatively coarse resin such as Diaion HP20 that allows high flow rates and to complete the capture and elution swiftly. For products associated with the biomass, capture and concentration are achieved by solvent extraction. In both cases, the net result is a solvent extract, which is dried under pressure (using large-scale rotary evaporators or even a thin film evaporator) to give a stable extract ready for the next purification step.

If the target compound is acidic or basic, then a capture column using an anion or cation exchange resin may be used (**Table 1**), with elution by pH adjustment or by increasing the ionic strength of the eluent. Alternatively, if the compound becomes hydrophobic on neutralization, ionic suppression by pH adjustment followed by use of hydrophobic resin as above may be more appropriate.

For hydrophobic compounds, it is possible to perform a direct extraction on unseparated fermentation broth using a water-immiscible solvent such as EtOAc, thereby omitting the need for separate clarification and product capture steps and potentially providing higher overall process efficiencies and product yields. The major challenges with this are handling

Table 1
Some Adsorption and Ion Exchange Resins Used by the Authors for Scale-Up Isolation of Natural Products

Resin type	Name	Strength/ selectivity[a]	Manufacturer
Adsorption	Amberlite XAD7	Coarse	Rohm and Haas
	Amberlite XAD1600	Fine	Rohm and Haas
	Diaion HP20	Coarse	Mitsubishi Chem. Corp.
	Diaion HP20SS	Fine	Mitsubishi Chem. Corp.
	Sepabeads SP20SS	Fine	Mitsubishi Chem. Corp.
	MCI Gel CHP20P	Fine	Mitsubishi Chem. Corp.
	Amberchrom CG-161	Fine	TosoHaas
	Amberchrom CG-1000	Fine	TosoHaas
Anion exchange	Amberlite IRA 900-Cl	Strong	Rohm and Haas
	Amberlite IRA 67	Weak	Rohm and Haas
	FPDA-13	Weak	Mitsubishi Chem. Corp
	Macro-Prep High Q	Strong	Bio-Rad Laboratories
Cation exchange	Amberlite IR120	Strong	Rohm and Haas
	Dowex MAC-3	Weak	Dow Chemical Co.

[a]Coarse, large particle size, low selectivity; fine, small particles, higher selectivity; strong, strong ion exchanger; weak, weak ion exchanger.

these solvents safely, and dealing with any emulsions that may form, at large scale. Another technique capable of capturing product directly from whole broth is expanded bed adsorption (EBA) chromatography, which utilizes a single pass through an adsorption resin. EBA is currently restricted to extracellular products, mostly proteins and peptides that bind to ion exchange and affinity resins, but has been used for the purification of immunomycin ***(15)*** and pneumocandins ***(16)***.

2.3.3. Chromatography

The net result of the product capture and concentration steps is a fairly crude extract requiring extensive purification that often needs multiple chromatographic separation steps comprising either a blend of lower resolution steps utilizing different separation modes or a smaller number of high-resolution steps. The chromatographic methods applied will depend on the physicochemical properties of the target compound and the impurities from which it must be separated. Chromatographic methods that are widely used in scale-up are adsorption chromatography using resins with high selectivity, anion and cation exchange chromatography, and preparative HPLC using reversed-phase stationary phases. For hydrophobic compounds, normal-phase chromatography has become much more straightforward to conduct at scale with the development of flash chromatography equipment using radially compressed prepacked cartridge columns ***(17,18)***. Chromatographic steps used first in the purification process will be intended for intermediate purification whereby further removal of bulk contaminants from the target compound and related analogs is achieved. A final high-resolution step, sometimes referred to as a polishing step, is used to remove trace contaminants and to isolate the target compound and related analogs as single entities. Preparative HPLC (*see* Chap. 8) is the most common method used for this step. The purification process can result in the isolation and characterization of minor related components if this is deemed to be useful—but the ultimate goal will be efficient, economically viable purification of the target compound itself. Chromatographic purification steps tend to be slow and expensive to operate, so they need to be kept to a minimum and be developed to be as efficient as possible.

2.3.4. Crystallization and Lyophilization

Once purity has been achieved, the compounds are usually in the form of an eluate from a chromatography column in an organic solvent/water mixture, or in a buffer containing salts. These solvents and salts require removal to stabilize the target compound in a usable form. Buffer salts can readily be removed by desalting columns or by diafiltration, whereas solvents can be evaporated leaving the compound in the aqueous phase before drying into a solid form by lyophilization (freeze-drying). Crystallization (*see* Chap. 11) or precipitation can expedite the recovery of pure product in crystalline or powdery solid forms, but appropriate methods need to be developed specifically for each new compound. A recently published example of a successful isolation scheme employing adsorption chromatography and preparative reversed-phase HPLC, and culminating in a crystallization step, is the large-scale isolation of epothilone D, heterologously expressed in 1000 L *Myxococcus xanthus* fermentations ***(19)***.

2.4. Natural Products From Nonmicrobial Sources

2.4.1. Plant Products

The scale-up challenges of working with plant products are well known and hence are only dealt with very briefly here. Raw material variability can be caused by different genotypes; seasonal, diurnal and geographical variation; and phenotypic differences between different parts of plants and plants of various ages *(20)*. It is therefore important to capture all the pertinent information relating to the plant material when initially sourced for screening, so that the chances of being able to obtain resupplied material at a later stage containing the same metabolites are maximized. These challenges also present opportunities for programs of raw material investigation analogous to some aspects of fermentation development. Thus, if a phytochemical lead compound is difficult to obtain and is not amenable to total chemical synthesis, alternative sourcing through collection and investigation of other members of the same genus, or other plants known to produce similar compounds can be initiated. Subsequent follow-up through semisynthetic modification of related, more abundant derivatives combined with alternative cultivation strategies such as establishment of plantations or plant tissue culture may also be considered *(21)*. The famous example is that of Taxol®, where concerns regarding the environmental impact of the original sourcing from the bark of the Pacific yew, *Taxus brevifolia*, led to the eventual development of a manufacturing process based on semisynthetic modification of 10-deacetylbaccatin III obtained from clippings from plantations of the European yew, *T. baccata* *(21)*. It should be noted that the decision to invest resources in finding alternative routes of manufacturing Taxol was only made after excellent activity was observed in clinical trials.

2.4.2. Marine Products

Of all the established sources of biologically active natural products, marine organisms present the greatest challenges in terms of scale-up for drug discovery and development *(22)*. Not only is the marine environment inherently more difficult for sample collection, but many of the problems associated with plants also apply, particularly variation in metabolite production between samples of the same organism collected from different geographical areas or even of varying depths *(23)*. Marine metabolites are attractive for drug discovery because of the remarkable potencies of their biological activities and their distinct and complex chemical structures.

The potencies are, however, generally reflected in low concentrations of these metabolites being present in the producing organisms. Consequently, collections of large volumes of the producing organisms may be required to even complete structure elucidation of metabolites of interest before starting to consider scale-up for further evaluation, leading to major concerns over environmental impact and sustainable harvesting. It is logical that scientists interested in discovery and development should focus on organisms that can be resampled easily in high volumes *(22)*. The structural complexity of marine metabolites usually makes total synthesis challenging on any scale, particularly one that needs to be commercially viable. Marine product scale-up and resupply have been addressed successfully by aquaculture of the bryozoan *Bugula neritina* for bryostatin 1, and the tunicate *Ecteinascidia tubinata* for ecteinascidin 743, both compounds with significant anticancer potential *(23)*. The details of a large-scale isolation of bryostatin 1 following good manufacturing practices have been described *(24)*. New possibilities are presented by the hypothesis that some metabolites found in marine invertebrates are actually produced by associated micro-organisms. If these symbionts can be cultured independently to produce metabolites of interest, then scale-up could become more straightforward, although the cultivation challenges here remain high *(22)*. Alternatively, the gene clusters for the biosynthesis of these products in the micro-organisms, or, eventually, in marine invertebrates could be identified and transferred by molecular genetic approaches into organisms more easily cultivated in the laboratory *(25)*.

3. Case Studies

So far, we have discussed general approaches and strategies rather than specific challenges and methods used to overcome them. Scale-up is challenging to describe as well as to practice. We think it best to illustrate the challenges, and some solutions adopted, through two case studies.

3.1. Xenovulene A

The production of XR368 (**Fig. 1**), a novel GABA-benzodiazepene receptor agonist *(26)*, in the fungus *Acremonium strictum*, was improved from 1 mg/L to 100 mg/L through the application of Placket Burman statistical experimental design methods to try to identify which carbon and nitrogen sources in the medium would improve the product titer. Maltose was identified as a key medium component for improving productivity, and this was further improved by doubling the overall C:N ratio of

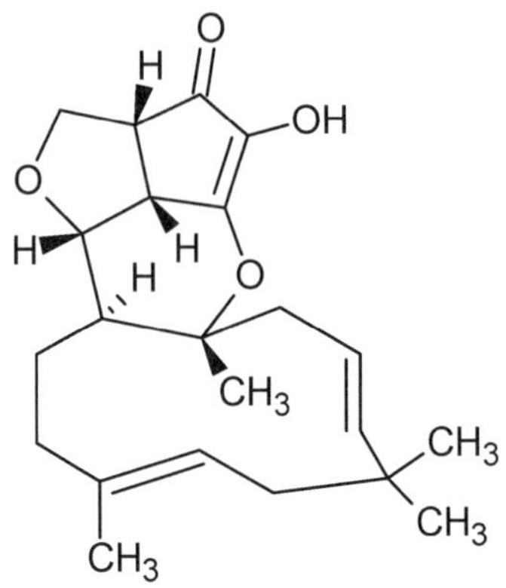

Fig. 1. Structure of xenovulene A.

the medium *(27)*. On initial discovery, xenovulene A was a minor component in a complex product mixture that required laborious purification principally by multiple reversed-phase HPLC steps. The titer improvement was accompanied by a highly significant improvement in the simplicity of the metabolite profile observed. Xenovulene A became the major fermentation product enabling much more straightforward DSP. This fermentation process was subsequently scaled-up directly to 3000 L with consistent performance to that observed in the lab-scale model.

The initial stage of any fermentation-based purification is product capture. On termination of the fermentation, xenovulene A was found distributed equally between the fermentation supernatant and the biomass, from which it was easily released by adding organic solvents. It would have been possible to process the supernatant and biomass as two separate product streams, but this would have been laborious—generally, it is always preferable to work on a single product stream. We opted in this instance to try to release the biomass-associated product into the supernatant before the fermentation broth was harvested from the tank. This was achieved by adding 1% Tween to the fermentation 1 h prior to harvest, which brought the product concentration in the supernatant to >90% (*see* **Note 1**).

Equipment, scale of operation, process limitations, and broth quality are some of the factors that govern which separation technique to use. For the *A. strictum* fermentation, despite the culture being pelleted, centrifugation resulted in such a sufficiently well-separated biomass and supernatant that this technique was suitable for direct scale-up to the 500 L scale using a Westfalia intermittent discharge disk stack centrifuge. The xenovulene was then captured from the supernatant onto a 40 L Diaion HP20 resin (Mitsubishi) column (30 cm × 60 cm and a linear flow rate 100 cm/h) by adsorption chromatography. After washing with water, the product was eluted from the column with acetone, and the eluate dried under reduced pressure to an aqueous concentrate using a cyclone thin film evaporator. This capture step onto HP20 was therefore a low-resolution chromatographic step, its main function was not only to trap the target compound and any closely related analogs in the supernatant, but also to facilitate a reduction in the volume of the product stream prior to the subsequent unit operations.

Further purification of xenovulene was achieved by back extracting the aqueous concentrate (15 L) with EtOAc (2 × 15 L). Xenovulene, a nonpolar compound, was partitioned into the organic phase leaving further polar impurities in the aqueous phase. Solvent partitioning (*see* Chap. 10) such as this is often used as a clean-up step in natural product isolation. Alteration of the pH and or the ionic strength of the aqueous layer may influence the partition coefficients and drive compounds in and out of the different phases resulting in a partial purification.

Initial screening experiments with thin layer chromatography (TLC) plates and a range of different solvent mixtures indicated that a fairly crude and fast semipurification could be achieved by a normal-phase chromatographic separation on silica, often referred to as flash chromatography ***(15,16)***. This was achieved using a 15 L SORBSIL™ silica column and an isocratic mobile phase consisting of hexane:EtOAc (2:1) plus 0.1% acetic acid; xenovulene-rich fractions were pooled following HPLC analysis. Conducting this separation step at this scale was facilitated by significant advances in flash chromatographic equipment (Biotage Flash™ radial compression technology), which had just become available at the time of this work. These laboratory-to-production scale flash chromatography systems using prepacked silica columns are available for purifying gram-to-kilogram quantities of compound. Xenovulene at this stage of purification is approx 70–80% pure by HPLC. High-purity xenovulene (>95%) was obtained by employing a final high-resolution polishing step involving preparative HPLC using a Biotage KP250 Kiloprep® HPLC system and a 100 mm × 600 mm BONDAPAK™ C_{18} column with an isocratic mobile phase of water:MeOH (25:75) and flow rate of 400 mL/min.

The compound-rich fractions were pooled and dried under reduced pressure with the final pool yielding 21.5 g of xenovulene with an overall process yield of 43% for a 5-step process, approximating to an average yield of 85% for each process step (**Fig. 2**).

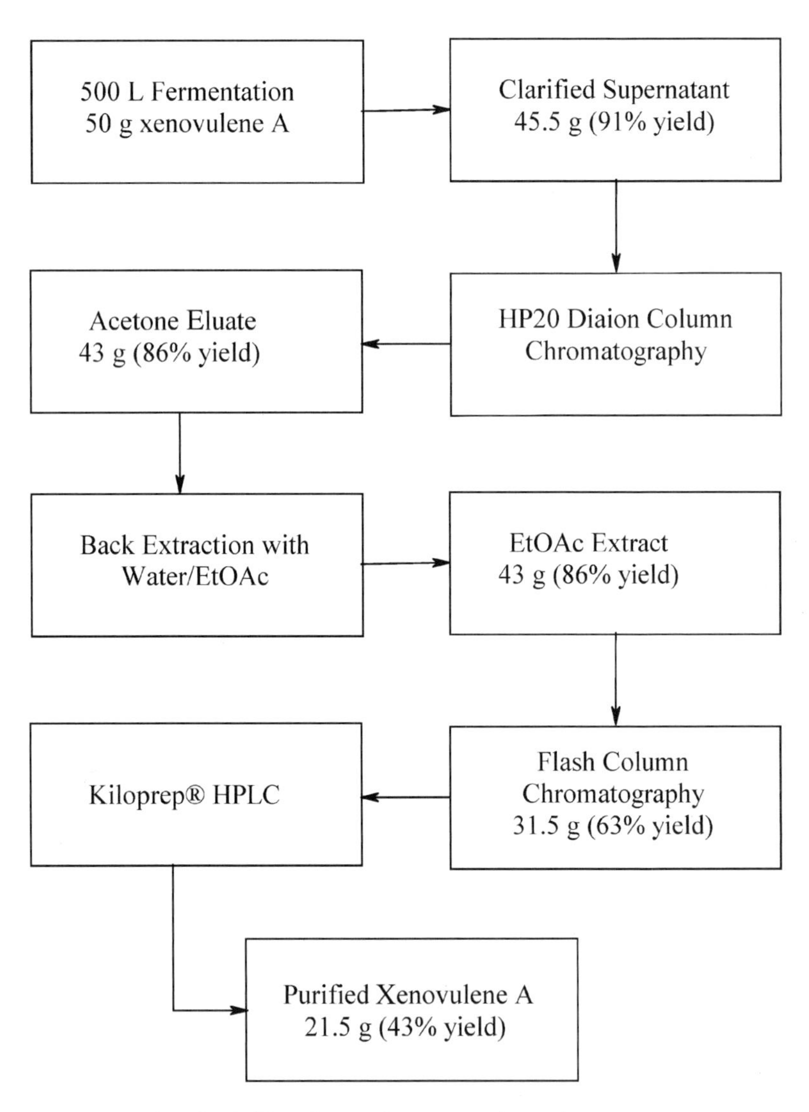

Fig. 2. DSP scheme for xenovulene A.

3.2. XR543 [(6S)-4,6-Dimethyldodeca-2E,4E-Dienoyl Phomalactone]

The value of a taxonomically characterized culture collection can be illustrated by the scale-up history for the production of the novel phomalactone derivative XR543 *(28)*. This was the most potent member of a series of macrophage activation inhibitors (novel 6-substituted 5,6-dihydro-α-pyrone esters) found in fermentations of the fungal strain *Phomopsis* sp. 22502 (**Fig. 3**). The first of the series to be identified was XR379, a novel ester of an acidic analog of the known fungal metabolite phomalactone. This exhibited only moderate biological activity but it was deemed worthwhile to investigate the complex mixture of metabolites produced in fermentations of this fungus for the presence of more active members of the series.

Initial scale-up work in fermenters focused mainly on increasing the volume from test tubes to shake flasks and then to 2 L and 10 L fermenters. As the process moved into fermenters, it was noted that the producing organism possessed a particularly challenging morphology. The fermentations tended to have an extended lag phase (up to 7 d), which was followed by a period of rapid growth leading to very high packed cell volumes (>70%) and a highly viscous oxygen-limited broth. While titers of XR379 were occasionally good at approx 300 mg/L, poor reproducibility meant that a high proportion of batches produced little or no product.

These issues were addressed initially by trying to control the biomass levels by reducing the medium concentration. While this did result in less biomass, it also enabled better control of the fermentation through avoidance of oxygen limitation and hence overall volumetric productivity was actually improved. The extended lag phase was eliminated by increasing

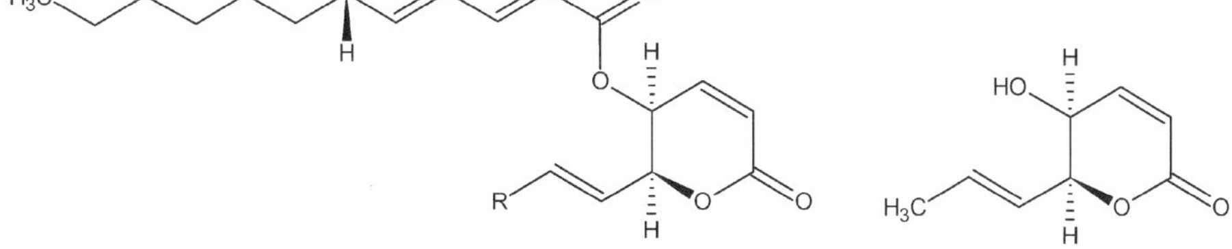

XR379: R = CO_2H

XR543: R = CH_3

Phomalactone

Fig. 3. Structures of XR379, XR543, and phomalactone.

the period of incubation of the inoculum from 3 d to 6 d to ensure that a consistently well-grown seed was transferred into each production phase. Solving these reproducibility issues was followed by some rapid medium optimization studies and the use of pH control that cumulatively resulted in process productivity reaching more than 800 mg/L. As the titer of XR379 increased, so did the those of other members in the series that had initially been difficult to isolate and characterize.

The effort paid off with the isolation and characterization of XR543, (6*S*)-4,6-dimethyldodeca-2*E*,4*E*-dienoyl phomalactone. XR543 was 25-fold more active than XR379 in the macrophage activation assay and performed well in secondary assays. As XR543 titers were much lower than those of XR379, a producer of phomalactone was sought to provide a template for both synthesis of XR543 itself, and to enable synthesis of analogs to probe the activity of this compound series further. This was facilitated by the discovery of another fungal strain (*Paecilomyces* sp. 3527), which produced phomalactone itself as a major metabolite on fermentation. Transfer of the *Paecilomyces* sp. into the fermentation process already optimized for the *Phomopsis* sp. yielded titers greater than 600 mg/L of product on the first attempt.

The purification strategy used here was similar to xenovulene except that XR543 is a cell-associated compound that is extractable with MeOH, and the first unit operation used was a combination of filtration and solvent extraction. On termination of the fermentation, 5 L of a 20% (v/v) solution of Hyflo Supercel celite in water was added to the broth (to act as a filter aid) and mixed prior to passing the mixture through a filter press. The retained biomass and celite mixture were then extracted *in situ* by pumping and recirculating 25 L of methanol through the filter press for 24 h. The solvent extract was recovered from the press and dried under reduced pressure to an aqueous concentrate (10 L) using a thin film evaporator. The aqueous fraction was then partitioned by repeated back extraction with an equal volume of EtOAc:hexane (1:1). XR543, being lipophilic in nature, extracts into the organic phase leaving the polar impurities in the aqueous phase. The solvent extract was dried under vacuum to a small volume (50 mL) and then fractionated by flash chromatography using a Biotage Flash 75 chromatography system with a 100 × 300 mm Hyperprep KPsil^{TM} 32–62 μm silica column and an isocratic mobile phase (EtOAc:hexane 1:1) with a flow rate 250 mL/min. XR543-rich fractions were pooled, and a final purification by preparative HPLC (Waters Delta Prep^{TM} HPLC system with a 25 × 200 mm Novapak^{TM} 5 μm C_{18} column and an isocratic mobile phase of 85% acetonitrile:15% water plus 0.1% v/v acetic acid and flow 50 mL/min) resulted in pure compound with an overall process yield of 20%.

Phomalactone was purified from the clarified fermentation broth using a three-step process. Initially, capture and concentration from the clarified broth was achieved by extraction into EtOAc and evaporation under reduced pressure. This was followed by a crude fractionation on silica gel using a Biotage Flash 75 chromatography system and an isocratic mobile phase (100% EtOAc). Phomalactone-rich fractions were pooled and then further purified by preparative HPLC using a Shandon Hyper prep HS BOS™ C_{18} (100 Å 12 μm) column (ID 10 × 30 cm length) and an isocratic mobile phase (85% water:15% acetonitrile, flow rate 170 mL/min) to yield pure phomalactone.

4. Note

1. The addition of surfactants and other chemicals to release compounds into the broth is well known, but care should be taken so that their addition does not interfere with the subsequent purification steps.

References

1. Verrall, M. S. and Warr, S. R. C. (1998) Scale-up of natural products isolation, in *Methods in Biotechnology, vol. 4: Natural Products Isolation* (Cannell, R. J. P., ed.), Humana, Totowa, NJ.
2. Katzer, W., Blackburn, M., Charman, K., Martin, S., Penn, J., and Wrigley, S. (2001) Scale-up of filamentous organisms from tubes and shake-flasks into stirred vessels. *Biochem. Eng. J.* **7,** 127–134.
3. Ignova, M., Montague, G. A., Ward, A. C., and Glassey, J. (1999) Fermentation seed quality analysis with self-organising neural networks. *Biotechnol. Bioeng.* **64,** 82–91.
4. Cunha, C. C., Glassey, J., Montague, G. A., Albert, S., and Mohan, P. (2002) An assessment of seed quality and its influence on productivity estimation in an industrial antibiotic fermentation. *Biotechnol. Bioeng.* **78,** 658–669.
5. Neves, A. A., Vieira, L. M., and Menezes, J. C. (2001) Effects of preculture variability on clavulanic acid fermentation. *Biotechnol. Bioeng.* **72,** 628–633.
6. Vinci, V. A. and Byng, G. (1999) Strain improvement by nonrecombinant methods, in *Manual of Industrial Microbiology and Biotechnology*, 2 ed. (Demain, A. L. and Davies, J. E., eds.), ASM, Washington, DC, pp. 103–113.

7. Baltz, R. H. (1997). Molecular approaches to yield improvements, in *Biotechnology of Antibiotics, 2 ed.* (Strohl, W. R., ed.), Marcel Dekker, New York, pp. 49–62.
8. Baltz, R. H. (2001) Genetic methods and strategies for secondary metabolite yield improvements in actinomycetes. *Antonie van Leeuwenhook* **79,** 251–259.
9. Baltz, R. H. (2003) Genetic engineering solutions for natural products in actinomycetes, in *Handbook of Industrial Cell Culture: Mammalian, Microbial, and Plant Cells* (Vinci, V. A. and Parekh, S. R., eds.), Humana, Totawa, NJ, pp. 137–170.
10. Rodriguez, E., Hu, Z., Ou, S., Volchegursky, Y., Hutchinson, C. R., and McDaniel, R. (2003) Rapid engineering of polyketide overproduction by gene transfer to industrially optimised strains. *J. Ind. Microbiol Biotechnol.* **30,** 480–488.
11. Hu, Z., Hopwood, D. A., and Hutchinson, C. R. (2003) Enhanced heterologous polyketide production in *Streptomyces* by exploiting plasmid co-integration. *J. Ind. Microbiol. Biotechnol.* **30,** 516–522.
12. Regentin, R., Cadapan, L., Ou, S., Zavala, S., and Licari, P. (2002) Production of a novel FK520 analog in *Streptomyces hygroscopicus*: improving titer while minimizing impurities. *J. Ind. Microbiol. Biotechnol.* **28,** 12–16.
13. Regentin, R., Kennedy, J., Wu, N., Carney, J. R., Licari, P., and Desai, R. (2004) Precursor-directed biosynthesis of novel triketide lactones. *Biotechnol. Prog.* **1,** 122–127.
14. Verall, M. S. (ed) (1996) *Downstream Processing of Natural Products—A Practical Handbook*. Wiley, Chichester, UK.
15. Gailliot, F. P., Gleason, C., Wilson, J. A., and Zwarich, J. (1990) Fluidised bed adsorption for whole broth extraction. *Biotechnol. Bioeng.* **44,** 922–929.
16. Schwartz, R. E., Sesin, D. F., Joshua, H. et al. (1992) Pneumocandins from *Zalerion arboricola*: I. Discovery and Isolation. *J. Antibiot.* **45,** 1853–1866.
17. Lawton, L. A., McElhiney, J., and Edwards, C. (1999) Purification of closely eluting hydrophobic microcystins (peptide cyanotoxins) by normal-phase and reversed-phase flash chromatography. *J. Chromatogr. A* **848,** 515–522.
18. Jarvis, A. P., Morgan, E. D., and Edwards, C. (1999) Rapid separation of triterpenoids from Neem seed extracts. *Phytochem. Anal.* **10,** 39–43.
19. Arslanian, R. L., Parker, C. D., Wang, P. K. et al. (2002) Large-scale isolation and crystallization of epothilone D from *Myxococcus xanthus* cultures. *J. Nat. Prod.* **65,** 570–572.
20. Verpoorte, R. (1998) Exploration of nature's chemodiversity: the role of secondary metabolites as leads in drug development. *Drug Discovery Today* **3**, 232–238.
21. Cordell, G. A. (1995) Changing strategies in natural products chemistry. *Phytochemistry* **40,** 1585–1612.

22. Faulkner, D. J. (2000) Marine pharmacology. *Antonie van Leeuwenhook* **77,** 135–145.
23. Mendola, D. (2003) Aquaculture of three phyla of marine invertebrates to yield bioactive metabolites: process developments and economics. *Biomol. Eng.* **20,** 441–458.
24. Schauffelberger, D. E., Koleck, M. P., Beutler, J. A. et al. (1991) The large-scale isolation of bryostatin 1 from *Bugula neritina* following current good manufacturing practices. *J. Nat. Prod.* **54,** 1265–1270.
25. Salomon, C. E., Magarvey, N. A., and Sherman, D. H. (2003) Merging the potential of microbial genetics with biological and chemical diversity: an even brighter future for marine natural product drug discovery. *Nat. Prod. Rep.* **21,** 105–121.
26. Ainsworth, A. M., Chicarelli-Robinson, M. I., Copp, B. R. et al. (1995) Xenovulene A, a novel GABA-benzodiazepine receptor binding compound produced by *Acremonium strictum*. *J. Antibiot.* **48,** 568–573.
27. Blackburn, M., Fauth, U., Katzer, W., Renno, D., and Trew, S. (1996) Optimization of fermentation conditions for the production of a novel GABA-benzodiazepine receptor agonist by *Acremonium strictum*. *J. Indust. Microbiol.* **17,** 36–40.
28. Wrigley, S. K., Sadeghi, R., Bahl, S. et al. (1999) A novel (6*S*)-4,6-dimethyl-dodeca-2*E*,4*E*-dienoyl ester of phomalactone and related α-pyrone esters from a *Phomopsis* sp. with cytokine production inhibitory activity. *J. Antibiot.* **52**, 862–872.

18

Follow-Up of Natural Product Isolation

Richard J. P. Cannell

Summary

Follow-up of natural product isolation means re-isolation of certain compounds in larger amounts for various reasons, e.g. for further biological testing, conclusive structure determination, structure modification for SAR studies, ecological or chemotaxonomic investigations, etc. Apart from conventional synthetic chemistry approaches, a number of other systems, which have been established in other chapters of this book, can be utilised in order to build on this initial natural products isolation protocols. This chapter summarises various strategies and methods involved in follow-up of natural product isolation, and presents a number of specific examples.

Key Words: Follow-up; squalestatin; blocked biosynthesis; maximising gene expression; enzyme inhibitors; mutasynthesis; biotransformation.

1. Introduction

What do we mean by "follow-up"? Let us assume that the natural product just isolated is of some interest, that is to say, it may have some biological activity worthy of further examination, it may represent a novel structure, or it may be of interest for ecological or chemotaxonomic reasons. In each case, we may want more of the compound, or analogs, biosynthetic precursors, and other related metabolites. If the compound is biologically active, we may look to these related compounds to provide structure-activity relationship data, for compounds that are more active, more chemically or

From: *Methods in Biotechnology, Vol. 20, Natural Products Isolation, 2nd ed.*
Edited by: S. D. Sarker, Z. Latif, and A. I. Gray © Humana Press Inc., Totowa, NJ

metabolically stable, or in commercial terms will strengthen the patent position of the original compound by describing the wider family of metabolites.

Apart from classical synthetic chemistry approaches, there are a number of ways that we as natural products scientists can utilize the systems that we have already established so far (e.g., organisms, growth/culture conditions, chromatographic systems and separation methods) in order to build on this initial natural product isolation.

2. Further Extraction

Perhaps the most obvious and most important approach is that of carrying out a repeat fermentation (or collection)—probably on a larger scale—in order to isolate either further quantities of the same compound or, any related metabolites that might exist within the same organism. It might also be productive to examine other strains of the organisms, or other related species, as these may yield greater levels of, or analogs of, the compound. Armed with knowledge of the original natural product, it is possible to look again at the extract with a better idea of how to find related structures, as follows.

2.1. Similar UV Spectrum

Comparison of UV profiles of the compound with those of materials corresponding to the other peak on a chromatogram can lead to identification of related metabolites. Related compounds often possess features or moieties that give rise to characteristic maxima—a handle by which some peaks can be picked out from the many of no interest.

With high-performance liquid chromatography (HPLC), this kind of analysis may be preformed "on-line" by use of a diode array detector. Alternatively, material corresponding to various peaks can be collected and their UV absorbance measured individually in a UV spectrophotometer.

The UV spectrum can therefore give a semi-identification that may at least point the extractor to a few selected peaks in an otherwise complex chromatogram.

2.2. Chemical Identification

This Principle of identification of compounds from the same structural or chemical family can also be carried out in other ways once the components of a mixture have been partially or fully separated. For example, a TLC plate or paper chromatogram may be sprayed or stained with reagents that react specifically with certain classes of chemical. An example of this was the detection of a number of different secondary amines in a culture broth of *Streptomyces luteogriseus* by staining TLC plates of extracts of the organism with phenothiazine perbromide *(1)*.

This approach of "chemical screening" can also be used to examine related organisms or different strains of the same organism, in order to detect those that exhibit related, or different chemical profiles, which may therefore be the focus of further isolation work.

This principle of detecting specific classes of related compound with relative ease is discussed at greater length as part of the dereplication procedures described in Chapter 12 and also in Chapter 9.

2.3. Mass Spectrometry and LC-MS

Mass spectrometry (MS) is powerful tool for identifying compounds that are related. Coupled to an LC system, mass spectrometry can often detect the similarities and relatedness of materials corresponding to chromatographic peaks by their characteristic fragmentation pattern and/or by the fact that they contain particular atoms, e.g., Cl and Br, that have characteristic isotope ratios. More simply, it may just be that two compounds have a similar molecular weight or molecular weights separated by expected or explicable differences; e.g., differences that are multiples of 16 could likely represent the same molecule with and without various oxygen groups.

2.4. Thorough Isolation

The most complete means of ensuring that all of the related metabolites present in an organism extract are isolated, is simply to isolate everything—or at least as many compounds as possible—from the extract. By this means, other unrelated secondary metabolites are often also isolated. Such metabolites may not be obviously structurally related to the original metabolite of interest but may comprise earlier biosynthetic precursors or provide clues as to the biosynthetic pathways in operation and, in addition, may help to build up a secondary metabolic profile of the organism. This process is facilitated by the fact that it is generally carried out on a larger scale repeat extraction, from which it is possible to examine properly the minor peaks of a Chromatogram. In the original small-scale separation, these minor peaks may have been overlooked or ignored in favor more readily obtainable material or may even have not been detectable. This process is often achieved by obtaining a good separation of the extract on preparative HPLC and monitoring the eluate with a highly-sensitivity detection system in order to pick out all the minor peaks, barely detectable humps, and perturbations on the chromatogram baseline that might represent minor components. This process—sometimes known as "looking in the grass"—can lead to interesting and surprising results. There is more than one case in which biological activity has been assumed to be solely the result of a major component of an extract only to find later that in fact, most or all of the activity is due to a potent compound present in the extract at very low levels and hence in danger of being overlooked. (*see* **Note 1**).

A subsequent fermentation might also lead to additional, different compounds because, for a number of possible reasons, the secondary metabolic profile of a repeat fermentation may be different. Such reasons might include the effects of scale-up (e.g., oxygenation, or shear forces), changes in the organism strain, subtle undetectable differences in the growth of the seed culture, inoculation into growth culture, and so on, all of which can lead to variations in levels and types of secondary metabolic expression.

2.5. Islolation of Squalestatin "Minors"

An example of this approach is given by a group of metabolites: the squalestains (also known as zaragozic acids), isolated from the fungus, sp. (*see* **Fig. 1**). Extracts of this organism were found to have squalene synthase inhibitory activity, and extraction resulted in the isolation of the major squalestatins of which there were approximately four *(2,3)*. As these compounds appeared to hold potential as lead for the development of cholesterol-lowering drugs, repeat fermentations were carried out on a larger scale. This enabled more of the previously characterized squalestatins to be isolated, but it also gave the opportunity and will to hunt for related compounds that might be present at much lower levels. By filtration, then adsorption onto a column of a nonionic adsorbent (styrene divinylbenzene; Whatman XAD-16), the concentrated initial extract (from a 500 L fermentation) was clarified and partially purified and then converted to a crude calcium salt, a fairly selective step for this group of tri-carboxylic acids (**Fig. 2**). Large-scale, open-column reverse phase chromatography (Whatman Partisil Prep P40) of acidified extracts of this calcium salt resulted in the isolation of the major squalestatins, some of them as crystalline tripotassium salts. The side fractions and the crystallization mother liquors were subjected to further preparative HPLC. By scrutinizing the chromatograms and attempting to isolate almost every component, an additional 24 related metabolites were isolated fom the same organism *(4)*.

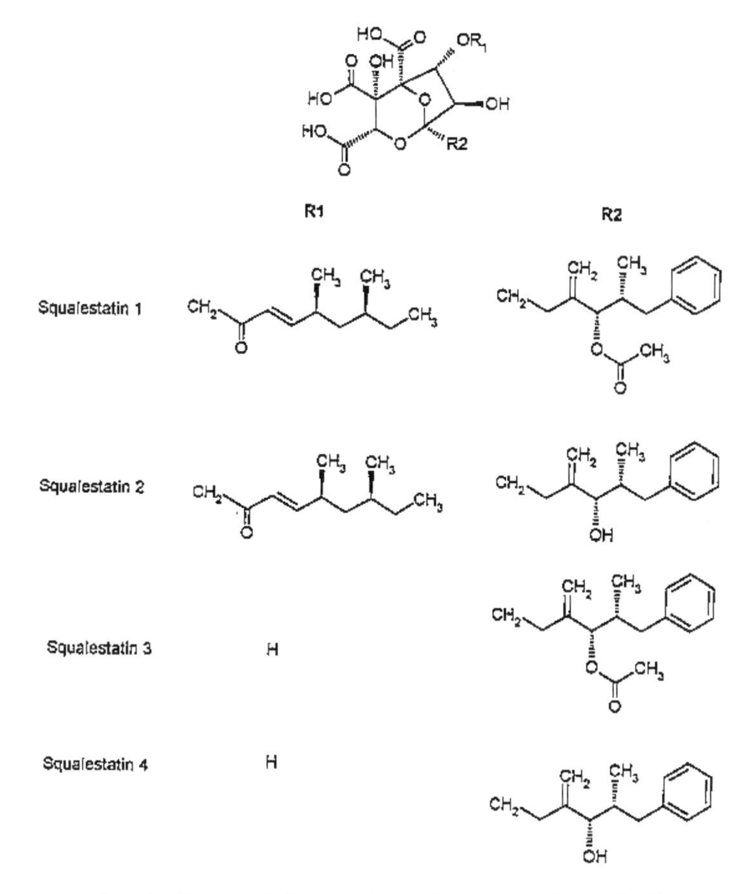

Fig. 1. Some of the squalestatin group of metabolites.

Some of these metabolites were extracted in relative abundance, but others were present only in minute quantities, such that it would have been virtually impossible to have isolated them from the initial 200 mL extract. Although this example is something of an extreme case, further related metabolites are frequently found from repeat extractions, even those involving no, or a much more modest, scale-up.

3. Maximizing Gene Expression

A more general method of generating analogs or related metabolites is to attempt to maximize secondary metabolic gene expression in order to maximize the range of products formed. Although there is no single unified

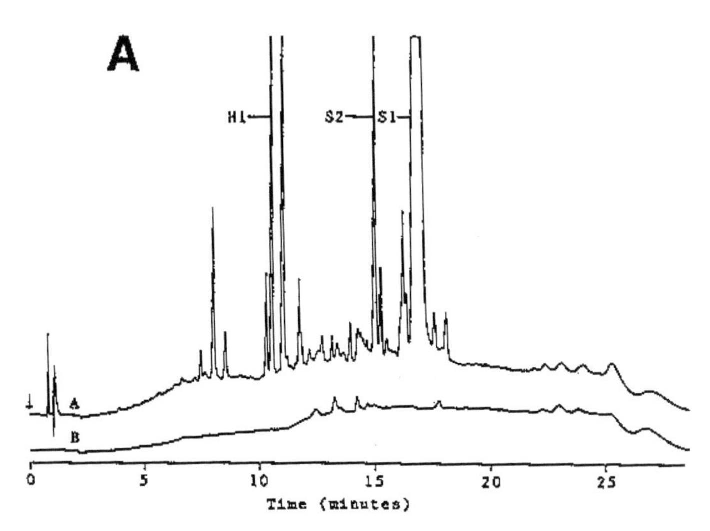

Column: Phase Separations Ltd Spherisorb C6, 5 μm (15×0.46 cm). Mobile Phases: A, H_2O-H_2SO_4, 1,000 : 0.15. B, MeCN - H_2O - H_2SO_4, 500 : 500 : 0.15. Linear gradient 0 to 100% B in 15 minutes, hold 10 minutes, 100 to 0% B in 1.5 minutes. Flow: 2 ml/minute. Detection: λ 210nm. Range: 0.1 AUFS. Chart speed: 1 cm/minute. Trace B is of a blank gradient.

B

Broth
- adjust to pH 10.5
- pass through rotary drum filter

Filtrate
- adjust to pH 6.5

Amberlite XAD-16
- wash with $(NH_4)_2SO_4$
- wash with EDTA (5,000 litre scale)
- wash with water
- elute with Me_2CO - H_2O (1 : 1)

Eluate
- $Ca(OAc)_2$
- filter (add filter aid on 5,000 litre scale)

Crude Ca salt

C

Squalestatin	Rt (minutes)	Squalestatin	Rt (minutes)
H7	5.8	V1 isomer a	13.6
H9	5.9	V1 isomer b	13.7
H2	6.4	S4	13.7
H6	6.9	W1	14.0
H1	9.4	S8	14.2
H5	9.8	S2	14.3
6-Deoxy H1	10.2	U1	14.5
7-Deoxy H5	11.2	X1	15.4
6-Deoxy H5	11.2	T1	15.6
V2	11.3	S1	15.8
W2	11.4	Y1	16.1
6,7-Dideoxy H5	11.8	S5	16.7
S3	12.6	7-Deoxy S1	16.8
U2	12.9		

HPLC conditions as for Fig. 2 but flow 3 ml/minute.

Fig. 2. Chromatogram of an acidified extract of crude calcium salt of the squalestatins. (**A**) Typical trace after HPLC of an acidified extract of crude calcium salt. (**B**) Isolation of crude calcium salt from fermentation of *Phoma* sp. C2932. (**C**) Squalestatins isolated in the course of this work and their retention times in a gradient HPLC system. (Reproduced with permission from **ref. 4.**)

concept of secondary metabolism, it is clear that metabolite production does vary according to an organism's chemical and physical environment. In microbial systems at least, the onset of secondary metabolism, generally coincides with the end of log phase growth and the onset of idiophase, which itself is generally the result of limitation of a specific nutrient. The nature of this limiting nutrient and the other nutrients in the medium can determine the secondary metabolites produced, through processes such as derepression, or inhibition, of secondary metabolic pathways. For instance, carbon catabolite control is exhibited by glucose, which represses phenoxazinone synthase expression and hence actinomycin production in *Streptomyces antibioticus*; high levels of ammonium suppress the formation of tylosin by *Streptomyces fradiae* and cephalosporin by *Streptomyces clavuligerus*; high inorganic phosphate levels repress enzymes involved in the synthesis of tetracyclins, candicidin, neomycin, and streptomycin; concentrations of trace metals (e.g., iron, zinc and cobalt) often have a profound effect on metabolite production *(5)*. Other compounds may induce or enhance the production of secondary metabolites. These include both commonly occurring molecules, such as various amino acids, which may or may not be components of the metabolites, or less prevalent compounds, such as the autoregulator molecules known to play a role in the related aspects of differentiation and quorum sensing, e.g., A-factor and pamamycin. Overall, the processes that regulate secondary metabolism are poorly understood, a situation that befits a whole form of metabolism whose biological role is still very unclear. Suffice it to say that many physical and chemical parameters can influence secondary metabolite production, and the way in which this can best be used to advantage is generally best determined empirically for any new metabolite or organism.

Hence, varying these factors can lead to a much wider range of metabolites than would be produced by an organism under a single set of conditions. This is most easily carried out for microorganisms, though secondary metabolite production by other organisms is also affected by their local environment. Much has been written about this aspect of microbiology (*see*, for example, **ref.** *6*), and will not be discussed in detail here, but some of the most important factors affecting metabolite production are:

1. Medium: carbon source, nitrogen source, phosphorus source, C/N/P ratios, levels of trace nutrients.
2. Autoregulators (e.g., A-factor and pamamycin).

3. Physical conditions: Shaken liquid culture, static liquid culture, solid state culture.
4. Oxygen levels.
5. Growth rate.
6. Temperature.
7. pH.

A general rule of thumb is that exposing an organism to various forms of "stress" will lead to an increase in the range of secondary metabolites produced. If the aim is to isolate from a microorganism the greatest possible number of metabolites, then the organism should always be grown in more than one medium and/or in more than one form (e.g., shaken liquid and solid state). Approaches for the more specific aim of maximizing the production of a particular metabolite are discussed in Chapter 17.

4. Blocked Biosynthesis

4.1. Biosynthetic Mutants

The generation from a producing organism of mutant strains that are blocked or altered in the biosynthetic pathway can lead to the isolation of related metabolites that could not otherwise be obtained. These metabolites may be intermediates from the blocked pathway that would normally be transient and detectable only in trace amounts, if at all, or they may represent "shunt" metabolites, whereby the intermediates have gone down a different biosynthetic route, resulting in novel compounds. The products of such mutants may be of interest in their own right, or they may be interest in biosynthetic studies, biotransformation experiments or as starting points for precursor-directed biosynthesis.

In its simplest form, this might involve mutating isolates of the organism and looking for changes in either the level or type of secondary metabolite produced in the resultant isolates. Mutation can be carried out by the use of UV light or chemical mutagens or by the use of selective media—nutrient-deficient, or toxin-containing—to isolate strains of an organism with a modified genetic make-up, able to survive on such media, and which may result in concomitant modification of the secondary metabolites. The same effect may be obtained by the use of enzyme inhibitors that block a biosynthetic pathway at particular steps.

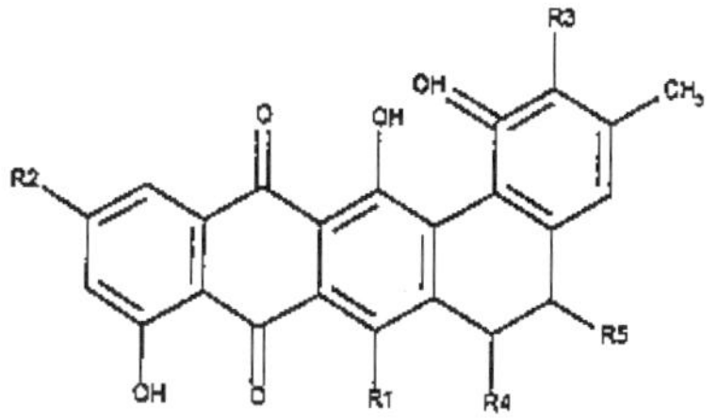

R1	R2	R3	R4	R5
OH	H	COOH	OH	OH
OCH_3	H	COOH	OH	OH
H	H	COOH	OH	OH
H	H	CONHCH(CH_3)COOH	OH	OH
OCH_3	H	CONHCH(CH_3)COOH	OH	OH
OH	CH_3	CONHCH(CH_3)COOH	OH	OH
H	H	COOH	H	OH
H	CH_3	COOH	OH	OH
H	CH_3	CONHCH(CH_3)COOH	OH	OH
OH	H	CONHCH(CH_3)COOH	OH	OH
OH	CH_3	CONHCH(CH_3)COOH	OH	
H	CH_3	CONHCH(CH_3)COOH	OH	

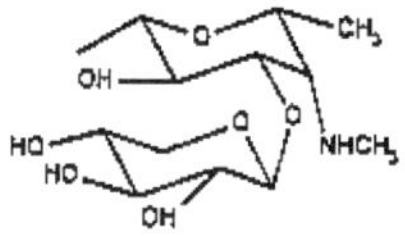

Fig. 3. Structures of pradimicins isolated from biosynthetic mutants.

4.1.1. Pradimicin Biosynthetic Mutants

Spores of *Actinomadura Verrucosospora* subsp. *neohibisca*, a producer of the dihydrobenzonaphthacene quinone antibiotic pradimicin, were mutagenized by UV light and/or *N*-methyl-*N*'-nitro-*N*-nitrosoguanidine. As pradimicin is a red pigment, mutants were initially selected from approximately 10,000 colonies on the basis of the production of colorless or non-red pigments, and these were assigned to a number of classes, each of which either accumulated intermediates at a particular stage in the pathway or produced novel shunt metabolites. From these strains were isolated eight novel metabolites and a number of known metabolites *(7,8)* (**Fig. 3**).

4.1.2. Aclacinomycin Biosynthetic Mutants

Similarly, mutants of *Streptomyces galilaeus*, which produces another class of glycosidic anthracyclines, aclacinomycins (**Fig. 4**), yielded a number

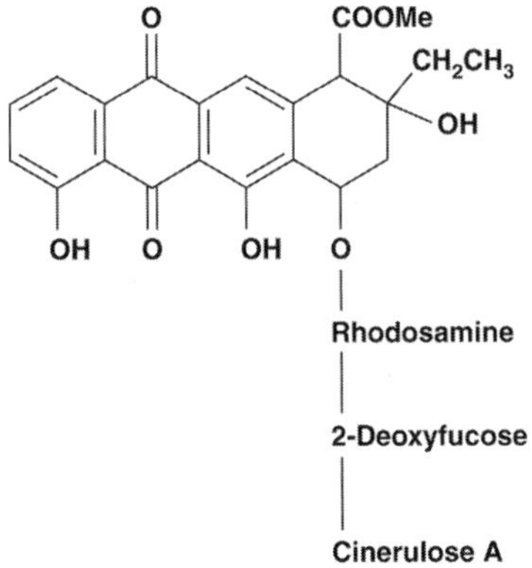

Fig. 4. Aclacinomycin A.

of novel metabolites. These included novel anthracyclinones devoid of sugars and others with combination of different sugars. Some were able to produce only one form of the aglycone, others were impaired in their ability to supply rhodosamine as they produced analogs lacking this sugars, others were unable to oxidize the terminal rhodinose to form cinerulose A, and others were unable to supply rhodinose at all ***(9–11)***. From the same mutagenesis program were isolated a number of strains capable of producing aclacinomycin A at a 30-fold higher concentration than that of the parent strain.

4.1.3. Tricothecene Analogs

The tricothecenes are a family of toxic fungal metabolites, and T-2 toxin is the major tricothecene produced by the fungus *Fusarium sporotrichioides* (**Fig. 5**). In studying the biosynthesis of this metabolite, Beremand et al. ***(12)*** generated a range of amino acid auxotrophs and found that the leucine auxotroph was blocked in the production of T-2 toxin. They found that this mutant, however, produced two analogs that were present only at barely detectable levels in the wild type. The relative concentrations of T-2 and its analogs could be manipulated by controlling the concentration of leucine in the medium, indicating that the metabolites were shunt metabolites and that all three compounds derive from a common intermediate.

Toxin T-2 R = $OCOCH_2CH(CH_3)_2$

Analogues R = $OCOCH(CH_3)_2$

R = $OCO(CH_3)_2$

Fig. 5. Trichothecene Toxin T-2 and shunt metabolites isolated from mutant.

4.2. Enzyme Inhibitors

Enzyme inhibitors can be used to produce the same effect as a genetic mutant, i.e., to block a biosynthetic pathway and lead to the buildup of otherwise transient intermediates or shunt products. Such inhibitors include cerulenin, which inhibits fatty acid synthase and the similar polyketide synthase, and sinefungin and ethionine that inhibit the transfer of methionine groups. Another useful group of inhibitors are those that inhibit cytochromes P450, a group of enzymes that perform oxidative reactions on a wide range of compounds. These inhibitors include ancymidol, metyrapone, and phenytoin. An example of this is the use of ancymidol in fermentations of the fungus *Gibberella pulicaris* in order to block the production of the tricothecenes and allow access to the otherwise transient intermediate trichodiene ***(13)***.

5. Directed Biosynthesis

There is a further, more elegant way of attempting to manipulate or "persuade" the organism to produce analogs or derivates. Most secondary metabolites are largely constructed for the most part from a number of standard building blocks (e.g., terpenoids riginate from 5-carbon isoprenyl units; polyketides originate from acetate units [in the case of fungi] or formyl, acetate or malonyl units [in the case of bacteria]). Aromatic groups tend to derive from one of the aromatic amino acids such as tyrosine, phenylalanine, or tryptophan, which themselves derive from shikimic acid or form acetate units. Alkaloids, too, derive in a large part from a number of amino acids such as tyrosine, phenylalanine, tryptophan, lysine, and ornithine, as well as from acetate units, mevalonate, and a number of other simple precursors.

Providing organisms with analogs of precursors or intermediates of a secondary metabolite can sometimes lead to these being incorporated by the organism's biosynthetic machinery into a novel secondary metabolite.

This approach assumes some knowledge, or at least some presumption, of the general biosynthetic pathway by which the metabolites is formed. It is usually possible to make a good guess at the biosynthetic origins of a molecule, particularly when biosynthetic information is available on a structurally similar molecule, and these theories cab be tested by the use of isotopically labeled precursors. It also relies on the fact that the biosynthetic enzymes are not so specific as to exclude all but the natural precursor. Fortunately, the enzymes of secondary metabolism tend to be less specific than those of primary metabolism.

5.1. Mutasynthesis

As mentioned in **Subheading 4.1.**, the products of a biosynthetic mutant may be of interest in their own right but, in addition, the mutant or its products can be utilized further. The mutant can be used as a reaction system to which can be fed a modified form of the first blocked intermediate, i.e., the product of the blocked enzyme, or subsequent building blocks. Assuming that this is the only blocked step in the pathway and that the remainder of the pathway's enzymes are functioning, the enzymes may well act on the modified intermediate free from the competition from the natural substrates to produce a derivatized final product. Mutasynthesis is the term generally used to refer to this process of "blocking and feeding."

The blocked intermediates may also be fed to other organisms in a rational manner—such as to those that usually modify a related precursor—in order to produce "hybrid" metabolites. This process of mutasynthesis usually requires some knowledge of the blocked step and that the biosynthetic enzymes are sufficiently unspecific to accept substrate analogs.

5.2. Methodology

An advantage of this approach is that, once a putative biosynthetic pathway has been elucidated and suitable precursor analogs obtained, there are very few additional experimental procedures, other than those already developed for the isolation of the original "parent" metabolite.

The major processes involved are feeding the precursor to the organism and the development of a separation method to analyze and isolate the products.

5.2.1. Feeding Precursor Analogs

The procedures involved in carrying out directed biosynthesis are largely based on educated guesses, and the best conditions for any given family of metabolites are best determined empirically. There is no right or wrong way, and in many cases it is not worth expending a great effort in adjusting the feeding conditions to obtain best product titer—it is often enough just to generate the modified metabolite at any level. However, there are a number of general guidelines that it makes sense to follow.

1. Addition of the precursor to a fermentation broth can be carried out at the beginning of the fermentation, i.e., as a medium component, or it can be added at the onset of the biosynthesis of the secondary metabolite, which is often at the beginning of stationary phase, the period when most secondary metabolites are formed. This latter method is generally preferred as it is more likely to avoid possible toxic effects of the precursor analogs on cell growth and primary metabolism as well as the possibilities of the precursor being catabolized and recycled during the processes of primary metabolism before secondary metabolite formation has begun.
2. The compound may be added as a single aliquot or it may be pulse-fed—perhaps once a day for 4 d from the end of lag phase, or in more sophisticated systems it may be added continuously.
3. The precursor is generally added at levels at which its natural relative might be found in the broth (although in many cases, the natural relative would not be "free" in the broth but would be intracellular and mobilized only in an activated form, e.g., phosphorylated or as a Coenzyme A ester). Obviously, this is a fairly vague figure but suffice to say that such "physiological" levels are generally assumed to be about 10 m*M*. An alternative approach is to add the precursor at much higher levels, say 50–100 m*M*, in order to outcompete the natural precursor for the biosynthetic enzymes' active site (for which the latter probably has greater specificity). The uptake of the unnatural precursor may only be low compared to that of the natural precursor, so by adding excess it may be possible to increase the final amount of a modified product. Disadvantages of this approach are that the high levels of the fed compound may have a toxic or inhibitory effect on the growth or metabolism of the organism and, more specifically, on the general biosynthesis of the secondary metabolite itself. It also requires a greater supply of the precursor analog.
4. The precursor analog is added aseptically, generally as a concentrated aqueous solution that has been filter-sterilized, autoclaved, or prepared under aseptic conditions. However, if the compound is insoluble in water, it may be added in a small volume of organic solvent or even as a solid. This may cause the compound to precipitate as soon as it enters the aqueous medium

but this does not make the experiment a lost cause. The compound may be sparingly soluble and may act in effect as a slow-release feed.

5. The fact that an artificial precursor is not incorporated may reflect problems relating to its uptake across the cell membrane and it may be necessary to modify conditions to create either a cell-free system or a resting cell culture. The former involves growing the culture to stationery phase, separating the cells from the liquid medium by centrifugation, then smashing the cells (e.g., in a sonicating bath) and resuspending in an isotonic buffer to which the precursor analogs are added. This removes the barrier of the cell walls, thus allowing access for exogenous precursor molecules to the biosynthesis enzymes.

In resting cell cultures the cell of the stationary phase culture are separated from the liquid medium by centrifugation, then washed in buffer, centrifuged again, and resuspended in buffer and/or minimal medium to which the precursor is added. Thus, removed of all exogenous substrates, the cells are essentially intact and viable but inert, and with few competing biosynthetic or metabolic pathways in operation, the likelihood of precursor uptake and incorporation is increased. An additional advantage is that the washed-cell system is cleaner than the original cell culture, making purification easier.

There may be problems relating to the conversion of the metabolite to an activated form (e.g., an acetyl coenzyme A ester), which it may require in order to be transported and/or recognized by the relevant enzymes. Other compounds may be rapidly metabolized and "chewed up" by the pathways of primary metabolism before they can be incorporated into the secondary metabolite.

5.2.2. Analysis of Directed-Biosynthesis Products

Presumably, an assay system will already have been established to monitor the natural product during the initial isolation, and this may allow for analysis of the modified product. By comparing the chromatogram of the precursor-fed organism extract with that of an unfed control organism, it should be possible to detect peaks in the former corresponding to the modified product and/or substrate that are not present in the control. If, however, a modified product is undetected by the established analytical system, this can mean either that the precursor has not been incorporated, or simply that the modified product has not been resolved from the natural product. This problem is compounded by the fact that there will be no standard of the compound available. Resolution of potentially coeluting compounds may require gradient HPLC or exploitation of further differences between the modified and natural products by means such as LC-MS—a technique ideal for such analysis. The presence of a particular atom such as a halogen or an additional oxygen can be readily detected by this method. The incorporation of a fluorine-containing precursor can be followed by the use of ^{19}F NMR. Any fluorine-containing metabolites will give a single peak with a characteristic shift on a ^{19}F NMR spectrum, and with no signals from any other molecules, this is a simple and unambiguous technique by which to follow the generation of fluorine-containing metabolites in a complex mixture (**Note 2**).

5.3. Precursor-Directed Biosynthesis of Squalestatins

The squalestatins, described in the previous section, also provide an example of how some knowledge of biosynthetic pathways can lead to interesting directed-biosynthetic products.

The molecule appeared at first sight to be polyketide in nature, and in order to test this assumption and ascertain the biosynthetic origin of the component parts of the molecule, the producing organism—*Phoma* sp.—was fed with isotopically labeled acetate units in the form of [1-^{13}C], [2-^{13}C], and [1,2-^{13}C2] acetate. These demonstrated that the backbone of the molecule was formed from two polyketide chains made up of acetate units. The remaining four carbons of the bicyclic ring structure (carbons ***21,3,4,22***) appeared from NMR coupling studies to be incorporated as adjacent intact acetate-derived units at a level lower than the others, suggesting metabolism of the acetate via the TCA cycle to a four-carbon unit; indeed, feeding [2,3-^{13}C succinate] resulted in incorporation of this double-label (albeit scrambled incorporation, suggesting that the succinate had also been metabolized to a large extent by the TCA cycle). Other carbons ***(19,20,32,33)*** appeared to derive from single carbon units in the form of *S*-adenosyl methionine. The aromatic portion of this molecule was investigated by feeding ^{13}C-labeled forms of phenylalanine and benzoic acid, which resulted in incorporation of both, but the particularly high incorporation of benzoic acid suggested its role as the starter unit of biosynthesis following its formation from phenylalanine ***(14)***.

To summarize, it had been established that the backbone of this family of molecules was built from the addition of acetate building blocks to an aromatic starter unit together with the condensation of this to a four-carbon α-keto dicarboxylic acid, followed by esterification/acylation of the tetraketide chain and methylation by methionine-derived carbons and probably hydroxylation at C7. All together this provides a body of information of interest not only for its own sake but as a means of "hitching a ride" on the organism's biosynthetic pathway to make analogs of the compound (**Note 3**).

It was felt, for a number of reasons, that the aromatic moiety represented an area of the molecule suitable for manipulation as a means of generating analogs. It derived from a small simple molecule but with a number of sites for modification that could potentially give rise to a large number of analogs. It had already been shown that feeding labeled benzoic acid resulted in high levels of incorporation into squalestatin—it was the starter unit and presumably relatively little was metabolized by other pathways—maximizing the chances of detecting incorporation. Finally, the methods involved in modifying specific sites on such aromatic groups are generally complex and difficult to carry out by chemical means, so the products were likely to be valuable.

A range of simple analogs of benzoic acid and phenylalanine, easily obtainable from standard chemical suppliers, were chosen as substrates. These included a large number of hydroxy- and dihydroxybenzoic acids, aminobenzoic acids, nitrobenzoic acids, fluoro-, difluoro-, trifluoro-, tetrafluoro- and pentafluorobenzoic acids, chlorobenzoic acids, iodobenzoic acids, and methoxybenzoic acids. A number of structures other than those with a six-membered aromatic ring were also tried, including pyridinecarboxaldehydes, alicyclic carboxylic acids, naphthalenecarboxylic acids, furancarboxylic acids, thiophenecarboxylic acids, nitro-, bromo-, and chlorothiophenecarboxylic acids.

5.3.1. Method

Feeding studies demonstrated that benzoic acid added at the time of inoculation was not incorporated but when added at d 3, 4, or 5 was incorporated at very significant levels. This presumably was because of the fact that the compound added at d 0 was metabolized by the processes of primary metabolism before the culture reached stationary phase where secondary metabolism for the most part takes over.

The samples were added as aqueous solutions (6.25 mg/mL), which were adjusted to neutral pH with sodium hydride and filter sterilized (**Note 4**). 2 mL aliquots of each of these were added to individual cultures (50 mL) to give a final concentration of precursor of 0.25 mg/mL and the cultures reincubated at 25°C with shaking, as they had been for the first part of the fermentation. Four days later (seven days after incubation), the cultures were harvested, analyzed, and the products isolated.

HPLC analysis: Samples of broth were mixed with an equal volume of acetonitrile containing sulfuric acid (5 mL/L), centrifuged, and the supernatant analyzed by reverse phase gradient HPLC (Spherisorb C6 [5 μm particle size, 150 × 4.5 mm] with a gradient of 0–50% acetonitrile/water with sulfuric acid [50 μL/L], flow rate 1 mL/min, detection 210 nm).

HPLC-MS analysis: Samples of broth were prepared for HPLC-MS analysis by mixing with an equal volume of acetonitrile containing trifluoroacetic acid (5 mL/L). The samples were centrifuged and the supernatant analyzed by an isocratic HPLC method (Spherisorb ODS2 [5μm particle size, 150 × 4.5 mm] acetonitrile/water [55:45] with trifluoroacetic acid [0.1%, v/v], flow rate 0.5 mL/min). This system was connected via a thermospray interface to a (Finnigan Mat TSQ 70B) mass spectrometer. Using the HPLC conditions described, it was not, in fact, possible to differentiate between squalestatin 1 and its fluorinated analogs as they coeluted but they could be detected by MS (*see* **Note 5**).

Purification of the analogs was carried out as before by solvent extraction, adsorption, and elution (solid phase extraction) from a column of Amberlite XAD16, followed by loading of the sample onto a preparative HPLC column (Spherisorb ODS2, 5 μm particle size), which was washed with 25% acetonitrile/water, then squalestatins 1 and analogs were eluted with 60% acetonitrile/water. A set of final preparative HPLC steps were carried out to resolve each of the analogs and the parent compound. The products isolated are shown in **Fig. 6** ***(15,16)*** (*see* **Note 6**).

So, in the above case, the production of these fluorinated and thiophenylated squalestatins necessitated the following:

1. Some idea of the biosynthesis in order to pick a sensible range of suitable precursor analogs.
2. Some simple optimization of feeding conditions. At its simplest, this might involve determination of whether to:
    a. Add compounds at inoculation, i.e., at the beginning of, or during, growth phase, or after several days growth when log phase growth has stopped (at the onset of, or during, secondary metabolism).
    b. Use whole cells or a washed-cell culture or a cell-free system for optimal incorporation.
3. A method for detection of incorporation, e.g., LC, LC-MS, GC-MS, TLC, NMR.
4. A method for purifying a number of very closely related compounds.

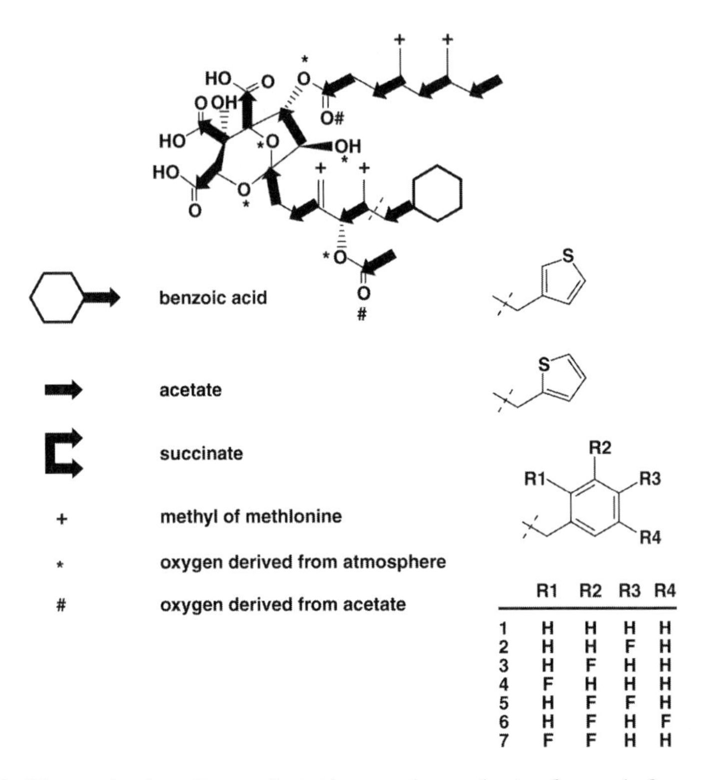

	R1	R2	R3	R4
1	H	H	H	H
2	H	H	F	H
3	H	F	H	H
4	F	H	H	H
5	H	F	F	H
6	H	F	H	F
7	F	F	H	H

Fig. 6. Biosynthesis of squalestatins and products formed from directed biosynthesis.

Arguably, the example described above represented a not particularly successful attempt to generate a wide range of metabolites, insofar as a relatively low proportion of the potential precursor analogs were incorporated into the final squalestatin molecule. Presumably, therefore, the active site of one or more of the biosynthetic enzymes exhibits fairly narrow specificity.

Why was it that only the fluorobenzoic acids and the thiophenes were apparently incorporated? In the case of the former, this may reflect the probability that fluorobenzoic acid is structurally the most similar of the analogs to the natural substrate. Fluorine is a fairly inert atom and isosteric with hydrogen. It is not so clear why the thiophenes should have been incorporated.

Alternative reasons may include:

1. Inability of the other precursors to cross the cell membrane. Had a cell-free system been used, more of the metabolites may have been incorporated (although it is unlikely that any squalestatins would be formed in a cell-free system).
2. Enzyme inhibitory effect of other precursors. (Addition of some of the compounds was associated with a reduction in levels of squalestatins.)
3. Some of the analogs were metabolized by different pathways.

As well as concentrating on the aromatic moiety, attention could also have focused on other parts of the molecule that derive from such as the 4-C unit that comprises part of the bicyclic core, or the various acetate units, by feeding analogs of these precursors. However, there is likely to be more success by using intermediates from later in the pathway.

5.4. Other Examples

There are numerous other examples of precursor-directed biosynthesis involving natural products of various biosynthetic origins ***(17)***. Many have utilized intermediates from much later stages in the biosynthetic pathway than the example of the squalestatins.

5.4.1. A54145

This group of lipopeptide antibiotics produced by *Streptomyces fradiae* spawned a wide number of analogs produced by feeding various fatty acids (**Fig. 7**). In addition, the ratios of the various natural metabolites could be controlled by addition of valine or isoleucine ***(18)***.

5.4.2. Mitomycins

This group of natural products derived from the shikimate pathway and produced by *Streptomyces caeapitosus* led to a large number of analogues by feeding the organism with a range of primary amines ***(19)***. These are summarized in (**Fig. 8**).

5.4.3. Cyclosporins

Cyclosporins are a class of cyclic peptides produced by *Beauvaria nivea*, that possess immunosuppressant activity and have spawned a large number of additional modified forms by feeding analogs of the natural amino acid building blocks. A large range of derivative have been formed via this method with modifications at almost all of the amino acid positions ***(20,21)***, and more were produced by carrying out the process in a cell-free system. These could not be formed by feeding the precursors to the whole

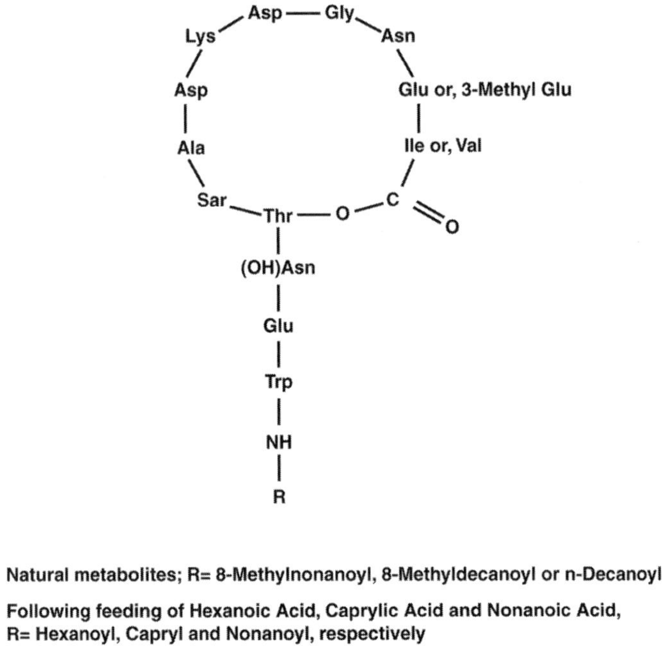
Asp
Gly
Lys
Asn
Asp
Glu or, 3-Methyl Glu
Ala
Ile or, Val
Sar
Thr
O
C
O
(OH)Asn
Glu
Trp
NH
R
Natural metabolites; R= 8-Methylnonanoyl, 8-Methyldecanoyl or n-Decanoyl
Following feeding of Hexanoic Acid, Caprylic Acid and Nonanoic Acid,
R= Hexanoyl, Capryl and Nonanoyl, respectively

Fig. 7. Directed biosynthesis products of A54145.

organism, probably because the amino acids involved were normally metabolized *(22)* (**Fig. 9**).

The range of examples that can be drawn from amino acid-derived natural products formed via amide synthases reflect the unspecificity of these enzymes in peptide biosynthesis compared to ribosomal protein synthesis.

5.4.4. Avermectins

One of the most successful attempts at generating a wide range of analogs of a natural product resulted from the isolation of a mutant of the avermectin-producing organism, *Streptomyces avermitilis*. Avermectins types a and b derive their C-25 substituents from isoleucine and valine, via incorporation of these compounds as coenzyme A derivatives of isobutyric and 2-methylbutyric acids, respectively. This mutant lacked the functional branched-chain 2-oxo acid dehydrogenase activity and could not therefore incorporate these likely biosynthetic starter *(23)*. When the fermentation was not supplemented with branched-chain carboxylic acids, no avermectins were produced. However, feeding a range of such acids to the

Mitomycin B: $R_1 = OCH_3$, $R_2 = H$, $R_3 = CH_3$
Mitomycin C: $R_1 = NH_2$, $R_2 = CH_3$, $R_3 = H$

Analogues: R_1 = Methylamine
Ethylamine
Propylamine
Propargylamine
Allylamine
2-Methylallyamine
2-Chloroethylamine
3-Chloropropylamine
Benzylamine

Type I Analogues - Amine incorporated intp Mitomycin C nucleus
Type II Analogues - Amine incorporated intp Mitomycin B nucleus

Fig. 8. Mitomycin analogs produced by feeding various amine precursors.

fermentation led to the production of a large number of new avermectins modified at the C-25 position with the corresponding acid substituent as shown in **Fig. 10** ***(24)***. The analog containing the cyclohexyl moiety at the C-25 position is now made commercially, by essentially this process, and is marketed as the antiparasitic, doramectin.

5.4.5. Feeding Natural Precursors

This general approach of feeding precursors can also be used just to increase levels of the naturally occurring metabolite by supplementing the culture with the natural precursors. This is illustrated by the simple example of pyrrolnitrin production by *Pseudomonas aureofaciens*, which was increased significantly by addition of exogenous tryptophan, the direct precursors of pyrrolnitrin (**Fig. 11**). Addition of tryptophan analogs also led to the corresponding analogs of pyrrolnitrin ***(25)***.

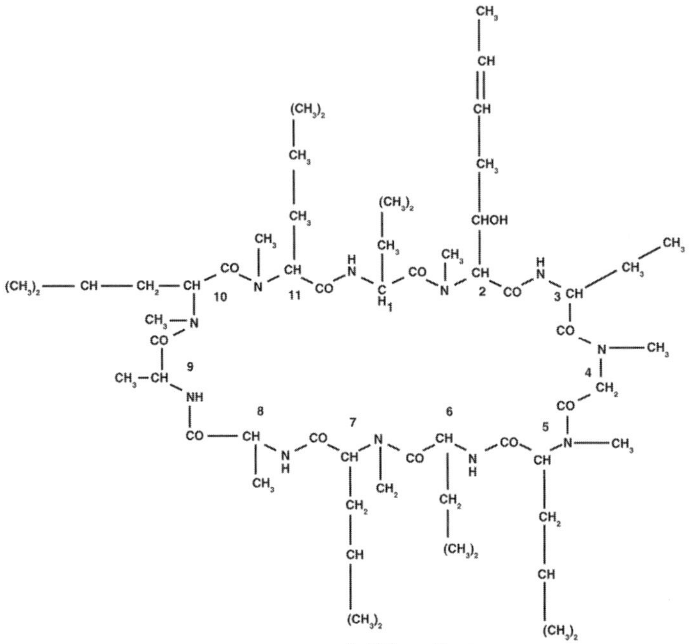

Cell Free Systems:

L-β-Cyclohexylalanine (1)	**N-Methyl-2-amino-3-hydroxy-4,4-dimethyloctanoic acid (1)**
DL-α-Allylglycine (2)	**N-Methyl-L-Norvaline (11), L-Noryvaline (5)**
D-Serine (8)	**N-Methyl-L-Norvaline (11), L-Noryvaline (2,5)**
DL-Threonine (2), D-Serine (8)	**L-*allo*-Isoleucine (5), N-Methyl-L-*allo*-Isolecine (11)**
DL-Valine (2), D-Serine (8)	**L-*allo*-Isoleucine (5,11)**
DL-Norvaline (2), D-Serine (8)	**D-2-Aminobutyric acid (8), β-Chloro-D-alanine (8)**
3-Fluoro-D-Alanine (8)	**2-Deufero-3-Fluoro-D-alanine (8)**

Fig. 9. Some amino acid analogs incorporated into Cyclosporin A by *Beauveria nivea* (numbers in brackets indicate position of incorporation).

5.4.6. Halogenation

Other than the method described in **Subheading 5.3.**, other halogens may be introduced or substituted more directly into a natural product by making use of the haloperoxidase activity of various organisms. Organisms producing chlorine-containing metabolites can often be "driven" to produce the same metabolites in a brominated form by addition of an inorganic source of bromine such as sodium bromide or potassium bromide. The haloperoxidases that carry out this reaction are fairly unspecific and can often readily utilize bromine in the place of chlorine. Examples include

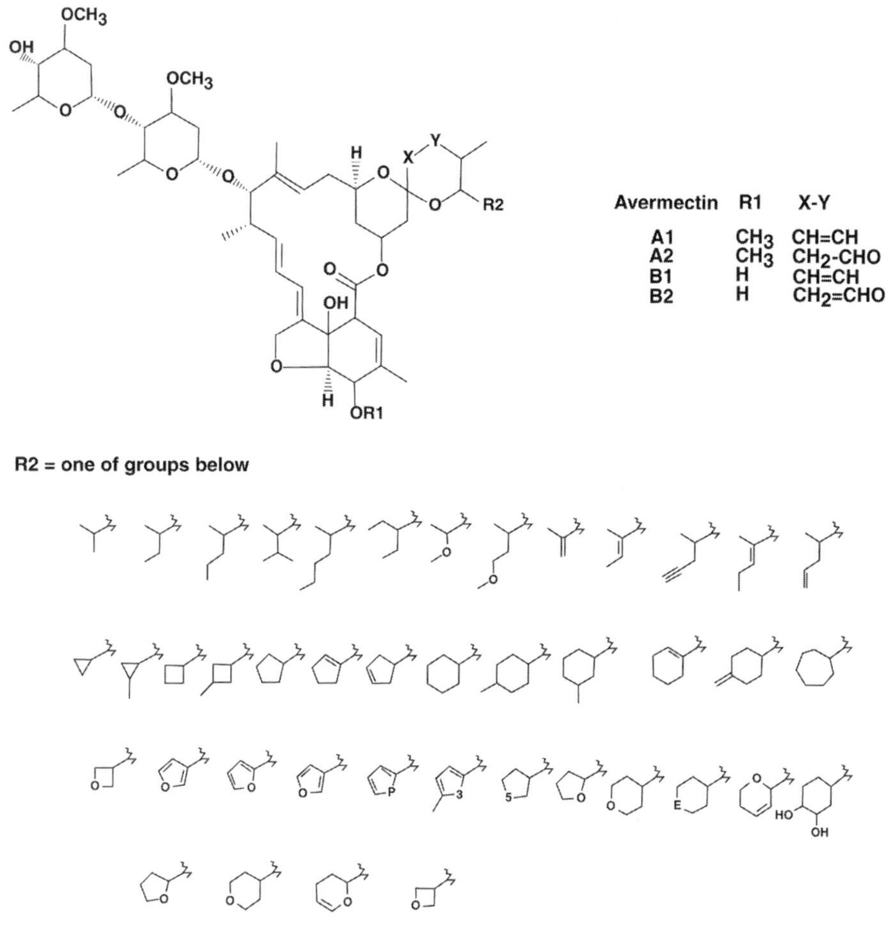
OCH3
OH
OCH3
H
X
Y
O
R2
OH
H
OR1
Avermectin R1 X-Y
A1 CH3 CH=CH
A2 CH3 CH2-CHO
B1 H CH=CH
B2 H CH2=CHO
R2 = one of groups below

Fig. 10. Structures of novel avermectins.

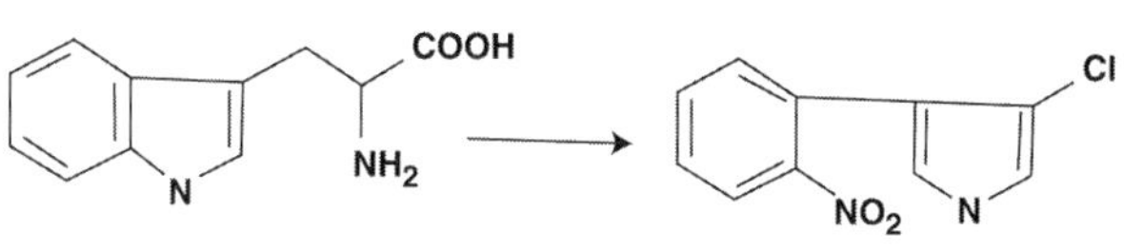
COOH
NH2
N
Cl
NO2
N

Fig. 11. Tryptophan–precursor of pyrrolnitrin.

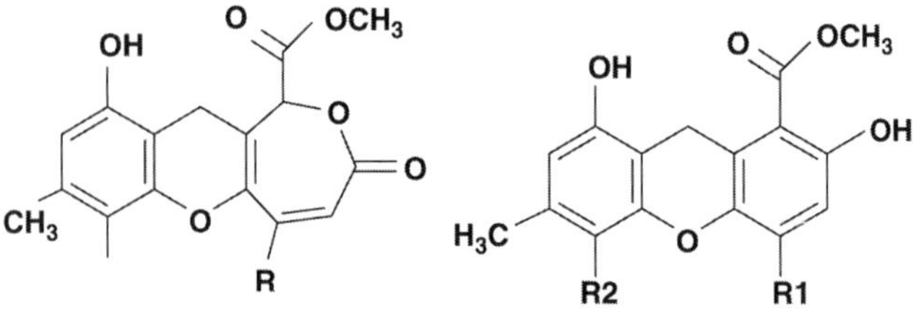

Chloromonilicin R = Cl
Bromomonilicin R = Br

4-Chloropinselin R1 = Cl, R2 = H
4-Bromopinselin R1 = Br, R2 = H
4,5-Dibromopinselin R1,R2 = Br

Fig. 12. Chloromonilicin and 4-Chloropinselin and their brominated forms isolated from *Monilinia fructicola*.

the metabolites of *Monilinia fructicola* (chloromonilicin and 4-chloropinselin), which were converted to their bromoanalogs following the addition to the medium of NaBr (1 g/L) (**Fig. 12**) *(26)*.

6. Biotransformation

Another method of maximizing diversity from a single natural product is the use of other biological systems, in the form of whole cells or isolated enzymes, to modify the molecule. In its simplest form, this process involves incubation of the isolated natural product with one or several microbial cultures, thus allowing the enzymes of each organism to act on the compound to produce modified forms.

This approach to generating chemical diversity from a single natural product is analogous to a program of chemical modification except that the chemists in this case are microorganisms and the reactions are enzymically mediated. The advantages of biotransformation lie in the fact that many organisms are able to carry out numerous site-specific and stereospecific reactions that are very challenging to a mere human chemist.

For example, chemical hydroxylation of an aromatic group, or at any specific point on a molecule, could involve many protection and deprotection steps and much complex chemistry, but it may well be possible to find an organism that performs the desired modification. This is particularly useful with natural products that, in the context of synthetic chemistry, are big molecules with many functionalities and chiral centers.

The main drawback of biotransformation systems is that the process is not very predictive (except sometimes in the case of purified enzymes).

Although it may be possible to build up groups of organisms that tend to carry out certain classes of reaction and to have some knowledge of the general modifications a given type of compound might undergo, such empirical rules are by no means absolute or predictive. In order to maximize the chances of obtaining a particular product, therefore, or of obtaining the widest range of products, it is often necessary to utilize a fairly large number of organisms in the initial screen.

To summarize, it is possible to make use of the various biosynthetic, secondary metabolic, and primary metabolic pathways and enzymes associated with particular organisms by giving them the natural product starting material to work with (*see* **Note 7**).

6.1. Methods

A general outline for carrying out biotransformations on a natural product consists of:

1. An initial screen to identify organisms that biotransform the compound.
2. Incubation of the compound with a range of organisms (or enzymes) alongside a set of controls (organisms with no compound added).
3. Analysis of extracts of both sets of cultures. Metabolites in the fed culture not seen in the control can be presumed to be related to the substrate.
4. Isolation of metabolites either from these or from larger scale cultures.

The issues associated with the practicalities of biotransformations are very much the same as those for precursor-directed biosynthesis. The basic analysis and preparative isolation procedures are likely to have been put in place for the initial isolation of the natural product.

The main aspects for consideration are in **Subheadings 6.1.1.–6.1.4.**

6.1.1. What to Feed

If the aim is to generate a particular derivative of a natural product, then it is necessary to feed either that natural product and/or related metabolites that are structurally closer to the target compound. If the aim is to produce the widest possible range of analogs, then clearly the natural product should be used as starting material. However, it is also worth considering the use of analogs as starting material if they are available (and especially if the natural product is itself in limited supply). The modifications to the main substrate may also occur to the analogs, thus giving a multiplicity of related variants on the original structure. If available, the use of radiolabeled substrate—perhaps with a ^{3}H or ^{14}C atom—can prove advantageous at the analysis stage, as this often allows for easy identification of related metabolities in a complex mixture.

6.1.2. Organisms

The organisms that can be used for biotransformation are limitless. The most circuitous approach is to isolate organisms for a specific biotransformation, perhaps based around the ability of different isolates to utilize particular substrates (e.g., the natural product in question) as a sole carbon source. However, in practice it is easier to use organisms that are already "proven" biotransformers. There exist hundreds of literature reports of specific reactions carried out by specific organisms (or specific enzymes), and many of these organisms are easily obtainable from standard microbial culture collections. Many commonly used microorganisms including, fungi, actinomycetes, and nonfilamentous bacteria, that have shown versatility in carrying out a range of biotransformations and that are available from major culture collections, are listed in **Table 1**. (This covers a wide range of reaction types and organism classes and is a good starting point for anyone establishing a biotransformation program.) Also to be included are organisms previously reported to carry out the types of reactions that are being aimed for. Other organisms that it would make sense to include in such a "panel" of potentially biotransforming organisms are those that are related to the producer of the natural product and those organisms that produce structurally related natural products, on the basis that these might possess biosynthetic/secondary metabolic enzymes that can act on these compounds. (Done in this rational way, this is essentially the same as some of the directed biosynthesis work described in **Subheading 5**.) Plant and animal cell cultures can also be used, though these are usually less convenient than microbial cultures.

Apart from the type of organisms to be used, there is also to be considered the form in which these cultures are used. Types of cultures include:

1. Growing cultures.
2. Stationary cultures.
3. Washed cell cultures/resting cell cultures—prepared by separating cells from a medium by centrifugation and then resuspending in dilute buffer. Washed cell suspensions are often used, as in many cases these behave in essentially the same manner as stationary cultures, but with the advantage that the mixture is much less complex, making analysis and purification more straightforward.

Table 1
A Selection of "Proven" Biotransformation Organisms

Organism	Strain
Actinoplanes sp.	ATCC 53771
Amycolatopsis orientalis	NRRL 2452
Aspergillus niger	ATCC 16404
Aspergillus oryzae	ATCC 9102
Bacillus sphaericus	ATCC 13805
Beauveria bassiana	ATCC 7159
Beauveria bassiana	IMI 012939
Cunninghamella bainierii	ATCC9244
Cunninghamella echinulata	IMI 199844
Cunninghamella echinulata var. *elegans*	ATCC 36112
Gibberella fujikuroi	ATCC 12616
Morteriella isabellina	ATCC 38063
Mucor circinelloides	IFO 4563
Mucor rouxii	ATCC 24905
Nocardia corallina	ATCC 31338
Penicillium patulum	IMI 039809
Rhizopus arrhizus	ATCC 11145
Rhizopus stolonifer	ATCC 6227B
Saccharomyces cerevisae	NCYC 1110
Streptomyces griseus	ATCC 13273
Streptomyces lavendulae	CBS 414.59
Streptomyces mashuensis	ISP 5221
Streptomyces punipalus	NRRL 3529
Streptomyces rimosus	NRRL 2234
Verticillium lecanii	IMI 68689

4. Broken cell system/cell-free system—By disrupting the cell wall, the substrate has ready access to the cellular enzymes, and problems of selective permeability are overcome. However, many enzymes will be rendered inactive by this damage to the cell wall.
5. Immobilized cultures—Cells can be removed from their medium as for resting cultures and then immobilized on or within a solid support to provide a system that is usually active for longer than with the above systems. Immobilization can be effected by entrapment in a polymer (alginate, polyacrylamide), adsorption onto a solid support (ion-exchange resins, silica), covalent attachment to a support such as cellulose, or chemical crosslinking with glutaraldehyde. The advantages of immobilization are that the system is usually active for relatively long periods and can be re-used for different batches, a system can be created with a high cell density, and the surrounding medium will be fairly clean, making purification of the product straightforward. Also, in some cases, the

biotransformation reaction may be affected by immobilization, e.g., in terms of the stereochemistry.

6. Spore suspension—Because spores contain many unique enzymes not expressed in other forms of the organism and because they are relatively stable and active in buffer over a long period of time, spore suspensions are often useful biotransformation systems (*see* **Note 8**).

6.1.3. Feeding Conditions

As with conditions for directed biosynthesis, there are no hard-and-fast predictive rules covering experimental conditions for biotransformations, but there are plenty of literature reports that help to give some idea of the best general practical approaches. Whereas there are limitless permutations of substrate concentrations with time of feeding with different organisms and so on, usually circumstances such as amount of available substrate, numbers of microorganisms, amount of shaker space, and time, all impose practical limitations.

Sample: The substrate itself is added to the culture to give typical concentrations of 0.1–0.5 mg/mL, with 0.25 mg/mL as a standard concentration. The aim is to add sufficient starting material to generate reasonable yields of biotransformed products at the same time remaining below concentrations that are toxic, inhibitory to biotransformation, or that waste valuable substrate. The sample can be added as a filter-sterilized or autoclaved aqueous solution or in a small volume of organic solvent. If the substrate is water-insoluble it may be necessary to dissolve it first in a small volume of organic solvent. As mentioned earlier, even if the compound is only sparingly soluble in water, it may still be transformed. The type and amount of solvent used should be considered as this may damage the cells and may permeabilize the cell walls, which may lead to reduced metabolic activity or to enhanced biotransformation by facilitating substrate uptake by the cells.

Time of feeding and incubation: Again, as with directed biosynthesis, the two points on a growth curve that are deemed most suitable for substrate addition are time of inoculation and at the onset of stationary phase. The latter is perhaps preferable, as by this time there is a high cell concentration, so the substrate cannot significantly inhibit culture growth, and there is the possibility that the substrate might be metabolized by the enzymes of both primary and secondary metabolism, thus maximizing the possible range of biotransformation products.

An alternative, possibly better method, is the pulse feed, or continuous feed. This lessens the potential for toxicity by the compound and can take the form of daily feeds starting from inoculation or from the end of log phase. Continuous feeding can be carried out with the help of advanced and expensive fermentation apparatus or by the use of a wick, with one end in the culture and the other in the solution of substrate. Once fed, the culture is left, typically for several days, to allow reactions to take place.

Typically, then, a bacterial culture might be inoculated from a seed culture and then incubated for approx 24 h, by which time it is likely to have reached the end of growth phase; then the substrate compound is added to the culture, followed by incubation of the mixture for another 3, 4, or even 7 d. Fungal cultures might take longer to reach stationary phase, perhaps 3 d, after which it is again advisable to leave for at least a few days in order to achieve maximum biotransformation.

6.1.4. Other Factors

As the processes and enzymes associated with such biotransformations are often those associated with secondary metabolism, any other experimental conditions that generally affect secondary metabolism should be brought into play whenever possible.

1. Medium: As discussed in **Subheading 2.**, the most important factor in this respect is probably the medium. There are numerous reports of the way in which the nutrient source affects the secondary metabolism. Such factors include carbon, phosphorus and nitrogen source, levels, and ratios; inorganics present in medium; levels of aeration/oxygenation; the physical state of the medium (liquid or solid). The diversity of metabolism elicited by a range of media should be considered insofar as the biotransformation reactions are a product of secondary metabolism.

   Ideally, the medium should be one that promotes dispersed growth such that the organism does not stick to the side of the vessel or, as is often the case with many fungi, grow in the form of a tight mycelial pellet. The physical form in which organisms grow can affect secondary metabolism, and many culture flasks contain internal baffles that promote dispersal of the organism. Whereas pelleting may be difficult to control and may not affect biotransformation, a pellet tends to be hydrophobic and generally less amenable to the diffusion of substrate.

   If only a single medium for each organism is to be used, this should be a liquid medium (to allow for dispersal and diffusion of substrate) that is sufficiently rich to result in a high biomass and contains Arkasoy to promote cytochrome P450 activity.

2. Induction of cytochromes P450: These enzymes are a family of monooxygenases ubiquitous in animals, plants, and many microorganisms, which perform many of the reactions that are carried out by an organism when challenged with a xenobiotic compound, i.e., a compound foreign to the organism, such as a drug. These reactions include many that are commonly associated with biotransformations (in both animals and microbes) such as hydroxylation and demethylation, as well as many reactions that involve conjugation of a metabolite with other groups, such as glucuronic acid, sugars, amino acids, and others.

   In most situations, the activity of cytochromes P450 is seen as a means by which the organism modifies a foreign compound in order to make it more polar and more water-soluble and therefore more readily excreted. As natural products scientists, we can view cytochrome P450 activity as a means of carrying out often subtle and site/stereo-specific reactions on a complex molecule to produce a range of closely related analogs, many of which would be difficult to carry out chemically. Cytochromes P450 have been shown to be expressed, either constitutively or inducibly, in many microorganisms that have been reported as capable of performing such reactions. Arkasoy (soya wheat flour) has been found to induce cytochrome P450 expression in some microbial species; an active component is genistein ***(27,28)***, so Arkasoy should be considered as a carbon source or at least as a medium component. (*see* **Note 9**).

3. Scale-up: Scaling up a fermentation for biotransformation purposes may bring with it the possibility of the biotransformation not reproducing. If a different, larger vessel is used, this may result in subtle differences in physicochemical conditions, such as oxygen limitation, mixing, and so on, that result in differences in metabolism. In such cases where it is practically possible, it is sometimes best to carry out the scale-up simply by increasing the number of original vessels.

4. Analysis: An analytical system is also required in much the same way as that described for directed biosynthesis work, i.e., as a means of comparing the biotransformation culture with a control culture to determine those components present only in the former, which are therefore presumably biotransformation-associated. This can take the form of gradient. HPLC, HPLC-MS, TLC, and so on. The aim of such an analytical system is to identify that biotransformation products have been produced and that there is something worth isolating from the cultures. The use of MS may give further information about the compounds; e.g., M+16 molecular ion would lead to a suspicion that oxygenation has occurred, but in order to properly identify and test the product, it is always necessary to isolate the compound.

5. Isolation of products: Again, as in the case of directed biosynthesis products, isolation is likely to be guided to a large extent by conditions used for the original isolation of the natural product. Many typical biotransformation reactions are oxidative, and it is therefore likely that a high proportion of the products will be more polar than the substrate. Therefore, a reverse phase system in which the substrate elutes after a reaonable length of time is probably desirable, though it is possible that some products will be less polar and will elute later from a reverse phase chromatography system.

6.2. Biotransformation of Squalestatins

Continuing with squalestatins as models of natural products that can be further exploited following their initial isolation, these molecules were also used as biotransformation substrates ***(29)***. The aim of generating biotransformation products from this class of compounds was to generate chemical diversity in the form of natural product analogs and for SAR studies, and to functionalize particular atoms to facilitate further chemistry.

In order to identify organisms that would biotransform the squalestatins, the compound was fed to a range of microbes isolated from soil samples, followed by HPLC analysis of crude extracts of the broths simply to look for a significant decrease in the substrate peak that would imply utilization of the substrate. Theses isolates were then analyzed more closely and compared with their control cultures (no squalestatin added) and the products identified.

Shake cultures were inoculated from seed cultures (2–3%, v/v) and incubated for 1 d (actinomycetes; 28°C) or 3 d (fungi; 25°C) before addition of the tripotassium salt of squalestatin 1 (0.1 and 0.5 mg/mL), followed by further incubation for up to 7 d.

Analysis and Product Isolation: Gradient HPLC analysis and product isolation could be carried out essentially as described for the squalestatin analogs described in the previous sections. The use of LC-MS allowed the identification of peaks corresponding to related metabolites with more certainty than LC alone. Additionally, MS often gives a strong indication of how the molecule has changed. In cultures in which squalestatin had apparently been metabolized, LC-MS indicated that in most cases the biotransformation had involved addition of oxygen and/or deacetylation. The fragmentation pattern suggested that these had occurred on the ester and the alkyl (C6 and C7) side chains, respectively. The isolation was carried out largely as described in **Subheading 5.3.1.**, except at the final stages when the resolution of compounds with hydroxyl groups on C23 and C24, respectively, required further HPLC.

The products generated by this particular biotransformation experiment are shown in **Fig. 13**. Six analogs were formed, but considering the complexity of the substrate and the number of potential reaction sites, this is perhaps a fairly meager tally compared to many instances of other simple molecules that generate a plethora of different metabolites. This highlights one of the problems of employing biotransformations as a means of modifying natural products—the unpredictability of such systems. For a given molecule, there are a number of reactions that would be predicted as being more likely than others, but this presumably indicates something about the active site specificity of the cytochrome P450 monooxygenases and the other enzymes involved. However, one of the advantages of biotransformations is exemplified in the products that represent extremely difficult targets for chemical modification.

Compound 7, while not a direct biotransformation product, may have been the result of the inhibition of squalene synthase leading to the buildup of farnesyl pyrophosphate, which was converted to 7 by one of the test organisms.

Possible reasons for the relatively poor "rate of return" of novel metabolites, for microbial cultures screened in the case of the squalestatins, may include the poor solubility of the compound resulting in only a small proportion of the substrate in solution at any one time, that the best organisms were not tested/selected, or that other conditions were unfavorable such as media, feeding time, and so on. These compounds are tricarboxylic acids, and this may have rendered them unsuitable substrates; feeding the compounds in their trimethylester forms may have led to more biotransformation. It may also have been that squalestatins, as antifungal agents, may have had some toxic effect on the organisms with which they were incubated.

6.3. Other Examples of Biotransformations

An example of a simple biotransformation of a small natural product is shown in **Fig. 14**, in which the sesquiterpene lactone, 7α-hydroxyfrullanolide from the plant *Sphaeranthus indicus* was acetylated and reduced by two different *Aspergillus* species to give the two products shown ***(30)***.

At the other extreme, Borghi et al. reported the biotransformation of the teicoplanin family of antibiotics. They found that these large, complex glycopeptides produced by *Actinoplanes teichomyceticus* could be

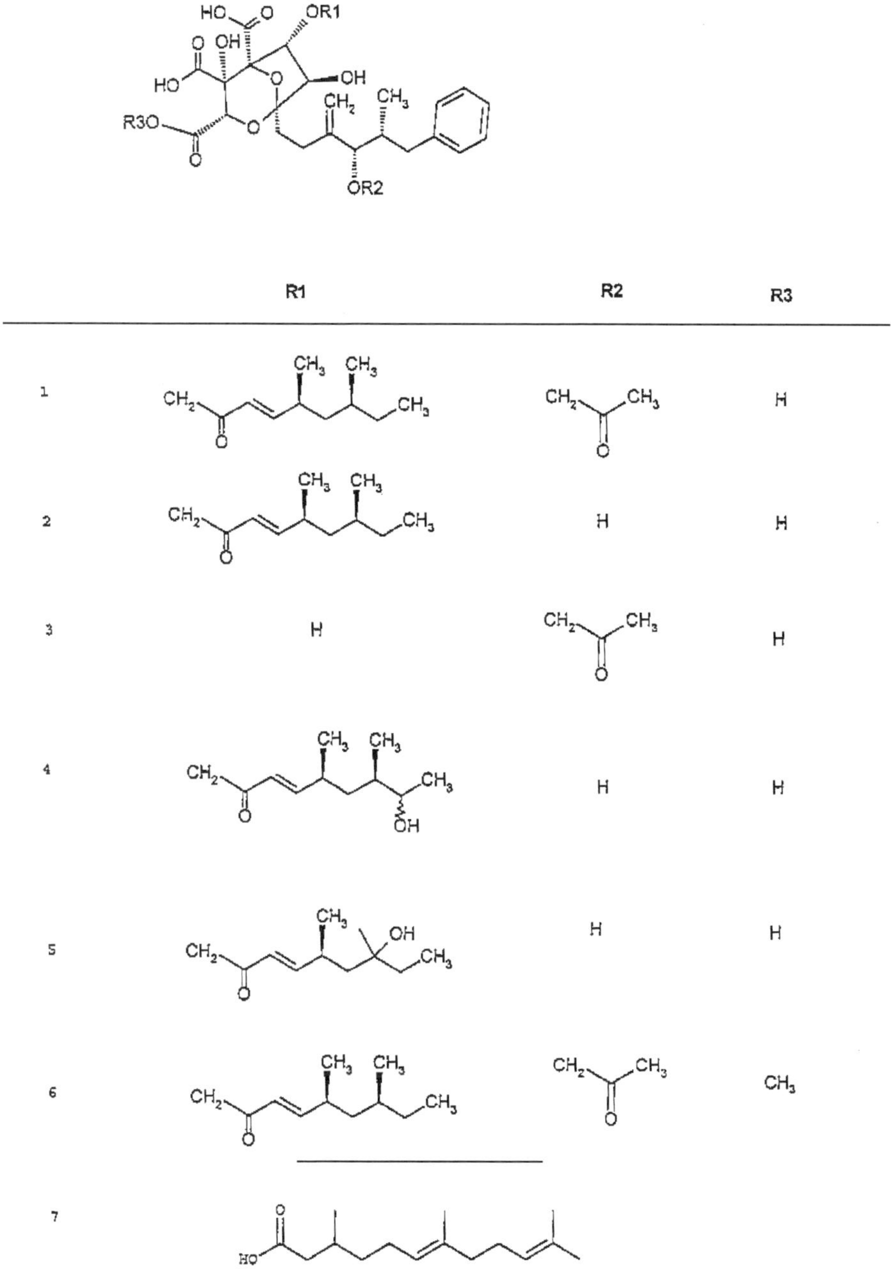

HO O OR1
O OH
HO OH
CH2 CH3
R3O O
O
OR2

	R1	R2	R3
1	CH2 CH3 CH3 CH3 O	CH2 CH3 O	H
2	CH2 CH3 CH3 CH3 O	H	H
3	H	CH2 CH3 O	H
4	CH2 CH3 CH3 CH3 O OH	H	H
5	CH2 CH3 OH CH3 O	H	H
6	CH2 CH3 CH3 CH3 O	CH2 CH3 O	CH3
7	O HO		

Fig. 13. Biotransformation of squalestatins.

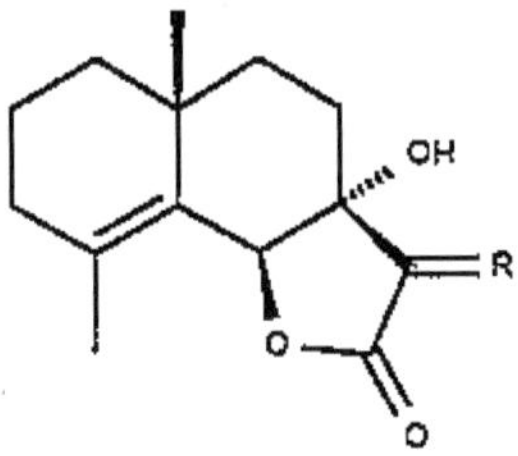

Fig. 14. Biotransformation of frullanolide ($R = CH_2$) to two products where $R = CH_3$ and $R = CH_2COCH_2$, respectively.

demannosylated by cultures of *Nocardia orientalis* NRRL 2450, or *Streptomyces candidus* NRRL 3218. Interestingly, they also found that the demannosylated teicoplanin and other de-mannosylated derivatives could be converted back to the mannosylated form by the original producing organism ***(31)***.

Chen et al. ***(32)*** reported that one of the cytochalasin family of natural products, fed to a culture of *Actinoplanes* sp. (4 × 50 mL) at a concentration of 0.05 mg/mL and incubation for a further 30 h, led to the isolation of six biotransformation products, resulting in functionalization at four new points on the molecule (**Fig. 15**).

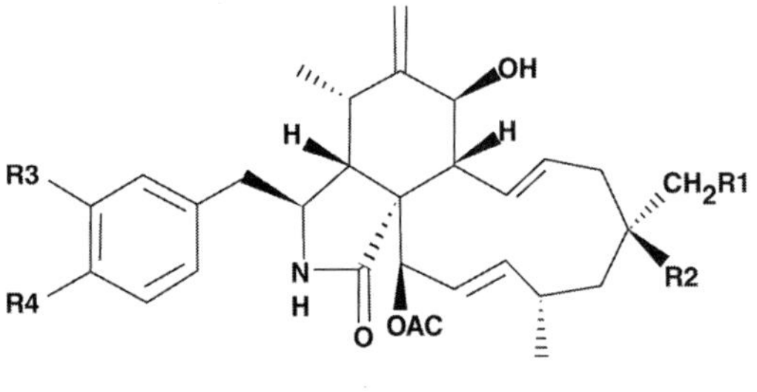

	R1	R2	R3	R4
L-696,474	H	H	H	H
Biotransformation products	OH	H	H	H
	OH	H	H	OH
	OH	H	OH	H
	OH	H	OH	OH
	OH	OH	H	H
	OH	OH	OH	H

Fig. 15. Biotransformation products of cytochalasin L-696, 474.

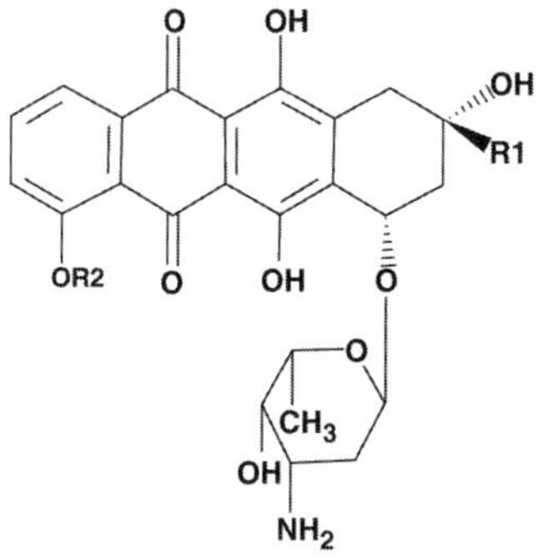

Daunomycin R1 = $COCH_3$, R2 = CH_3
Adriamycin R1 = $COCH_2OH$, R2 = CH_3
13 -Dihydrodaunomycin R1 = $CHOHCH_3$, R2 = CH_3
Carminomycin I R1 = $COCH_3$ R2 = H

Fig. 16. Anthracycline antibiotics.

6.3.1. Anthracycline Antibiotics

As with many commercially important groups of natural products, anthracycline antibiotics and their analogs have undergone a whole range of microbial transformations including oxidation, reduction, acylation, and alkylation *(33)*. These main reaction types can be illustrated with one of this family of compounds, daunomycin (**Fig. 16**). The methyl group α to the side-chain ketone of daunomycin is oxygenated by a mutant of *Streptomyces peuceticus* to give adriamycin *(34)*. This was also found to be the final step in the biosynthesis of adriamycin. This ketone group can also be reduced by a variety of organisms, including both filamentous and nonfilamentous bacteria, and fungi, to give 13-dihydrodaunomycin *(35)*. This reaction is also the first step in the mammalian metabolism of daunomycin (*see* **Note 9**). A *Streptomyces peuceticus* strain is also capable of alkylating carminomycin I to produce daunomycin *(36)*.

Many of these transformations are also carried out on the analogous anthracyclinones—the aglycone portions of these molecules. Others occur on the glycosides, such as the *N*-acetylation of the daunosamine moiety of daunomycin and daunomycinol by *Bacillus subtilis* var. *mycoides (36)*.

Daunomycinone and 13-dihydrodaunomycinone were also converted to glycosylated forms by cultures of *Streptomyces coeruleorubidus* ***(37)***.

6.3.2. Milbemycins

The milbemycins are a group of 16-membered macrolides produced by *Streptomyces hygroscopicus* subsp. *aureolacrimosus*, structurally related to the avermectins, which also have potent antihelminthic and insecticidal activities. As part of a program to prepare new analogs of this family of compounds for further derivatization and as standards of potential animal metabolites, Nakagawa et al. ***(38)*** tested several hundred strains of ctinomycetes, nonfilamentous bacteria, and fungi for the ability to biotransform milbemycin A_4. Organisms were obtained from culture collections and as isolates from soil samples, then cultured on a scale of 20 mL medium in a 100-mL Erlenmeyer flask for 2–3 d, following which milbemycin A_4 was added (5% [w/v] in 1,4-dioxane) to give a final concentration of 0.5 mg/mL. Samples were taken from the cultures at intervals following this, extracted with ethyl acetate, and analyzed for conversion of milbemycin A_4 by TLC with staining with ammonium molybdate. Many strains of actinomycetes and zygomycetes were found to convert the compound, and some of the most efficient converters were selected to carry out larger scale conversions and for testing with other milbemycin analogs. Some of the conversion products of milbemycin A_4 (**Fig. 17**) include

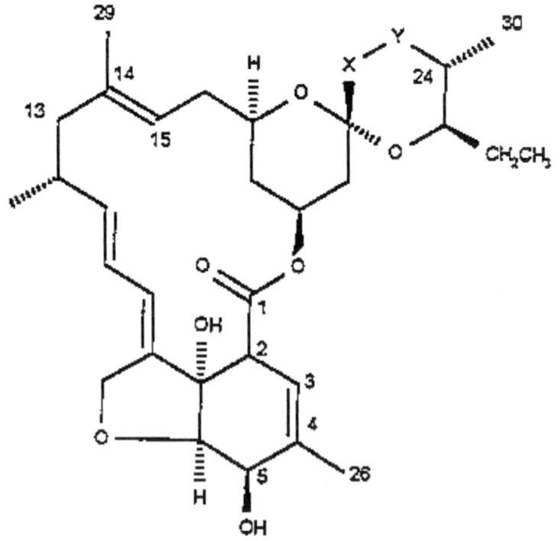

Fig. 17. Milbemycin A_4 showing sites of modification.

30-hydroxymilbemycin A_4, 26,30-dihydroxymilbemycin A_4 and milbemycin A_4 30-oic acid (all by *Amycolata autotrophica* subsp. *amethystina*) ***(38)***, 29-hydroxymilbemycin A_4 (*Syncephalastrum* sp.) ***(39)***, 13β-hydroxymilbemycin A_4, 13β,24-dihydroxymilbemycin A_4, 13β,30-dihydroxymilbemycin A_4 *(Cunninghamella echinulata)* ***(40)***, and 13β,29-dihydroxymilbemycin A_4 *(Streptomyces cavourensis)* ***(41)***. Other biotransformations of this compound that have been reported include 13β-hydroxylation and 14, 15-epoxidation by *Streptomyces violascens* ***(42)***. Many of these conversions, and others, were also carried out on many of the closely related analogs of this compound resulting in a multiplicity of derivatives from a relatively small number of starting compounds.

Although these strategies of mutasynthesis, precusor-directed biosynthesis, and biotansformation have been discussed separately, one of their main strengths lies in the way in which they can be overlapped and can complement each other. Mutants can provide novel analogs or biosynthetic intermediates, which can be fed in a rational manner to other organisms to provide further chemical modification, or they themselves can be fed other precursors.

7. Combinatorial Biosynthesis

Modification of genetic material can be used on a number of levels to exploit the isolation of a novel natural product. The production of mutants can be used to generate organisms with blocked or altered biosynthetic pathways, which can be used for feeding experiments with precursor analogs (as with the avermectins, **Subheading 5.4.4.**). Mutation programs and more precise molecular genetic methods can also be used to improve levels of production of particular metabolites ***(43)***.

More recently, however, genetic engineering for the production of secondary metabolites has involved not so much the follow-up and improvement of an initial natural product, but more the creation almost from "scratch" of novel, or "unnatural" natural products, but still using the organism's secondary metabolite apparatus.

The biosynthesis of polyketides, for example, is carried out by polyketide synthases that operate in a production-line mode, each one carrying out one of a series of condensations of regular small building blocks such as acetyl- and malonyl coenzyme A, with subsequent reductions to make the final molecule. The enormous variety of polyketides is essentially the result of different combinations of a relatively small number of reactions with a relatively small number of building blocks, which includes such important secondary metabolites as erythromycin, avermectin, actinorhodin, and rapamycin. It is now becoming possible to shuffle the genes that code for these enzymes in a myriad of ways to produce many different combinations of synthases resulting in a whole new range of polyketides ***(44–46)***.

Modification of enzymes involved in the synthesis from universal precursors of other secondary metabolites has also been demonstrated. The sesquiterpene synthases are a large family of enzymes that act by a common mechanism and catalyze the cyclization of farnesyl diphosphate, each to form a distinct sesquiterpene. The active site of the synthase that produces trichodiene, the precursor of the trichothecenes in *Fusarium sporotrichioides*, was modified by site-directed mutagenesis, resulting in altered cyclization products as indicated by the isolation of additional sesquiterpenes not previously seen in this organism ***(47)***. In the future, therefore, it is likely that engineering of terpenoid synthases and other classes of enzymes will lead to the generation of many novel metabolites.

8. Combinatorial Synthesis

Combinatorial chemistry has revolutionized the means by which a relatively small number of compounds can be converted into a large number of new compounds. Apart from classical chemistry techniques, the use of natural products as templates and monomers in combinatorial chemistry programs is likely to become one of the major methods for optimizing and maximizing the diversity generated around one natural product. Seven libraries that are based around natural products have been created, such as those comprising a set of modified *Rauwolfia* alkaloids ***(48)***. The benefits of expending the effort to isolate novel natural products, with their advantages of unpredictable structures, usually in a chiral form, are likely to be greatly enhanced if they can be coupled to the power of combinatorial chemistry to produce a large family of derivatives.

9. Notes

1. This highlights another reason why it is always desirable to account for the total amount of biological activity present in the extract in terms of the level and activity of the metabolite deemed to be responsible. The fact that an isolated compound is active may not mean that it can account for all the observed biological activity of the extract, and it may be that there exists in the extract a more potent minor compound.

2. Such atoms, which are relatively uncommon in biological molecules, often make useful "handles" by which to monitor or follow metabolism of compounds that contain them.
3. As mentioned earlier, this process relies on the fact that the enzymes of secondary metabolism are generally less specific than those of primary metabolism and often accept a relatively wide range of substrates. This may reflect the less immediately important role of secondary metabolism in the organism's survival than the more direct and critical role of primary metabolites. Secondary metabolite biosynthetic systems can afford to be more "relaxed" with regard to substrate specificity as the resultant metabolites may play a more indirect, longer term role (or no role at all) than primary metabolites; indeed, it may even be beneficial to the fitness of the species to have secondary-metabolic machinery with rather broad specificity in order to maximize the number of molecules that may be formed by this process and hence to maximize the possibility of producing a compound to protect the organism from any future challenge.
4. The feed sample can be sterilized by autoclaving or filtration, if it is aqueous. Samples in organic solvents can be regarded as being sterile without treatment. Although it is preferable to add a sterile solution, in many cases, any contamination is likely to be insignificant as the culture will already have grown substantially and the biosynthesis is likely to occur before growth of the contaminant becomes significant.
5. The HPLC systems used with and without MS were different. The difference in acetonitrile concentration may reflect the fact that as trifluoroacetic acid is a weaker acid than sulfuric acid, as well as the fact that a C18 (ODS) column is generally more retentive than a C6 column, slightly more acetonitrile was required to elute all the compounds in the sample.
6. During the process of generating each of the products, a total of three different HPLC systems was used. A gradient analytical system was employed initially to analyze whole-broth extracts, because these contained a wide range of metabolites of greatly differing polarities and a gradient separation maximized the chances of detecting compounds closely related to the parent molecule. A gradient system could probably have been used for the purification, but once the analog had been detected and its isolation commenced, many of the unwanted components could be fairly simply removed and an isocratic preparative HPLC system developed to resolve the parent and product. The LC-MS system was different again—no gradient was used, but the possible loss of resolution in which this might result is offset by the power of the MS (to which the LC system is coupled). In most cases the molecular weight of the expected compounds is known, and MS is a powerful means of detecting these compounds even if they coelute with other compounds. Chromatography coupled to MS must also take account of the fact that many of the

salts and ions commonly used in mobile phases will build up on the MS probe and interfere with the spectrometry. Therefore, only volatile buffer salts and acids can be used; commonly used LC-MS mobile phase components include sodium or ammonium acetate or formate with trifluoroacetic acid, formic acid, or tetrahydrofuran as modifiers.

7. This process does not, of course, have to be confined to natural products—synthetic compounds are just as amenable to this approach and indeed the use of enzymes in organic synthesis is now fairly commonplace.
8. The advantage of using purified enzymes as opposed to whole cells is that enzymes are specific for certain reactions so that, for the production of a specific compound, enzymes might be the method of choice. Also, the reaction mixture is likely to be cleaner, thus facilitating purification. Disadvantages of using enzymes in this context are that they are often unstable outside the cell, often require inconvenient and expensive cofactors, and they tend to be expensive. For the purposes of generating a wide diversity of analogs, whole cells are preferable, as they are capable of a multiplicity of reactions. The use of isolated enzymes as tools of organic chemistry is now widespread and increasing and will not be considered at length here. (For Further reading, *see* **refs. *49–51*.**)
9. An additional feature of using microbes to metabolize compounds—be they natural or synthetic—is that many of the biotransformation reactions carried out by microbes are the same as those carried out in humans and animals on xenobiotic compounds. Thus microbial biotransformation systems can be used as a means of preparing quantities of mammalian metabolites of a drug—metabolites that may be very difficult to generate from animals in any reasonable amount—and can even be used in a limited sense as a method by which to predict the metabolic fate of a drug in an animal (***52***).

References

1. Grabley, S., Hammann, P., Kluge, H., Wink, J., Kricke, P., and Zeeck, A. (1991) Secondary metabolites by chemical screening 4. Detection, isolation and biological activities of chiral synthons from streptomyces. *J. Antibiot.* **44,** 797–800.
2. Dawson, M. J., Farthing, J. E., Marshall, P. S., et al. (1992) The squalestatins, novel inhibitors of squalene synthase produced by a species of *Phoma* I. Taxonomy, fermentation, isolation, physico-chemical properties and biological activity. *J. Antibiot.* **45,** 639–647.
3. Sidebottom, P. J., Highcock, R. M., Lane, S. J., Procopiou, P. A., and Watson, N. S. (1992) The squalestatins, novel inhibitors of squalene synthase produced by a species of *Phoma* II. Structure elucidation. *J. Antibiot.* **45,** 648–658.

4. Blows, W. M., Foster, G., Lanes, S. J., et al. (1994) The squalestatins, novel inhibitors of squalene synthase produced by a species of *Phoma* V. Minor metabolites *J. Antibiot.* **47,** 740–754.
5. Trilli, A. (1990) Kinetics of secondary metobolite production, in *Microbial Growth Dynamics* (Poole R. K., Bazin M. J., Keevil C. W., eds.), IRL, Oxford, pp. 103–126.
6. Hutter, R. (1982) Design of culture media capable of provoking wide gene expression, in *Bioactive Microbial Products: Search and Discovery* (Bu'Lock J. D., Nisbet L. J., Winstanley D. J., eds.,) Academic, London, pp. 37–50.
7. Furumai, T., Kakinuma, S., Yamamoto, H., et al. (1993) Biosynthesis of the predimicin family of antibiotics I. Generation and selection of pradimicin non-producing mutants. *J. Antibiot.* **46,** 412–419.
8. Tsuno, T., Yamamoto, H., Narita, Y., et al. (1993) Biosynthesis of the pradimicin family of antibiotics II. Fermentation, isolation and structure determination of metabolites associated with predimicins biosynthesis. *J. Antibiot.* **46,** 420–429.
9. Yoshimoto, A., Matsuzawa, Y., Oki, T., Takeuchi, T., and Umezawa, H. (1981) New anthracycline metabolites from mutant strains of *Streptomyces galilaeus* MA144-MI. I. Isolation and characterization of various blocked mutants. *J. Antibiot.* **34,** 951–958.
10. Matsuzawa, Y., Yoshimoto, A., Shibamoto, N., et al. (1981) New anthracycline metabolites from mutant strains of *Streptomyces galilaeus* MA144-MI. II. Structure of 2-hydroxyaklavinone and new aklavinone glycosides. *J. Antibiot.* **34,** 959–964.
11. Tobe, H., Yoshimoto, A., Ishikur, T., Naganawa, H., Takeuchi, T., and Umezawa, H. (1982) New anthracycline metabolites from two blocked mutants of *Streptomyces galilaeus* MA144-MI. *J. Antibiot.* **35,** 1641–1645.
12. Beremand, M. N., VanMiddlesworth, F., Taylor, S., Plattner, R., and Weisleder, D. (1988) Leucine auxotrophy specifically alters the pattern of tricothecene production in a T-2 Toxin-producing strain of *Fusarium sporotrichioides. Appl. Env. Microbiol.* **54,** 2759–2766.
13. VanMiddlesworth, F., Desjardins, A., Taylor, S., and Plattner, R. (1986) Trichodiene accumulation by ancymidol treatment of *Gibberella pulicaris. J. Chem. Soc. Chem. Comm.* 1156,1157.
14. Jones, C. A., Sidebottom, P. J., Cannell, R. J. P., Noble, D., and Rudd, B. A. M. (1992) The squalestatins, novel inhibitors of squalene synthase produced by a species of *Phoma* III. Biosynthesis. *J. Antibiot.* **45,** 1492–1498.
15. Cannell, R. J. P., Dawson, M. J., Hale, R. S., et al. (1993) The squalestatins, novel inhibitors of squalene synthase produced by a species of *Phoma* IV. Preparation of fluorinated squalestatins, by directed biosynthesis. *J. Antibiot.* **46,** 1381–1389.

16. Cannell, R. J. P., Dawson, M. J., Hale, R. S., et al. (1994) Production of additional squalestatin analogues by directed biosynthesis. *J. Antibiot.* **47,** 247–249.
17. Thiericke, R. and Rohr, J. (1993) Biological variation of microbial metabolites by precursor-directed biosynthesis. *Nat. Prod. Rep.* **10,** 265–289.
18. Boeck, L. D. and Betzel, R. W. (1990) A54145, a new lipopeptide antibiotic complex: factor control through precursor directed biosynthesis. *J. Antibiot.* **43,** 607–615.
19. Claridge, C., Bush, J. A., Doyle, T. W., et al. (1986) New mitomycin analogs produced by directed biosynthesis. *J. Antibiot.* **39,** 437–446.
20. Traber, R., Hofmann, H., and Kobel, H. (1989) Cyclosporins-new analogues by precursor directed biosynthesis. *J. Antibiot.* **42,** 591–597.
21. Hensens, O. D., White, R. F., Goegelman, R.T., Inamine, E. S., and Patchett, A. A. (1992) The preparation of [2-deutero-3-fluoro-D-ala^{8}]cyclosporin A by directed biosynthesis. *J. Antibiot.* **45,** 133–135.
22. Lawen, A., Traber, R., Geyl, D., Zocher, R., and Kleinkauf, H. (1989) Cell-free biosynthesis of new cyclosporins. *J. Antibiot.* **42**, 1283–1289.
23. Hafner, E. W., Holley, B. W., Holdom, K. S., et al. (1991) Branched-chain fatty acid requirement for avermectin production by a mutant of *Streptomyces avermitilis* lacking baranched-chain 2-oxo acid dehydrogenase activity *J. Antibiot.* **44,** 349–356.
24. Dutton, C. J., Gibson, S. P., Goudie, A. C., et al. (1991) Novel avermectins produced by mutational biosynthesis. *J. Antibiot.* **44,** 357–365.
25. Hamill, R. L., Elander, R. P., Mabe, J. A., and Gorman, M. (1970) Metabolism of tryptophan by *Pseudomonas aureofaciens* III. Production of substituted pyrrolnitrins from tryptophan analogues. *Appl. Microbiol.* **19,** 721–725.
26. Kachi, H., Hattori, H., and Sassa, T. (1986) A new antifungal substance, bromomonilicin, and its precursor produced by *Monilinia fructicola. J. Antibiot.* **39,** 164–166.
27. Sariaslani, F. S. and Kunz, D. A. (1986) Induction of cytochrome P-450 in *Streptomyces griseus* by soybean flour. *Biochem. Biophys. Res. Comm.* **141,** 405–410.
28. Trower, M. K., Sariaslani, F. S., and Kitson, F. S. (1988) Xenobiotic oxidation by cytochrome P-450-enriched extracts of *Streotimyces griseus. Biochem. Biophys. Res. Comm.* **157,** 1417–1422.
29. Middleton, R. F., Foster, G., Cannell, R. J. P., et al. (1995) Novel squalestatins produced by biotransformation. *J. Antibiot.* **48,** 311–316.
30. Atta-ur-Rahman, Choudhary, M. I., Ata, A., et al. (1994) Microbial transformations of 7α-hydroxyfrullanolide. *J. Nat. Prod.* **57**, 1251–1255.

31. Borghi, A., Ferrari, P., Gallo, G. G., Zanol, M., Zerilli, L. F., and Lancini, G. C. (1991) Microbial de-mannosylation and mannosylation of teicoplanin derivatives *J. Antibiot.* **44,** 1444–1451.
32. Chen, T. S., Doss, G. A., Hsu, A., et al. (1993) Microbial transformation of L-696, 474, a novel cytochalasin as an inhibitor of HIV-1 protease. *J. Nat. Prod.* **56,** 755–761.
33. Marshall, V. P. (1985) Microbial transformation of anthracycline antibiotics and their analogs. *Dev. Ind. Microbiol.* **26,** 129–142.
34. Oki, T., Takatsuki, Y., Tobe, H., Yoshimoto, A., Takeuchi, T., and Umezawa, H. (1981) Microbial conversion of daunomycin, carminomycin I and feudomycin A to adriamycin. *J. Antibiot.* **34,** 1229–1231.
35. Aszalos, A. A., Bachur, N. R., Hamilton, B. K., et al. (1977) Microbial reduction of the side-chain carbonyl of daunorubicin and N-acetyl daunorubicin *J. Antibiot.* **30,** 50–58.
36. Hamilton, B. K., Sutphin, M. S., Thomas, M. C., Wareheim, D. A., and Aszalos, A. A. (1977) Microbial N-acetylation of daunorubicin daunorubicinol *J. Antibiol.* **30,** 425–426.
37. Blumauerova, M., Kralovcova, E., Mateju, J., Jizba, J., and Vanek, Z. (1979) Biotransformations of anthracyclinoness in *Streptomyces coeruleorubidus* and *Streptomyces galilaeus*. *Folia Microbiol.* **24,** 117–127.
38. Nakagawa, K., Torikata, A, Sato, K., and Tsukamoto, Y. (1990) Microbial convesion of milbemycins: 30-Oxidation of milbemycin A_4 and related compounds by *Amycolata autotrophica* and *Amycolatopsis mediterranei*. *J. Antibiot.* **43,** 1321–1328.
39. Nakagawa, K., Sato, K., Tsukamoto, Y., and Torikata, A. (1992) Microbial conversion of milbemycins: 29-Hydroxylation of milbemycins by genus Syncephalastrum *J. Antibiot.* **45,** 802–805.
40. Nakagawa, K., Miyakoshi, S, Torikata, A., Sato, K., and Tsukamoto, Y. (1991) Microbial conversion of milbemycins: Hydroxylation of milbemycin A_4 and related compounds by *Cunninghamella enchinulata* ATCC 9244. *J. Antibiot.* **44,** 232–240.
41. Nakagawa, K., Sato, K., Okazaki, T., and Torikata, A. (1992) Microbial conversion of milbemycins: 13β, 29-Dihydroxylation of milbemycins by soil isolate *Streptomyces cavourensis*. *J. Antibiot.* **44,** 803–805.
42. Ramos Tombo, G. M., Ghisalba, O., Schar, H.-P., Frei, B., Maienfisch, P., and O'Sullivan, A. C. (1989) Diastereoselective microbial hydroxylation of milbemycin derivatives. *Agric. Biol. Chem.* **53,** 1531–1535.
43. Baltz, R. H. and Hosted, T. J. (1996) Molecular genetic methods for improving secondary-metabolite production in actinomycetes. *Trends. Biotechnol.* **14,** 245–250.
44. Tsoi, C. J. and Khosla, C. (1995) Combinatorial biosynthesis of "unnatural" natural products: The polyketide example. *Chem. Biol.* **2,** 355–362.

45. Khosla, C. and Zawada, R. (1996) Generation of polyketide libraries via combinatorial biosynthesis. *Trends. Biotechnol.* **14,** 335–341.
46. Hopwood, D. A. (1993) Genetic engineering of *Streptomyces* to create hybrid antibiotics. *Curr. Opin. Biotechnol.* **4,** 53–537.
47. Cane, D. E. and Xue, Q. (1996) Trichodiene synthase. Enzymatic formation of multiple sesquiterpenes by alteration of the cyclase active site. *J. Am. Chem. Soc.* **118,** 1563,1564.
48. Atuegbu, A., Maclean, D., Nguyen, C., Gordan, E., and Jacobs, J. (1996) Combinatorial modification of natural products: preparation of unencoded and encoded libraries of Rauwolfia alkaloids. *Biorg. Med. Chem.* **4,** 1097–1106.
49. Davies, H. G., Green, R.,H., Kelly, D.R., Roberts, S.M. (1989) *Biotransformations in Preparative Organic Chemistry: The Use of Isolated Enzymes and Whole Cell Systems in Synthesis.* Academic, London, UK.
50. Faber, K. (1977) *Biotransformations in Organic Chemistry* (3rd ed.), Springer-Verlag, Berlin, Germany.
51. Hanson, J. R. (1995) *An Introduction to Biotransformations in Organic Chemistry*, W. H. Freeman, Oxford, UK.
52. Cannell, R. J. P., Knaggs, A. R., Dawson, M. J., et al. (1995) Microbial biotransformation of the angiotensin II antagonist GR117289 by *Streptomyces rimosus* to identify a mammalian metabolite. *Drug. Metab. Dispos.* **23,** 724–729.

Index